Compendium of Plant Genomes

Series Editor

Chairman, Prof. Chittaranjan Kole, Foundation for Science & Society (PCKFSS)
President, Intl. Climate Resilient Crop Genomics Consortium (ICRCGC)
President, Intl. Phytomedomics & Nutriomics Consortium (IPNC)
President, Genome India International (GII)
Kolkata, India

Whole-genome sequencing is at the cutting edge of life sciences in the new millennium. Since the first genome sequencing of the model plant *Arabidopsis thaliana* in 2000, whole genomes of about 100 plant species have been sequenced and genome sequences of several other plants are in the pipeline. Research publications on these genome initiatives are scattered on dedicated web sites and in journals with all too brief descriptions. The individual volumes elucidate the background history of the national and international genome initiatives; public and private partners involved; strategies and genomic resources and tools utilized; enumeration on the sequences and their assembly; repetitive sequences; gene annotation and genome duplication. In addition, synteny with other sequences, comparison of gene families and most importantly potential of the genome sequence information for gene pool characterization and genetic improvement of crop plants are described.

Roberto Tuberosa • Luigi Cattivelli
Aldo Ceriotti • Assaf Distelfeld
Curtis Pozniak
Editors

The Durum Wheat Genome

Sequencing, Mining and Leveraging the Tetraploid Wheat Genome

Springer

Editors
Roberto Tuberosa
Department of Agricultural
and Food Sciences
University of Bologna
Bologna, Italy

Aldo Ceriotti
Institute of Agriculture Biology
and Biotechnology (IBBA)
Milano, Italy

Curtis Pozniak
Crop Development Centre
University of Saskatchewan
Saskatoon, SK, Canada

Luigi Cattivelli
CREA Research Centre for Genomics
Fiorenzuola d'Arda, Italy

Assaf Distelfeld
Institute of Evolution
University of Haifa
Mount Carmel, Haifa, Israel

ISSN 2199-4781 ISSN 2199-479X (electronic)
Compendium of Plant Genomes
ISBN 978-3-031-84688-5 ISBN 978-3-031-84689-2 (eBook)
https://doi.org/10.1007/978-3-031-84689-2

This Springer imprint is published by the registered company Springer Nature Switzerland AG
The registered company address is: Gewerbestrasse 11, 6330 Cham, Switzerland

This book series is dedicated to my wife Phullara and our children Sourav and Devleena

Chittaranjan Kole

Preface to the Series

Genome sequencing has emerged as the leading discipline in the plant sciences coinciding with the start of the new century. For much of the twentieth century, plant geneticists were only successful in delineating putative chromosomal location, function, and changes in genes indirectly through the use of a number of "markers" physically linked to them. These included visible or morphological, cytological, protein, and molecular or DNA markers.

Among them, the first DNA marker, the RFLPs, introduced a revolutionary change in plant genetics and breeding in the mid-1980s, mainly because of their infinite number and thus potential to cover maximum chromosomal regions, phenotypic neutrality, absence of epistasis, and codominant nature. An array of other hybridization-based markers, PCR-based markers, and markers based on both facilitated construction of genetic linkage maps, mapping of genes controlling simply inherited traits, and even gene clusters (QTLs) controlling polygenic traits in a large number of model and crop plants. During this period, a number of new mapping populations beyond F2 were utilized and a number of computer programs were developed for map construction, mapping of genes, and mapping of polygenic clusters or QTLs. Molecular markers were also used in the studies of evolution and phylogenetic relationship, genetic diversity, DNA fingerprinting, and map-based cloning. Markers tightly linked to the genes were used in crop improvement employing the so-called marker-assisted selection. These strategies of molecular genetic mapping and molecular breeding made a spectacular impact during the last one-and-a-half decades of the twentieth century. But still they remained "indirect" approaches for elucidation and utilization of plant genomes since much of the chromosomes remained unknown and the complete chemical depiction of them was yet to be unraveled.

Physical mapping of genomes was the obvious consequence that facilitated the development of the "genomic resources" including BAC and YAC libraries to develop physical maps in some plant genomes. Subsequently, integrated genetic–physical maps were also developed in many plants. This led to the concept of structural genomics. Later on, emphasis was laid on EST and transcriptome analysis to decipher the function of the active gene sequences leading to another concept defined as functional genomics. The advent of techniques of bacteriophage gene and DNA sequencing in the 1970s was extended to facilitate sequencing of these genomic resources in the last decade of the twentieth century.

As expected, sequencing of chromosomal regions would have led to too much data to store, characterize, and utilize with the then available computer software could handle. But the development of information technology made the life of biologists easier by leading to a swift and sweet marriage of biology and informatics, and a new subject was born—bioinformatics.

Thus, the evolution of the concepts, strategies, and tools of sequencing and bioinformatics reinforced the subject of genomics—structural and functional. Today, genome sequencing has traveled much beyond biology and involves biophysics, biochemistry, and bioinformatics!

Thanks to the efforts of both public and private agencies, genome sequencing strategies are evolving very fast, leading to cheaper, quicker, and automated techniques right from clone-by-clone and whole-genome shotgun approaches to a succession of second-generation sequencing methods. The development of software for different generations facilitated this genome sequencing. At the same time, newer concepts and strategies were emerging to handle sequencing of the complex genomes, particularly the polyploids.

It became a reality to chemically—and so directly—define plant genomes, popularly called whole-genome sequencing or simply genome sequencing.

The history of plant genome sequencing will always cite the sequencing of the genome of the model plant *Arabidopsis thaliana* in 2000 that was followed by sequencing the genome of the crop and model plant rice in 2002. Since then, the number of sequenced genomes of higher plants has been increasing exponentially, mainly due to the development of cheaper and quicker genomic techniques and, most importantly, the development of collaborative platforms such as national and international consortia involving partners from public and/or private agencies.

As I write this preface for the first volume of the new series "Compendium of Plant Genomes," a net search tells me that complete or nearly complete whole-genome sequencing of 45 crop plants, 8 crop and model plants, eight model plants, 15 crop progenitors and relatives, and 3 basal plants is accomplished, the majority of which are in the public domain. This means that we nowadays know many of our model and crop plants chemically, i.e., directly, and we may depict them and utilize them precisely better than ever. Genome sequencing has covered all groups of crop plants. Hence, information on the precise depiction of plant genomes and the scope of their utilization are growing rapidly every day. However, the information is scattered in research articles and review papers in journals and dedicated Web pages of the consortia and databases. There is no compilation of plant genomes and the opportunity of using the information in sequence-assisted breeding or further genomic studies. This is the underlying rationale for starting this book series, with each volume dedicated to a particular plant.

Plant genome science has emerged as an important subject in academia, and the present compendium of plant genomes will be highly useful to both students and teaching faculties. Most importantly, research scientists involved in genomics research will have access to systematic deliberations on the plant genomes of their interest. Elucidation of plant genomes is of interest not only for the geneticists and breeders but also for practitioners of an array of plant science disciplines, such as taxonomy, evolution, cytology, physiology, pathology, entomology, nematology, crop production, biochemistry, and obviously bioinformatics. It must be mentioned that information regarding each plant genome is ever-growing. The contents of the volumes of this compendium are, therefore, focusing on the basic aspects of the genomes and their utility. They include information on the academic and/or economic importance of the plants, description of their genomes from a molecular genetic and cytogenetic point of view, and the genomic resources developed. Detailed deliberations focus on the background history of the national and international genome initiatives, public and private partners involved, strategies and genomic resources and tools utilized, enumeration on the sequences and their assembly, repetitive sequences, gene annotation, and genome duplication. In addition, synteny with other sequences, comparison of gene families, and, most importantly, the potential of the genome sequence information for gene pool characterization through genotyping by sequencing (GBS) and genetic improvement of crop plants have been described. As expected, there is a lot of variation of these topics in the volumes based on the information available on the crop, model, or reference plants.

I must confess that as the series editor, it has been a daunting task for me to work on such a huge and broad knowledge base that spans so many diverse plant species. However, pioneering scientists with lifetime experience and expertise on the particular crops did excellent jobs editing the respective volumes. I myself have been a small science worker on plant genomes since the mid-1980s and that provided me the opportunity to personally know several stalwarts of plant genomics from all over the globe. Most, if not all, of the volume editors are my longtime friends and colleagues. It has been highly comfortable and enriching for me to work with them on this book series. To be honest, while working on this series I have been and will remain a student first, a science worker second, and a series editor last. And, I must express my gratitude to the volume editors and the chapter authors for providing me the opportunity to work with them on this compendium.

I also wish to mention here my thanks and gratitude to Springer staff, particularly Dr. Christina Eckey and Dr. Jutta Lindenborn, for the earlier set of volumes and presently Ing. Zuzana Bernhart for all their timely help and support.

I always had to set aside additional hours to edit books besides my professional and personal commitments—hours I could and should have given to my wife, Phullara, and our kids, Sourav and Devleena. I must mention that they not only allowed me the freedom to take away those hours from them but also offered their support in the editing job itself. I am really not sure whether my dedication of this compendium to them will suffice to do justice to their sacrifices for the interest of science and the science community.

Kolkata, India — Chittaranjan Kole

Preface

Tetraploid wheat encompasses a diverse group of species and subspecies all grouped within the *Triticum turgidum* group, characterized by the presence of the AA and BB genomes (AABB). Among these, *Triticum durum*—commonly known as durum or pasta wheat—is the most widely cultivated and economically significant. Other notable members include wild emmer (*Triticum dicoccoides*), a valuable and underutilized reservoir of genetic diversity, and various *Triticum turgidum* subspecies and domesticated emmer (*Triticum dicoccum*) adapted to various agroecological zones.

Global production of durum wheat exceeds 35 million metric tons annually, with the European Union (particularly Italy), Canada, and Turkey leading its cultivation. Durum wheat is primarily used in the production of pasta, couscous, and bulgur, and its global demand is currently rising, driven by population growth and evolving dietary preferences.

Despite its importance, tetraploid durum wheat accounts for only about 7–8% of total global wheat production, which makes it a relatively minor crop when compared to hexaploid bread wheat (*Triticum aestivum*), which provides 92–93% of production. Durum wheat is distinguished from bread or common wheat by its hard kernels and high protein content, making it particularly suitable for semolina-based products.

Importantly, durum wheat is gaining renewed attention as a source of unique haplotypes and allelic diversity, especially within the AA and BB genomes partially depleted in haplotype diversity in bread wheat consequent to the additional domestication bottleneck, thus positioning durum wheat as a strategic asset for wheat improvement and resilience breeding. It has now been three decades since the mapping of a Quantitative Trait Locus (QTL) for agronomic traits was first reported in durum using RFLP markers to map QTLs controlling grain yield and grain protein content in a population derived from crossing the durum wheat cv. 'Messapia' with the wild emmer accession 'MG4343'. Following this pioneering and groundbreaking work, the pace of progress in durum wheat genomics and its breeding applications has been nothing short of impressive. This progress was accelerated by the publication in 2015 of the most complete consensus tetraploid wheat linkage map, then followed in 2019 by the first sequenced durum wheat genome of cv. 'Svevo'. These two milestone papers prompted and paved the way to a wealth of manuscripts on the mapping of many genes/QTLs with a relevant role in durum wheat genomics.

Based upon this premise, the volume builds on such knowledge and provides a valuable glimpse into some of the most recent advances in the study and characterization of durum wheat genome structure and function. Importantly, this information can be harnessed to enhance the effectiveness of genomics-assisted breeding, gene/QTL cloning, and editing. Suitable platforms, genetic materials, and databases now provide valuable opportunities to best leverage forward- and reverse-genetics approaches while allowing for an unprecedented level of genetic resolution and molecular information, particularly for quantitative traits.

Durum wheat genomics now provides breeders with an impressive toolbox for tailoring novel cultivars better equipped to cope with the challenges posed by climate change. This will hopefully ensure environmentally sustainable production and a stronger and synergic genetic bridge between durum and bread wheat, a bridge now still in its infancy. In short, in the past decade, the collaborative and concerted international research efforts and action promoted durum wheat from 'orphan' to a higher genomics research asset and toward better informed breeding.

The 16 chapters of the volume present valuable and up-to-date information/examples on the latest achievements in durum wheat genomics/sequencing and how the entire wheat community and other stakeholders, including consumers, will eventually benefit from such knowledge. Overall, the chapters in this volume emphasize the importance to identify rare haplotypes with beneficial effects that are not yet represented in elite wheat germplasm, including also the generation of synthetic wheat, an increasingly cultivated hexaploid wheat derived from the cross between durum wheat with *Aegilops tauschii* as donor of the DD genome.

Large-scale resequencing coupled with an equally deep analysis of the transcriptome, proteome, and metabolome will accelerate the cloning of agronomically valuable loci, paving the way to a more effective harnessing of durum wheat biodiversity. We hope that this volume will allow wheat scientists to better appreciate the complexity underpinning phenotypic variability as influenced by the QTLome, while stimulating their curiosity and interest in undertaking new studies to further enhance our understanding of such complexity.

The editors are grateful to the authors of the different chapters for reviewing the published research work in their area of expertise. We also appreciate their cooperation in meeting the deadlines and revising their manuscripts as required.

Bologna, Italy — Roberto Tuberosa
Fiorenzuola d'Arda, Italy — Luigi Cattivelli
Milano, Italy — Aldo Ceriotti
Haifa, Israel — Assaf Distelfeld
Saskatoon, SK, Canada — Curtis Pozniak

Contents

The Wild Emmer Wheat Genome

Leah Oren, Hanan Sela, and Assaf Distelfeld

Abstract

Wild emmer wheat (WEW), the progenitor of most cultivated wheat species, evolved in the Fertile Crescent about 0.5 million years ago. The large size of the wheat genome, its polyploidy, and its high content of repetitive elements delayed the construction of a comprehensive reference genome for many years. However, despite the difficulties, a high-quality polyploid genome assembly of the wild emmer wheat accession "Zavitan" was published in 2017, presenting 14 chromosome-scale assemblies (pseudomolecules) with a total size of 10.1 Gb, as well as one group of unanchored scaffolds (0.4 Gb). Integration of an optical map enabled improvement of assembly parameters and the annotation of 97,908 genes. In this chapter, we aligned exome-capture data of 134 WEW accessions from the Fertile Crescent to the Zavitan genome assembly and describe the population structure. Admixture and PCA (principal component analysis) analyses revealed three genetically separate groups. We defined these three groups according to their geographical and morphological characteristics as narrow-spike WEW from Turkey, Iraq, and Iran (northeastern [NE]-Horanum group); narrow-spike WEW from the southwestern (SW) Fertile Crescent (SW-Horanum group); and wide-spike WEW from SW Fertile Crescent (SW-Judaicum group). We found that the SW-Horanum group exhibited the highest diversity, thus representing the core of WEW's diversity. Moreover, we demonstrated the genetic uniqueness of the SW-Judaicum group by focusing on a region close to the flowering induction *VRN1-A* gene. Finally, we review the use of WEW's genomic resources and genome reference in recent years. The WEW gene pool has been used in wheat breeding programs to improve disease resistance and other desirable agricultural traits. It has also been used to study wheat's genetic evolution and domestication processes. These advances improve our food security as well as increase basic and applied knowledge. Considering that only the Zavitan genome, belonging to the unique SW-Judaicum group, was sequenced, we propose that more reference genomes, preferably from the other groups, should be assembled and become publicly available.

1 Introduction

Wild emmer wheat (WEW; *Triticum turgidum* subspecies [ssp.] *dicoccoides* [Körn] Thell [$2n = 4x = 28$], AABB genomes) is the progenitor

L. Oren · H. Sela · A. Distelfeld (✉)
Department of Evolutionary and Environmental Biology, Faculty of Natural Sciences and the Institute of Evolution, University of Haifa, Haifa, Israel
e-mail: adistelfe@univ.haifa.ac.il

R. Tuberosa et al. (eds.), *The Durum Wheat Genome*, Compendium of Plant Genomes,
https://doi.org/10.1007/978-3-031-84689-2_1

of most domesticated wheat species. It evolved about 0.5 million years ago from hybridization between the diploid species *Triticum urartu* (genome AA) and a diploid *Aegilops* species from the *Sitopsis* section (genome BB) in the Fertile Crescent (Dvorak et al. 1993; Dvorak and Akhunov 2005; Dvorak and Zhang 1990; Huang et al. 2002; Özkan et al. 2005; Sarkar and Stebbins 1956; Li et al. 2024a). WEW is an annual, predominantly self-pollinated plant with relatively large and elongated grains. WEW spikes are brittle, meaning that upon maturity, the spikelets disarticulate and penetrate the soil via their unique arrow-shaped form and long awns. Thereby, WEW seeds can remain safe until their forthcoming germination (Elbaum et al. 2007; Harlan and Zohary 1966).

Wild emmer wheat, initially known as "wild two-grained wheat," was first recognized in 1873 by the botanist Friedrich August Körnicke in the herbarium of the National Museum of Vienna. He found it among specimens of wild barley that had been collected by the botanist Theodor Kotschy on the northwestern slopes of Mount Hermon 18 years earlier. Körnicke described these specimens, which contained no viable seeds, as the wild progenitor "prototype" of cultivated wheat, and named the species *Triticum vulgare* Vill. var. *dicoccoides* Körn (Aaronsohn 1910; Lack and Slageren 2020; Poyarkova 1988).

Years later, the wild two-grained wheat was identified in its natural habitat by Aaron Aaronsohn at Rosh-Pinna, Eastern Galilee, Israel, on June 18, 1906 (Aaronsohn and Schweinfurth 1906), a revelation that triggered substantial progress in wheat research. Aaronson envisioned WEW serving as a genetic source for wheat improvement due to its natural genetic variation and adaptation to semi-arid conditions (Aaronsohn 1910). Indeed, there are many examples of WEW's use in wheat improvement programs and research. Schweinfurth and Aaronsohn named the wild two-grained wheat "*T. dicoccoides*" (Aaronsohn and Schweinfurth 1906), but it was later modified to *T. turgidum* ssp. *dicoccoides* to reflect the AABB genomic constitution of all *Triticum turgidum* subspecies (Lack and Slageren 2020; Thellung 1920; van Slageren and Payne 2013).

Since the discovery of WEW in nature in 1906, it has been identified in many locations in the Fertile Crescent at a range of altitudes (from 150 m below sea level up to 1800 m above sea level) and in various geological conditions (Harlan and Zohary 1966; Özkan et al. 2011; Schweinfurth 1908; Zohary et al. 2012). Several of these populations were later identified as *Triticum araraticum*, a separate species of wild tetraploid wheat (Badaeva et al. 2022). In most cases, WEW grows in sporadic, isolated patches and thin scattered stands often in association with wild oats, wild einkorn, and wild barley. It is quite common for these other wild cereals to predominate in nature, rendering WEW's identification difficult. Also, plowing and overgrazing have probably reduced WEW from its natural habitats over time (Özkan et al. 2011).

Early studies of WEW indicated the existence of "another form of wild emmer wheat," characterized by larger spikelets and glumes. The variety of morphological characteristics that WEW exhibits were described by Aaronsohn as follows: "The most astonishing thing about it was the large number of forms it displayed… in some, the form of the glume resembled that of *Triticum durum*, in others the development of the secondary nerve was similar to that of *T. monococcum*" (Aaronsohn 1910). In 1926, an expedition of Russian geneticists and breeders, headed by N. Vavilov, searched for wild relatives of domesticated plants. The expedition visited numerous sites, including the Gilboa and Carmel mountain ranges. Following that expedition, Vavilov divided the local (southern Levant) wild wheat into narrow-spike and wide-spike forms (Vavilov et al. 1931). Following Vavilov's work, Jakubziner named the narrow-spike form "Horanum" and the wide-spike form "Judaicum" (Jakubziner 1932a, 1932b).

Since the late twentieth century to the present day, with the development of molecular genetics techniques, genome-wide estimates of diversity were recognized as an essential tool for discovering the dynamic processes of natural populations (Martin and Salamini 2000; Mazzucotelli et al. 2024). The first studies of global tetraploid wheat population structure were conducted on wild and domesticated populations via the AFLP technique, and revealed two widely differing genetic populations: the southwestern (SW) Fertile Crescent population colonizing habitats in Israel, Syria, Lebanon, and Jordan; and the northeastern (NE) Fertile Crescent population, colonizing habitats in Turkey, Iran, and Iraq (Özkan et al. 2002, 2005, 2011). Accordingly, Mori et al. (2003) examined the variation of wild and domesticated wheat using chloroplast microsatellite markers, and similarly identified two WEW taxa: the "western Palestine" strain and the "central-eastern Turkish-Iraqi" strain. The division into two main populations was later confirmed using single nucleotide polymorphism (SNP) data from 115 WEW accessions (Maccaferri et al. 2019). The NE population was further subdivided into several populations: Iran and Iraq population, west of Diyarbakir population, and an additional population that grows 36.2 km west of Diyarbakir (Luo et al. 2007). Likewise, the SW Fertile Crescent genotypes were further subdivided according to spike morphology and genetic markers into two populations corresponding to Horanum and Judaicum (Luo et al. 2007; Özkan et al. 2002, 2005, 2011; Poyarkova et al. 1991; Sela et al. 2014). The morphological differences between Horanum and Judaicum genotypes are evident, as can be seen from a comparison of spikes belonging to these contrasting groups (Fig. 1).

Fig. 1 Representative spike morphology of wild emmer wheat genotypes. (**a**) WEW collected from Mt. Hermon, Israel, belongs to the southwestern (SW) Fertile Crescent population (Horanum). (**b**) WEW collected from Urfa, Turkey, belongs to the northeastern (NE) Fertile Crescent population. (**c**) WEW accession "Zavitan," collected from the Yehudia Nature Reserve, Israel, belongs to the Judaicum population with a more robust spike than those of the other two genotypes. The three WEW genotypes were grown under common garden conditions at Aaron Aaronsohn's experimental farm in Atlit, Israel, during the winter growing season of 2022

2 Wild Emmer Wheat Genome Assembly

Wheat has a large genome size (~5 Gb per diploid genome), a polyploidy nature, and a high content of repetitive elements, all of which delayed for many years the construction of any comprehensive reference genome. However, the first high-quality polyploid-genome assembly was published in 2017, derived from WEW accession "Zavitan" (Avni et al. 2017). Zavitan was chosen as a reference-genome to make use of the extensive genetic data already collected for this accession (Avni et al. 2014; Nave et al. 2016).

The first version of Zavitan genome (v.1) was constructed by whole-genome shotgun (WGS) sequencing, combined with an innovative assembly algorithm (NRGene, Ltd.) that generated large sequence-scaffolds (N_{50} = ~7 Mb). Altogether, the genome assembly pipeline used a total of 2.1 Tb of raw sequence data (equivalent to 200× coverage, based on an estimated genome size of 10.5 Gb). The scaffolds were validated with genetic data and combined with three-dimensional (3D) chromosome conformation capture sequencing (HiC) data (Lieberman-Aiden et al. 2009) enabling construction of 14 chromosome-scale assemblies (pseudomolecules) with a total size of 10.1 Gb as well as one group of unanchored scaffolds (0.4 Gb).

The v.1's utility and quality was demonstrated via the cloning of the *Btr1* genes that control spike shattering in barley and wheat (Avni et al. 2017; Pourkheirandish et al. 2015, 2018). This was achieved because the two main quantitative trait loci (QTLs) for this trait on chromosomes 3A and 3B, were fully represented in the genome reference. Many domesticated wheat genome assemblies followed a similar approach (Guo et al. 2020; Maccaferri et al. 2019; The International Wheat Genome Sequencing Consortium (IWGSC) et al. 2018; Walkowiak et al. 2020). Later, an improved WEW reference (v.2) was constructed by integrating the v.1 scaffolds with 10.4 Gb optical mapping of Zavitan (Zhu et al. 2019). The analyses presented in this chapter are based on Zavitan genome assembly v.2.

3 Wild Emmer Wheat Genome Structure

Originally, 62,813 genes were positioned along the chromosomes (Avni et al. 2017). Gene density was higher toward the distal regions compared to the pericentromeric regions of chromosome arms (up to 14-fold higher), while toward the proximal regions there was an increased abundance of housekeeping genes (Avni et al. 2017). Here, we integrated two annotation versions (Maccaferri et al. 2019; NCBI release 100) to one comprehensive gene annotation that includes a total of 97,908 genes with 79,138 protein coding genes, of which 40% are uncharacterized. Gene density levels range from 1 to 37 genes per Mbp (Fig. 2a, Table 1) and the improved annotation can be found on GrainGenes.

To further explore the WEW genome and its population structure, we computed population genetic indices using exome-capture data of 134 wild wheat accessions (Avni et al. 2017; Fu et al. 2019). This enabled us to estimate Tajima's D (Tajima 1989) (Fig. 2b, Table 1), nucleotide diversity (π) (Fig. 2c, Table 1), marker density (Fig. 2d, Table 1), and linkage disequilibrium (LD) levels (Fig. 2e, Table 1). The nucleotide diversity levels depend upon mutation rate, population size (Kimura 1968), and recombination (Sandhu and Gill 2002). In wheat, there is generally a high recombination level in distal regions of wheat and a low recombination level in proximal regions (Able et al. 2007; Akhunov et al. 2003; Sandhu and Gill 2002). As such, several studies have demonstrated that wheat polymorphism increases with physical distance from the centromere, a phenomenon that has been attributed to the effects of local recombination rates on mutation factors and indirect selection events (Able et al. 2007; Dvorak et al. 1998, 2004; Huang et al. 2002; Thuillet et al. 2004). In our analysis, nucleotide diversity, marker density, and gene density exhibited similar patterns: lower toward the centromeric regions, and higher toward the telomeric sites. Similarly, Tajima's D values, which represent the difference between the number of segregating sites and the number of nucleotide differences estimated from pairwise comparisons (Tajima 1989), exhibited a pattern wherein the highest values were found in telomeric regions. While these high telomeric Tajima's D values may reflect balancing selection, since the overall mean of Tajima's D is highly positive (3.84), we suggest that it is due to population shrinkage that is expected due to loss of habitats.

Apart from the effect on nucleotide diversity, recombination has an impact also on linkage disequilibrium (LD), a statistic that quantifies nonrandom associations among alleles at various loci. Likewise, our analysis shows that WEW's LD values are higher in the centromeric regions than in the telomeric regions (Fig. 2e). This is likely due to recombination, as well as other factors that influence LD, such as selection, mutation, and genetic drift (Flint-Garcia et al. 2003; Slatkin 2008). The plots also reveal outlier regions that need further detailed investigation such as the overall low LD in chromosome 7A, a segment with unique values across all indices in chromosome 5A near 500 Mbp; and a segment with very high LD and low Tajima D values in chromosome 1B in the 200–300 Mbp range.

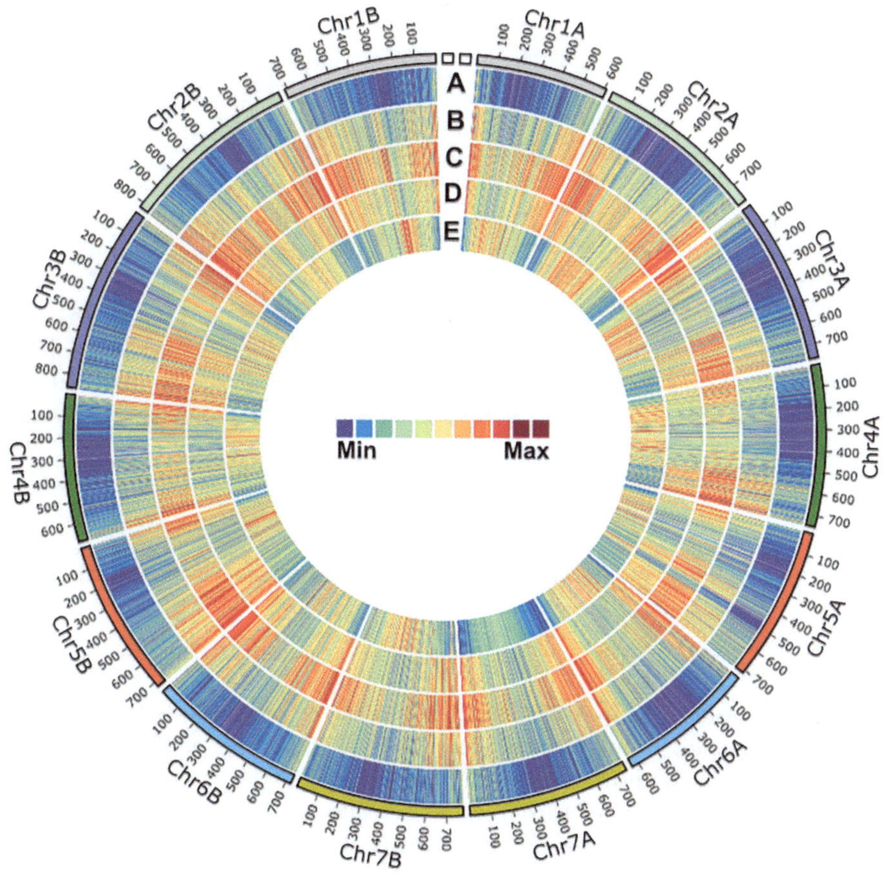

Fig. 2 Gene density and population genetic indices of 134 wild emmer wheat accessions. Estimates were averaged in a 1 Mbp sliding window. (**a**) Gene density levels (genes/Mbp); (**b**) Marker density levels (Markers/ Mbp); (**c**) Tajima's D; (**d**) Nucleotide diversity (π); (**e**) LD levels (r^2)

Table 1 Summary statistics of population genetic indices for 134 wild emmer accessions

	Gene density (Genes/1Mbp)	Tajima's D	Nucleotide diversity (π)	Marker density (Markers/1Mbp)	LD (r^2)
Min.	1.000	0.765	0.0000002	1.000	0.0002197
Max.	37.000	6.055	0.0001633	339.000	0.9904320
Mean	6.469	3.841	0.0000092	21.340	0.3922857
Std. dev.	4.851	1.096	0.0000126	27.958	0.1967827

4 Wild Emmer Population Structure

The admixture analysis of 134 WEW accessions shows a clear separation into three groups, with a partial admixture between them (Fig. 3). These three groups are defined herein as per their geographical and morphological characteristics as: (1) "NE-Horanum," which are narrow-spike WEW from Turkey, Iraq, and Iran (NE Fertile Crescent); (2) "SW-Horanum," which are narrow-spike WEW from SW Fertile Crescent; and (3) "SW-Judaicum," which are wide-spike WEW from SW Fertile Crescent. The SW-Judaicum accessions (40 accessions including Zavitan, all from Israel) display the least admixture, whereas the SW-Horanum (77 accessions, 72 from Israel, 3 from Lebanon, and 2 from Syria) and

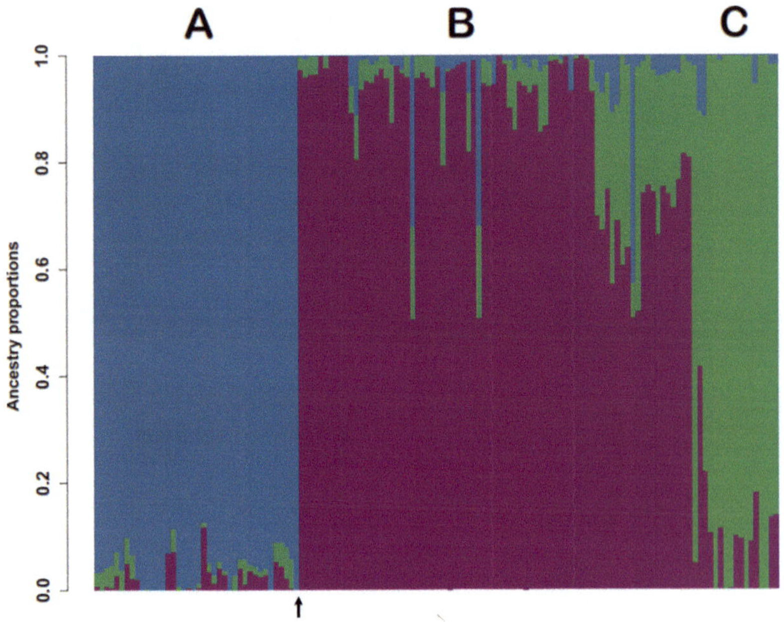

Fig. 3 Admixture analysis of 134 wild wheat accessions based on 197,648 SNPs at $K = 3$. Each accession is divided into colored sections, and the longest section of each individual is associated with its ancestry group. (**a**) Blue, (**b**) purple, and (**c**) green are attributed to the SW-Judaicum, SW-Horanum, and NE-Horanum groups, respectively. WEW accession "Zavitan" is indicated by the black arrow

NE-Horanum (17 accessions, 13 from Turkey, 2 from Iraq, 1 from Iran, and 1 from Lebanon) are more admixed. Despite being distributed over a smaller geographical area, accessions from the SW Fertile Crescent (SW-Horanum and SW-Judaicum groups) are more diverse than are accessions from NE Fertile Crescent genotypes.

Further classification based on principal component analysis (PCA) supports the division into three groups. The SW-Judaicum group represents a distinct group based on principal component (PC) 1, which explains 81.7% of the genetic variation. The two Horanum groups show a continuum diffusion of diversity with respect to geographic proximity based on PC 2 (10.2% of the genetic variation), where values between −100 and 85 represent the SW-Horanum genotypes; and values higher than 85 represent the NE-Horanum genotypes (Fig. 4). Overall, both ancestry and PCA analyses showed the genetic uniqueness of the SW-Judaicum group with the Zavitan accession to be representative of this group. Also, the highest genetic diversity of the SW genotypes (both Horanum and Judaicum) supports the idea that the diversity center, or WEW's origin, lies within the SW Fertile Crescent, as previously suggested (Fahima et al. 1999, 2002; Harlan and Zohary 1966; Yahiaoui et al. 2008).

While the 197,648 SNPs used for the diversity analysis spread across the 14 chromosomes of WEW (Fig. 2), to underscore the genetic differentiation of the 3 wild wheat groups, we created a haplotype map of a specific region on chromosome 5A (582,845,128–583,211,189 bp) (Fig. 5). This region exceeded the 99th percentile of the total genome-wide pairwise fixation index (Fst) values, as measured by comparing the SW-Judaicum group to the SW-Horanum and NE-Horanum groups. In the resulting map, the SW-Horanum differed from the SW-Judaicum in 30/30 SNPs, and the NE-Horanum differed from the Judaicum in 27/30 SNPs. By contrast, only 3/30 SNPs varied between the SW-Horanum and the NE-Horanum groups. Remarkably, this region is close to the *VRN1-A* gene (located on chromosome 5A at 594 Mbp), a key gene regulating flowering time, and one of the factors determining winter and spring growth habits (Distelfeld et al. 2009; Murai et al. 2003; Yan et al. 2004). The SW-Judaicum genotypes' uniqueness in an area near the *VRN1-A* may be

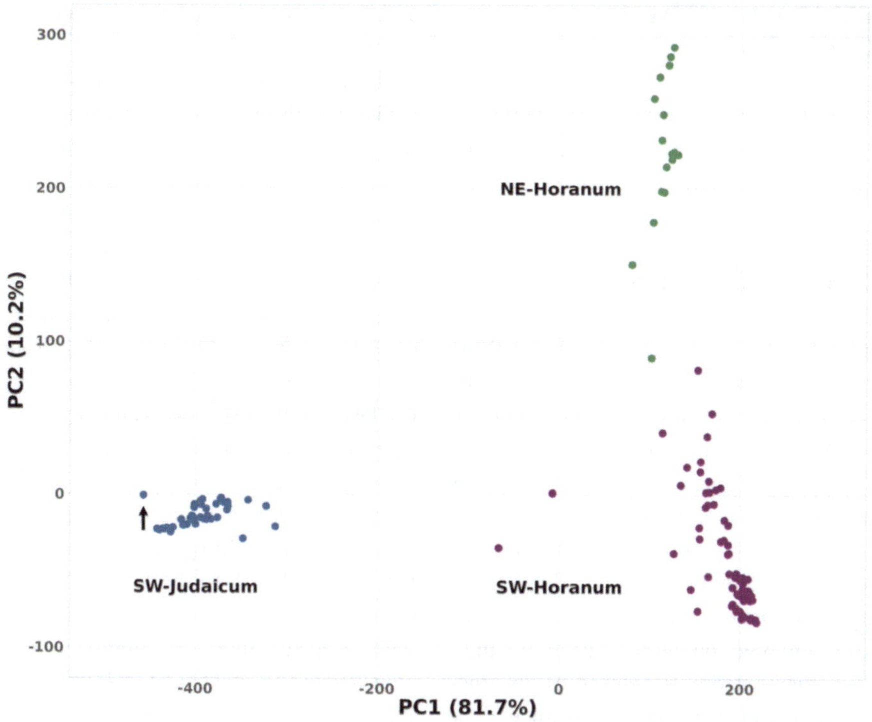

Fig. 4 PCA of 134 wild wheat accessions based on 197,648 SNPs representing the wild emmer wheat population structure. The colors correspond to the admixture analysis: blue, purple, and green are attributed to the SW-Judaicum, SW-Horanum, and NE-Horanum groups, respectively. WEW accession "Zavitan" is indicated by the black arrow

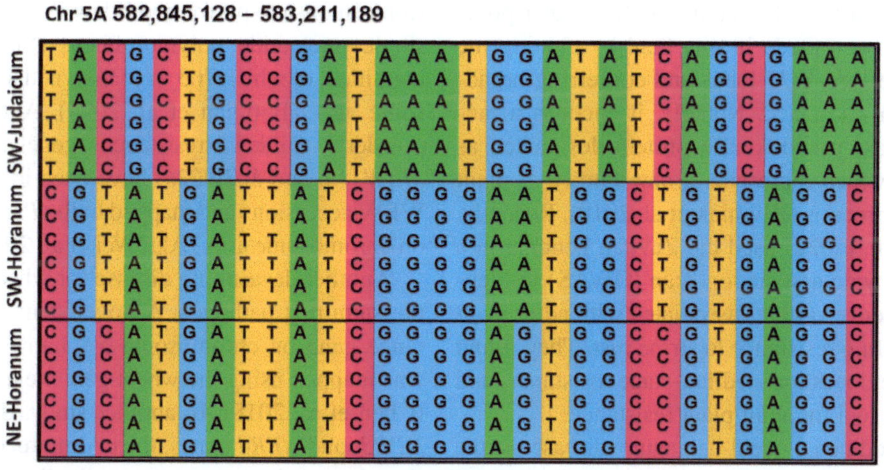

Fig. 5 Schematic presentation of the sequence polymorphism in groups of WEW at chromosome 5A, 582,845,128–583,211,189 bp

associated with various alleles that contribute to the preservation of the Horanum and Judaicum as relatively distinct groups despite their geographic proximity. The Judaicum group heads two weeks before the Horanum group on average, thus preventing cross pollination and hybridization (Poyarkova et al. 1991).

5 Wild Emmer Wheat Genomics

The following section describes some examples of the use of WEW genetic resources and genome assembly in the identification of various genes.

5.1 Disease Resistance

Yellow (stripe) rust is a prevalent disease striking temperate cereals, leading to reduced yields and low-quality grains (Wellings 2011). Prior to WEW's sequencing, most markers were not effective for mapping genes in regions of low gene density or low recombination. The WEW genome and other assemblies enabled efficient high-throughput genetic analysis based on sequencing. For instance, the newly developed genomic tools enabled the identification of candidate genes in the centromeric region of chromosome 1B that includes the rust resistance gene *YR26* (Wu et al. 2018). Another study identified four stable QTLs for yellow rust resistance on chromosomes 1B, 4A, and 6A and identified candidate genes using the bread wheat and WEW genome assemblies to retrieve genes between the flanking markers (Ma et al. 2019). Further progress has been made in this field by cloning the WEW-derived *Yr15* locus and conferring broad-spectrum resistance. Notably, it encodes a tandem kinase-pseudokinase protein designated as WTK1, which represents a new family of plant defense proteins (Klymiuk et al. 2018). Recently, Tene et al. (2022) used the Zavitan genome in a genome-wide association scan (GWAS) to identify sequence variants associated with adult plant and seedling resistance to stripe rust. They found one novel locus for seedling-stage resistance, and 12 novel loci for adult plant resistance.

An additional yield-damaging disease is the plant parasitic nematode pathogen (Jones and Fosu-Nyarko 2014). The root lesion nematodes are migratory endoparasites that penetrate the root tissues, release cell wall degrading enzymes, and migrate intracellularly (Davis et al. 2011). Consequently, plants are unable to uptake nutrients and water from the soil properly (Jones and Fosu-Nyarko 2014). To dissect the disease's genetic basis, an accurate mapping was conducted, and QTLs for disease resistance were found on chromosomes 6D and 2B. The Zavitan reference, along with other genomes, was used for BLAST searches to identify the SNPs' physical genomic positions, and to further identify the candidate genes underlying these QTLs (Rahman et al. 2020).

Powdery mildew is a devastating disease that causes severe yield losses (Morgounov et al. 2012; Singh et al. 2016). In a study by Wu et al. (2019), a high-resolution linkage map of the powdery mildew resistance *Pm52* locus was constructed, and candidate genes were identified using the genome references. A similar approach was used for mapping the powdery mildew resistance gene *Pm68* to the terminal portion of 2BS (He et al. 2021). In addition, the WEW genome was used to map the powdery mildew gene *PmG16* and to identify it as *Pm60* allele (Li et al. 2021). Recently, a powdery mildew resistance gene *Pm36* derived from WEW was identified and mapped on chromosome arm 5BL of the tetraploid WEW-durum introgression line (IL) (Li et al. 2024b). A bulked RNA-Seq and additional qRT-PCR verification revealed the profile expression response to powdery mildew. A total of 6196 single nucleotide polymorphisms (SNPs) were identified, of which 763 were assigned to chromosome arm 7B (Qian et al. 2024). Additionally, powdery mildew resistance genes (*MlIW6*, *MlIW113*, and *MlIW226*) have been identified in WEW accessions and mapped to the *PM60* region on chromosome arm 7AL (Wei et al. 2024).

Another devastating disease for wheat yield is the stem rust pathogen. A novel stem rust resistance gene on chromosome 5A known as *Sr60* was mapped using the wheat genome assemblies (Chen et al. 2018). In another study, the *NPR1*, *NPR3*, and *NPR4* genes, the coding receptors for the plant defense hormone salicylic acid, were investigated (Zhang and Li 2019). These genes, which were originally discovered and characterized in *Arabidopsis* (Ding et al. 2018), were identified in wheat species owing to the progress in wheat genomics (Liu et al. 2019).

5.2 Yield Components

Several agronomical traits contribute to the improvement of wheat yield, including thousand-grain weight, spikelet number per spike, tillering, and others. A recent study identified candidate genes for the spikelet number per spike trait on chromosome 7A using the WEW genome and other assemblies (Kuzay et al. 2019). Regarding the thousand-grain weight, a QTL on chromosome 4AL was validated using WEW and bread wheat genome resources (Guan et al. 2019). In addition, the *Growth-Regulating Factor4* gene, *GRF4*, was suggested as being responsible for a WEW-originating QTL for grain weight (Avni et al. 2018). Notably, two positively epistatic major QTLs were reported in durum wheat for thousand-grain weight (Graziani et al. 2014). Their positional cloning would eventually provide a valuable opportunity to better explore haplotype diversity of wild emmer for thousand-grain weight.

Another investigated agronomical trait is tillering, a special branching phenomenon occurring in monocot plants (Kebrom et al. 2013) that affects branch structures, flowering, plant height, and seed development (Kuraparthy et al. 2007). It also manifests in wheat yield when it is calculated as spike number per hectare (Yang et al. 2013). Several QTLs for the tillering trait were tested in multiple environments, and a stable QTL for the effective tiller number trait was identified on chromosome 1B. BLAST searches against the bread wheat and WEW genome assemblies retrieved the genes between the flanking markers for further study (Liu et al. 2020).

5.3 Diverse Agronomical Traits

There may be differences in the importance of agronomical traits across various geographical regions. For instance, in wetter, more humid environments such as those in northern and central Europe, the absence of awns has been selected, likely as they would not necessarily provide an adaptive advantage there (Börner et al. 2005; Rebetzke et al. 2016). To identify the gene causing awn inhibition, a fine-resolution mapping of a segregating biparental durum wheat population was conducted, along with an examination of bread wheat mutants. The awn inhibition gene *B1* has been identified on chromosome arm 5AL and annotated as a *C2H2 zinc finger* gene that may function as a transcriptional repressor. To further characterize the *B1* gene, full-length ORFs were retrieved from the genomes of WEW and other *Triticeae* species (Huang et al. 2020).

The extracellular matrix of plant organs connects the plant with its surrounding environment. Accordingly, the cuticular layer is crucial to maintaining fitness under varying environmental conditions. Among the main components of plant cuticular waxes are aliphatic compounds (hydrocarbons, esters, aldehydes, and alcohols), triterpenoids, flavonoids, and phenolic lipids. Cuticular waxes of several *Triticeae* species contain high quantities of aliphatic β-diketones and related compounds, which generates a distinct blue/gray hue (Adamski et al. 2013; von Wettstein-Knowles 1976). A study by Hen-Avivi et al. (2016) discovered a cluster of metabolic genes on chromosome 2B associated with the biosynthesis of β-diketones in wheat using the Zavitan assembly. According to their research, genes in the cluster encode proteins from several families, including type-III polyketide synthases, hydrolases, and cytochrome P450s.

Phenotypic plasticity is another important agricultural trait that refers to the ability to adapt to changes in the environment by altering the phenotype (Bloomfield et al. 2014). A QTL mapping of recombinant inbred lines under water-limited conditions revealed 79 QTLs associated with phenotypic plasticity. The favorable alleles for most of the plasticity QTLs were contributed by wild emmer wheat, demonstrating WEW's high potential for wheat improvement (Fatiukha et al. 2021). Another drought-induced form of phenotypic plasticity limits shoot growth by allocating more carbon to roots, resulting in a higher root-to-shoot ratio (Correa et al. 2019). Bacher et al. (2021) evaluated a large set of wild emmer wheat introgression lines (ILs) under contrasting water conditions and identified an adaptive IL that exhibited a significant shift in root-to-shoot ratio in response to dry conditions. Genetic analysis and root transcriptome profiling were used to identify candidate genes in Svevo and Zavitan reference genomes that may be linked to root-to-shoot carbon reallocation under water stress. Moreover, Bacher et al. (2022) characterized the physiological mechanism associated with whole-plant water flux, and showed that drought-induced shift in root-to-shoot ratio can improve plant water uptake potential in certain situations. In a study conducted by Amini et al. (2023), drought priming and seed priming were examined for the potential to improve drought tolerance in diverse wheat germplasm. Results indicated that wheat plants pre-exposed to primary and moderate drought stress retained a drought memory that induced a more efficient and faster stress recovery response to severe drought stress (Amini et al. 2023). In another research, 76 quantitative trait loci (QTLs) were identified for yield components and for several morpho-physiological traits associated with the adaptive response to water and heat stresses. Two major epistatic QTLs on chromosomes 2BL and 3BS showed consistent effects on grain yield (Graziani et al. 2014).

Health-related issues were also investigated, including the genomic regions harboring *α-gliadin* genes, which have been found to cause severe human health problems such as celiac disease. A comparison of chromosomal regions containing *α-gliadin* genes in WEW and the homologous sequences in hexaploid wheat revealed rapid evolution of these regions (Huo et al. 2019). WEW's genome was used also for population genetics studies such as that of Dahan-Meir et al. (2022), that found high genetic stability over 30 years in the in situ WEW conservation site at Ami'ad, Israel.

6 Conclusions and Perspectives

Wild emmer wheat genome is the origin of most domesticated wheat varieties, including durum and bread wheat. Currently, there are two published reference assemblies that represent only one WEW genotype: v.1 and v.2 of "Zavitan" (Avni et al. 2017; Zhu et al. 2019). The annotated and ordered reference genome sequences enabled a significant improvement in the ability to detect loci (genes and/or QTLs) of interest, to obtain information on regulatory sequences, to evaluate genome diversity, to clone and conduct genome editing, and to identify candidate genes associated with the studied traits (Fatiukha et al. 2021; Rai et al. 2019; Sánchez-Martín and Keller 2019; Maccaferri et al. 2022) hence contributing to (i) identify and eventually select valuable tetraploid wild wheat haplotypes for Mendelian and/or quantitative traits, hence the QTLome (Salvi and Tuberosa 2015), and (ii) leverage still untapped biodiversity for a more effective breeding of cultivated wheat hence better bridging tetraploid and hexaploid wheat (Maccaferri et al. 2015), and accessing haplotype diversity of tetraploid wheat germplasm somehow lost during bread wheat domestication. Yet, the assignment of the Zavitan genome to the relatively small and distant SW-Judiacum group points to the need for additional WEW genome assemblies from diverse accessions of the NE-Horanum and SW-Horanum groups. These additional genomic resources will improve our capacity to leverage WEW genetic resources for wheat improvement at the tetraploid and hexaploid levels, providing also a more effective bridging for the A and B genomes, key components for the selection of improved synthetic hexaploid wheat (Amini et al. 2023; Vasudevan et al. 2023; Haugrud et al. 2024; Li et al. 2024a). Therefore, exploring the WEW gene pool will not only improve our food security but also increase our knowledge about wheat evolution and domestication, as well as basic and functional plant genomics.

References

Aaronsohn A (1910) Agricultural and botanical explorations in Palestine. US Government Printing Office

Aaronsohn A, Schweinfurth G (1906) Die auffindung des wilden emmers (*Triticum dicoccum*) in Nordpalästina. Altneuland 3:213–216

Able JA, Langridge P, Milligan AS (2007) Capturing diversity in the cereals: many options but little promiscuity. Trends Plant Sci 12:71–79

Adamski NM, Bush MS, Simmonds J, Turner AS, Mugford SG, Jones A, Findlay K, Pedentchouk N, von Wettstein-Knowles P, Uauy C (2013) The inhibitor of *wax 1* locus (*Iw1*) prevents formation of β- and OH-β-diketones in wheat cuticular waxes and maps to a sub-cM interval on chromosome arm 2BS. Plant J 74:989–1002

Akhunov ED, Goodyear AW, Geng S, Qi L-L, Echalier B, Gill BS, Miftahudin, Gustafson JP, Lazo G, Chao S, Anderson OD, Linkiewicz AM, Dubcovsky J, Rota ML, Sorrells ME, Zhang D, Nguyen HT, Kalavacharla V, Hossain K, Kianian SF, Peng J, Lapitan NLV, Gonzalez-Hernandez JL, Anderson JA, Choi D-W, Close TJ, Dilbirligi M, Gill KS, Walker-Simmons MK, Steber C, McGuire PE, Qualset CO, Dvorak J (2003) The organization and rate of evolution of Wheat genomes are correlated with recombination rates along chromosome arms. Genome Res 13:753–763

Amini A, Majidi MM, Mokhtari N, Ghanavati M (2023) Drought stress memory in a germplasm of synthetic and common wheat: antioxidant system, physiological and morphological consequences. Sci Rep 13:8569

Avni R, Nave M, Eilam T, Sela H, Alekperov C, Peleg Z, Dvorak J, Korol A, Distelfeld A (2014) Ultra-dense genetic map of durum wheat × wild emmer wheat developed using the 90K iSelect SNP genotyping assay. Mol Breed 34:1549–1562

Avni R, Nave M, Barad O, Baruch K, Twardziok SO, Gundlach H, Hale I, Mascher M, Spannagl M, Wiebe K, Jordan KW, Golan G, Deek J, Ben-Zvi B, Ben-Zvi G, Himmelbach A, MacLachlan RP, Sharpe AG, Fritz A, Ben-David R, Budak H, Fahima T, Korol A, Faris JD, Hernandez A, Mikel MA, Levy AA, Steffenson B, Maccaferri M, Tuberosa R, Cattivelli L, Faccioli P, Ceriotti A, Kashkush K, Pourkheirandish M, Komatsuda T, Eilam T, Sela H, Sharon A, Ohad N, Chamovitz DA, Mayer KFX, Stein N, Ronen G, Peleg Z, Pozniak CJ, Akhunov ED, Distelfeld A (2017) Wild emmer genome architecture and diversity elucidate wheat evolution and domestication. Science 357:93–97

Avni R, Oren L, Shabtay G, Assili S, Pozniak C, Hale I, Ben-David R, Peleg Z, Distelfeld A (2018) Genome based meta-QTL analysis of grain weight in tetraploid Wheat identifies rare alleles of GRF4 associated with larger grains. Genes 9:636

Bacher H, Zhu F, Gao T, Liu K, Dhatt BK, Awada T, Zhang C, Distelfeld A, Yu H, Peleg Z, Walia H (2021) Wild emmer introgression alters root-to-shoot growth dynamics in durum wheat in response to water stress. Plant Physiol 187:1149–1162

Bacher H, Sharaby Y, Walia H, Peleg Z (2022) Modifying root-to-shoot ratio improves root water influxes in wheat under drought stress. J Exp Bot 73:1643–1654

Badaeva ED, Konovalov FA, Knüpffer H, Fricano A, Ruban AS, Kehel Z, Zoshchuk SA, Surzhikov SA, Neumann K, Graner A, Hammer K, Filatenko A, Bogaard A, Jones G, Özkan H, Kilian B (2022) Genetic diversity, distribution and domestication history of the neglected GGAtAt genepool of wheat. Theor Appl Genet 135:755–776

Bloomfield JA, Rose TJ, King GJ (2014) Sustainable harvest: managing plasticity for resilient crops. Plant Biotechnol J 12:517–533

Börner A, Schäfer M, Schmidt A, Grau M, Vorwald J (2005) Associations between geographical origin and morphological characters in bread wheat (*Triticum aestivum* L.). Plant Genetic Resourc 3:360–372

Chen S, Guo Y, Briggs J, Dubach F, Chao S, Zhang W, Rouse MN, Dubcovsky J (2018) Mapping and characterization of wheat stem rust resistance genes *SrTm5* and *Sr60* from *Triticum monococcum*. Theor Appl Genet 131:625–635

Correa J, Postma JA, Watt M, Wojciechowski T (2019) Soil compaction and the architectural plasticity of root systems. J Exp Bot 70:6019–6034

Dahan-Meir T, Ellis TJ, Mafessoni F, Sela H, Manisterski J, Avivi-Ragolsky N, Raz A, Feldman M, Anikster Y, Nordborg M, Levy AA (2022) Stability of the genetic structure of a wild wheat population across microhabitats over 36 years. bioRxiv:2022.01.10.475641

Davis EL, Haegeman A, Kikuchi T (2011) Degradation of the plant Cell Wall by nematodes. In: Jones J, Gheysen G, Fenoll C (eds) Genomics and molecular genetics of plant-nematode interactions. Springer Netherlands, Dordrecht, pp 255–272

Ding Y, Sun T, Ao K, Peng Y, Zhang Y, Li X, Zhang Y (2018) Opposite roles of salicylic acid receptors NPR1 and NPR3/NPR4 in transcriptional regulation of plant immunity. Cell 173:1454–1467.e15

Distelfeld A, Li C, Dubcovsky J (2009) Regulation of flowering in temperate cereals. Curr Opin Plant Biol 12:178–184

Dvorak J, Akhunov ED (2005) Tempos of gene locus deletions and duplications and their relationship to recombination rate during diploid and polyploid evolution in the Aegilops-Triticum alliance. Genetics 171:323–332

Dvorak J, Zhang H-B (1990) Variation in repeated nucleotide sequences sheds light on the phylogeny of the wheat B and G genomes. Proc Natl Acad Sci 87:9640–9644

Dvorak J, di Terlizzi P, Zhang H-B, Resta P (1993) The evolution of polyploid wheats: identification of the a genome donor species. Genome 36:21–31

Dvorak J, Luo MC, Yang ZL (1998) Restriction fragment length polymorphism and divergence in the genomic regions of high and low recombination in self-fertilizing and cross-fertilizing aegilops species. Genetics 148:423–434

Dvorak J, Yang Z-L, You FM, Luo M-C (2004) Deletion polymorphism in wheat chromosome regions with contrasting recombination rates. Genetics 168:1665–1675

Elbaum R, Zaltzman L, Burgert I, Fratzl P (2007) The role of Wheat awns in the seed dispersal unit. Science (New York, N.Y.) 316:884–886

Fahima T, Sun GL, Beharav A, Krugman T, Beiles A, Nevo E (1999) RAPD polymorphism of wild emmer wheat populations, *Triticum dicoccoides*, in Israel. Theor Appl Genet 98:434–447

Fahima T, Röder MS, Wendehake K, Kirzhner VM, Nevo E (2002) Microsatellite polymorphism in natural populations of wild emmer wheat, Triticum dicoccoides, in Israel. Theor Appl Genet 104:17–29

Fatiukha A, Deblieck M, Klymiuk V, Merchuk-Ovnat L, Peleg Z, Ordon F, Fahima T, Korol A, Saranga Y, Krugman T (2021) Genomic architecture of phenotypic plasticity in response to water stress in tetraploid Wheat. Int J Mol Sci 22:1723

Flint-Garcia SA, Thornsberry JM, Buckler ES (2003) Structure of linkage disequilibrium in plants. Annu Rev Plant Biol 54:357–374

Fu Y-B, Peterson GW, Horbach C, Konkin DJ, Beiles A, Nevo E (2019) Elevated mutation and selection in wild emmer wheat in response to 28 years of global warming. Proc Natl Acad Sci 116:20002–20008

Graziani M, Maccaferri M, Royo C, Salvatorelli F, Tuberosa R (2014) QTL dissection of yield components and morpho-physiological traits in a durum wheat elite population tested in contrasting thermo-pluviometric conditions. Crop Pasture Sci 65:80–95

Guan P, Di N, Mu Q, Shen X, Wang Y, Wang X, Yu K, Song W, Chen Y, Xin M, Hu Z, Guo W, Yao Y, Ni Z, Sun Q, Peng H (2019) Use of near-isogenic lines to precisely map and validate a major QTL for grain weight on chromosome 4AL in bread wheat (*Triticum aestivum* L.). Theor Appl Genet 132:2367–2379

Guo W, Xin M, Wang Z, Yao Y, Hu Z, Song W, Yu K, Chen Y, Wang X, Guan P (2020) Origin and adaptation to high altitude of Tibetan semi-wild wheat. Nat Commun 11:5085

Harlan JR, Zohary D (1966) Distribution of wild wheats and barley. Science 153:1074–1080

Haugrud PAR, Achilli AL, Martínez-Peña R, Klymiuk V (2024) Future of durum wheat research and breeding: insights from early career researchers. Plant Genome e20453

He H, Liu R, Ma P, Du H, Zhang H, Wu Q, Yang L, Gong S, Liu T, Huo N, Gu YQ, Zhu S (2021) Characterization of *Pm68*, a new powdery mildew resistance gene on chromosome 2BS of Greek durum wheat TRI 1796. Theor Appl Genet 134:53–62

Hen-Avivi S, Savin O, Racovita RC, Lee W-S, Adamski NM, Malitsky S, Almekias-Siegl E, Levy M, Vautrin S, Bergès H, Friedlander G, Kartvelishvily E, Ben-Zvi G, Alkan N, Uauy C, Kanyuka K, Jetter R, Distelfeld A, Aharoni A (2016) A metabolic gene cluster in the Wheat W1 and the barley *Cer-cqu* loci determines β-Diketone biosynthesis and Glaucousness. Plant Cell 28:1440–1460

Huang X, Börner A, Röder M, Ganal M (2002) Assessing genetic diversity of wheat (*Triticum aestivum* L.) germplasm using microsatellite markers. Theor Appl Genet 105:699–707

Huang D, Zheng Q, Melchkart T, Bekkaoui Y, Konkin DJF, Kagale S, Martucci M, You FM, Clarke M, Adamski NM, Chinoy C, Steed A, McCartney CA, Cutler AJ, Nicholson P, Feurtado JA (2020) Dominant inhibition of awn development by a putative zinc-finger transcriptional repressor expressed at the *B1* locus in wheat. New Phytol 225:340–355

Huo N, Zhu T, Zhang S, Mohr T, Luo M-C, Lee J-Y, Distelfeld A, Altenbach S, Gu YQ (2019) Rapid evolution of α-gliadin gene family revealed by analyzing *Gli-2* locus regions of wild emmer wheat. Funct Integr Genomics 19:993–1005

Jakubziner MM (1932a) Contribution to the knowledge of wild wheat in Transcaucasia. Tud Prikl Bot Genet Selek 5:147–198

Jakubziner MM (1932b) The wheats of Syria, Palestine and Transjordania, cultivated and wild. Trudy prikl Bot Genet Selek Suppl, 53

Jones MGK, Fosu-Nyarko J (2014) Molecular biology of root lesion nematodes (*Pratylenchus* spp.) and their interaction with host plants. Ann Appl Biol 164:163–181

Kebrom TH, Spielmeyer W, Finnegan EJ (2013) Grasses provide new insights into regulation of shoot branching. Trends Plant Sci 18:41–48

Kimura M (1968) Evolutionary rate at the molecular level. Nature 217:624–626

Klymiuk V, Yaniv E, Huang L, Raats D, Fatiukha A, Chen S, Feng L, Frenkel Z, Krugman T, Lidzbarsky G, Chang W, Jääskeläinen MJ, Schudoma C, Paulin L, Laine P, Bariana H, Sela H, Saleem K, Sørensen CK, Hovmøller MS, Distelfeld A, Chalhoub B, Dubcovsky J, Korol AB, Schulman AH, Fahima T (2018) Cloning of the wheat Yr15 resistance gene sheds light on the plant tandem kinase-pseudokinase family. Nat Commun 9:3735

Kuraparthy V, Sood S, Dhaliwal HS, Chhuneja P, Gill BS (2007) Identification and mapping of a tiller inhibition gene (tin3) in wheat. Theor Appl Genet 114:285–294

Kuzay S, Xu Y, Zhang J, Katz A, Pearce S, Su Z, Fraser M, Anderson JA, Brown-Guedira G, DeWitt N, Peters Haugrud A, Faris JD, Akhunov E, Bai G, Dubcovsky J (2019) Identification of a candidate gene for a QTL for spikelet number per spike on wheat chromosome arm 7AL by high-resolution genetic mapping. Theor Appl Genet 132:2689–2705

Lack HW, Slageren MV (2020) The discovery, typification and rediscovery of wild emmer wheat, *Triticum turgidum* subsp. *dicoccoides* (Poaceae). Willdenowia 50:207–216

Li Y, Wei Z-Z, Fatiukha A, Jaiwar S, Wang H, Hasan S, Liu Z, Sela H, Krugman T, Fahima T (2021) TdPm60 identified in wild emmer wheat is an ortholog of *Pm60* and constitutes a strong candidate for *PmG16* powdery mildew resistance. Theor Appl Genet 134:2777–2793

Li H, Zhu L, Fan R, Li Z, Liu Y, Shaheen A, Nie F, Li C, Liu X, Li Y, Liu W, Yang Y, Guo T, Zhu Y, Bu M, Li C, Liang H, Bai S, Ma F, Guo G, Zhang Z, Huang J, Zhou Y, Song C-P (2024a) A platform for whole-genome speed introgression from *Aegilops tauschii* to wheat for breeding future crops. Nat Protoc 19:281–312

Li M, Zhang H, Xiao H, Zhu K, Shi W, Zhang D, Wang Y, Yang L, Wu Q, Xie J, Chen Y, Qiu D, Guo G, Lu P, Li B, Dong L, Li W, Cui X, Li L, Tian X, Yuan C, Li Y,

Yu D, Nevo E, Fahima T, Li H, Dong L, Zhao Y, Liu Z (2024b) A membrane associated tandem kinase from wild emmer wheat confers broad-spectrum resistance to powdery mildew. Nat Commun 15:3124

Lieberman-Aiden E, van Berkum NL, Williams L, Imakaev M, Ragoczy T, Telling A, Amit I, Lajoie BR, Sabo PJ, Dorschner MO, Sandstrom R, Bernstein B, Bender MA, Groudine M, Gnirke A, Stamatoyannopoulos J, Mirny LA, Lander ES, Dekker J (2009) Comprehensive mapping of Long-range interactions reveals folding principles of the human genome. Science 326:289–293

Liu X, Liu Z, Niu X, Xu Q, Long Y (2019) Genome-wide identification and analysis of the *NPR1*-like gene family in bread Wheat and its relatives. Int J Mol Sci 20:5974

Liu J, Tang H, Qu X, Liu H, Li C, Tu Y, Li S, Habib A, Mu Y, Dai S, Deng M, Jiang Q, Liu Y, Chen G, Wang J, Chen G, Li W, Jiang Y, Wei Y, Lan X, Zheng Y, Ma J (2020) A novel, major, and validated QTL for the effective tiller number located on chromosome arm 1BL in bread wheat. Plant Mol Biol 104:173–185

Luo M-C, Yang Z-L, You FM, Kawahara T, Waines JG, Dvorak J (2007) The structure of wild and domesticated emmer wheat populations, gene flow between them, and the site of emmer domestication. Theor Appl Genet 114:947–959

Ma J, Qin N, Cai B, Chen G, Ding P, Zhang H, Yang C, Huang L, Mu Y, Tang H, Liu Y, Wang J, Qi P, Jiang Q, Zheng Y, Liu C, Lan X, Wei Y (2019) Identification and validation of a novel major QTL for all-stage stripe rust resistance on 1BL in the winter wheat line 20828. Theor Appl Genet 132:1363–1373

Maccaferri M, Ricci A, Salvi S, Milner SG, Noli E, Martelli PL, Casadio R, Akhunov E, Scalabrin S, Vendramin V, Ammar K, Blanco A, Desiderio F, Distelfeld A, Jorge DJ, Fahima T, Faris J, Korol A, Massi A, Mastrangelo AM, Morgante M, Pozniak C, N'Diaye A, Xu S, Tuberosa R (2015) A high-density, SNP-based consensus map of tetraploid wheat as a bridge to integrate durum and bread wheat genomics and breeding. Plant Biotechnol J 13(5):648–663

Maccaferri M, Harris NS, Twardziok SO, Pasam RK, Gundlach H, Spannagl M, Ormanbekova D, Lux T, Prade VM, Milner SG, Himmelbach A, Mascher M, Bagnaresi P, Faccioli P, Cozzi P, Lauria M, Lazzari B, Stella A, Manconi A, Gnocchi M, Moscatelli M, Avni R, Deek J, Biyiklioglu S, Frascaroli E, Corneti S, Salvi S, Sonnante G, Desiderio F, Marè C, Crosatti C, Mica E, Özkan H, Kilian B, De Vita P, Marone D, Joukhadar R, Mazzucotelli E, Nigro D, Gadaleta A, Chao S, Faris JD, Melo ATO, Pumphrey M, Pecchioni N, Milanesi L, Wiebe K, Ens J, MacLachlan RP, Clarke JM, Sharpe AG, Koh CS, Liang KYH, Taylor GJ, Knox R, Budak H, Mastrangelo AM, Xu SS, Stein N, Hale I, Distelfeld A, Hayden MJ, Tuberosa R, Walkowiak S, Mayer KFX, Ceriotti A, Pozniak CJ, Cattivelli L (2019) Durum wheat genome highlights past domestication signatures and future improvement targets. Nat Genet 51:885–895

Maccaferri M, Bruschi M, Tuberosa R (2022) Sequence-based marker assisted selection in Wheat. In: Reynolds MP, Braun H-J (eds) Wheat improvement: food security in a changing climate. Springer International Publishing, Cham, pp 513–538. Available at: https://doi.org/10.1007/978-3-030-90673-3_28. Accessed 22 July 2024

Martin W, Salamini F (2000) A meeting at the gene. EMBO Rep 1:208–210

Mazzucotelli E, Mastrangelo AM, Desiderio F, Barabaschi D, Maccaferri M, Tuberosa R, Cattivelli L (2024) Gene flow between tetraploid and Hexaploid Wheat for breeding innovation. In: The Wheat genome. Springer, Cham, pp 135–163. Available at: https://link.springer.com/chapter/10.1007/978-3-031-38294-9_8. Accessed 19 July 2024

Morgounov A, Tufan HA, Sharma R, Akin B, Bagci A, Braun H-J, Kaya Y, Keser M, Payne TS, Sonder K, McIntosh R (2012) Global incidence of wheat rusts and powdery mildew during 1969–2010 and durability of resistance of winter wheat variety Bezostaya 1. Eur J Plant Pathol 132:323–340

Mori N, Ishii T, Ishido T, Hirosawa S, Watatani H, Kawahara T, Nesbitt M, Belay G, Takumi S, Ogihara Y (2003) Origin of domesticated emmer and common wheat inferred from chloroplast DNA fingerprinting. In: 10th International wheat genetics symposium, 1–6

Murai K, Miyamae M, Kato H, Takumi S, Ogihara Y (2003) WAP1, a Wheat APETALA1 homolog, plays a central role in the phase transition from vegetative to reproductive growth. Plant Cell Physiol 44:1255–1265

Nave M, Avni R, Ben-Zvi B, Hale I, Distelfeld A (2016) QTLs for uniform grain dimensions and germination selected during wheat domestication are co-located on chromosome 4B. Theor Appl Genet 129:1303–1315

Özkan H, Brandolini A, Schäfer-Pregl R, Salamini F (2002) AFLP analysis of a collection of tetraploid wheats indicates the origin of emmer and hard Wheat domestication in Southeast Turkey. Mol Biol Evol 19:1797–1801

Özkan H, Brandolini A, Pozzi C, Effgen S, Wunder J, Salamini F (2005) A reconsideration of the domestication geography of tetraploid wheats. Theor Appl Genet 110:1052–1060

Özkan H, Willcox G, Graner A, Salamini F, Kilian B (2011) Geographic distribution and domestication of wild emmer wheat (*Triticum dicoccoides*). Genet Resour Crop Evol 58:11–53

Pourkheirandish M, Hensel G, Kilian B, Senthil N, Chen G, Sameri M, Azhaguvel P, Sakuma S, Dhanagond S, Sharma R, Mascher M, Himmelbach A, Gottwald S, Nair SK, Tagiri A, Yukuhiro F, Nagamura Y, Kanamori H, Matsumoto T, Willcox G, Middleton CP, Wicker T, Walther A, Waugh R, Fincher GB, Stein N, Kumlehn J, Sato K, Komatsuda T (2015) Evolution of the grain dispersal system in barley. Cell 162:527–539

Pourkheirandish M, Dai F, Sakuma S, Kanamori H, Distelfeld A, Willcox G, Kawahara T, Matsumoto T, Kilian B, Komatsuda T (2018) On the origin of the non-brittle rachis trait of domesticated einkorn Wheat. Front Plant Sci 8

Poyarkova H (1988) Morphology, geography and infraspecific taxonomics of *Triticum dicoccoides* Körn. A retrospective of 80 years of research. Euphytica 38:11–23

Poyarkova H, Gerechter-Amitai ZK, Genizi A (1991) Two variants of wild emmer (*Triticum dicoccoides*) native to Israel: morphology and distribution. Can J Bot 69:2772–2789

Qian Z, Liu R, Liu X, Qie Y, Wang J, Yin Y, Xin Q, Yu N, Zhang J, Li Y, Li J, Dai Y, Liu C, Jin Y, Ma P (2024) Bulked segregant RNA-seq reveals complex resistance expression profile to powdery mildew in wild emmer wheat W762. Front Plant Sci 15. Available at: https://www.frontiersin.org/journals/plant-science/articles/10.3389/fpls.2024.1387427/full. Accessed 20 July 2024

Rahman MS, Linsell KJ, Taylor JD, Hayden MJ, Collins NC, Oldach KH (2020) Fine mapping of root lesion nematode (Pratylenchus thornei) resistance loci on chromosomes 6D and 2B of wheat. Theor Appl Genet 133:635–652

Rai A, Yamazaki M, Saito K (2019) A new era in plant functional genomics. Curr Opin Syst Biol 15:58–67

Rebetzke GJ, Bonnett DG, Reynolds MP (2016) Awns reduce grain number to increase grain size and harvestable yield in irrigated and rainfed spring wheat. J Exp Bot 67:2573–2586

Salvi S, Tuberosa R (2015) The crop QTLome comes of age. Curr Opin Biotechnol 32:179–185

Sánchez-Martín J, Keller B (2019) Contribution of recent technological advances to future resistance breeding. Theor Appl Genet 132:713–732

Sandhu D, Gill KS (2002) Gene-containing regions of Wheat and the other grass genomes. Plant Physiol 128:803–811

Sarkar P, Stebbins GL (1956) Morphological evidence concerning the origin of the B genome in wheat. Am J Bot 43:297–304

Schweinfurth GA (1908) Über die von A. Aaronsohn ausgeführten Nachforschungen nach dem wilden Emmer (Triticum dicoccoides Kcke)

Sela H, Ezrati S, Ben-Yehuda P, Manisterski J, Akhunov E, Dvorak J, Breiman A, Korol A (2014) Linkage disequilibrium and association analysis of stripe rust resistance in wild emmer wheat (*Triticum turgidum* ssp. *dicoccoides*) population in Israel. Theor Appl Genet 127:2453–2463

Singh RP, Singh PK, Rutkoski J, Hodson DP, He X, Jørgensen LN, Hovmøller MS, Huerta-Espino J (2016) Disease impact on wheat yield potential and prospects of genetic control. Annu Rev Phytopathol 54:303–322

Slatkin M (2008) Linkage disequilibrium — understanding the evolutionary past and mapping the medical future. Nat Rev Genet 9:477–485

Tajima F (1989) Statistical method for testing the neutral mutation hypothesis by DNA polymorphism. Genetics 123:585–595

Tene M, Adhikari E, Cobo N, Jordan KW, Matny O, del Blanco IA, Roter J, Ezrati S, Govta L, Manisterski J, Yehuda PB, Chen X, Steffenson B, Akhunov E, Sela H (2022) GWAS for stripe rust resistance in wild emmer Wheat (*Triticum dicoccoides*) population: obstacles and solutions. Crops 2:42–61

The International Wheat Genome Sequencing Consortium (IWGSC), Appels R, Eversole K, Stein N, Feuillet C, Keller B, Rogers J, Pozniak CJ, Choulet F, Distelfeld A, Poland J, Ronen G, Sharpe AG, Barad O, Baruch K, Keeble-Gagnère G, Mascher M, Ben-Zvi G, Josselin A-A, Himmelbach A, Balfourier F, Gutierrez-Gonzalez J, Hayden M, Koh C, Muehlbauer G, Pasam RK, Paux E, Rigault P, Tibbits J, Tiwari V, Spannagl M, Lang D, Gundlach H, Haberer G, Mayer KFX, Ormanbekova D, Prade V, Šimková H, Wicker T, Swarbreck D, Rimbert H, Felder M, Guilhot N, Kaithakottil G, Keilwagen J, Leroy P, Lux T, Twardziok S, Venturini L, Juhász A, Abrouk M, Fischer I, Uauy C, Borrill P, Ramirez-Gonzalez RH, Arnaud D, Chalabi S, Chalhoub B, Cory A, Datla R, Davey MW, Jacobs J, Robinson SJ, Steuernagel B, van Ex F, Wulff BBH, Benhamed M, Bendahmane A, Concia L, Latrasse D, Bartoš J, Bellec A, Berges H, Doležel J, Frenkel Z, Gill B, Korol A, Letellier T, Olsen O-A, Singh K, Valárik M, van der Vossen E, Vautrin S, Weining S, Fahima T, Glikson V, Raats D, Číhalíková J, Toegelová H, Vrána J, Sourdille P, Darrier B, Barabaschi D, Cattivelli L, Hernandez P, Galvez S, Budak H, Jones JDG, Witek K, Yu G, Small I, Melonek J, Zhou R, Belova T, Kanyuka K, King R, Nilsen K, Walkowiak S, Cuthbert R, Knox R, Wiebe K, Xiang D, Rohde A, Golds T, Čížková J, Akpinar BA, Biyiklioglu S, Gao L, N'Daiye A, Kubaláková M, Šafář J, Alfama F, Adam-Blondon A-F, Flores R, Guerche C, Loaec M, Quesneville H, Condie J, Ens J, Maclachlan R, Tan Y, Alberti A, Aury J-M, Barbe V, Couloux A, Cruaud C, Labadie K, Mangenot S, Wincker P, Kaur G, Luo M, Sehgal S, Chhuneja P, Gupta OP, Jindal S, Kaur P, Malik P, Sharma P, Yadav B, Singh NK, Khurana JP, Chaudhary C, Khurana P, Kumar V, Mahato A, Mathur S, Sevanthi A, Sharma N, Tomar RS, Holušová K, Plíhal O, Clark MD, Heavens D, Kettleborough G, Wright J, Balcárková B, Hu Y, Salina E, Ravin N, Skryabin K, Beletsky A, Kadnikov V, Mardanov A, Nesterov M, Rakitin A, Sergeeva E, Handa H, Kanamori H, Katagiri S, Kobayashi F, Nasuda S, Tanaka T, Wu J, Cattonaro F, Jiumeng M, Kugler K, Pfeifer M, Sandve S, Xun X, Zhan B, Batley J, Bayer PE, Edwards D, Hayashi S, Tulpová Z, Visendi P, Cui L, Du X, Feng K, Nie X, Tong W, Wang L (2018) Shifting the limits in wheat research and breeding using a fully annotated reference genome. Science 361:eaar7191

Thellung A (1920) 1918. Neuere Wege und Ziele der botanischen Systematik, erläutert am Beispiele unserer Getreidearten. Z Indukt Abstamm Vererbungsl 22:293–295

Thuillet AC, Bataillon T, Sourdille P, David JL (2004) Factors affecting polymorphism at microsatellite loci in bread wheat [*Triticum aestivum* (L.) Thell]: effects of mutation processes and physical distance from the centromere. Theor Appl Genet 108:368–377

van Slageren M, Payne T (2013) Concepts and nomenclature of the Farro wheats, with special reference to Emmer, *Triticum turgidum* subsp. *dicoccum* (Poaceae). Kew Bull 68:477–494

Vasudevan A, Lévesque-Lemay M, Edwards T, Cloutier S (2023) Global transcriptome analysis of allopolyploidization reveals large-scale repression of the D-subgenome in synthetic hexaploid wheat. Commun Biol 6:426

Vavilov NI, Fortunatova OK, Jakubziner MM, Palmova ET, Nikolaenko EI, Stoletova EA, Verkhovskaya KA, Shreiber LL, Syrovatsky SG (1931) The wheats of Abyssinia. Trudy prikl Bot Genet Selek Suppl, 51

von Wettstein-Knowles P (1976) Biosynthetic relationships between β-diketones and esterified alkan-2-ols deduced from epicuticular wax of barley mutants. Mol Gen Genet MGG 144:43–48

Walkowiak S, Gao L, Monat C, Haberer G, Kassa MT, Brinton J, Ramirez-Gonzalez RH, Kolodziej MC, Delorean E, Thambugala D, Klymiuk V, Byrns B, Gundlach H, Bandi V, Siri JN, Nilsen K, Aquino C, Himmelbach A, Copetti D, Ban T, Venturini L, Bevan M, Clavijo B, Koo D-H, Ens J, Wiebe K, N'Diaye A, Fritz AK, Gutwin C, Fiebig A, Fosker C, Fu BX, Accinelli GG, Gardner KA, Fradgley N, Gutierrez-Gonzalez J, Halstead-Nussloch G, Hatakeyama M, Koh CS, Deek J, Costamagna AC, Fobert P, Heavens D, Kanamori H, Kawaura K, Kobayashi F, Krasileva K, Kuo T, McKenzie N, Murata K, Nabeka Y, Paape T, Padmarasu S, Percival-Alwyn L, Kagale S, Scholz U, Sese J, Juliana P, Singh R, Shimizu-Inatsugi R, Swarbreck D, Cockram J, Budak H, Tameshige T, Tanaka T, Tsuji H, Wright J, Wu J, Steuernagel B, Small I, Cloutier S, Keeble-Gagnère G, Muehlbauer G, Tibbets J, Nasuda S, Melonek J, Hucl PJ, Sharpe AG, Clark M, Legg E, Bharti A, Langridge P, Hall A, Uauy C, Mascher M, Krattinger SG, Handa H, Shimizu KK, Distelfeld A, Chalmers K, Keller B, Mayer KFX, Poland J, Stein N, McCartney CA, Spannagl M, Wicker T, Pozniak CJ (2020) Multiple wheat genomes reveal global variation in modern breeding. Nature 588:277–283

Wei W, Liu N, Zhang S, Zhang J, Pan W, Xie X, Yang Z, Sun J, Ma J, Hu Z, Guo W, Luo Q, Xie J, He F, Li Y, Xie C, Sun Q (2024) Mapping of powdery mildew resistance genes transferred to common wheat from wild emmer wheat revealed three functional *Pm60* haplotypes. Crop J 12:540–548

Wellings CR (2011) Global status of stripe rust: a review of historical and current threats. Euphytica 179:129–141

Wu J, Zeng Q, Wang Q, Liu S, Yu S, Mu J, Huang S, Sela H, Distelfeld A, Huang L, Han D, Kang Z (2018) SNP-based pool genotyping and haplotype analysis accelerate fine-mapping of the wheat genomic region containing stripe rust resistance gene *Yr26*. Theor Appl Genet 131:1481–1496

Wu P, Hu J, Zou J, Qiu D, Qu Y, Li Y, Li T, Zhang H, Yang L, Liu H, Zhou Y, Zhang Z, Li J, Liu Z, Li H (2019) Fine mapping of the wheat powdery mildew resistance gene *Pm52* using comparative genomics analysis and the Chinese spring reference genomic sequence. Theor Appl Genet 132:1451–1461

Yahiaoui N, Kaur N, Keller B (2008) Independent evolution of functional *Pm3* resistance genes in wild tetraploid wheat and domesticated bread wheat. Plant J 57:846–856

Yan L, Helguera M, Kato K, Fukuyama S, Sherman J, Dubcovsky J (2004) Allelic variation at the VRN-1 promoter in polyploid wheat. Theor Appl Genet 109:1677–1686

Yang L, Shao H, Wu QX, Yu J, Rang CF, Li LQ, Li X (2013) QTLs mapping and epistasis analysis for the number of tillers and spike number per plant in wheat. J Tritic Crops 33:875–882

Zhang Y, Li X (2019) Salicylic acid: biosynthesis, perception, and contributions to plant immunity. Curr Opin Plant Biol 50:29–36

Zhu T, Wang L, Rodriguez JC, Deal KR, Avni R, Distelfeld A, McGuire PE, Dvorak J, Luo M-C (2019) Improved genome sequence of wild emmer Wheat Zavitan with the aid of optical maps. G3 (Bethesda) 9:619–624

Zohary D, Hopf M, Weiss E (2012) Domestication of plants in the Old World: the origin and spread of domesticated plants in Southwest Asia, Europe, and the Mediterranean Basin. OUP Oxford

The Durum Wheat Genome

Francesca Desiderio, Marco Maccaferri,
Roberto Tuberosa, and Luigi Cattivelli

Abstract

The durum wheat reference genome (cv. Svevo) assembly has become an essential tool to study durum wheat evolution and breeding, as well as for acquiring new knowledge on gene and genome function and organization. The modern durum wheat cultivars and its progenitor, the wild emmer, are separated by about 10,000 years of domestication and subsequent evolution under domestication characterized by a two-step domestication process and one century of knowledge-based plant breeding. The availability of wild emmer accession Zavitan and of the durum wheat cv. Svevo genome sequences allowed for a direct comparison of the two genomes to better understand the changes that occurred during the recent domestication and selection processes. The durum wheat genome has paved the way to a deeper analysis and, eventually, exploitation of tetraploid wheat germplasm diversity for studying the mechanisms of polyploidy evolution, domestication, genetic and epigenetic regulation of homoeologous expression, as well as defining and mapping the wide chart of genetic diversity present in the tetraploid wheat germplasm and assisting breeding strategies for the improvement of wheat as a whole, including durum and bread wheat.

F. Desiderio · L. Cattivelli (✉)
CREA Research Centre for Genomics and Bioinformatics, Fiorenzuola d'Arda, Italy
e-mail: francesca.desiderio@crea.gov.it; luigi.cattivelli@crea.gov.it

M. Maccaferri · R. Tuberosa (✉)
Department of Agricultural and Food Sciences, University of Bologna, Bologna, Italy
e-mail: marco.maccaferri@unibo.it; roberto.tuberosa@unibo.it

1 Introduction

Durum wheat (*Triticum turgidum* L. subsp. *durum* [Desf.] Husn, genome BBAA) was domesticated from wild emmer (*T. turgidum* subsp. *dicoccoides*) through an intermediate step represented by cultivated emmer (*T. turgidum* subsp. *dicoccum*) and then subjected to further genetic modifications during the selection of the modern cultivars. Cultivated emmer was domesticated from wild emmer in the Near-Eastern Fertile Crescent ~10,000 years BP, via selection of plants carrying non-functional mutations on the two homoeologous genes conferring the non-brittle rachis phenotype (Dubcoksky and Dvorak 2007). Cultivated emmer, the result of the first and most important domestication step, is characterized by non-brittle spikes, but, like its wild progenitor, retains hulled kernels with strong glumes with only two fertile florets per spikelet. Then the transition from emmer to durum wheat involved both the acquisition of free threshing

R. Tuberosa et al. (eds.), *The Durum Wheat Genome*, Compendium of Plant Genomes,
https://doi.org/10.1007/978-3-031-84689-2_2

and the appearance of the naked kernel phenotype (Faris 2014). Naked tetraploid wheat forms, including *Triticum turgidum* ssp. *turgidum*, *carthlicum*, *aethiopicum*, *turanicum*, *polonicum*, and *durum*, gradually replaced emmer in most wheat-growing areas thanks to its larger grains and higher productivity.

Archaeological evidence suggests a sporadic appearance of durum-like wheat about 7000 years BP in the Near East, but it was only about 2500 years BP when durum wheat became a major crop in the Mediterranean Basin and in the eastern Europe and west/central Asia (Feldman and Millet 2001; Faris 2014). Today, durum is the most widely grown tetraploid wheat and provides the essential ingredients for making pasta, a main component of the now global Mediterranean diet.

The expansion of emmer cultivation promoted the hybridization of domesticated and semi-domesticated forms with *Aegilops tauschii* (genome DD) and the emergence of the hexaploid bread wheat (*Triticum aestivum* L. subsp. *aestivum*, genome BBAADD) about 6000 years BP (Dubcoksky and Dvorak 2007). Domesticated emmer being one of the two progenitors of bread wheat also represents a major reservoir of genetic diversity to be exploited for bread wheat improvement. Domesticated emmer populations expanded from the Fertile Crescent with northern–southern and eastern–western directions and developed subsequently expanded populations and sub-populations adapted to a range of new environments, recently colonized by cultivated forms through the already known classical human migration routes.

Researchers were faced with the challenge of understanding how much basic and applied knowledge gain they could recover from knowing of the nature and effects that domestication and selection impacted a germplasm to eventually lead to modern cultivars (Katamadze et al. 2023). Wild progenitors of modern crop plants provide valuable genetic diversity, for traits ranging from disease resistance (as an example: *Sbcmv1*, *Yr15*, *Pm41*, and *Lr53*; Bruschi et al. 2024; Klymiuk et al. 2018; Li et al. 2009; Li et al. 2020; Marais et al. 2005) to (i) quality attributes (as *Gpc-B1*, the gene responsible for high grain protein and mineral content: Uauy et al. 2006; Tabbita et al. 2017) and (ii) phenology and plant development (as *TaEps*, Faricelli et al. 2010); and comparative genomic analysis with wild progenitors is a key strategy for detecting structural variation that can constrain breeding efforts, dissecting the mechanisms underlying domestication traits, producing new extreme phenotypes, and understanding the genetic consequences of selection history (Dubcoksky and Dvorak 2007; Gaut 2015; Brozynska et al. 2016; Jordan et al. 2020; Yuan et al. 2022).

In recent years the main wheat genomes (tetraploid and hexaploid) were all sequenced. In 2017, the total assembly size of 10.5 Gb resulting in the 14 chromosomes of wild emmer (*Triticum turgidum* ssp. *dicoccoides*, accession Zavitan) was published (Avni et al. 2017) and later revised in a second version improved using optical maps (Zhu et al. 2019). In 2018, the 21 chromosomes of bread wheat (17 Gb; cultivar Chinese Spring) were released (IWGSC 2018). In 2019, an international consortium reported a high-quality reference sequence of the modern durum wheat cultivar Svevo (10.5 Gb; Maccaferri et al. 2019). These genome sequences represent a major milestone for wheat biology providing resources for functional wheat genomics.

2 The Svevo Genome

The durum wheat genome (cv. Svevo) assembly was published by Maccaferri et al. (2019). It has become an essential and pivotal tool to study durum wheat evolution and breeding, as well as for acquiring new knowledge both on the gene function and on the organization, at the genomic level, of the quantitative trait loci (QTLs) related to the relevant agronomic traits. The Svevo genome assembly was assembled from the 3.35 Tb of Illumina sequencing data using the software package DenovoMAGIC2™ (NRGene, Nes Ziona, Israel) and the resulting scaffolds (N50 length = 6.0 Mb; N50 number of scaffolds = 493) were ordered and oriented by genetic mapping and chromosome conformation capture (HiC) sequencing. The genome resulted in a set of pseudomolecules (9.96 Gb) corresponding to the 14 chromosomes and one group of unassigned scaffolds (Chr_Un, 499 Mb). The pseudomolecules encompass 95.3% of the assembled sequence and have 90% of the scaffolds oriented (Maccaferri et al. 2019).

Annotation of the Svevo genome, supported with RNA-seq data generated from 57 different combinations of tissues, treatments, and developmental stages, led to the identification of 66,559 high-confidence genes (HC: a coding sequence with annotated start and stop codon and representative hit to a reference protein sequence), 90.5% of which exhibited detectable evidence of expression. Furthermore, the annotation identified 303,404 low-confidence genes (LC: a coding sequence that was not annotated completely or showed an insufficient homology to reference proteins or was a likely candidate for transposons). The presence of low-confidence genes should not be underestimated, as genes that are low confidence in a cultivar could be high confidence in others.

The main annotation statistics are reported in Table 1. The annotation also identified 774 unique miRNA loci grouped into 67 families.

Beside the identification of the genes, the annotation process also highlighted a vast number of pseudogenes. A homology search with a combined query set of durum wheat canonical HC genes resulted in the annotation of ~280,000 gene-like sequences. About 90% of them are short gene fragments. Taking only those that cover at least 80% of their parent reduces the numbers to ~28,000 (Table 2). The number of pseudogenes is higher compared to barley, and this is likely to be a consequence of the polyploidy of the tetraploid genome that could represent a sign of an ongoing gene-loss process facilitated by complementation through copies of homologous genes.

Table 1 Annotation statistics with high-confidence (HC) and low-confidence (LC) genes

	HC	LC
Number of genes	66,559	303,404
Number of genes on chr_Un	2566	12,145
Mean loci size (bp)	6681	1089
Number of single transcript genes	31,283	282,546
Number of multi transcript genes	35,276	20,858
Number of transcripts	196,153	341,975
Mean transcripts per gene	2.95	1.13
Mean CDS size (bp)[a]	1241	520
Mean exons per transcript[a]	4.6	1.2
Median exons per transcript[a]	3	1
Number of single exon transcripts[a]	17,250	273,063
Number of multi exon transcripts[a]	49,309	30,341

Source: Modified from Maccaferri et al. (2019)
[a]Numbers are for one representative transcript per gene

The alignment of the durum wheat genome with high-density SNP genetic maps showed highly recombinogenic distal chromosome regions (approximately 22% of the genome) with an average recombination rate of 1.8 Mb/cM and an approximately linear relationship between genetic distance and physical distance. In contrast, large pericentromeric regions nearly devoid of recombination represent approximately 44% of the genome, with a mean recombination rate of 107 Mb/cM. The 42.4% (27,109) of all HC genes fall in the highly recombinogenic distal chromosome regions while 23.1% (14,805) of all HC genes were located in the low recombinogenic pericentromeric regions. A clear example of the unequal distribution of the recombination across the chromosome is presented in Fig. 1 for chromosomes 1A and 1B, a very similar result to what found in all other chromosomes (Maccaferri et al. 2019). The presence of large low recombinogenic region has some practical consequence on plant breeding: as an example, a resistance gene located in a pericentromeric region has a minimal possibility to be introgressed in an elite cultivar without the large linkage drag represented by the low recombinogenic associated region. Furthermore, a selective sweep during the domestication process with causal genes located in pericentromeric regions led to strong diversity depletions, close to fixation, in wide chromosome regions, as for chromosomes 1A, 2A, 3A, 4A, 6A, 1B, 4B, 5B, 6B (an example is reported in Fig. 2). Interestingly, the extent of the regions subjected to diversity depletion was much higher in chromosomes belonging to the AA genomes as compared to those of the BB genome, indicating that domestication sweeps and bottlenecks preferentially targeted the AA genome.

Table 2 Pseudogenes basic metrics

Pseudogene features					Pseudogene types (%)				
No.	Mean length (no.)	Mean identity to parent (no.)	Mean coverage of parent (%)	Parent genes (no.)	Duplicated	Processed	Mono exon Parent	Fragmented	Chimeric
All pseudogenes and gene fragments									
279,773	289	89.2	27.7	36,314	24.1	2.6	17.4	54.8	1
Pseudogenes with ≥ 80% coverage of their parent gene									
28,106	756	92.7	94.7	9952	47.2	1.6	45.8	3.8	1.5

Source: Modified from Maccaferri et al. (2019)

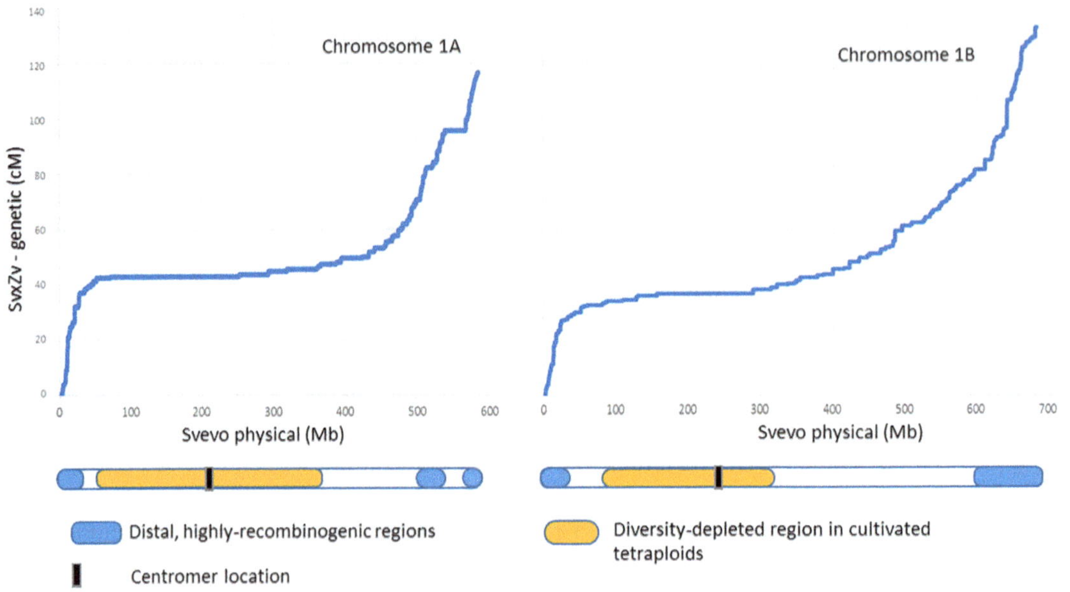

Fig. 1 Recombination rate, location of high- and low-recombinogenic regions, and LD pattern (focal SNP) for chromosomes 1A and 1B of the Svevo genome. (Modified from Maccaferri et al. 2019)

A vast majority (82.2%) of the durum wheat assembly was annotated as repetitive elements with about 7.3 Gb of long terminal repeat retrotransposons (LTR-RTs) and 1.2 Gb of transposon elements (TE, Table 3). The detection of full-length LTR-RTs resulted in a final set of 51,077 high quality full-length LTR-RTs. A large part of long non-coding RNAs (lncRNAs) were associated with transposable elements, particularly with Type I-CACTA and Type II-Gypsy family repeats.

A high-quality genome assembly offers the unique opportunity to fully describe large gene families either spread across the genome or concentrated in one of few clusters. The best example of a large and relevant gene family spread across the entire genome is represented by the nucleotide-binding leucine-rich repeat domains (NB-NLRs), a gene family with an important role in plant resistance mechanisms and plant innate immune systems (Borrelli et al. 2018). Notably, 2442 NB-NLR loci were predicted in the genome of the durum wheat cv. Svevo; 1462 (61%) of them were complete, 857 were pseudogenes or partial genes, while 101 were complete genes but included in the unassigned contigs on Chr_Un.

Prolamins are the main seed storage proteins of wheat, and their characteristics have a strong influence on the technological properties of wheat flour (Ruiz and Giraldo 2021). Prolamins are a perfect example of a gene family organized in few clusters with many members sharing a

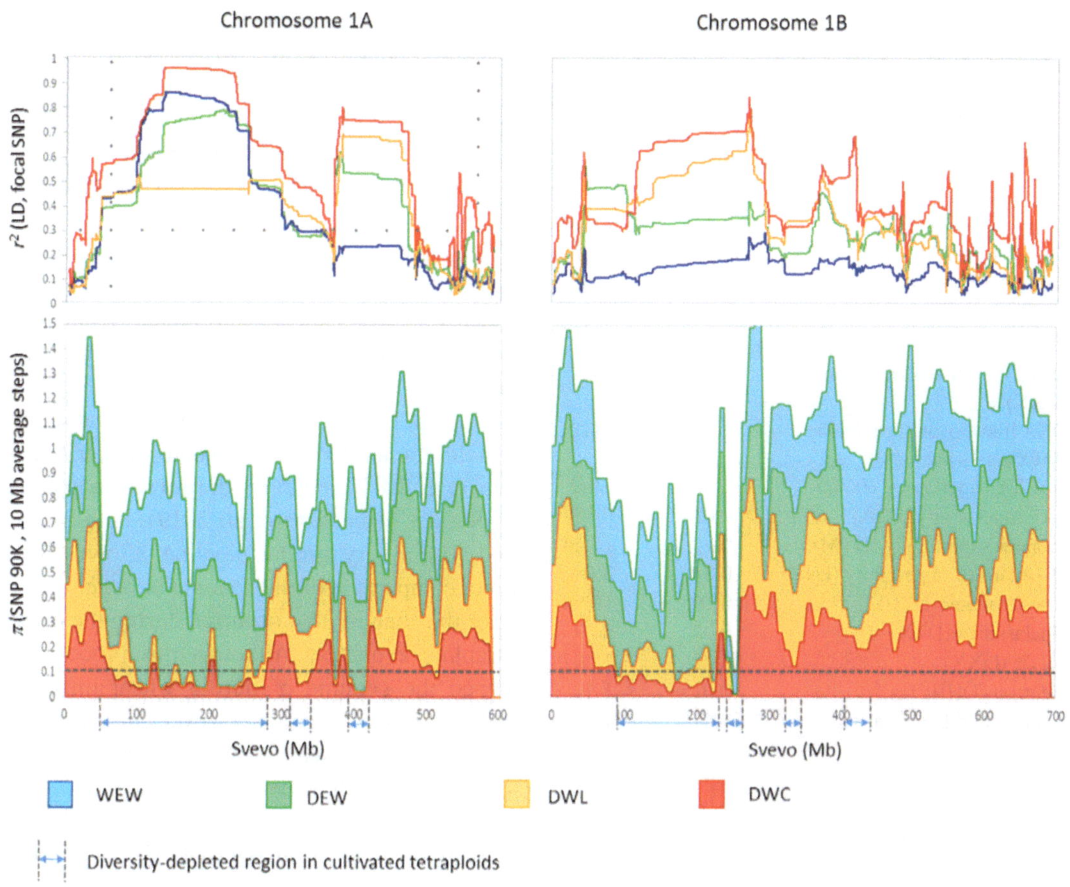

Fig. 2 Example of extended diversity depletions in *T. turgidum* landraces and in durum wheat cultivars in chromosomes 1A and 1B. Local linkage disequilibrium (r^2) and staked histograms of SNP diversity (Pi) in wild emmer wheat (WEW), domesticated emmer wheat (DEW), durum wheat landraces (DWL), and durum wheat cultivars (DWC) representative germplasm. (Modified from Maccaferri et al. 2019)

repetitive sequence, a feature that makes genome assembly very problematic. Wheat prolamins are classified as high-molecular weight (HMW) glutenin subunits, low-molecular weight (LMW) glutenin subunits, and α, γ, ω, and δ-gliadins. Prolamine-encoding genes belong to multigene families, in which individual members are organized in tandem and interspersed by repetitive elements. In total, 102 prolamine sequences were identified in the Svevo genome correctly mapped to four expected chromosome regions: the short and long arms of chromosome 1A/1B (*Glu-3*, *Gli-1* and *Glu-1* loci), the short arm of chromosome 6A/6B (*Gli-2* loci), the long arms of chromosome 4A, and the short arm of chromosome 7A (avenin-like loci) (Table 4). Additional 18 prolamine sequences were included in the unassigned contigs (Chr_Un). The assembly also captured a specific feature of the Gli-2 loci with all α-gliadin sequences located between glutamate-like receptor genes, a signature present already in the ancestral wheat genomes (Huo et al. 2017).

The Svevo genome and its annotations are accessible on two scientific open access repositories (GrainGenes Durum Wheat (cv. Svevo) RefSeq Release 1.0 at GrainGenes | GrainGenes (usda.gov) and Ensemble plants *Triticum_ turgidum*-Ensembl Genomes 54). Furthermore, the durum wheat cv. Svevo GFF3 and VCF annotation files can be downloaded at figshare, https://doi.org/10.6084/m9.figshare.6984035. Finally, all information related to RNA-seq and microRNA

Table 3 Transposon composition of durum wheat

	% of genome	Mb
Mobile element (TXX)	**82.2**	**8596**
Class I: Retroelement (RXX)	**70.3**	**7358**
LTR retrotransposon (RLX)	70.0	7320
Ty1/copia (RLC)	16.4	1713
Ty3/gypsy (RLG)	32.5	3396
Unclassified LTR (RLX)	21.1	2210
Non-LTR retrotransposon (RXX)	0.36	37.49
LINE (RIX)	0.34	36.08
SINE (RSX)	0.01	1.41
Class II: DNA transposon (DXX)	**11.4**	**1194**
DNA transposon superfamily (DTX)	11.3	1179
CACTA superfamily (DTC)	10.9	1143
hAT superfamily (DTA)	0.01	0.55
Mutator superfamily (DTM)	0.15	15.44
Tc1/mariner superfamily (DTT)	0.04	3.68
PIF/harbinger (DTH)	0.10	10.94
Unclassified (DTX)	0.05	5.59
MITE (DXX)	0.12	12.07
Helitron (DHH)	0.01	1.12
Unclassified DNA transposon (DXX)	0.01	1.23
Unclassified element (TXX)	0.42	44.32
Retro-TE/DNA-TE ratio	6.20	
Gypsy/Copia ratio	2.00	

Source: Modified from Maccaferri et al. (2019)
Note: The three-letter code given in brackets refers to official transposon classification

Table 4 Prolamin genes identified in the genome of the durum wheat Svevo code

Protein family	Family type	Chromosomal location	Number of genes
Gliadin	α	6AS	35
		6BS	24
	γ	1AS	5
		1BS	6
	ω	1AS	6
		1BS	7
	δ	1AS	2
		1BS	1
Glutenin	HMW	1AL	2
		1BL	2
	LMW	1AS	3
		1BS	1
Gliadin-like avenin		4AL	3
		7AS	5
Total			106

Source: Modified from Maccaferri et al. (2019)

(miRNA) datasets can be downloaded at the SRA database (under accession SRP149116).

3 A Comparison Between Svevo and Zavitan Genomes

Wild emmer and durum wheat are separated by about 10,000 years of evolution characterized by a two-step domestication process (from wild emmer to cultivated emmer and from cultivated emmer to durum wheat) and one century of knowledge-based plant breeding. The availability of wild emmer sequence genome (acc. Zavitan; Avni et al. 2017) and of the durum wheat genome cv. Svevo (Maccaferri et al. 2019), both assembled with a very similar pipeline, allowed for a direct comparison of the genome to better understand the changes that occurred during the recent tetraploid evolution. Although this process has shaped the genome of durum wheat imposing many selective signatures (Maccaferri et al. 2019), only minimally structural variations can be found between the two genomes.

The genomes comparison produced very similar numbers in terms of assembly size and repetitive element content (10.5 Gb and 82.2% for both genotypes) and number of genes (65,559 vs. 65,102 Svevo and Zavitan, respectively). Additionally, the number of full-length LTR-RTs resulted very similarly (Svevo: 51,077; Zavitan: 53,295). Notably, about 8000 full-length LTRs were still found in syntenic positions between the genomes of Svevo and Zavitan (Fig. 3) because the estimated divergence time was too short to lead to the usually almost complete turnover of the intergenic space observed between different *Triticeae* species (Dubcoksky and Dvorak 2007).

Very similar numbers of NB-NLR genes were predicted in Svevo and Zavitan genomes (2442 and 2420 NB-LRR genes, respectively) among which 172 and 136 NB-NLR sequences are specific for Svevo and Zavitan, respectively (Maccaferri et al. 2019). The syntenic analysis of the complete set of NB-NLR genes showed that these genes are more conserved as collinearity and number of copies between Svevo and Zavitan in genome A as compared to genome B.

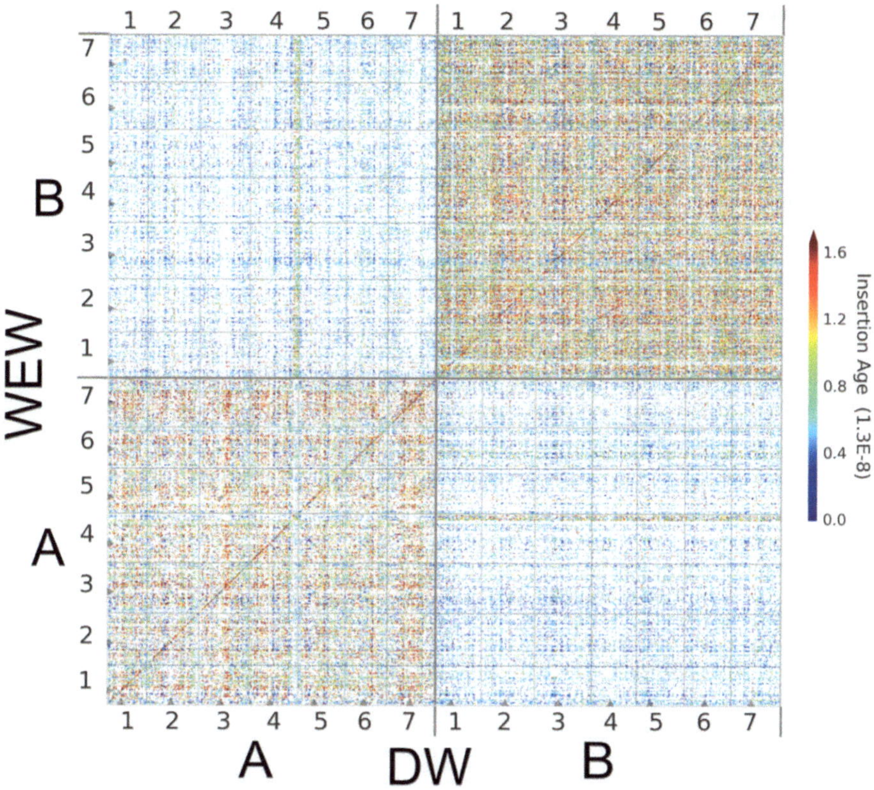

Fig. 3 Genome-wide comparison of Svevo and Zavitan full-length LTR-RTs subfamilies. The dotplot depicts the insertion age (color scale) and location of the full-length LTR subfamilies. The clearly visible diagonal represents about 8000 full-length LTRs with still conserved syntenic positions between durum wheat (DW) and wild emmer wheat (WEW). (Modified from Maccaferri et al. 2019)

NB-NLR loci were found more frequently at the distal regions of the chromosome arms and, often, in overlapping position with confidence intervals of known disease resistance QTLs (Maccaferri et al. 2019, Fig. 4).

A small mutation in the coding region of a gene can introduce a functional variant (e.g., a stop codon) that may impact significantly on gene function and eventually on plant phenotype. Single-base mutations or small indels are responsible for domestication-associated traits (e.g., brittle rachis) as well as major breeding-related traits such as reduced plant height (Nave et al. 2019; Peng et al. 1999). The comparison between two genomes separated by a short evolutionary process can offer the opportunity to make an atlas of putative functional variances capable to impact on the function of the corresponding genes. When the gene sequences of the Svevo genome were compared with the orthologous genes of Zavitan, 597 putative high-impact variants were identified (Maccaferri et al. 2019). High-impact variants identified in Svevo vs. Zavitan could be simply genotype-specific (variation between two individuals) rather than domestication or breeding-specific related polymorphisms (variation attributable to domestication or specifically selected by breeders); nevertheless, it is significant that hundreds of genes carrying functional variants are found when two genomes of fully interfertile plants are compared.

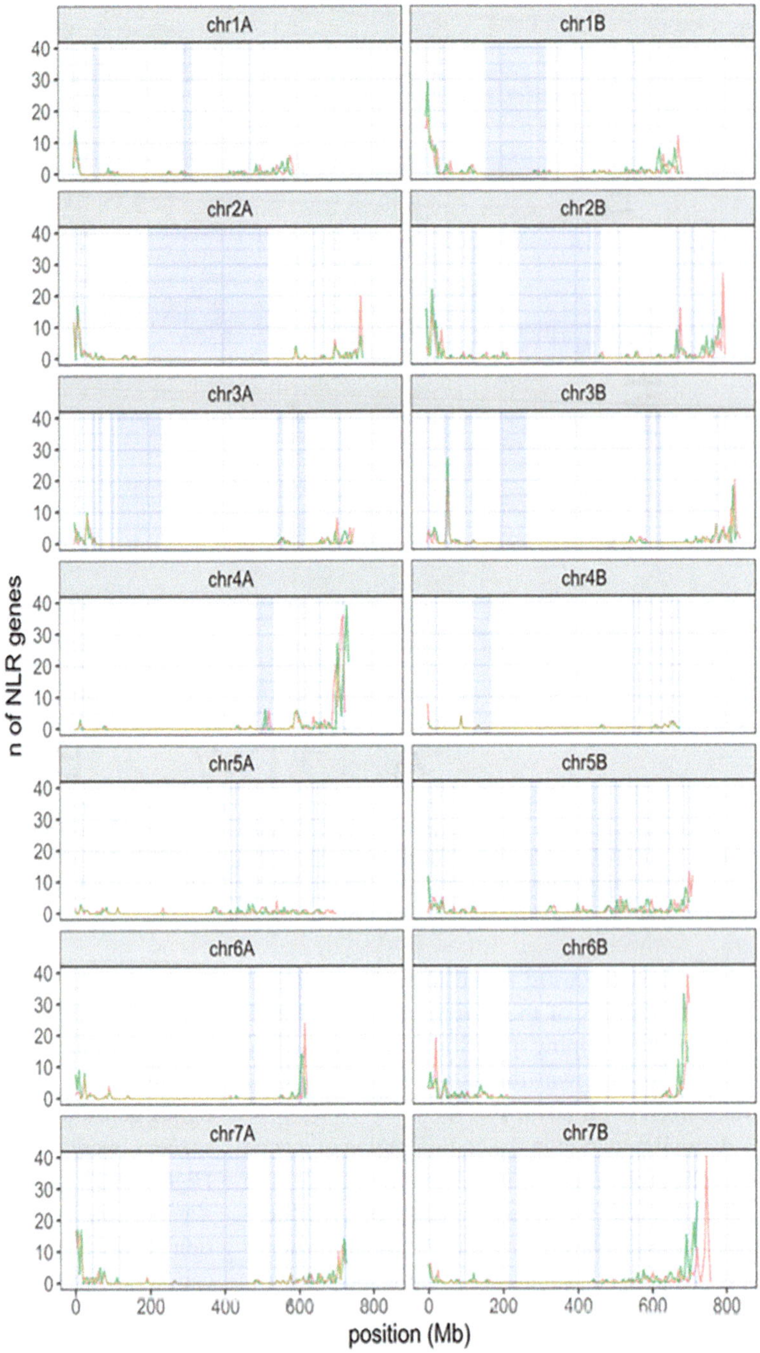

Fig. 4 Whole genome NB-NLR gene density graph at 5 Mb window. Blue transparent area represents confidence intervals of published disease resistance QTLs. Green color represents Svevo; red, Zavitan. (Modified from Maccaferri et al. 2019)

4 The Svevo Genome as a Tool to Study Evolution in Wheat and Clone Genes: Some Examples from the Literature

The release of genome sequences and corresponding gene annotations, genetic data linking the phenotypic traits to specific genomic regions, and information on gene expression provide fundamental tools for gene discovery, gene cloning, and the application of genomics-assisted breeding (Maccaferri et al. 2022). Furthermore, armed with these genomics tools, wheat (both durum and bread wheat) will re-establish itself as a model for polyploid plants in studying the mechanisms of polyploidy evolution, domestication, genetic and epigenetic regulation of homoeolog expression, as well as defining its genetic diversity and breeding at the genome level (Guan et al. 2020). The publication of the reference genome for durum wheat has boosted the fine mapping and cloning of genes and QTLs governing important traits such as phenology, disease resistance, grain yield and quality, and their plasticity, accelerating durum wheat genomics-based breeding and allowing for a more effective haplotype bridging of tetraploid and hexaploid wheat for the A and B genomes (Maccaferri et al. 2015; Peters Haugrud et al. 2024).

Already in the paper where the Svevo genome was presented (Maccaferri et al. 2019), the new genome information was instrumental to study a locus on chromosome 5B that carries a gene encoding a metal transporter (*TdHMA3-B1*) with a non-functional variant causing high accumulation of cadmium in the grain. The cloning of *TdHMA3-B1* (locus *TRITD5Bv1G197370*) rescued a wild beneficial allele and demonstrated the practical use of the Svevo genome for wheat breeding. The 17 bp duplication in the first exon of *TRITD5Bv1G197370* creates two alternative alleles namely *TdHMA3-B1a* in Zavitan (functional; low-grain Cd) and *TdHMA3-B1b* in Svevo (non-functional; high-grain Cd). The diagnostic marker X*usw*59 clearly identified the alternative alleles discriminating low- and high-Cd accumulators in a tetraploid wheat panel. The functional *TdHMA3-B1a* allele was genetically fixed in wild emmer accessions, while the *TdHMA3-B1b* allele showed an increasing frequency from emmer wheat to durum landraces and durum cultivars. The increase in frequency of *TdHMA3-B1b* during durum wheat breeding could be due to the presence of a selective sweep for another sequence (gene or non-coding sequence) present in the linkage disequilibrium or, alternatively, the *TdHMA3-B1b* allele could pleiotropically exert some beneficial effects on plant fitness, particularly for functionally complex quantitative traits controlled by the QTLome (Salvi and Tuberosa 2015; Roselló et al. 2018; Condorelli et al. 2022).

Coupling the Svevo genome and the sequenced mutant population of the tetraploid wheat variety Kronos, Glenn et al. (2022) isolated a natural *FT-A2* allele on the short arm of chromosome 3A that increases grain number per spike in both durum and bread wheat with limited effect on heading time. This natural mutation, resulting in an aspartic acid to alanine change, was associated with significant increases in spikelet number per spike and no negative effects on fertility. The positive effects of this natural allele (indicated as *A10*) on spikelet number per spike, grain number, and grain yield per spike was validated in near-isogenic tetraploid wheat lines and in hexaploid winter wheat population. This allele is present at very low frequency in durum wheat and at much higher frequency in hexaploid wheat, particularly in winter wheat and in fall-planted spring wheat varieties. These results suggest that the *FT-A2 A10* allele may be particularly useful for improving grain yield in durum wheat and fall-planted common wheat spring varieties (Glenn et al. 2022).

The combined used of Zavitan and Svevo genomes allowed to clone and functionally characterize *Pm41*, a powdery mildew resistance gene derived from wild emmer, which encodes a coiled-coil, nucleotide-binding site, and leucine-rich repeat protein (CNL) (Li et al. 2020). The non-functional allele (*pm41*), identified in domesticated emmer, durum, and bread wheat, such as in some wild emmer accessions, has been originated by the insertion of transposons in the promoter and in the UTR regions. These findings suggest that the ancestral *Pm41* was restricted to its place of origin and support evidence that multiple gene flow events occurred during domestication from wild emmer to domesticated tetraploid wheat as well as during hexaploidization of common wheat (Matsuoka 2011).

The wheat stem sawfly, *Cephus cinctus* Norton (Hymenoptera: Cephidae), is one of the most destructive insect pests of wheat and other grain crops. Stem solidness is an important agronomic trait of durum and bread wheat that provides resistance to the wheat stem sawfly (Nilsen et al. 2017). The causal gene that controls stem solidness in durum wheat (*TdDof*) was recently identified (Nilsen et al. 2020). The authors showed that copy number gain of *TdDof* correlates with its increased expression and the solid-stem phenotype in durum wheat suggesting that *TdDof* could function as a key regulator of genes involved in programmed cell death (PCD) of the pith parenchyma cells. Hollow-stemmed wheat cultivars carry a single copy of *TdDof*, whereas solid-stemmed cultivars carry multiple identical copies of the gene. Deletion of all *TdDof* copies from solid stemmed cultivars resulted in the loss of stem solidness, whereas the transgenic overexpression of *TdDof* restored stem solidness in the *TdDof* deletion mutant *pithless1* (a mutant exhibiting a complete loss of phenotypic expression of stem solidness) and conferred stem solidness in hollow-stemmed cultivars. In solid-stemmed cultivars, increased *TdDof* expression was correlated with the downregulation of genes whose orthologs have been implicated in programmed cell death (PCD). Anatomical and histochemical analyses revealed that hollow-stemmed lines had stronger PCD-associated signals in the pith cells compared to solid-stemmed lines, which suggests copy number dependent expression of *TdDof* could be directly or indirectly involved in the negative regulation of PCD. These findings provide opportunities to manipulate stem architecture in wheat to enhance wheat stem sawfly resistance, modify carbon partitioning and water-soluble carbohydrate remobilization in plants under drought and temperature stress, and enhance bioenergy production (Nilsen et al. 2020).

The tetraploid wheat genomes are instrumental to precisely define the wild emmer introgressions into a durum wheat elite variety associated with vegetative growth capacity under water-limited conditions (Bacher et al. 2021) as well as for fine mapping and cloning of genes in emmer, durum wheat, and sometime also in bread wheat. The comparative genomic analysis performed using both the bread wheat pangenomes and the tetraploid genomes reveals that the region associated with hybrid necrosis on chromosome 2B and carrying *Ne2*, a typical CNL gene responsible for hybrid necrosis in wheat, was highly conserved (Si et al. 2021).

Several studies have used the Svevo reference genome to mine for possible candidate genes supporting the expression of specific QTLs. For instance, a QTL mapping carried out in a segregating population generated by a cross between a domesticated emmer landrace and a durum wheat variety coupled with the information provided by the Svevo genome has allowed to identify candidate genes for major QTLs controlling seed size (Valladares-Garcıa et al. 2023).

All available wheat genomes, and their developed resources, are widely used in comparative genomics studies to analyze and elucidate the mechanisms of polyploidization, domestication, and evolution of wheat. Yuan et al. (2022) explored the potential role of open chromatin interactions in gene expression and the effect of structure variations during polyploidization and evolution of wheat. As well documented, the organization of chromatin affects cellular processes by controlling access to the genomic regions involved in the regulation of transcription, recombination, replication, and DNA repair (Klemm et al. 2019; Bell et al. 2011). By comparing hexaploid wheat with tetraploid and diploid relatives, a positive correlation between chromatin interaction changes and expression changes was found, suggesting that chromatin interactions are also involved in gene regulation during wheat polyploidization (Yuan et al. 2022). More chromatin interactions were observed in the D subgenome compared with the A and B subgenomes, which is consistent with higher chromatin accessibility and lower levels of repressive histone marks in the D subgenome (Jordan et al. 2020; Ramirez-Gonzalez et al. 2018). Although chromatin interaction changes were widely observed in hexaploid wheat compared with tetraploid and diploid relatives, only limited triads (that define the homoeolog groups with only one gene copy in each subgenome) showed chromatin interaction changes in all three homoeologs, indicating chromatin interaction changes may contribute to functional differentiation of homoeologs after polyploidization by expression regulation (Yuan et al. 2022).

He et al. (2022) revealed the importance of genomic variants affecting homoeolog expression dosage in shaping agronomic phenotypes and points at their potential value for improving yield in wheat and other polyploid crops. This study highlights the importance of inter-genomic trans-effects in the regulation of genes that lost their *cis*-regulatory diversity in polyploid wheat. The loss of *cis*-acting variants in the A and B genomes is likely associated with domestication and improvement of selective sweeps, which affected a significant portion of the wheat genome (Avni et al. 2017; He et al. 2019; Maccaferri et al. 2019; Walkowiak et al. 2020; Zhou et al. 2020). Furthermore, on the D genome are located half of the genes whose expression is largely explained by trans-genomic effects. These genes also show evidence of reduced genetic diversity suggesting that the loss of *cis*-acting diversity occurred consequent to hybridization between tetraploid wheat *Triticum tauschii*, the ancestor of the D genome. The cumulative effects of *trans*-acting variants on the expression variance of genes that lost *cis*-regulatory diversity were comparable to the effects of *cis*-acting variants on genes showing no evidence of *trans*-regulation. These findings suggest that inter-genomic interactions could play a significant role in regulating the expression of loci controlling domestication and quantitative trait improvement in wheat. By investigating the distribution of *cis*- and *trans*-acting eQTL across genomic regions showing distinct epigenetic marks and chromatin architecture, it was revealed that eQTLs are more frequent in the regions of active chromatin. Finally, it was demonstrated that the frequency of genomic variants associated with the relative expression dosage of homoeologous genes is predictive of variations in productivity traits in allopolyploid wheat and appears to be targeted by improvement selection (Li et al. 2019; Jordan et al. 2020; Concia et al. 2020).

Following the early work of the transcriptome of cv. Svevo (Vendramin et al. 2019), the first translatome and in vivo RNA structurome of polyploid wheat was recently obtained providing the context and basis for future studies of polyploid evolution and for developing new strategies for wheat molecular breeding (Yang et al. 2021). The work has revealed a novel contribution of RNA structure in subgenome translation asymmetry identifying 3564 single-nucleotide variations (SNVs) across the transcriptomes between the two subgenomes of tetraploid wheat, which induce large RNA structure disparities. These SNVs are highly conserved within durum wheat cultivars but are divergent in both domesticated and wild emmer wheat. This finding revealed that RNA structure serves as an important modulator of translational subgenome expression asymmetry in polyploids. The integration of multi-omics, transcriptional network with genome information and genetic analyses could open new perspective in the understanding of the molecular bases of crop yield (Lin et al. 2024).

The Svevo genome was also instrumental for the interpretation of metabolic studies. A comparative metabolomic analysis of kernels, performed by Ben-Abu and Itsko (2022), was applied to investigate the tetraploid wheat evolution along the primary and secondary domestication events. This study revealed that many metabolites in wheat had undergone significant changes in each step of domestication. Only a limited number of metabolites steadily increased or decreased in association with both domestication events, suggesting that these metabolites may have contributed to the improvement in the agricultural fitness of wheat. Amino acid metabolism was significantly altered during wheat domestication, demonstrating an increased proportion of several amino acids except for the tyrosine, the direct precursor of p-coumaric acid, a component of defense mechanisms of wheat against biotic and abiotic stresses.

The transcriptomic, metabolomic, and physiologic approaches were combined to better understand the differences in wheat defense mechanisms during the domestication (Batyrshina et al. 2020). Chemical defenses are based on the accumulation of toxic chemicals such as benzoxazinoids, a class of specialized metabolites acting as deterrents, while mechanical defense, including physical barriers, depends on the increasing density of thorns, spikes, or trichomes (Howe and Jander 2008). Zavitan, Svevo, and Chinese Spring were analyzed to assess the effect of domestication on plant resistance against aphid (*Rhopalosiphum padi*). Differences between the abundance of defense mechanisms in the wild and domesticated wheats were observed: wild emmer possesses more effective physical defenses, while the domesticated wheat genotypes have more active chemical defenses. Measurement of benzoxazinoid levels revealed high levels in the two domesticated genotypes than in wild emmer wheat; in contrast, the trichome density was much higher in the wild emmer than in the domesticated wheat. The results show that the wild emmer, which possessed a high number of trichomes and a low benzoxazinoid level, is more susceptible to the *R. padi* aphid than Chinese Spring and Svevo. This suggested that under insect pressure, wheat plants might have undergone evolutionary convergence, which resulted in similarities in defense mechanisms via the biosynthesis of defense metabolites and, to a lesser extent, trichome formation (Batyrshina et al. 2020).

5 Perspectives

During the recent years the unprecedented development of the sequencing technologies has allowed to achieve the sequence and assembly of the large wheat genomes. The availability of these genomic resources has boosted the isolation of agronomically relevant genes that will significantly impact on wheat breeding. Currently few high-quality tetraploid wheat genomes are available, besides Zavitan, Svevo, and Chinese spring, while a pangenome is already available for hexaploid wheat (Walkowiak et al. 2020; N'Diaye et al. 2023) and many more high-quality wheat genome assemblies will be available soon. A tetraploid wheat pangenome is underway. In few years there will be plenty of high-quality wheat genomes that will significantly change wheat genomics and breeding. As results, wheat breeding will trust more and more on genomics information and selection will became more efficient, while the focus of genomics will progressively move to functional genomics and epigenomics. The combination of advanced genomics knowledge, genome editing, and speed breeding (Li et al. 2021) will soon enhance the precision and the speed of the breeding process for improving grain quality and quantity of pasta wheat, making durum wheat more adapted to the new climatic scenario (Broccanello et al. 2023; Peters Haugrud et al. 2024).

References

Avni R, Nave M, Barad O et al (2017) Wild emmer genome architecture and diversity elucidate wheat evolution and domestication. Science 357:93–97. https://doi.org/10.1126/science.aan0032

Bacher H, Zhu F, Gao T et al (2021) Wild emmer introgression alters root-to-shoot growth dynamics in durum wheat in response to water stress. Plant Physiol 187:1149–1162. https://doi.org/10.1093/plphys/kiab292

Batyrshina ZS, Yaakov B, Shavit R et al (2020) Comparative transcriptomic and metabolic analysis of wild and domesticated wheat genotypes reveals differences in chemical and physical defense responses against aphids. BMC Plant Biol 20:19. https://doi.org/10.1186/s12870-019-2214-z

Bell O, Tiwari VK, Thomä NH et al (2011) Determinants and dynamics of genome accessibility. Nat Rev Genet 12:554–564. https://doi.org/10.1038/nrg3017

Ben-Abu Y, Itsko M (2022) Metabolome dynamics during wheat domestication. Sci Rep 12:8532. https://doi.org/10.1038/s41598-022-11952-9

Borrelli GM, Mazzucotelli E, Marone D et al (2018) Regulation and evolution of NLR genes: a close interconnection for plant immunity. Int J Mol Sci 19:1662. https://doi.org/10.3390/ijms19061662

Broccanello C, Bellin D, DalCorso G et al (2023) Genetic approaches to exploit landraces for improvement of *Triticum turgidum* ssp. *durum* in the age of climate change. Frontiers. Plant Sci 14:1101271. https://doi.org/10.3389/fpls.2023.1101271

Brozynska M, Furtad A, Henry RJ (2016) Genomics of crop wild relatives: expanding the gene pool for crop improvement. Plant Biotech J 14:1070–1085. https://doi.org/10.1111/pbi.12454

Bruschi M, Bozzoli M, Ratti C et al (2024) Dissecting the genetic basis of resistance to soil-borne cereal mosaic virus (SBCMV) in durum wheat by bi-parental mapping and GWAS. Theor Appl Gen 137(9):213. https://doi.org/10.1007/s00122-024-04709-7

Concia L, Veluchamy A, Ramirez-Prado JS et al (2020) Wheat chromatin architecture is organized in genome territories and transcription factories. Genome Biol 21:104. https://doi.org/10.1186/s13059-020-01998-1

Condorelli GE, Newcomb M, Groli EL et al (2022) Genome wide association study uncovers the QTLome for osmotic adjustment and related drought adaptive traits in durum wheat. Genes 13(2). https://doi.org/10.3390/genes13020293

Dubcoksky J, Dvorak J (2007) Genome plasticity a key factor in the success of polyploid wheat under domestication. Science 316:1862–1866. https://doi.org/10.1126/science.1143986

Faricelli ME, Valarick M, Dubcovsky J (2010) Control of flowering time and spike development in cereals: the earliness per se Eps-1 region in wheat, rice, and *Brachypodium*. Funct Integr Genomics 10:293–306 https://doi.org/10.1007/s10142-009-0146-7

Faris JD (2014) Wheat domestication: key to agricultural revolution past and future. In: Tuberosa R, Graner A, Frison E (eds) Genomics of plant genetic resources. Springer, Dordrecht, pp 439–464. https://doi.org/10.1007/978-94-007-7572-5_18

Feldman M, Millet E (2001) The contribution of the discovery of wild emmer to an understanding of wheat evolution and domestication and to wheat improvement. Isr J Plant Science 49:25–36. https://doi.org/10.1560/JCMX-WGXM-D40G-BG4P

Gaut BS (2015) Evolution is an experiment: assessing parallelism in crop domestication and experimental evolution. Mol Biol Evol 32:1661–1671. https://doi.org/10.1093/molbev/msv105

Glenn P, Zhang J, Brown-Guedira G et al (2022) Identification and characterization of a natural polymorphism in FT-A2 associated with increased number of grains per spike in wheat. Theor Appl Genet 135:679–692. https://doi.org/10.1007/s00122-021-03992-y

Guan J, Garcia DF, Zhou Y et al (2020) The battle to sequence the bread wheat genome: a tale of the three kingdoms. Genomics Proteomics Bioinformatics 18:221–229. https://doi.org/10.1016/j.gpb.2019.09.005

He F, Pasam R, Shi F et al (2019) Exome sequencing highlights the role of wild relative introgression in shaping the adaptive landscape of the wheat genome. Nat Genet 51:896–904. https://doi.org/10.1038/s41588-019-0382-2

He F, Wang W, Rutter WB et al (2022) Genomic variants affecting homoeologous gene expression dosage contribute to agronomic trait variation in allopolyploid wheat. Nat Commun 13:826. https://doi.org/10.1038/s41467-022-28453-y

Howe GA, Jander G (2008) Plant immunity to insect herbivores. Annu Rev Plant Biol 59:41–66. https://doi.org/10.1146/annurev.arplant.59.032607.092825

Huo N, Dong L, Zhang S et al (2017) New insights into structural organization and gene duplication in a 1.75-Mb genomic region harboring the α-gliadin gene family in *Aegilops tauschii*, the source of wheat D genome. Plant J 92:571–583. https://doi.org/10.1111/tpj.13675

IWGSC (2018) Shifting the limits in wheat research and breeding using a fully annotated reference genome. Science 361:7191. https://doi.org/10.1126/science.aar7191

Jordan KW, He F, de Soto MF et al (2020) Differential chromatin accessibility landscape reveals structural and functional features of the allopolyploid wheat chromosomes. Genome Biol 21:176. https://doi.org/10.1186/s13059-020-02093-1

Katamadze A, Vergara-Díaz O, Uberegui E et al (2023) Evolution of wheat architecture, physiology, and metabolism during domestication and further cultivation: lessons for crop improvement. Crop J 11:1080–1096. https://doi.org/10.1016/j.cj.2023.06.006

Klemm SL, Shipony Z, Greenleaf WJ (2019) Chromatin accessibility and the regulatory epigenome. Nat Rev Genet 20:207–220. https://doi.org/10.1038/s41576-018-0089-8

Klymiuk V, Yaniv E, Huang L et al (2018) Cloning of the wheat *Yr15* resistance gene sheds light on the plant tandem kinase-pseudokinase family. Nat Commun 9:3735. https://doi.org/10.1038/s41467-018-06138-9

Krasileva KV, Vasquez-Gross HA, Howell T et al (2017) Uncovering hidden variation in polyploid wheat. Proc Natl Acad Sci USA 114:E913–E921. https://doi.org/10.1073/pnas.1619268114

Li G, Fang T, Zhang H et al (2009) Molecular identification of a new powdery mildew resistance gene *Pm41* on chromosome 3BL derived from wild emmer (*Triticum turgidum* var. *dicoccoides*). Theor Appl Genet 119:531–539. https://doi.org/10.1007/s00122-009-1061-y

Li Z, Wang M, Lin K et al (2019) The bread wheat epigenomic map reveals distinct chromatin architectural and evolutionary features of functional genetic elements. Genome Biol 20:139. https://doi.org/10.1186/s13059-019-1746-8

Li M, Dong L, Li B et al (2020) A CNL protein in wild emmer wheat confers powdery mildew resistance. New Phytol 228:1027–1037. https://doi.org/10.1111/nph.16761

Li S, Zhang C, Li J et al (2021) Present and future prospects for wheat improvement through genome editing and advanced technologies. Plant Commun 2:100211. https://doi.org/10.1016/j.xplc.2021.100211

Lin X, Xu Y, Wang D et al (2024) Systematic identification of wheat spike developmental regulators by integrated multi-omics, transcriptional network, GWAS, and genetic analyses. Mol Plant 17:438–459. https://doi.org/10.1016/j.molp.2024.01.010

Maccaferri M, Ricci A, Salvi S et al (2015) A high-density, SNP-based consensus map of tetraploid wheat as a bridge to integrate durum and bread wheat genomics and breeding. Plant Biotech J 13(5):648–663. https://doi.org/10.1111/pbi.12288

Maccaferri M, Harris NS, Twardziok SO et al (2019) Durum wheat genome highlights past domestication signatures and future improvement targets. Nat Genet 51:885–895. https://doi.org/10.1038/s41588-019-0381-3

Maccaferri M, Bruschi M, Tuberosa R (2022) Sequence-based marker assisted selection in wheat. Wheat improvement. In: Wheat improvement: food security in a changing climate. Springer International Publishing, Cham, pp 513–538. https://doi.org/10.1007/978-3-030-90673-3_28

Marais GF, Pretorius ZA, Wellings CR et al (2005) Leaf rust and stripe rust resistance genes transferred to common wheat from *Triticum dicoccoides*. Euphytica 143:115–123. https://doi.org/10.1007/s10681-005-2911-6

Matsuoka Y (2011) Evolution of polyploid *Triticum* wheats under cultivation: the role of domestication, natural hybridization and allopolyploid speciation in their diversification. Plant Cell Physiol 52:750–764. https://doi.org/10.1093/pcp/pcr018

N'Diaye A, Walkowiak S, Pozniak C (2023) Wheat sequencing: the pan-genome and opportunities for accelerating breeding. In: The wheat genome. Compendium of plant genomes. Springer, Cham, pp 273–288. https://doi.org/10.1007/978-3-031-38294-9_14

Nave M, Avni R, Çakır E et al (2019) Wheat domestication in light of haplotype analyses of the *brittle rachis 1* genes (*BTR1-a* and *BTR1-B*). Plant Sci 285:193–199. https://doi.org/10.1016/j.plantsci.2019.05.012

Nilsen KT, N'Diaye A, MacLachlan PR et al (2017) High density mapping and haplotype analysis of the major stem-solidness locus *SSt1* in durum and common wheat. PLoS One 12:e0175285. https://doi.org/10.1371/journal.pone.0175285

Nilsen KT, Walkowiak S, Xiang D et al (2020) Copy number variation of *TdDof* controls solid-stemmed architecture in wheat. Proc Natl Acad Sci USA 117:28708–28718. https://doi.org/10.1073/pnas.2009418117

Peng J, Richards DE, Hartley NM et al (1999) 'Green revolution' genes encode mutant gibberellin response modulators. Nature 400:256–261. https://doi.org/10.1038/22307

Peters Haugrud AR, Achilli AL, Martínez-Peña R, Klymiuk V (2024) Future of durum wheat research and breeding: insights from early career researchers. Plant Genome e20453. https://doi.org/10.1002/tpg2.20453

Ramirez-Gonzalez RH, Borrill P, Lang D et al (2018) The transcriptional landscape of polyploid wheat. Science 361:eaar6089. https://doi.org/10.1126/science.aar6089

Roselló M, Royo CÁlvaro F et al (2018) Pasta-making quality QTLome from Mediterranean durum wheat landraces. Front Plant Sci 9:1512. https://doi.org/10.3389/fpls.2018.01512

Ruiz M, Giraldo P (2021) The influence of allelic variability of prolamins on gluten quality in durum wheat: an overview. J Cereal Sci 101:103304. https://doi.org/10.1016/j.jcs.2021.103304

Salvi S, Tuberosa R (2015) The crop QTLome comes of age. Curr Opin Biotechnol 32:179–185. https://doi.org/10.1016/j.copbio.2015.01.001

Si Y, Zheng S, Niu J et al (2021) *Ne2*, a typical CC–NBS–LRR-type gene, is responsible for hybrid necrosis in wheat. New Phytol 232:279–289. https://doi.org/10.1111/nph.17575

Tabbita F, Pearcec S, Barneix AJ (2017) Breeding for increased grain protein and micronutrient content in wheat: ten years of the *GPC-B1* gene. J Cereal Sci 73:183–191. https://doi.org/10.1016/j.jcs.2017.01.003

Uauy C, Distelfeld A, Fahima T et al (2006) A *NAC* gene regulating senescence improves grain protein, zinc and iron content in wheat. Science 314:1298–1300. https://doi.org/10.1126/science.1133649

Valladares-Garcıa AP, Desiderio F, Simeone R et al (2023) QTL mapping for kernel-related traits in a durum wheat x *T. dicoccum* segregating population. Front Plant Sci 14:1253385. https://doi.org/10.3389/fpls.2023.1253385

Vendramin V, Ormanbekova D, Scalabrin S et al (2019) Genomic tools for durum wheat breeding: de novo assembly of Svevo transcriptome and SNP discovery in elite germplasm. BMC Genomics 20(1). https://doi.org/10.1186/s12864-019-5645-x

Walkowiak S, Gao L, Monat C et al (2020) Multiple wheat genomes reveal global variation in modern breeding. Nature 588:277–283. https://doi.org/10.1038/s41586-020-2961-x

Yang X, Yu H, Sun W et al (2021) Wheat in vivo RNA structure landscape reveals a prevalent role of RNA structure in modulating translational subgenome expression asymmetry. Genome Biol 22:326. https://doi.org/10.1186/s13059-021-02549-y

Yuan J, Sun H, Wang Y et al (2022) Open chromatin interaction maps reveal functional regulatory elements and chromatin architecture variations during wheat evolution. Genome Biol 23:34. https://doi.org/10.1186/s13059-022-02611-3

Zhou Y, Zhao X, Li Y et al (2020) *Triticum* population sequencing provides insights into wheat adaptation. Nat Genet 52:1412–1422. https://doi.org/10.1038/s41588-020-00722-w

Zhu T, Wang L, Rodriguez JC et al (2019) Improved genome sequence of wild emmer wheat Zavitan with the aid of optical maps. G3 (Bethesda) 9:619–624. https://doi.org/10.1534/g3.118.200902

Genome Assembly Completeness of the Durum Wheat Genome as a Function of PacBio HiFi Sequencing Coverage

Md Nafis Ul Alam, Marco Maccaferri, Roberto Tuberosa, Rod Wing, and Dario Copetti

Abstract

Durum wheat is a commercially valuable cereal crop with a large 10.45 gigabase allotetraploid genome. The PacBio Revio's highly accurate long reads allow to sequence genomes of this scale at a much more affordable cost. At present, there have been no comprehensive studies investigating the effects of sequencing coverage on de novo genome assemblies for large plant genomes. To evaluate the impact of sequencing coverage on assembly completeness, we developed 14 genome assemblies with PacBio's HiFi reads across a gradient of coverage ranging from 3.05× to 34.05×. We compiled and evaluated assembly statistics, gene features, and benchmarks of universal single-copy orthologs across the genome assemblies. The results highlight key differences in assembly features that result from coverage variation. These analyses will help researchers make informed decisions in determining a target coverage range for large germplasm resequencing projects. We demonstrate that aside from contig numbers and N and L metrics, durum wheat assemblies prepared at v8–10× coverage display genomic features nearly identical to the highly contiguous assemblies obtained from 15–34× coverage. These observations mark the 8–10× coverage as an optimal range for applications where a reference-grade reference assembly is available, thus enabling the resequencing of broad germplasm collections and maximizing sequencing outcomes.

Keywords

Durum wheat Svevo genome · PacBio HiFi · Optimal coverage · Assembly completeness · PacBio Revio

M. N. U. Alam
Arizona Genomics Institute, School of Plant Sciences, University of Arizona, Tucson, AZ, USA

Plant Biotechnology Lab, Department of Biochemistry and Molecular Biology, University of Dhaka, Dhaka, Bangladesh

M. Maccaferri · R. Tuberosa
Department of Agricultural and Food Sciences (DISTAL), Alma Mater Studiorum – Università di Bologna, Bologna, Italy

R. Wing
Arizona Genomics Institute, School of Plant Sciences, University of Arizona, Tucson, AZ, USA

Center for Desert Agriculture, Biological and Environmental Sciences & Engineering Division (BESE), King Abdullah University of Science and Technology (KAUST), Thuwal, Saudi Arabia

D. Copetti (✉)
Arizona Genomics Institute, School of Plant Sciences, University of Arizona, Tucson, AZ, USA
e-mail: dcopetti@arizona.edu

R. Tuberosa et al. (eds.), *The Durum Wheat Genome*, Compendium of Plant Genomes,
https://doi.org/10.1007/978-3-031-84689-2_3

1 Introduction

The tetraploid durum wheat (*Triticum turgidum* L. ssp. *durum* [Desf.] Husn.) is the result of human selection of the wild emmer wheat and has a large genome size of 10.45 Gb (Maccaferri et al. 2019). Maccaferri et al. (2019) released the first highly accurate assembly of the modern Svevo cultivar (cv.). They developed four sets of Illumina paired-end libraries of varying insert sizes, assembling them into contigs and anchored in 14 pseudomolecules by chromosome conformation capture (Hi-C) sequencing (Maccaferri et al. 2019). Though it contained most of the estimated bases in the genome, this assembly was also highly fragmented with 129,464 total scaffolds, and an N50 of ~6 Mb (Maccaferri et al. 2019). Based on present-day assembly standards (Grewal et al. 2024; Liu et al. 2024), the very large number of short-length scaffolds raises valid concerns regarding both missing sequences and errors in pseudomolecule construction. For the cv. Svevo, a highly complete assembly (that we will hereafter refer to as a platinum-grade) provided of Iso-Seq and RNA-seq based annotation is presently under development. The availability of a new genome assembly will allow for new research avenues and sequencing efforts focused on characterizing the past evolutionary history and present genetic diversity of major durum wheat varieties.

Thanks to its higher accuracy (mean quality value (QV) score around 30), the PacBio HiFi technology developed in 2019 (Wenger et al. 2019) has drastically improved the de novo assembly process for large, complex genomes. In 2023, the deployment of the PacBio Revio sequencer has allowed the acquisition of HiFi reads for a fraction of the cost per base of the previous platforms (PacBio 2022), enabling cost-effective sequencing of large genomes. There are few bread wheat genome assemblies assembled from long reads (Alonge et al. 2020; Grewal et al. 2024; Liu et al. 2024; Sato et al. 2021; Zhu et al. 2021; Zimin et al. 2017) and there is limited information on the quantity of sequencing data that can economically maximize information gain for sequencing plants with similarly sized genomes (Kersey 2019; Kong et al. 2023). Furthermore, there has been no study comprehensively testing the de novo assembly capabilities of data generated from this new technology pertaining to large polyploid genomes. Up to now, long-read sequencing and assembling performance has been addressed by a single study, limited to the diploid barley genome only, which is not affected by the issues typically associated with polyploidy (Mascher et al. 2021).

To elucidate the effects of HiFi sequencing coverage on assembly quality, we assembled and evaluated completeness metrics for 14 genome assemblies across a coverage gradient ranging from 3.05× to 34.05×. We compared assembly properties, gene features, and benchmarks of universal single-copy orthologs (SCOs) of assemblies prepared at lower coverages to those prepared at higher coverages. Our goal was to compile the metrics necessary to evaluate a coverage range where the tradeoff between genomic feature representation and sequencing costs could be optimized. We also were able to estimate the limited gain that additional coverage has on assembly contiguity above 34.05×. Our findings will help determine the most efficient allocation of resources in a potential large-scale resequencing effort of multiple durum wheat accessions. This work can additionally serve as a proxy for similarly structured plant genomes of comparable sizes, such as other tetraploid grasses.

2 Materials and Methods

2.1 DNA Extraction, Library Preparation, and Sequencing

PacBio SMRTbell libraries were sequenced by Corteva Agriscience™ with the PacBio Sequel II platform, based on their own Corteva/PacBio high-throughput optimized workflow including tissues disruption, DNA extraction, DNA shearing, library prep, size selection, and sequencing.

2.2 Quality Control and Assembly

Raw read statistics were assessed using FastQC version 0.12.1 and NanoPlot version 1.40.0 (Andrews 2010; De Coster et al. 2018). The bam files were converted to fastq with the SAMtools version 1.6 running with HTSlib 1.6 (Li et al. 2009). Reads with a QV higher than 20 were selected using the BBDuk script from BBTools version 38.18 with the maq = 20 option (Bushnell 2015). Different combinations of read sets from 13 single molecule, real-time (SMRT) cells were used to prepare raw read files spanning a range of coverage values. Properties of the filtered read files were verified using FastQC and NanoPlot. The total size of the durum wheat cultivar Svevo assembly, 10.50 Gb (Maccaferri et al. 2019), was used to calculate coverage. Reads were assembled using Hifiasm version 0.19.6-r597 with the default parameters (Cheng et al. 2021). Only the primary contigs were used in all subsequent analyses.

2.3 Assembly Evaluation

Assembly statistics were generated using assembly-stats version 1.0.1 with the -t option for tabular output (Source code available at https://github.com/sanger-pathogens/assembly-stats). Illumina paired-end reads from 800 bp libraries were sourced from NCBI SRA accessions ERR3309119, ERR3309722, ERR3309739, and ERR3310010 for k-mer comparisons. K-mer completeness tests and comparison plots were made using KAT version 2.4.1 (Mapleson et al. 2017) with the KAT comp command using the total 466.39 Gb of SRA Illumina data. Counts of single-copy orthologs were carried out using Compleasm version 0.2.2–0 for the poales_odb10 lineage (Huang and Li 2023). Primary transcripts from Maccaferri et al. (2019) were mapped to their source assembly using minimap2 2.26-r1175 under the —split-prefix -ax splice:hq -uf options (Li 2018). Transcripts that mapped with 100% coverage and accuracy were selected to be aligned to the assembled genomes following the same method. The stats_from_bam program from the Pomoxis software suite version 0.3.13 (source code available at https://github.com/nanoporetech/pomoxis) was used to calculate query coverage and accuracy to the reference. From manual observation, a minimum cutoff of 99% coverage and 97% accuracy was selected to consider a reference transcript properly mapped and thus a gene model "present" in the reference assembly. Assembly-to-assembly mappings were done using minimap2 2.26-r1175 with the asm5 option. Dotplots were visualized using the dotPlotly scripts from https://github.com/tpoorten/dotPlotly with a minimum query length cutoff of 5000 and minimum alignment length cutoff of 20,000.

2.4 Organellar Contig Removal

The 135.88 kb plastid genome of the Chinese Spring cultivar (GenBank CM022232.1) and the 452.53 kb mitochondrial genome of the Chinese Yamai cultivar (NCBI RefSeq NC_036024.1) of *Triticum aestivum* were used to screen for organellar contigs in the output assemblies. Complete contigs were first aligned as queries to the chloroplast genome using the blastn command from BLAST+ with default settings and tabular output with the -outfmt 6 option (Camacho et al. 2009). Contigs having alignments that covered 60% or more of the entire contig while having at least 98% sequence identity were labeled as chloroplast and removed. The remaining contigs were aligned to the mitochondrial genome with blastn and alignments covering at least 60% of the query contig with over 75% sequence identity were removed.

2.5 Numerical Trends and Curve Fitting

In Fig. 1, ascending metrics are plotted as the percentile of the value at the 34.05× coverage assembly. For metrics expected to descend with increasing coverage, the formula $\frac{\min}{x}\%$ was used to obtain the percentile value, where min is the value of metric

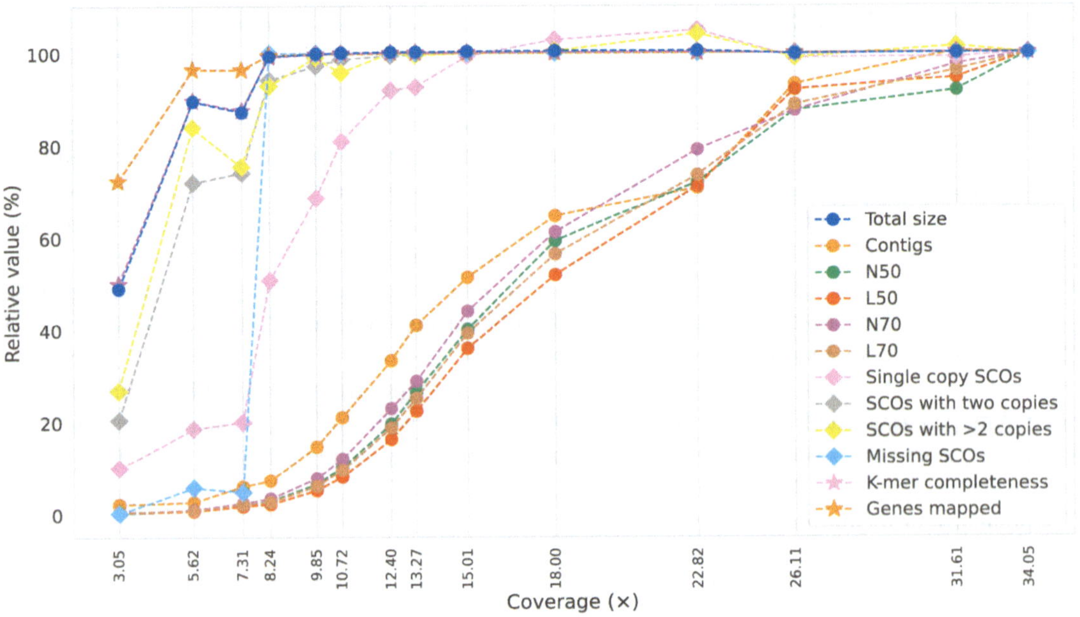

Fig. 1 Variation in genomic feature proportion in assemblies at increasing read coverage. Completeness in terms of assembly statistics, universal single-copy orthologs, k-mer comparisons, and putative genes mapped are shown as a relative fraction of the value observed in the assembly at highest coverage. Over 90% k-mers and genes are already present at 8.24× coverage. Single copies of universal orthologs are slower to converge at 15.01×. Quantities like N50, N70, L50, and L70, which describe the fraction of bases in large fragments and level of fragmentation, slowly develop in a sigmoidal pattern

at the highest coverage assembly. In Fig. 4, curves on the trend of the natural log of N50 and contig number as a function of coverage were fit using the formula $y = a(1 - e^{(b-cx)})$ and $y = a(1 + e^{(-cx)})$, respectively, where the log metric is the response and a, b, and c are estimated parameters. Parameter values were estimated using the nls function from the stats 4.3.1 library with starting values of 5, 0.1, and 0.1 in R (Team 2010).

3 Results

3.1 Raw Read Statistics

Genomic DNA libraries were sequenced using the PacBio SequelII in 13 SMRT cells. All the results that follow are based only on reads that had a mean QV value of 20 or more. The total yield was 357.55 Gb and the mean yield per SMRT cell was 27.50 Gb. The mean number of reads per SMRT cell, mean read length, and mean N50 were 1.53 million, 17.91 kb, and 17.94 kb, respectively.

3.2 Assembly Statistics

Fourteen subsets of reads, ranging in coverage from 3.05× to 34.05×, were assembled (Table 1). The coverage values were selected with the aim of maximizing the resolution of completeness statistics at focal coverages. The smallest nuclear assembly at 3.05× coverage was 5.11 Gb in length with 58,623 contigs and an N50 of 103 kb. At the next coverage level of 5.62×, assembly size increased to 9.35 Gb, which was over 80% of the estimated genome size (Fig. 1). However, the number of contigs was still similarly high and changes in N50 and L50 were marginal (Fig. 2a). When coverage was increased to 8.24×, assembly size improved to 10.36 Gb (over 99% of the expected size, Fig. 1). At this coverage, a significant decrease in the number of contigs and a ten-

Table 1 Assembly completeness statistics for Svevo HiFi assemblies

Coverage (×)	3.05	5.62	7.31	8.24	9.85	10.72	12.40	13.27	15.01	18.00	22.82	26.11	31.61	34.05	Maccaferri et al. (2019)
Assembly statistics for nuclear contigs															
Total size (Gb)	5.11	9.35	9.1	10.36	10.42	10.44	10.46	10.46	10.47	10.48	10.49	10.42	10.43	10.44	9.83
Contigs	58,623	48,009	21,154	17,423	8745	6103	3857	3133	2502	1988	1816	1379	1281	1285	3,30,244
N50 (Kb)	103	301	733	1048	2367	3678	7074	9578	14,453	21,353	25,960	31,544	33,084	36,046	57
L50	16,163	9604	3854	3037	1325	844	424	308	192	133	97	75	73	69	50,988
N70 (Kb)	74	193	470	665	1497	2295	4386	5503	8409	11,672	15,083	16,692	18,606	19,088	35
L70	27,874	17,345	6950	5506	2433	1556	798	596	382	266	204	169	156	150	94,081
miniBUSCO (%)															
Single	57.10	30.90	28.66	11.30	8.37	7.10	6.25	6.20	5.78	5.58	5.47	5.80	5.80	5.74	10.19
Duplicated	19.30	67.70	69.71	88.60	91.44	92.80	93.59	93.70	94.06	94.26	94.36	94.00	94.10	94.10	89.48
>2 Copies	1.16	3.64	3.27	4.02	4.29	4.15	4.33	4.31	4.33	4.35	4.51	4.29	4.39	4.33	3.70
Fragmented	2.50	0.30	0.39	0.10	0.12	0.10	0.10	0.10	0.10	0.10	0.10	0.10	0.10	0.10	0.14
Missing	21.10	1.04	1.25	0.06	0.06	0.06	0.06	0.06	0.06	0.06	0.06	0.06	0.06	0.06	0.18
K-mer statistics (%)															
Completeness	49.51	88.80	86.77	98.23	98.85	99.03	99.16	99.20	99.24	99.27	99.29	99.01	99.04	99.10	97.51
Found 0× at 30×	51.53	11.73	14.00	1.29	–	–	–	–	–	–	–	–	–	–	2.18
Found 1× at 30×	48.46	87.35	85.73	98.21	99.53	99.53	99.54	99.54	99.54	99.53	99.53	99.56	99.56	99.55	97.61
Gene models (%)															
Mapped	72.25	96.37	96.33	99.49	99.73	99.80	99.89	99.91	99.95	99.97	99.97	99.94	99.95	99.96	100.00

Note: To avoid inflation caused by the multiple organellar contigs, the contiguity statistics presented are derived from nuclear contigs only

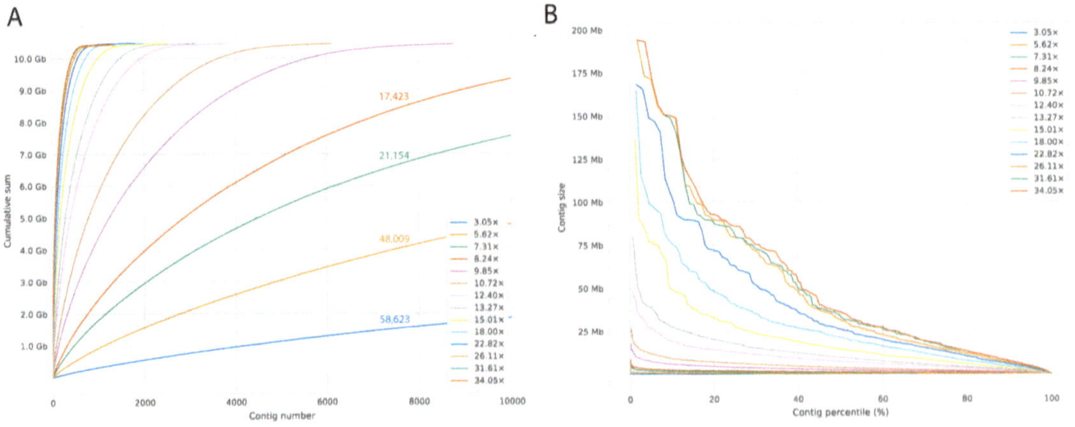

Fig. 2 Effect of input read coverage on assembly contiguity. (**a**) Cumulative number of bases as a function of contig number. Spans of the curves that exceed the x-axis range are shown with numbers. Decreases are dramatic at coverages below 9.85× relative to higher coverages. (**b**) Changes in contig size across the total number of contigs

fold increase in N50 were observed. With higher coverages, a marginal yet consistent improvement in contiguity was maintained up to the highest value (Fig. 1). The 15.01× assembly had about a 14-fold increase in contiguity compared to the 8.24×, with an N50 of 14.45 Mb and an L50 of 192. The assembly at the highest coverage of 34.05× had the highest N50 and N70 at 36.04 Mb and 19.05 Mb, respectively, and the lowest L50 and L70 at 69 and 150, respectively. Beyond 8.24×, the total size of the assemblies was consistent and not affected by a >14-fold drop in contig number (Table 1, Figs. 1 and 2a). The minor (0.2%) increase in assembly size observed between 9.85× and 34.05× coverage was accompanied by significant gains in contiguity, with 14- and 12-fold increases in N50 and N70, respectively (Table 1). For the higher coverage assemblies, the differences are better visualized in Fig. 2b where descending contig lengths are shown. Above 18× coverage, over 40% of the contigs in the assemblies are longer than 25 Mb and above 30× this fraction increases to over 60% (Fig. 2b).

3.3 miniBUSCO and Gene Space Completeness

Given that there are two subgenomes in durum wheat, exactly two copies of conserved universal single-copy orthologs were expected to be found in each assembly. The 3.05× assembly had a single copy for 57.10% of the universal single-copy orthologs (SCOs, Table 1), whereas 19.30% were found in multiple copies, and within that fraction 1.20% had more than two copies. The percentage of SCOs found once in the assembly gradually decreased with increasing coverage, reaching a stable minimum at 15.01×. The duplicated SCO fraction reached a plateau sooner at around 10× (Fig. 1 and Table 1). Already at 8.24×, miniBUSCO completeness of the HiFi assembly was comparable to the Svevo assembly reported in Maccaferri et al. (2019), with further improvements observed at higher coverages (Table 1, discussed in Sect. 4.1). The missing orthologs were 21.10% at 3.05×, but at 5.62× the amount already decreased to 1.00% (Table 1). This trend in representation of the gene space was validated by mapping putative primary transcripts from the Svevo assembly (Maccaferri et al. 2019). Upon aligning the 54,419 models that displayed a perfect mapping to their reference, 72.25% of the genes mapped to the 3.05× assembly. A more generous 96.37% of the genes could be found in the 5.62×. From 8.24×, 99.50% or more of the genes consistently mapped to the assemblies and beyond 13.27×, this fraction exceeds 99.90%.

3.4 K-Mer Comparisons

In the 3.05× assembly, the fraction of assembly k-mers found in the unassembled reads was 49.51% (Table 1). At 5.62× coverage, this increased to 88.80%. At 8.24×, over 98% of the assembly k-mers were found and starting from 10.72×, less than 1% of the read k-mers were missing. The original Svevo assembly (Maccaferri et al. 2019) had a notably lower fraction of assembly k-mers, at 97.51%, as discussed in Sect. 4.1. The completeness metric was reflected in the k-mer completeness plots shown in Fig. 3. The 3.05× assembly had 51.53% missing k-mers at the coverage peak. In the 5.62× assembly, the fraction of k-mers not found in the peak reduced to 11.73%. At the next coverage level of 8.24×, 1.29% of k-mers were clearly missing the peak. Beyond 8.24× coverage, there was a consistent, near-complete representation (99.53% or more) of the read k-mers in the assemblies (Table 1).

3.5 Prediction of Assembly Contiguity at Higher Coverage

The general trend for the Nx and Lx metrics against coverage was sigmoidal (Fig. 1). After applying a log transformation to the N50 values to make the range of values more comparable, an ascending nonlinear trend that reached a plateau was observed (Fig. 4). The estimated nonlinear regression curve had the formula $10.79(1 - e^{(-0.13 - 0.12x)})$. All three parameters were statistically significant under the null t distribution, with p-values 1.08×10^{-14}, 0.04, and 4.38×10^{-7}, respectively. Based on the observed data points, the N50 value theoreti-

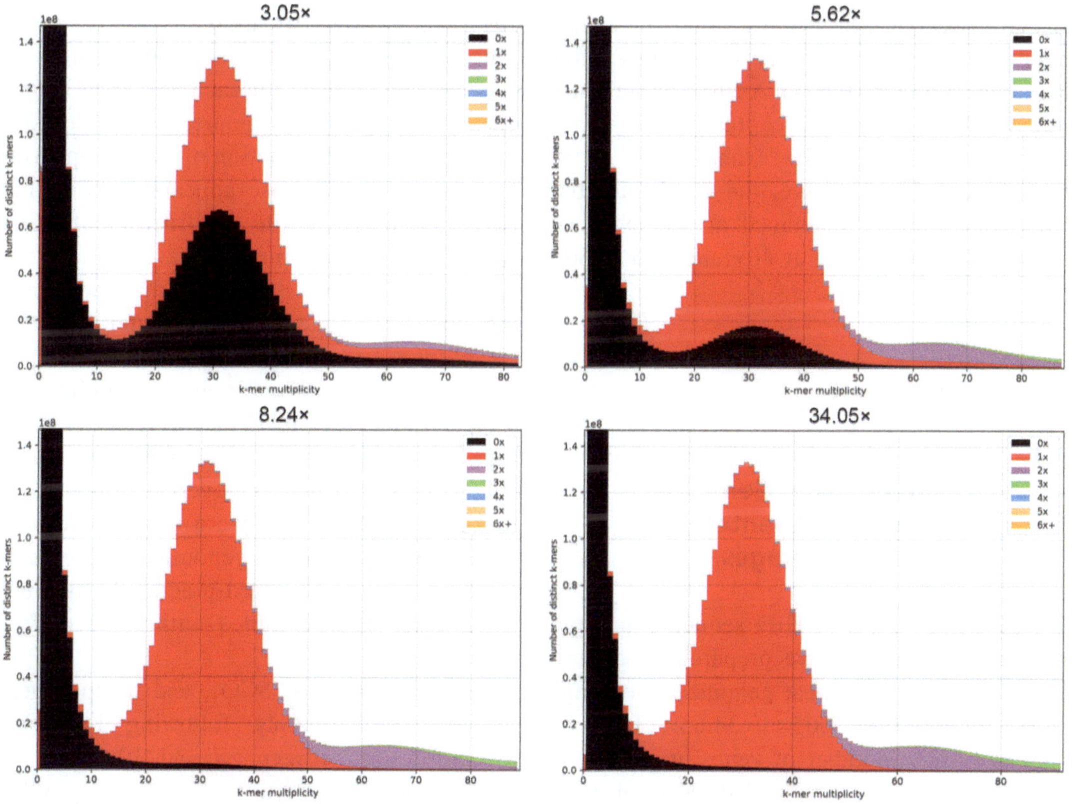

Fig. 3 K-mer comparison plots for assemblies at 3.05×, 5.62×, 8.24×, and 34.05×. The unassembled fraction of k-mers shown in black quickly recedes with increasing coverage, reaching 98.23% completeness at 8.24×

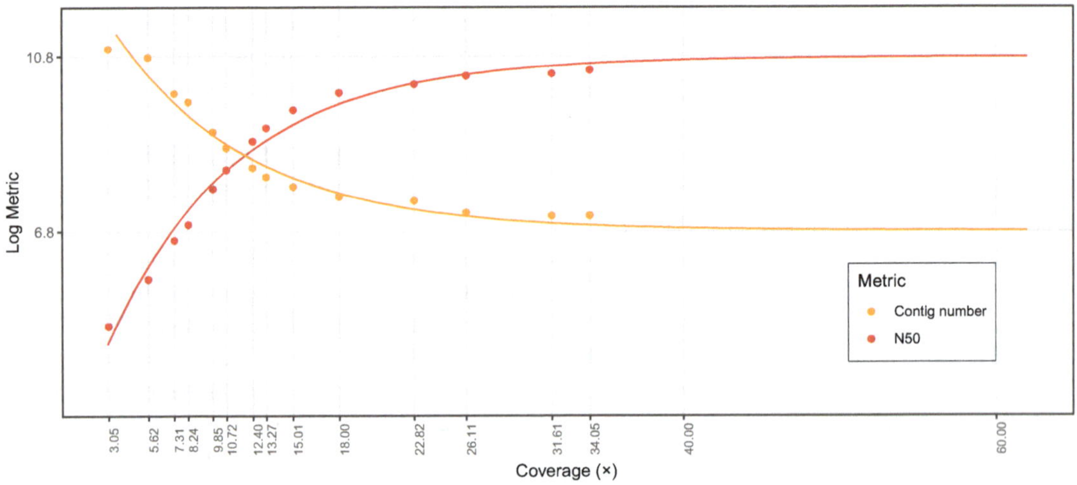

Fig. 4 Projections for N50 and contig number values at higher sequencing coverage. The curves are theoretical and may deviate from practical outcomes. The asymptote for N50 is equivalent to 48.53 Mb and for contig number is equivalent to 904

cally reached a plateau near 48.53 Mb with an estimated N50 of 44.89 Mb at 40× and 48.19 Mb at 60× coverage. The general trend for contig number was asymptotically descending after log transformation (Fig. 4). A curve of the function $6.80(1 + e^{(-0.1143x)})$ captured its general trend (p-values 5.65×10^{-13}, and 1.01×10^{-5}). Based on the fitted curve, the contig number value theoretically plateaued at 903 with an estimated value of 969 at 40× and 910 at 60× coverages.

4 Discussion

4.1 Assembly K-Mers, Single-Copy Orthologs, and Gene Space Are Well Represented at 8–10× Coverages

To study the effects of HiFi sequencing depth on assembly metrics, we prepared 14 assemblies of the durum wheat genome at coverage values ranging from 3.05× to 34.05× (Table 1). Assembly size, number of genes mapped, and k-mer completeness exceeded 90% of their respective best values already at 8.24× (Table 1 and Fig. 1), while higher coverages progressively developed contiguity (Fig. 2a, b). The reported results are similar to what reported by Mascher et al. (2021) for barley, where a 5× coverage allowed to align 80.5% of genes. Since durum wheat is an allotetraploid, we expected two copies for each universal single-copy ortholog in its genome. With increasing input coverage, the fraction of SCOs appearing exactly once and those appearing more than twice decreased, while the fraction of SCOs appearing in two copies increased (Fig. 1 and Table 1). For both SCOs and mapped genes, the lowest coverage value for greater than 90% representation was observed to be 8.24×. Below 8.24× input coverage, assemblies had large amounts of k-mers that were either missing or appeared at unexpected multiplicity (Fig. 3). At the 8–10× coverage range, even though the assemblies were fragmented with low N and L values, all genomic completeness features were represented practically as much as in assemblies obtained with three to four more reads.

Assembly k-mers, SCOs, and mapped gene numbers rose sharply following increasing assembly size between 5× and 10× (Fig. 1). Among these, the fraction of single-copy orthologs was the slowest to reach its plateau. This was unexpected since the assembly size at 9.85× was already 99.80% of the 34.05× assembly and

98.85% of the putative genes could be mapped to it. One possible hypothesis is that the two homoeologous gene copies collapsed during assembly because of low input read coverage. Higher coverages provided the necessary contrast between the two subgenomes to both assemble the homoeologs separately and to extend contigs across repetitive regions.

4.2 Assembly Contiguity Develops Markedly Between 10× and 30× Coverage

The general trend for N50, N70, L50, and L70 values representing the level of fragmentation in the assemblies was observed to be sigmoidal (Fig. 1). Assemblies with coverage lower than 8.24× were more fragmented and were also significantly shorter than the estimated genome size (Table 1). As coverage increased beyond 8.24×, the minimal gains in contiguity were counterbalanced by the substantial increase in assembly size. This was manifested as a stagnation in the N and L values in Fig. 1. Once the assembly size stabilized near the expected genome size at 8.24× coverage, contiguity metrics (contig number, N and L values) increased steeply until plateauing again above 30× coverage. This suggests that beyond 30×, it is likely that additional HiFi coverage may lead to marginal gains.

4.3 General Trends and Predictive Models for N50 and Contig Number

A polynomial curve to contig N50 was fitted as a function of read coverage from the double-haploid ryegrass assembly (Frei et al. 2021). Data from Table 1 were suitable for similar analyses. The N50 and contig number values varied drastically between 3.05× and 34.05×, making it necessary to apply a log transformation to compare the numerical differences across the coverage range. Although our curve fit was probabilistically robust, because these response metrics are generated by a stochastic assembly process, it can only be assumed that practical outcomes will reliably follow the trend, especially outside the lower and upper bounds of the coverage range. Furthermore, this trend should be generalized with caution, as genomes will vary in size, ploidy, chromosome counts, genome features, and read profiles. However, most HiFi experiments target a coverage value within our studied range and these curves could provide an approximate expectation when considering the prospects of generating more read data.

4.4 High-Coverage HiFi Assemblies Are Highly Contiguous

The 34.05× assembly had the highest coverage value and produced only 1285 nuclear contigs. Whole genome alignments of the 3.05× and 8.24× assemblies to the 34.05× showed that the less complete and more fragmented 3.05× assembly had visible gaps between fragments that were not seen for the near-full-size 8.24× assembly (Fig. 5). The haploid durum wheat genome is composed of 14 chromosomes, mostly with metacentric configuration (Feldman and Levy 2023). A simple interpretation of assembly contiguity can be obtained by assuming roughly equal lengths for the 28 chromosome arms, each about 375 Mb long. For the most contiguous 34.05× assembly, if we considered the contigs being homogeneously distributed across the 28 chromosome arms, the assembly would only have 5 gaps per chromosome arm. Furthermore, there were only 589 contigs longer than 1 Mb, for a combined size of 10.35 Gb. Considering only these 589 contigs spread across the 28 chromosome arms, an assembly containing over 99% of the bases would have only 21 gaps per chromosome arm. From a historical perspective of genome sequencing and assembly technology, this level of contiguity for a genome of this scale is highly impressive.

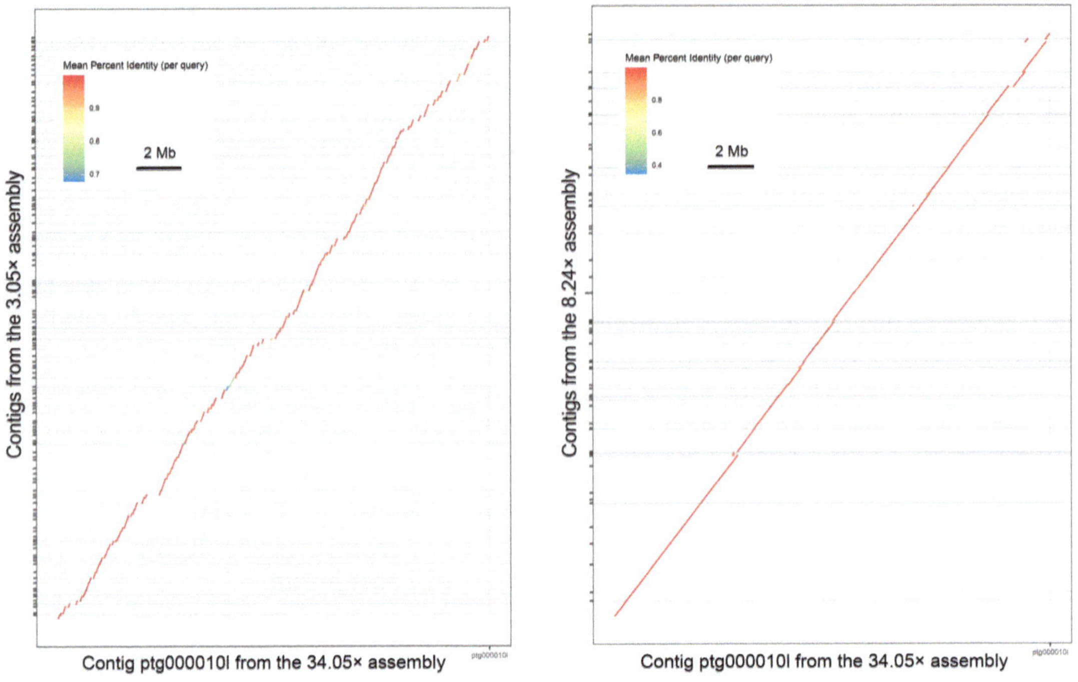

Fig. 5 Genomic dotplot alignments for 3.05× and 8.24× assemblies mapped to the 34.05× assembly. Mappings for a single reference contig are shown. The 3.05× assembly shows higher levels of fragmentation and more missing bases compared to 8.24×

5 Conclusion

De novo assembly of large, complex plant genomes has become more financially feasible in recent years. Using the 10.50 Gb tetraploid durum wheat genome as an experimental model, we extensively analyzed the effects of PacBio HiFi read coverage on assembly completeness metrics. The presented results highlight notable differences in assembly properties resulting from coverage variation that will help researchers make informed decisions when setting a target coverage range for sequencing projects. We demonstrate that highly fragmented durum wheat assemblies prepared at 8–10× coverage display genomic features nearly equivalent to highly contiguous assemblies prepared at 15–34× coverage (Fig. 1).

At input coverages above 10×, assembly contiguity develops markedly up to roughly 30× coverage, after which the improvements in N and L metrics begin to slow down. The predictive model suggests that beyond the 30–40x coverage range, small improvements in N values and contig numbers are possible but financially restrictive. Increasing the sequencing coverage from 40× to 60× would increase data and cost by an additional 50%; yet, according to our models, it would result in only a 9% increase in N50 and a 6% decrease in contig number.

Overall, these observations mark the 8–10× coverage range as an optimal choice for applications where contiguity is not prioritized, as in the case of genomes considered at a silver-level sequencing in the pangenome pyramid. We conclude that when a platinum-grade reference assembly is available, resequencing of broad germplasm collections at this coverage range would economically and scientifically maximize sequencing outcome.

Acknowledgments The authors would like to thank Cassandra Trier and Tiffany Bledsoe for their helpful comments. The authors acknowledge the funding support from the Italian projects "Grain Pangenomics for Durum Wheat Sustainable Production," PANWHEATGRAIN PRIN-2020 project, "National Center for Technology in

Agriculture" AGRITECH PNRR project, and the Agritech National Research Center, Spoke1, as well as funding received from the European Union Next-GenerationEU (PIANO NAZIONALE DI RIPRESA E RESILIENZA [PNRR]—MISSIONE 4 COMPONENTE 2, INVESTIMENTO 1.4—D.D. 1032 17/06/2022, CN00000022). This manuscript reflects only the authors' views and opinions, and neither the European Union nor the European Commission can be considered responsible for them.

References

Alonge M, Shumate A, Puiu D, Zimin AV, Salzberg SL (2020) Chromosome-scale assembly of the bread wheat genome reveals thousands of additional gene copies. Genetics 216(2):599–608. https://doi.org/10.1534/genetics.120.303501

Andrews S (2010) FastQC: a quality control tool for high throughput sequence data. In: Babraham bioinformatics. Babraham Institute, Cambridge, UK

Bushnell B (2015) BBMap short-read aligner, and other bioinformatics tools. University of California, Berkeley

Camacho C, Coulouris G, Avagyan V, Ma N, Papadopoulos J, Bealer K, Madden TL (2009) BLAST+: architecture and applications. BMC Bioinform 10:1–9

Cheng H, Concepcion GT, Feng X, Zhang H, Li H (2021) Haplotype-resolved de novo assembly using phased assembly graphs with hifiasm. Nat Methods 18(2):170–175

De Coster W, D'hert S, Schultz DT, Cruts M, Van Broeckhoven C (2018) NanoPack: visualizing and processing long-read sequencing data. Bioinformatics 34(15):2666–2669

Feldman M, Levy AA (2023) Genome structure of Triticeae species. In: Feldman M, Levy AA (eds) Wheat evolution and domestication. Springer International Publishing, pp 43–70. https://doi.org/10.1007/978-3-031-30175-9_3

Frei D, Veekman E, Grogg D, Stoffel-Studer I, Morishima A, Shimizu-Inatsugi R, Yates S, Shimizu KK, Frey JE, Studer B, Copetti D (2021) Ultralong Oxford Nanopore reads enable the development of a reference-grade perennial ryegrass genome assembly. Genome Biol Evol 13(8). https://doi.org/10.1093/gbe/evab159

Grewal S, Yang C-Y, Scholefield D, Ashling S, Ghosh S, Swarbreck D, Collins J, Yao E, Sen TZ, Wilson M, Yant L, King IP, King J (2024) Chromosome-scale genome assembly of bread wheat's wild relative *Triticum timopheevii*. Sci Data 11(1):420. https://doi.org/10.1038/s41597-024-03260-w

Huang N, Li H (2023) Compleasm: a faster and more accurate reimplementation of BUSCO. Bioinformatics 39(10):btad595

Kersey PJ (2019) Plant genome sequences: past, present, future. Curr Opin Plant Biol 48:1–8

Kong W, Wang Y, Zhang S, Yu J, Zhang X (2023) Recent advances in assembly of complex plant genomes. Genomics Proteomics Bioinformatics 21(3):427–439

Li H (2018) Minimap2: pairwise alignment for nucleotide sequences. Bioinformatics 34(18):3094–3100

Li H, Handsaker B, Wysoker A, Fennell T, Ruan J, Homer N, Marth G, Abecasis G, Durbin R, Subgroup GPDP (2009) The sequence alignment/map format and SAMtools. Bioinformatics 25(16):2078–2079

Liu Z, Yang F, Deng C, Wan H, Tang H, Feng J, Wang Q, Yang N, Li J, Yang W (2024) Chromosome-level assembly of the synthetic hexaploid wheat-derived cultivar Chuanmai 104. Sci Data 11(1):670

Maccaferri M, Harris NS, Twardziok SO, Pasam RK, Gundlach H, Spannagl M, Ormanbekova D, Lux T, Prade VM, Milner SG, Himmelbach A, Mascher M, Bagnaresi P, Faccioli P, Cozzi P, Lauria M, Lazzari B, Stella A, Manconi A et al (2019) Durum wheat genome highlights past domestication signatures and future improvement targets. Nat Genet 51(5):885–895. https://doi.org/10.1038/s41588-019-0381-3

Mapleson D, Garcia Accinelli G, Kettleborough G, Wright J, Clavijo BJ (2017) KAT: a K-mer analysis toolkit to quality control NGS datasets and genome assemblies. Bioinformatics 33(4):574–576

Mascher M, Wicker T, Jenkins J, Plott C, Lux T, Koh CS et al (2021) Long-read sequence assembly: a technical evaluation in barley. Plant Cell 33(6):1888–1906

PacBio (2022) PacBio Announces Revio, a Revolutionary New Long Read Sequencing System Designed to Provide 15 Times More HiFi Data and Human Genomes at Scale for Under $1,000. www.pacb.com. Retrieved 11/15/2023 from https://www.pacb.com/press_releases/pacbio-announces-revio-a-revolutionary-new-long-read-sequencing-system-designed-to-provide-15-times-more-hifi-data-and-human-genomes-at-scale-for-under-1000/

Sato K, Abe F, Mascher M, Haberer G, Gundlach H, Spannagl M, Shirasawa K, Isobe S (2021) Chromosome-scale genome assembly of the transformation-amenable common wheat cultivar 'fielder'. DNA Res 28(3). https://doi.org/10.1093/dnares/dsab008

Team, R. D. C. (2010) R: a language and environment for statistical computing. (No Title)

Wenger AM, Peluso P, Rowell WJ, Chang P-C, Hall RJ, Concepcion GT, Ebler J, Fungtammasan A, Kolesnikov A, Olson ND, Töpfer A, Alonge M, Mahmoud M, Qian Y, Chin C-S, Phillippy AM, Schatz MC, Myers G, DePristo MA et al (2019) Accurate circular consensus long-read sequencing improves variant detection and assembly of a human genome. Nat Biotechnol 37(10):1155–1162. https://doi.org/10.1038/s41587-019-0217-9

Zhu T, Wang L, Rimbert H, Rodriguez JC, Deal KR, De Oliveira R, Choulet F, Keeble-Gagnère G, Tibbits J, Rogers J, Eversole K, Appels R, Gu YQ, Mascher M, Dvorak J, Luo M-C (2021) Optical maps refine the bread wheat *Triticum aestivum* cv. Chinese spring genome assembly. Plant J 107(1):303–314. https://doi.org/10.1111/tpj.15289

Zimin AV, Puiu D, Hall R, Kingan S, Clavijo BJ, Salzberg SL (2017) The first near-complete assembly of the hexaploid bread wheat genome, Triticum aestivum. Gigascience 6(11). https://doi.org/10.1093/gigascience/gix097

Gene Expression Analysis in Durum Wheat

Matteo Bozzoli, Riccardo Aiese Cigliano, and Cristian Forestan

Abstract

The technologies used for transcriptome studies have rapidly evolved during the last two decades, becoming more powerful and less expensive. In this post-genomic era, deep understanding of gene expression, regulation and co-expression in tissues during developmental processes or in response to biotic and/or abiotic constraints as related to changing environmental conditions is crucial. Moreover, transcriptomics represents also a yet-underexplored source of molecular information for developing molecular markers, particularly useful in species with large genomes rich in repetitive sequences.

In this chapter we briefly review the history and state of transcriptomic technologies and data analysis tools, with particular emphasis on the challenges posed by the tetraploid durum wheat, and we review some original publications that used transcriptomic analysis for deciphering gene expression during seed development, stress response or for the identification of varietal-specific SNPs.

Keywords

Wheat · NGS technologies · Transcriptomics · Wheat breeding · Plant development regulation

1 Introduction

Durum wheat (*Triticum turgidum L.* ssp. *durum*), although it represents only 5%–8% of global wheat production, is a major cereal and staple in the semi-arid regions of the Mediterranean Basin and cultivated also in Canada, the USA, Mexico and Australia. It is a tetraploid (genome BBAA) species of wheat, originated in the Fertile Crescent from the domesticated emmer wheat (*T. turgidum* spp. *dicoccum*), which in turn is derived from the wild tetraploid emmer wheat (*T. turgidum* spp. *dicoccoides*) (Dubcovsky and Dvorak 2007). Natural hybridisation and allopolyploid speciation therefore significantly impacted on the diversification within the species evolved in up to 11 *T. turgidum* subspecies, including durum wheat landraces and modern cultivars (Matsuoka 2011).

Thus, the phases of domestication from wild to domesticated emmer wheat, continuous evolution of landraces from the domesticated emmer

M. Bozzoli · C. Forestan (✉)
Department of Agricultural and Food Sciences, University of Bologna, Bologna, Italy
e-mail: matteo.bozzoli2@unibo.it; cristian.forestan@unibo.it

R. A. Cigliano
Sequentia Biotech SL, Barcelona, Spain
e-mail: raiesecigliano@sequentiabiotech.com

R. Tuberosa et al. (eds.), *The Durum Wheat Genome*, Compendium of Plant Genomes,
https://doi.org/10.1007/978-3-031-84689-2_4

and the breeding of improved cultivars resulted in the definition of different germplasm groups of tetraploid wheat, exhibiting a very large genetic diversity (Maccaferri et al. 2019). In particular, hundreds of landraces originating from the continuous transition from the domesticated form of emmer were selected by ancient farmers for being adapted to different environmental conditions and could be fundamental for breeding programmes aimed at developing new improved varieties with desirable traits (Mazzucotelli et al. 2020).

Traditionally grown under rainfed conditions in the Mediterranean Basin, durum wheat is more adapted to drought and heat stresses than bread wheat; however, the dramatic effects of climate changes are constituting a main limiting factor also in durum wheat productivity, affecting both yield and quality (Shewry 2009; De Santis et al. 2021). It is therefore fundamental to investigate and elucidate the molecular basis involved in traits responsible for the adaptation and resilience of this crop to climate changes, along with other challenges posed by the need of a modern, more sustainable agriculture (Habash et al. 2009 ; Habash et al. 2014). Moreover, being the donor of the A and B genomes of hexaploid bread wheat (DDAABB), domesticated tetraploid wheat represents a valuable source of genetic variability and beneficial alleles also for bread wheat breeding programmes, hence more effectively bridging the two species (Maccaferri et al. 2015; Klymiuk et al. 2019).

The whole genome sequencing coupled with the high-quality assembly of the Zavitan emmer wheat (Avni et al. 2017) and durum wheat cultivar Svevo (Maccaferri et al. 2019) represented therefore an important milestone for genetic researches and breeding efforts. In both subspecies, annotation of ~10 Gbp assembly revealed roughly 65,000 high-confidence genes, equally distributed between the A and B subgenomes. In addition, about 45,000 and 300,000 low-confidence genes were annotated in Zavitan and Svevo respectively. The presence of homoeologous genes, derived from each chromosome set, represents indeed a challenge in identifying its differential contribution to plant phenotypes (Leach et al. 2014; Ramírez-González et al. 2018). For example, gene expression studies in emmer wheat under 20 different conditions indicated that, with 30.4% of the genes expressed in all conditions and 48% in at least one condition, the number of genes expressed in each subgenome was very similar, with only a slightly higher expression on average for genes of the A subgenome (Avni et al. 2017). Similarly, 90.5% of the high-confidence genes annotated in the Svevo genome were expressed in at least one of the 21 transcriptome samples analysed.

In this post-genomic context, a deeper understanding of gene expression patterns and dynamics, gene regulation and co-expression in tissues, during developmental processes or in response to biotic or abiotic constraints, is essential for advancing our understanding of durum wheat biology, crop improvement and responses to changing environmental conditions. These transcriptomic studies will help and support the identification of candidate genes underlining quantitative trait locus (QTL) effects, but, even more importantly, will identify the metabolic and signalling pathways involved in the different biological responses and will provide valuable data that can be applied to enhance crop yields, develop stress-tolerant varieties and address global food security challenges.

The methods used for studying and measuring the transcriptome have changed and improved repeatedly over the last decades (Lowe et al. 2017), from the earliest Sanger sequencing of concatenated random transcript fragments (SAGE—serial analysis of gene expression) (McIntosh et al. 2007; Poole et al. 2008), to the hybridisation-based microarrays (Foroud et al. 2012; Pilcher et al. 2018) concluding with the use of high-throughput, next-generation sequencing (NGS) technologies for sequencing the RNA (cDNA) molecules (RNA-Seq). RNA-Seq allows to estimate with precision and sensitivity the abundance in expression of a gene from the counts of RNA sequences obtained from a specific tissue or sample, without the requirement of predefined, transcript-specific probes.

In addition to the gene expression studies, RNA-Seq has other more specific applications:

taking advantage of its sequencing step RNA-Seq allows *de novo* gene and transcript identification, study of small RNAs with regulatory functions, prediction of non-coding RNAs (Di Marsico et al. 2022) and it also allows to obtain variant information at the RNA level, as source of Single Nucleotide Polymorphisms (SNPs) in the coding regions, that could be combined with gene expression to study the variation that affects gene-expression levels. RNA-Seq is a powerful technology to identify such expression quantitative trait locus (eQTL) through Genome-Wide Association Studies (GWAS) mapping, as recently done in bread wheat (He et al. 2022).

In this chapter, we describe the new advances in NGS transcriptomic technologies, in particular their potentialities and their drawbacks studying and analysing the autopolyploid durum wheat transcriptome, and discuss some seminal recent publications, specifically those concerning with the study of gene regulation during developmental processes or in response to biotic and biotic stress. Finally, the potentiality of RNA-Seq in transcriptome assembly and SNP detection in coding regions for developing molecular markers useful for breeding purposes will be presented.

2 NGS Sequencing and RNA-Seq Applications

The last 10 years have seen a dramatic increase in the application of next-generation sequencing (NGS) to the analysis of transcriptomes, including both long and small RNAs, through what is simply called RNA-Seq. Three main technologies are currently being used for sequencing of RNA: sequencing-by-synthesis (Illumina), Single-molecule real-time (SMRT) sequencing (PacBio) and Nanopore sequencing (ONT) (McCombie et al. 2019). Illumina short-read sequencing technology is currently the most used platform for transcriptomics analyses being characterised by lower error rates, higher throughput and lower costs. The main limitation in the Illumina technology is the relatively short size of the obtained sequences that can reach 300 bp using a MiSeq or 150 bp using a NovaSeq 6000 (Stark et al. 2019). On the other hand, long read technologies such as PacBio and ONT are characterised by much longer read lengths capable to cover entire transcripts but also by higher error rate and lower throughput that, in turn, leads to increased costs per sample (Stark et al. 2019).

The Sequence Read Archive (SRA), the largest publicly available repository of high-throughput sequencing data from all branches of life, currently contains transcriptomics datasets from 978 *Triticum durum* samples as of February 2024. Interestingly, all these datasets are produced with the Illumina technology. The samples deposited in SRA are equally distributed between transcriptome RNA-Seq experiments (50.1%) and miRNA-Seq experiments (49.9%). Of the RNA-Seq sample deposited, 66% of them are obtained from paired-end sequencing (see Table 1). On average, 4.2 Gbp of data per sample were produced, mostly from NovaSeq 6000 (40.1%) and HiSeq (55.3%) instruments. The most analysed tissues are flag leaf (27.8%) and developing head (17.1%) from control individuals (34.4%) but also heat stressed (15.2%) and water or water/heat stressed plants (50.3%). The transcriptomic resources generated on durum wheat and the corresponding scientific insights gained into abiotic and biotic stress responses have been recently described and reviewed by Curci and colleagues (Zuluaga et al. 2023). On the other hand, we were unable to retrieve any publication or datasets showing the application of PacBio or ONT sequencing to the study of *T. durum* transcriptome, whereas more than 150 PacBio samples were found for common wheat (*Triticum aestivum*).

The production of sequencing data has become a relatively easy process thanks to commercial kits that can produce high-quality RNA from multiple tissues in order to generate the libraries for sequencing. On the other hand, correct processing and analysis of the data is still a challenge for many researchers (Attwood et al. 2019). Over the years, more than 2000 bioinformatics tools have been developed and published for the analysis of transcriptomics data (Ison et al. 2016). The main approaches to analyse mRNA-Seq data have been already reviewed extensively in previ-

Table 1 Summary of the publicly available durum wheat RNA-Seq data

BioProject	Description	Tissue(s)	Cultivar/genotype	Additional information	References
PRJNA491827	Transcriptome of *Triticum turgidum* ssp. *durum* cultivar with high and low Cd	Root, shoot	12 commercial cv. and 2 Near Isogenic Lines (NILs)	Cadmium accumulation	Vergine et al. (2017) https://doi.org/10.1021/acs.jafc.7b01946
PRJEB35112	Genetic and transcriptional dissection of resistance to *Claviceps purpurea* in the durum wheat cultivar Greenshank	Flower	Greenshank × AC Avonlea, Recombinant Inbred Lines (RILs)and Doubled Haploids (DH)	Resistant vs. susceptible to *Claviceps purpurea*	Gordon et al. (2020) https://doi.org/10.1007/s00122-020-03561-9
PRJNA259173	Regulation of Zn and Fe transporters by the *GPC1* gene during early wheat monocarpic senescence	Flag leaf	Kronos mutant for *GPC1* and *GPC2*	Zn and Fe content	Pearce et al. (2014) https://doi.org/10.1186/s12870-014-0368-2
PRJNA473404	Durum wheat genome highlights past domestication signatures and future improvement targets	Grain, caryopses, root, coleoptile, spike, rachis, glumes	Svevo		Maccaferri et al. (2019) https://doi.org/10.1038/s41588-019-0381-3
PRJNA773969	Transcriptomic and physiological response of durum wheat grain to short-term heat stress during early grain filling	Grain	Queule-INIA cultivar	Control vs. heat stress	Arenas-M et al. (2022) https://doi.org/10.3390/plants11010059
PRJNA715857	RNA-Seq analysis of durum wheats inoculated with *Fusarium graminearum*	Spikes 72 h after inoculation	Langdon (S) and derived (R) line	Inoculum with *Fusarium*	Soresi et al. (2023) https://doi.org/10.1080/07060661.2022.2091663
PRJNA780180	Transcriptional regulation of metabolic and cellular processes in durum wheat in the face of temperature increasing	Root, shoot	CIRNO C2008	Thermal adaptation	Chaparro-Encinas et al. (2021) https://doi.org/10.3390/plants10122792
PRJNA719092	Small RNA, transcriptome and degradome analysis of the transgenerational heat stress response network in durum wheat	Flag leaf, grain	DBA Aurora and DBA Artemis	Control vs. heat stress	Liu et al. (2021b) https://doi.org/10.3390/ijms22115532

PRJNA624178	Comparative transcriptomic and metabolic analyses of wild and domesticated wheat genotypes reveal differences in chemical and physical defence responses against aphids	Leaf	Svevo, Zavitan	Aphid infection vs. control	Shavit et al. (2022) https://doi.org/10.1016/j.plantsci.2021.111171
PRJNA746118	Transcriptome changes induced by arbuscular mycorrhizal symbiosis in leaves of durum wheat (*Triticum durum* Desf.) promote higher salt tolerance	Root	Anco Marzio	Salt with or without mycorrhizal symbiosis	Puccio et al. (2023) https://doi.org/10.1038/s41598-022-26903-7
PRJNA644573	Multi-omics analysis of small RNA, transcriptome and degradome in *T. turgidum*—regulatory networks of grain development and abiotic stress response	Grain	DBA Aurora (water tolerant), L6 (sensitive)	Control vs. heat, water or heat + water stress, including miRNA	Liu et al. (2020a) https://doi.org/10.3390/ijms21207772
PRJNA680308	Small RNAs and their targets are associated with the transgenerational effects of water-deficit stress in durum wheat	Grain, flag leaf	DBA Artemis (heat tolerant), DBA Aurora (water tolerant)	Control vs. water stress	Liu et al. (2021b) https://doi.org/10.1038/s41598-021-83074-7
PRJNA673987 PRJNA681687	Leaf RNA-Seq of Kronos plus *tgw1* and *kat2* mutants	Leaf	Kronos (WT), *tgw1* and *kat2* mutant plus *Kat2* overexpression line		Chen et al. (2020) https://doi.org/10.1038/s41467-020-20133-z
PRJNA680890 PRJNA681099 PRJNA681097 PRJNA681067 PRJNA681036 PRJNA681032 PRJNA681065 PRJNA681027	Interactions between SQUAMOSA and SHORT VEGETATIVE PHASE MADS-box proteins regulate meristem transitions during wheat spike development	Apical meristem	Kronos *vrn1* and *vrn1 ful2* mutants	Stages W1.0, W2.0, W3.0, W3.25 and W3.5 of meristem and spike development—*vrn1* and *vrn1ful2* mutants	Li et al. (2021a) https://doi.org/10.1093/plcell/koab243
PRJNA795036	Transcriptional signatures of wheat inflorescence development	Apical meristem	Kronos	Stages W1.0, W2.0, W3.0, W3.25 and W3.5 of meristem and spike development	VanGessel et al. (2022) https://doi.org/10.1038/s41598-022-21571-z

(continued)

Table 1 (continued)

BioProject	Description	Tissue(s)	Cultivar/genotype	Additional information	References
PRJNA470527	Comparative transcriptome profile during grain filling of cultivated wheat genotypes and their progenitors	Grain	AA, BB, DD and AABBDD	Evolutionary study with different ploidy species	Kaushik et al. (2020) https://doi.org/10.3390/genes11050509
PRJNA641098	Integrated analysis of small RNA, transcriptome and degradome sequencing reveals the water-deficit and heat-stress response network in durum wheat	Flag leaf	DBA Aurora (water tolerant), L6 (sensitive)	Control vs. heat, water or heat + water stress, including miRNA	Liu et al. (2020b) https://doi.org/10.3390/ijms21176017
PRJNA532455	The evolution of gene expression in polyploid wheats and their diploid ancestors during embryogenesis	Endosperm, embryo, pericarp	Strongfield, AC Barrie, diploid progenitors	Different embryo developmental stages: two cells, octant, transition, leaf early, leaf middle, leaf late, mature, early endosperm, late endosperm	Xiang et al. (2019) https://doi.org/10.1105/tpc.19.00397
PRJNA600522	Transgenerational effects of water-deficit and heat stress on germination and seedling vigour—new insights from durum wheat microRNAs	Seed	DBA Aurora (water tolerant), L6 (sensitive)	Four different seed groups from control and stressed plants	Liu et al. (2020c) https://doi.org/10.3390/plants9020189
PRJNA630287	Copy number variation of *TdDof* controls solid-stemmed architecture in wheat	Stem		Solid and hollow stem phenotype	Nilsen et al. (2020) https://doi.org/10.1073/pnas.2009418117
PRJNA642414	Alternative splicing dynamics and evolutionary divergence during embryogenesis in wheat species	Embryo, endosperm, pericarp	CDC Commander-CM + Chinese Spring	Seven stages of embryo development, two of endosperm and pericarp	Gao et al. (2021) https://doi.org/10.1111/pbi.13579
PRJNA541573	Functional characterisation of the *B1* gene through transcriptome analysis of transgenic *B1* overexpression durum and bread wheat lines	Lemma, florets and spike	Strongfield × Glossy Huguenot		Huang et al. (2020) https://doi.org/10.1111/nph.16154
PRJNA503549	Transcriptome data of cultivated tetraploid and hexaploid wheat variety during grain development	Grain	PBW343 and PDW233	14–35 days after anthesis	Mandal et al. (2019) https://doi.org/10.1016/j.dib.2018.12.058

PRJNA322418	Genomic imprinting was evolutionarily conserved during wheat polyploidisation	Endosperm	Diploid, tetraploid and hexaploid wheat	15, 20 and 25 DAP	Yang et al. (2018) https://doi.org/10.1105/tpc.17.00837
PRJNA244006	mRNA and sRNA transcriptomes reveal insights into dynamic homoeologue regulation for allopolyploid heterosis in germinating hexaploid wheat	Seed, spike, seedling	*T. turgidum* × *Aegilops tauschii*		Li et al. (2014) https://doi.org/10.1105/tpc.114.124388
PRJNA343594	*Triticum turgidum* ssp. durum transcriptome *de novo* assembly and SNP call	Root, leaf, grains, anthers	Svevo, Meridiano, Claudio, Kofa, Strongfield, Creso and other cvs.		Vendramin et al. (2019) https://doi.org/10.1186/s12864-019-5645-x
PRJNA579337	RNA-Seq of hybrid hexaploid wheat from progenitor species *Triticum turgidum* (durum wheat, AABB) cv. HOH501 and two *Aegilops tauschii* lines (Clae23 and ENT336, DD) transcriptome	Leaf			
PRJNA272886	Transcriptome asymmetry in synthetic and natural allotetraploid wheats, revealed by RNA-sequencing	Young leaves and young inflorescences	Korn, a newly synthesised allotetraploid wheat		Wang et al. (2016) https://doi.org/10.1111/nph.13678
PRJNA288606	Transcriptome profiling of wheat glumes in wild emmer, hulled landraces and modern cultivars	Glumes of the central spikelet at heading time	Inbar and Svevo		Zou et al. (2015) https://doi.org/10.1186/s12864-015-1996-0
PRJNA191054	Separating homoeologues by phasing in the tetraploid wheat transcriptome	Root and shoot, 3 weeks after germination; spike and grain tissue from mature plant	Kronos		Krasileva et al. (2013) https://doi.org/10.1186/gb-2013-14-6-r66
PRJNA285254	Genome-wide identification of microRNAs in leaves and the developing head of four durum genotypes during water-deficit stress	Flag leaf and developing head	Two drought-tolerant and two drought-susceptible genotypes	Water limiting stress treatment and control	Liu et al. (2015b) https://doi.org/10.1371/journal.pone.0142799

(continued)

Table 1 (continued)

BioProject	Description	Tissue(s)	Cultivar/genotype	Additional information	References
PRJEB20904	Root precursors of microRNAs in wild emmer and modern wheats show major differences in response to drought stress	Root	Tolerant and sensitive genotypes to water stress	Drought treatment	Akpinar et al. (2015) https://doi.org/10.1007/s10142-015-0453-0
PRJEB2672	Combining SNP discovery from next-generation sequencing data with bulked segregant analysis (BSA) to fine-map genes in polyploid wheat	5th leaf	Recombinant Substitution Lines (RSL), lines with high/low protein content	SNP discovery	Trick et al. (2012) https://doi.org/10.1186/1471-2229-12-14
PRJNA314909	RNA-Seq studies using wheat PHYTOCHROME B and PHYTOCHROME C mutants reveal shared and specific functions in the regulation of flowering and shade-avoidance pathways	3rd leaf	*phyB-A* and *phyB-B* null mutations		Pearce et al. (2016) https://doi.org/10.1186/s12870-016-0831-3
PRJNA353753	Transcriptomic response of durum wheat to nitrogen starvation	Roots, leaves/stems, flag leaf, spikes	Svevo	N normal and N deficient	Curci et al. (2017) https://doi.org/10.1038/s41598-017-01377-0
PRJNA892720	Gene expression profile of the developing endosperm in durum wheat provides insight into starch biosynthesis	Endosperm	Kronos	Eight timepoints (6, 8, 10, 13, 15, 18, 20, 30 days post anthesis)	Chen et al. (2023) https://doi.org/10.1101/2022.10.21.513215
PRJNA834779	Leaf RNA-Seq of *wise1* mutant and Kronos infected by CYR32	Leaf	Kronos WT, mutant with resistance, sensitive line	CYR32 inoculation (14 dpi)	
PRJNA767375	Transcriptome-based identification and validation of heat stress transcription factors in wheat progenitor species *Aegilops speltoides*	Leaf	Two cultivated wheat varieties susceptible to heat stress, diploid wild wheat heat tolerant		Seni et al. (2021) https://doi.org/10.1038/s41598-021-01596-6
PRJNA790482	RNA sequencing of the whole grains at early development from SHWs and their parents	Whole grains	Synthetic Hexaploid Wheat lines (SHWs) and their tetraploid and diploid parents	Six timepoints (0, 2, 4, 6, 8 and 10 days after pollinations)	

PRJNA706013	Nitrogen starvation-responsive microRNAs are affected by transgenerational stress in durum wheat seedlings	Shoot	DBA Aurora (water tolerant), L6 (sensitive)		Liu et al. (2021a) https://doi.org/10.3390/plants10050826
PRJNA733166	RNA-Seq raw reads of durum wheat (*Triticum turgidum* ssp. *durum*) seedlings inoculated by *Bacillus paralicheniformis* TRQ65	Seed	CIRNO C2008	Inoculated with *B. paralicheniformis* (Plant Growth-Promoting Rizhobacteria - PGPR) and not inoculated	Chaparro-Encinas et al. (2022) https://doi.org/10.1007/s00425-022-03837-y
PRJNA531693	Weighed gene co-expression network analysis unveils gene networks associated with the *Fusarium* head blight resistance in tetraploid wheat	Head 48 h after inoculation	R and S to *Fusarium*	DH line of the SF/BB population carrying alternative alleles at the reported Fusarium Head Blight (FHB) resistance QTL on chromosomes 1A, 2B, 3A and 6B	Sari et al. (2019) https://doi.org/10.1186/s12864-019-6161-8
PRJNA591632	Effect of *phyB* and *phyC* loss-of-function mutations on the wheat transcriptome under short- and long-day photoperiods	Last fully expanded leaf	*phyB*-null and *phyC*-null mutants	Eight-weeks of age, short-day and long-day conditions	Kippes et al. (2020) https://doi.org/10.1186/s12870-020-02506-0
PRJNA910827	RNA-Seq of Triticeae pollen and stigma samples	Stigma, pollen			
PRJEB34302	Transcriptome analysis of durum wheat under heat stress and elevated CO_2	Flag leaf	*Triticum durum* Desf. cv. Regallo	Different CO_2 and temperature conditions	Vicente et al. (2019) https://doi.org/10.3389/fpls.2019.01605
PRJNA686121	Wild emmer introgression alters root-to-shoot growth dynamics in durum wheat in response to water stress	Roots	Svevo and RIL	Well-watered (WW, 80% field capacity) and water-limited (WL, 30% field capacity)	Bacher et al. (2021) https://doi.org/10.1093/plphys/kiab292
E-MTAB-7266	Response to the presence of Cd	Root, shoot	Creso, Svevo	50-day exposure to 0.5 μM cadmium	Aprile et al. (2018) https://doi.org/10.1186/s12870-018-1473-4
PRJEB19929	Wild emmer wheat genome sequencing project	Leaf, root, lemma, grain, spike, glume, flower, flag leaf	Zavitan		Avni et al. (2017) https://doi.org/10.1126/science.aan0032

(continued)

Table 1 (continued)

BioProject	Description	Tissue(s)	Cultivar/genotype	Additional information	References
PRJNA905376	RNA-Seq data for synthetic hexaploid wheat lines and their corresponding tetraploid and diploid parents	Shoot, root, spikes, glumes, pistils, anthers	SHW lines and their tetraploid (*Triticum turgidum*) and diploid (*Aegilops tauschii*) parents		Vasudevan et al. (2023) https://doi.org/10.1038/s42003-023-04781-7
PRJNA908799	Transcriptomes of synthetic wheat allohexaploid and related natural species	Leaf, seed	Langdon, Joyau		
PRJNA795708	Mutation analysis of *Triticum dicoccoides* accession carrying *Pm* gene	Leaf			
PRJNA977839	Oligogalacturonide (OG) sensing increases resistance to *Fusarium* head blight in durum wheat	Seedling		Inoculation with oligogalacturonides (OGs) at different concentrations, *Fusarium graminearum* spores or a co-treatment of both OGs and fungus spores	Bigini et al. (2024) https://doi.org/10.1093/jxb/erae050
PRJNA917254	RNA-Seq and Whole-Genome Bisulfite Sequencing (WGBS) of diploid and tetraploid wheat roots	Root	Diploid wheats and their synthetic allotetraploid wheat lines		Miao et al. (2024) https://doi.org/10.1111/nph.19593

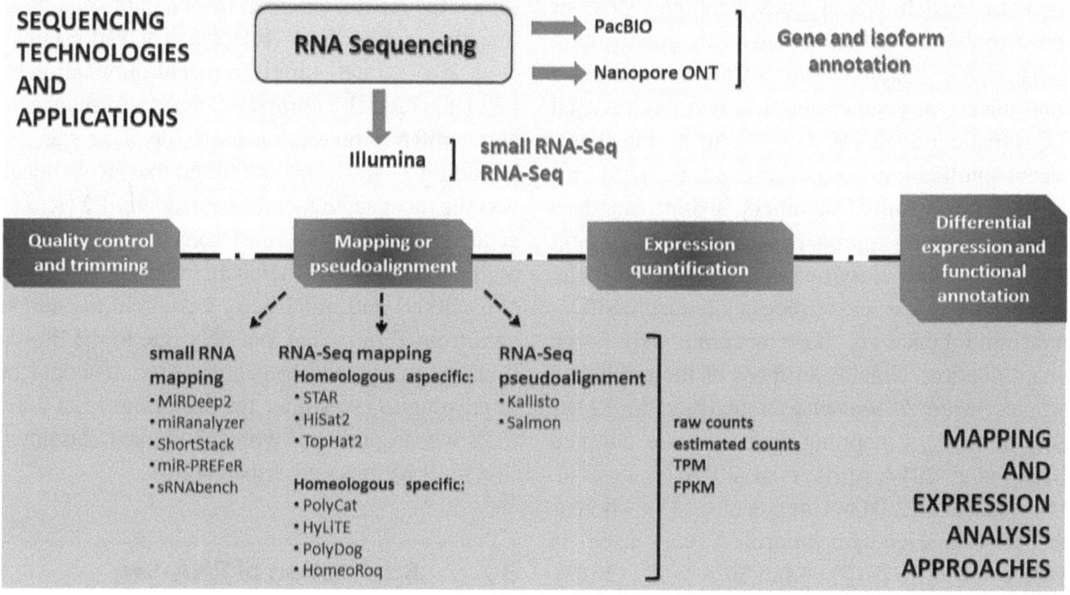

Fig. 1 Overview of RNA Sequencing application and analysis workflow. Upper panel reports the potential applications of the available sequencing technologies; in the middle, the schematic workflow for Illumina-based expression analysis is represented, while in bottom panel the methods and algorithms for read alignment and pseudoalignment are summarised

ous publications (Conesa et al. 2016; Stark et al. 2019) and can be summarised in the following steps (Fig. 1):

- Quality control and trimming of the raw sequencing data
- Mapping or pseudo-alignment against the reference genome or transcriptome
- (Optional) *De novo* or reference-guided transcriptome assembly
- Gene/Transcript expression quantification
- Statistical analysis of the data
- Functional annotation and biological interpretation of the results.

In addition to the intrinsic complexity of bioinformatics data analysis, the data produced from *T. durum* poses some additional challenges. Durum wheat is an allotetraploid (genome BBAA) species and the presence of homoeologous sequences in the genome represents a challenge especially for mapping and gene expression quantification. In fact, most of the mapping algorithms, such as STAR (Dobin et al. 2013), HiSAT2 (Kim et al. 2015) or TopHat2 (Trapnell et al. 2009), were not specifically designed to align reads coming from highly repetitive or duplicated sequences. As a consequence, for the most similar homoeologous genes, reads will be considered multiple mapping for the impossibility to assign them to the correct locus. In the following steps, such as read summary with featureCounts (Liao et al. 2014), those multiple mapping reads should be discarded to avoid unreliable gene expression quantification. In addition to 'standard' read mapping, alternative approaches can be used to mitigate this problem, such as those based on pseudo-alignment such as Kallisto (Bray et al. 2016) or Salmon (Patro et al. 2017), which implement statistical models to properly distribute multiple mapping reads to the most probable genes. Interestingly, tools specifically designed to map reads from allopolyploid species have been developed in recent years. Tools like PolyCat (Page et al. 2013), HyLiTE (Duchemin et al. 2015), PolyDog (Page and Udall 2015), EAGLE-RC or HomeoRoq (Akama et al. 2014) use SNPs to classify reads and assign them to each homoeologous genome. Most of these tools can be combined with downstream tools such as featureCounts in order to obtain a final matrix of gene expression that can be used for statistical analyses. It is worth noting that a recent publication from Kuo et al. (2018) compared the results obtained using standard mapping-based approaches, pseudo-alignment and subgenome classification, showing that the latter lead to the least amount of misclassifications and hence to more accurate expression quantification. Notably in most of the published articles where *T. durum* was analysed by RNA-Seq, a standard mapping approach was adopted (Curci et al. 2017; Aprile et al. 2018; Xiang et al. 2019; Liu et al. 2020a; Gong et al. 2021), whereas the pseudo-alignment approach was used in Pearce et al. (2015), Mandal et al. (2019), Chaparro-Encinas et al. (2021) and Arenas-M et al. (2022) and the subgenome classification approach was used only in Wang et al. (2016).

The challenges associated with the analysis of mRNA-Seq dataset from durum wheat are also faced by those interested in analysing small RNA-Seq data. Besides, the shorter length of small RNAs such as microRNAs (miRNAs), may actually increase the difficulty in discriminating the subgenome from which those sequences are being produced. Even if more than 40 bioinformatics tools are available for the analysis of plant small RNA-Seq data (Lukasik et al. 2016) including MiRDeep2 (Friedländer et al. 2012), miRanalyzer (Hackenberg et al. 2009), ShortStack (Axtell 2013), miR-PREFeR (Lei and Sun 2014) and sRNAbench (Aparicio-Puerta et al. 2019), most of the publications from *T. durum* rely on in-house generated pipelines. Most of these can be summarised as a two-step process in which known miRNA sequences are found based on sequence similarity search against databases of known miRNAs, that is, miRBase (Griffiths-Jones et al. 2006), whereas new sequences are detected by mapping the small RNA-Seq reads against the reference genome followed by prediction of secondary structure of the identified regions in order to identify putative precursors of small RNAs (De Paola et al. 2016; Zuluaga et al. 2017, 2018; Liu et al. 2020a, b, 2021a). Only two publications presented the results obtained from a published pipeline, that is, ShortStak (Fileccia et al. 2017; Giusti et al. 2017). Interestingly, a recent publication by Li et al. (2021b) compared the performance of eight miRNA prediction tools for plant species, including *T. aestivum*, and found that sRNAbench was the most accurate while miRDeep-P2 (Kuang et al. 2019) was the most sensitive. Out of the eight tools, only sRNABench, miRExpress (Wang et al. 2009) and miRDeep2 were able to analyse data from *T. aestivum*, possibly due to the limitations given by the big genome size. It would be interesting to investigate the performance of these tools when combined with subgenome classification tools as those described above.

3 Application of RNA-Seq to Study the Transcriptional Regulation During Seed and Endosperm Development in Durum Wheat

The high throughput, improved precision and cost-effectiveness of NGS technologies allowed for transcriptome studies in wheat for dissecting important processes occurring during plant development or to study at the molecular level the plant responses and adaptation mechanisms to biotic or abiotic stresses (Baloglu et al. 2014; Ramírez-González et al. 2018; Buerstmayr et al. 2021; De Santis et al. 2021; Azameti et al. 2022; Backhaus et al. 2022; Langridge et al. 2022; Rasool et al. 2022). Many studies have been initially focused on hexaploid bread wheat (*Triticum aestivum*), while relatively fewer resources exist for tetraploid durum wheat, even if durum wheat is gaining much interest due to its resilience to warm, dry climates and its use as an experimental model for wheat research and breeding (Roy et al. 2023; Groli et al. 2024; Mazzucotelli et al. 2024; Tabbita et al. 2024).

One of the developmental processes that has received main attention in the last years is the development of grains, under normal and stress conditions. Understanding the wheat grain transcriptome is crucial for unravelling the molecular processes involved in grain development, nutrient storage and responses to environmental factors, hence yield and quality of wheat production. Generating a transcriptome dataset for grain development could indeed provide insights into several key processes, such as the transcriptional dynamics across different stages of development, from double fertilisation events to early grain filling (GF) stages and later stages when starch and storage proteins accumulate. Within this developmental framework, numerous biological processes could be investigated, starting with the regulation of genes for the translocation of nutrients from the plant to the developing grain, nutrients transport and mobilisation within the grain nutrients, to genes encoding enzymes for starch biosynthesis and storage lipid metabolism or encoding storage proteins such as gliadins and glutenins. Understanding the gene transcriptional dynamics and their response to environmental cues is therefore essential for improving wheat grain quality and nutritional content.

To explore gene expression patterns that underpin key processes of endosperm development, Chen and colleagues (2023) recently published a high-quality transcriptomics dataset throughout grain development in durum wheat, cultivar Kronos. Developing grains, from plants grown in controlled environment chambers, were harvested at 8 timepoints during grain development, from 6 to 30 days post anthesis (DPA). RNA-Seq reads were analysed using the pseudoaligner Kallisto (Bray et al. 2016) to estimate transcripts per million (TPM) expression values for each of the 66,559 annotated canonical genes and the more than 300,000 low-confidence durum wheat transcripts. Using the expression values of high-confidence transcripts, in a principal component analysis (PCA), the authors showed that early timepoints (6, 8, 10 and 13 DPA) distinctly clustered, well separated from each other, were an indication that the largest changes in gene expression are observable during early grain development when the basic endosperm structure is forming, rather than later on development. Differential expression analysis revealed 29,925 genes significantly differentially expressed (DEGs) between at least 2 timepoints, with a larger number of DEGs detectable when comparing early timepoints, confirming that most changes in the endosperm transcriptome happen during the formation of the endosperm rather than during grain filling. Clustering analysis of DEGs over endosperm development revealed nine distinct gene expression profiles, some of them indicating general up- or down-regulation pattern during the developmental time-course and others more dynamic, potentially associated with processes that act during a specific stage (Chen et al. 2023).

This dataset was inspected to extensively study the expression patterns of known genes related to starch metabolism, subcategorised by their role in starch metabolism: ADP-glucose metabolism, amylopectin synthesis, amylose synthesis, starch phosphorylation, starch degradation, malto-oligosaccharide (MOS) metabolism, plastidial sugar transporters and starch granule initiation. The expression of the genes belonging to these pathways was then compared to RNA-Seq data obtained from a similar study in hexaploid wheat (Gu et al. 2021), revealing similarities but also species-specific differences in expression profiles of some genes that could be responsible for the biological differences between durum and bread wheat endosperm composition (Chen et al. 2023).

This work demonstrated the potential to provide new insights and hypotheses on starch metabolism but could be further explored to investigate other biological processes involved in grain development such as protein and lipid biosynthesis, cell wall biosynthesis, micronutrient transport, as well as development of endosperm structure and subsequent maturation in general (Chen et al. 2023).

4 Application of RNA-Seq Datasets to Study the Transcriptional Regulation of Heat Response in Durum Wheat

Plant developmental processes are highly interconnected and influenced by external environmental cues, such as drought, high temperatures, salinity or pathogens. Studying the transcriptomic regulation of responses to these stresses could provide insights into the different mechanisms of stress response and tolerance. Heat and drought stresses are among the main determinants of yield in durum wheat and account for 30%–50% of the overall output variability (Liu et al. 2016). Temperature increase has been estimated to reduce the durum wheat yield from 4.1% to 6.4% per Celsius degree increase, impacting yield-related traits, such as the reproductive tiller number and thousand-kernel weight (Quint et al. 2016; Ni et al. 2018), with the anthesis and filling stages as the phenological stages most sensitive to heat stress (HS). Similarly, water-deficit from the booting stage until harvest causes a reduction in grain number per plant by up to 64.6%, and reduction in grain yield per plant by up to 61.6% (Liu et al. 2015a). Although water-deficit stress (WS) and high temperature are commonly coupled in the field, their combined effects can be quite different from their individual impact (Barnabás et al. 2008; Liu et al. 2019).

A recent work studied the response of durum wheat to short-term heat stress (HS) treatment at transcriptomic and physiological level during early grain filling (Arenas-M et al. 2022). Short-term HS applied during early grain filling stage (GF, 10–14 DAP) significantly affects grain weight and quality. To gain new insights into the molecular factors underlying the early heat-stress responses, the authors conducted an RNA-Seq analysis on grains subjected to HS for 3 h. The RNA-Seq data were pseudo-aligned to the *Triticum turgidum* transcriptome (annotation version Svevo.v1) using Kallisto (Bray et al. 2016) and then used in differential expression analysis, resulting in the identification of 1590 DEGs between control and HS grains (671 HS upregulated and 919 repressed DEGs). Functional enrichment analysis revealed that 'response to heat stress', 'photosynthesis', 'carbon-nitrogen lyase activity' and 'oxidoreductase activity' were among the most enriched GO terms in upregulated genes. These results suggested that heat stress induces the expression of genes associated with photosynthesis and photosynthetic electron transport system, potentially involved in acclimation of photosynthetic CO_2 fixation and photoprotection during short-term high-temperature stress, as previously shown on bread wheat (Law and Crafts-Brandner 2001; Wan et al. 2008; Rangan et al. 2020) and also *Arabidopsis* (Wang et al. 2020).

Additionally, gene regulatory network analysis of HS-responsive genes allowed to identify novel transcription factors (TFs) controlling the expression of genes involved in abiotic stress response and grain quality, such as a member of the DNA-Binding with One Finger (DOF) family predicted to regulate glycogen and starch biosynthetic processes in response to HS in grains (Arenas-M et al. 2022). The identification of this potential key regulatory factor associated with grain quality points at the huge potentialities of transcriptomic studies and integrates data analysis to facilitate developing stress-resilient crops in the future.

In this framework, integration of different omics approaches would provide an even more comprehensive description of the molecular networks in response to environmental cues. To systematically elucidate the mechanisms underlying the response of durum wheat to single and combined water-deficit and heat stress (WH), a combination of RNA-Seq and small RNA-Seq was recently conducted. Single and combined water-deficit and heat stress were applied to stress-tolerant and stress-sensitive durum wheat genotypes to study their response at multiple timepoints during reproduction stages (Liu et al. 2020b). The multi-omics analysis was conducted on flag leaves of plants subjected to pre-anthesis water-deficit stress (WS), post-anthesis heat stress (HS), and pre-anthesis water-deficit plus post-anthesis heat stress (WH), plus the control samples.

Analysis of 120 sRNA libraries (including more than 1.2 billion raw reads) allowed for the identification of 523 microRNAs (miRNAs), 468 conserved and 55 novel ones, showing different genotype- and treatment-dependent expression. In parallel, transcriptome sequencing allowed for the identification of thousands of stress-responsive DEGs (based on pairwise comparisons between the control and each stress samples within each genotype), functionally including a wide range of transcription factors, plant growth regulators, hormone signalling receptors and key metabolic enzymes. Genotypic DEGs were instead identified comparing gene expression between stress-tolerant and stress-sensitive genotypes within each treatment, revealing an always higher number of genes highly expressed in the stress-tolerant genotype, irrespective of treatment (Liu et al. 2020b).

Finally, throughout an integrated omics analysis, the authors identified key miRNA–mRNA pairs antagonistically regulated, in particular modules composed by novel, differentially expressed miRNAs and oppositely regulated target transcription factors, in response to different stresses. These results provide valuable resources for deciphering molecular mechanism of plant growth and stress adaption for breeding programmes aimed at enhancing crop resilience to specific stresses.

5 Application of RNA-Seq Datasets to *De Novo* Gene Annotations and Varietal SNP Comparisons

The application of NGS techniques in plant breeding resulted in several improvements as to the data output availability and access to new sources of information. In addition to study developmental and environmental regulation of gene expression, RNA-Seq analysis was exploited in durum wheat to dissect candidate genes underlying agronomic traits QTL and phenotypic differences for important agronomic traits between modern and ancient durum wheat cultivars, as for example salt stress tolerance (Kim et al. 2021), *Fusarium* head blight resistance (Kumar et al. 2020), cadmium exposure (Aprile et al. 2018), drought stress and many more.

Marker-assisted selection (MAS) had a huge impact on wheat breeding with the analysis of durum and bread wheat panels to discern QTL responsible for agronomic traits of interest by using molecular markers such as Amplified fragment length polymorphism (AFLP), Simple Sequence Repeats (SSR), Cleaved Amplified Polymorphic Sequence (CAPS) and/or SNP (Lotti et al. 2000; Mantovani et al. 2008; Maccaferri et al. 2015). The development of SNP chip array like 35K and 820K (Allen et al. 2017) and 90K (Wang et al. 2014) paved the way to high-throughput analysis of wheat panels or population for association mapping analysis. Furthermore, other techniques for SNP detection in wheat panels include exome or whole genome sequencing, particularly useful for different samples of ancient wheat due to the large genome size and sequencing costs (Michikawa et al. 2019). However, RNA-Seq analysis can be a very efficient and cost-effective method to detect informative SNPs on a wide (IWGS Consortium et al. 2013; Vendramin et al. 2019) or narrow number of wheat varieties (Zou et al. 2015; Michikawa et al. 2019; Kim et al. 2021) in order to assemble de novo transcriptomes or design molecular markers such as Kompetitive Allele Specific PCR (KASP) useful for breeding and MAS (Maccaferri et al. 2022). On the other hand, only SNP on exons can be detected through the RNA-Seq data analysis and no information occurring on promoter or introns can be acquired (Michikawa et al. 2019).

One of the major challenges in assembling de novo transcriptome in a polyploid species relies on the separation of homoeologous chromosomes that merge together in the annotation steps, generating chimerical genomes (Duan et al. 2012; IWGS Consortium et al. 2013; Schreiber et al. 2012). Earlier than the publication of the assembled *Triticum turgidum* cv. Svevo wheat genome (Maccaferri et al. 2019), Krasileva and colleagues (IWGS Consortium et al. 2013) presented a method with established pipelines and software to de novo assemble the transcriptome of tetraploid durum wheat cultivar Kronos. Even using up to date de novo different assemblers, a large portion of contigs were expected to be merged into chimeric chromosomes between A and B genomes. This was resolved with a post-processing pipeline, which first consisted in the realignment of all the reads to already assembled contigs and further proceeded to the SNP calling using Freebayes bioinformatic tool (Garrison and Marth 2012). The analysis reported that a huge proportion of the contigs did not reveal any polymorphism; on the other hand, about 50% of the contigs contained at least one polymorphism resulting in chimeric A/B contigs. These latter contigs were subjected to a further process through HapCut that allowed to phase almost 90% of the detected SNP. This methodology for de novo transcriptome annotation of *T. turgidum* cv. Kronos durum wheat reported by Krasileva and colleagues was a major source of data and methodology approach for de novo transcriptome assembly in wheat and polyploid species.

A similar approach was indeed used to annotate de novo *T. turgidum* cv. Svevo transcriptome and to carry out SNP calling analysis among 13 modern durum wheat cultivars representing a wide dataset useful for breeding purposes (Vendramin et al. 2019). The Svevo de novo transcriptome annotation was performed similarly to what was carried out with Kronos by Krasileva and colleagues (IWGS Consortium et al. 2013) but with a different annotation approach, resulting in a higher gene reconstruction and implementing available transcriptome information on durum wheat. Furthermore, Vendramin and colleagues sequenced the transcriptome of 13 modern durum wheat cultivars coming from the Mediterranean Basin to provide a high-throughput analysis of the differential SNPs between the different varieties. The analysis revealed almost 800k SNP; the major part of them were characterised as rare SNPs probably as a consequence of the stringent parameters using the calling step of the analysis. A subset of these SNPs was included in the *T. aestivum* 90K SNPChip (Wang et al. 2014) already known and used in association mapping studies and this implemented SNP array was used to analyse a modern durum panel of 288 varieties. As expected, the polymorphic SNPs detected came, for the major part, from allelic variations implemented from the RNA-Seq transcriptome analysis, although 30% of the SNP originally derived from *T. aestivum* revealed to be polymorphic in durum wheat panel as well (Vendramin et al. 2019).

These results provide a useful dataset of additional SNP interesting for breeding purposes and to develop molecular markers on durum wheat. Nevertheless, it should be borne in mind that the implemented dataset was developed from 13 elite cultivar and is not as valuable to analyse wild or ancient wheat durum wheat accessions. In addition, the paper highlights a further interesting application of RNA-Seq data to dissect cultivar-specific differences in gene expression levels of the A and B subgenomes. Aligning the Svevo RNA-Seq reads on *Triticum urartu* (progenitor of the A genome) annotated genes, allowed the authors to differentiate the transcripts perfectly matching to the reference, and thus derived from the A genome to the homoeologues derived from the B genome. The analysis resulted in a higher percentage of transcripts overexpressed in the A genome in comparison to the B genome, confirming the results reported for bread wheat (Harper et al. 2016) although the lack of a reference B genome could have led to some biased results. Furthermore, the same pattern was visible also analysing a subset of genes between the 13 durum elite varieties. In particular, the authors show how a particular gene had an expression pattern biased towards the A or B genome based on the origin of each variety. It was indeed shown that varieties belonging to the CYMMYT germplasm have a higher transcription level of the B genome in contrast to the varieties related to the Italian and North America region.

The above discussed studies show different methods to use RNA-Seq data for gene de novo annotation and SNP calling from transcripts, useful for phasing of chimeric genomes but also to differentiate modern wheat varieties generating a broad-spectrum dataset useful for breeding purposes. Even if these works predate the durum and bread wheat whole genome sequencing, assembly and annotation, recent researches apply similar methods to specific case-studies, such as to discern genomic differences in ancient durum wheat ancestors (Michikawa et al. 2019) or to identify molecular markers associated with abiotic stress resistance (Kim et al. 2021).

Focusing on salt tolerance, a major problem for durum wheat cultivated in the Mediterranean region (Yousfi et al. 2016), Kim and colleagues tested different Tunisian durum wheat cultivars under strong salt stress application. For each cultivar, resistance/susceptibility was assessed based on growth parameters and chlorophyll content (Kim et al. 2021), and RNA-Seq analysis was carried out. Differential expressed genes detected in stressed condition were mainly connected to secondary metabolism pathways. In order to create molecular markers specific on the different expressed genes for salinity tolerance, SNP calling was therefore performed on the aligned RNA-Seq reads. Obviously, not all the called SNPs are valuable through this methodology for MAS. In fact, out of 159 designed primers only 17 worked and 14 resulted in missing data or unspecific results, but the 3 resulting molecular markers represented a valuable source for MAS on salt-tolerant durum varieties.

RNA-Seq analysis found major application also on evolutionary studies in wheat, allowing to discern differences between modern wheat relatives and compare differences between polyploid genome donor species. RNA-Seq data were used to study differences between the same genomes of diploid ancient durum species and to construct a linkage map for association studies (Michikawa et al. 2019). The authors focused on einkorn wheat (*T. monococcum*) widespread in Mediterranean region; in particular, they performed RNA-Seq analysis on different accessions composed of *T. monococcum* ssp. *monococcum* with Am genome (cultivated form), *T. monococcum* ssp. *speltoides*, which is the wild cultivar (Am genome), and *T. urartu* progenitor of the A genome of durum wheat. Authors detected SNP specific for Am genome and A genome also comparing the SNP with the available bread wheat genome (The International Wheat Genome Sequencing Consortium (IWGSC) et al. 2018), understanding allelic variant differences in closely related species. The set of SNPs detected could be a source of information useful to differentiate different accessions of diploid wheat and understanding the process of evolution towards the modern durum wheat and designing CAPS markers able to discriminate among all the accessions which one carried the A and Am genome. Furthermore, these molecular markers detected by RNA-Seq analysis were used to construct a linkage mapping on accessions of *T. monococcum*, detecting QTL responsible for agronomical traits like heading date in einkorn wheat.

These results clearly show that RNA-Seq analysis has different application both on research and breeding sides. In particular, the transcriptome datasets can be used to annotate de novo reference durum transcriptomes representing a reference dataset of information for research purposes. Furthermore, aligning transcriptome data from different cultivars leads to the identification of SNPs useful both for phasing of chromosome in de novo annotation to avoid chimeric homoeologous genomes, and to design informative molecular markers valuable for MAS analysis on different agronomic traits. However, it should be kept in mind that SNPs detected through RNA-Seq analysis can target only the exons of expressed genes and not introns or promoter regions, requiring further validation steps during the designing of molecular markers to exclude monomorphic and uninformative SNPs.

6 Conclusion

This chapter described the recent expansion of transcriptomic applications in durum wheat research. Thanks to the recent sequencing, assembly and annotation of the durum wheat genome (Maccaferri et al. 2019) combined with the technological progress in sequencing technologies and the resulting decrease of sequencing costs over the last years, the durum wheat community has been able to apply global gene expression profiling to wheat genetics research. Combining and integrating transcriptomics with other experimental data and 'omics', genome-wide transcriptional profiling sheds light on many key issues related to developmental programmes, stress response and adaptation, evolutionary and breeding-pushed genomic diversification in this crop species.

While RNA-Seq has revolutionised wheat genetics through insights into gene expression patterns, functional genomics and molecular mechanisms underlying various biological processes in durum wheat, the dataset and the knowledge produced so far are still too fragmented and not easily accessible to the durum wheat community. The development of comprehensive database, modelled on that of bread wheat such as the Wheat Expression Browser (http://www.wheat-expression.com/) or the WheatOmics platform (Ma et al. 2021; Sehgal et al. 2023), will eventually provide researchers with user-friendly tools to explore and analyse durum wheat transcriptomic data to more effectively dissect and leverage its QTLome (Salvi and Tuberosa 2015), hence facilitating genetic and functional data integration for candidate gene mapping and the identification of key regulators for trait-specific breeding.

Competing Interests The authors declare no competing financial and/or non-financial interests in relation to the content of the book.

References

Akama S, Shimizu-Inatsugi R, Shimizu KK, Sese J (2014) Genome-wide quantification of homeolog expression ratio revealed nonstochastic gene regulation in synthetic allopolyploid Arabidopsis. Nucleic Acids Res 42(6):e46

Akpinar BA, Kantar M, Budak H (2015) Root precursors of microRNAs in wild emmer and modern wheats show major differences in response to drought stress. Funct Integr Genomics 15(5):587–598

Allen AM, Winfield MO, Burridge AJ, Downie RC, Benbow HR, Barker GLA, Wilkinson PA, Coghill J, Waterfall C, Davassi A, Scopes G, Pirani A, Webster T, Brew F, Bloor C, Griffiths S, Bentley AR, Alda M, Jack P, Phillips AL, Edwards KJ (2017) Characterization of a Wheat Breeders' Array suitable for high-throughput SNP genotyping of global accessions of hexaploid bread wheat (*Triticum aestivum*). Plant Biotechnol J 15(3):390–401

Aparicio-Puerta E, Lebrón R, Rueda A, Gómez-Martín C, Giannoukakos S, Jaspez D, Medina JM, Zubkovic A, Jurak I, Fromm B, Marchal JA, Oliver J, Hackenberg M (2019) sRNAbench and sRNAtoolbox 2019: intuitive fast small RNA profiling and differential expression. Nucleic Acids Res 47(W1):W530–W535

Aprile A, Sabella E, Vergine M, Genga A, Siciliano M, Nutricati E, Rampino P, De Pascali M, Luvisi A, Miceli A, Negro C, De Bellis L (2018) Activation of a gene network in durum wheat roots exposed to cadmium. BMC Plant Biol 18(1):238

Arenas-M A, Castillo FM, Godoy D, Canales J, Calderini DF (2022) Transcriptomic and physiological response of durum wheat grain to short-term heat stress during early grain filling. Plants 11(1):59

Attwood TK, Blackford S, Brazas MD, Davies A, Schneider MV (2019) A global perspective on evolving bioinformatics and data science training needs. Brief Bioinform 20(2):398–404

Avni R, Nave M, Barad O, Baruch K, Twardziok SO, Gundlach H, Hale I, Mascher M, Spannagl M, Wiebe K, Jordan KW, Golan G, Deek J, Ben-Zvi B, Ben-Zvi G, Himmelbach A, MacLachlan RP, Sharpe AG, Fritz A, Ben-David R, Budak H, Fahima T, Korol A, Faris JD, Hernandez A, Mikel MA, Levy AA, Steffenson B, Maccaferri M, Tuberosa R, Cattivelli L, Faccioli P, Ceriotti A, Kashkush K, Pourkheirandish M, Komatsuda T, Eilam T, Sela H, Sharon A, Ohad N, Chamovitz DA, Mayer KFX, Stein N, Ronen G, Peleg Z, Pozniak CJ, Akhunov ED, Distelfeld A (2017) Wild emmer genome architecture and diversity elucidate wheat evolution and domestication. Science 357(6346):93–97

Axtell MJ (2013) ShortStack: comprehensive annotation and quantification of small RNA genes. RNA (New York, NY) 19(6):740–751

Azameti MK, Ranjan A, Singh PK, Gaikwad K, Singh AK, Dalal M, Arora A, Rai V, Padaria JC (2022) Transcriptome profiling reveals the genes and pathways involved in thermo-tolerance in wheat (Triticum aestivum L.) genotype Raj 3765. Sci Rep 12(1):14831

Bacher H, Zhu F, Gao T, Liu K, Dhatt BK, Awada T, Zhang C, Distelfeld A, Yu H, Peleg Z, Walia H (2021) Wild emmer introgression alters root-to-shoot growth dynamics in durum wheat in response to water stress. Plant Physiol 187(3):1149–1162

Backhaus AE, Lister A, Tomkins M, Adamski NM, Simmonds J, Macaulay I, Morris RJ, Haerty W, Uauy C (2022) High expression of the MADS-box gene *VRT2* increases the number of rudimentary basal spikelets in wheat. Plant Physiol 189(3):1536–1552

Baloglu MC, Inal B, Kavas M, Unver T (2014) Diverse expression pattern of wheat transcription factors against abiotic stresses in wheat species. Gene 550(1):117–122

Barnabás B, Jäger K, Fehér A (2008) The effect of drought and heat stress on reproductive processes in cereals. Plant Cell Environ 31(1):11–38

Bigini V, Sillo F, Giulietti S, Pontiggia D, Giovannini L, Balestrini R, Savatin DV (2024) Oligogalacturonide application increases resistance to Fusarium head blight in durum wheat. J Exp Bot 75(10):3070–3091

Bray NL, Pimentel H, Melsted P, Pachter L (2016) Near-optimal probabilistic RNA-seq quantification. Nat Biotechnol 34(5):525–527

Buerstmayr M, Wagner C, Nosenko T, Omony J, Steiner B, Nussbaumer T, Mayer KFX, Buerstmayr H (2021) Fusarium head blight resistance in European winter wheat: insights from genome-wide transcriptome analysis. BMC Genomics 22(1):470

Chaparro-Encinas LA, Santoyo G, Peña-Cabriales JJ, Castro-Espinoza L, Parra-Cota FI, de los Santos-Villalobos S (2021) Transcriptional regulation of metabolic and cellular processes in durum wheat (*Triticum turgidum* subsp. *durum*) in the face of temperature increasing. Plants 10(12):2792

Chaparro-Encinas LA, Parra-Cota FI, Cruz-Mendívil A, Santoyo G, Peña-Cabriales JJ, Castro-Espinoza L, de los Santos-Villalobos S (2022) Transcriptional regulation of cell growth and reprogramming of systemic response in wheat (Triticum turgidum subsp. durum) seedlings by *Bacillus paralicheniformis* TRQ65. Planta 255(3):56

Chen Y, Yan Y, Wu T-T, Zhang G-L, Yin H, Chen W, Wang S, Chang F, Gou J-Y (2020) Cloning of wheat keto-acyl thiolase 2B reveals a role of jasmonic acid in grain weight determination. Nat Commun 11(1):6266

Chen J, Watson-Lazowski A, Kamble NU, Vickers M, Seung D (2023) Gene expression profile of the developing endosperm in durum wheat provides insight into starch biosynthesis. BMC Plant Biol 23(1):363

Conesa A, Madrigal P, Tarazona S, Gomez-Cabrero D, Cervera A, McPherson A, Szcześniak MW, Gaffney DJ, Elo LL, Zhang X, Mortazavi A (2016) A survey of best practices for RNA-seq data analysis. Genome Biol 17(1):13

Curci PL, Aiese Cigliano R, Zuluaga DL, Janni M, Sanseverino W, Sonnante G (2017) Transcriptomic response of durum wheat to nitrogen starvation. Sci Rep 7(1):1176

De Paola D, Zuluaga DL, Sonnante G (2016) The miRNAome of durum wheat: isolation and characterisation of conserved and novel microRNAs and their target genes. BMC Genomics 17(1):505

De Santis MA, Soccio M, Laus MN, Flagella Z (2021) Influence of drought and salt stress on durum wheat grain quality and composition: a review. Plants 10(12):2599

Di Marsico M, Paytuvi Gallart A, Sanseverino W, Aiese Cigliano R (2022) GreeNC 2.0: a comprehensive database of plant long non-coding RNAs. Nucleic Acids Res 50(D1):D1442–D1447

Dobin A, Davis CA, Schlesinger F, Drenkow J, Zaleski C, Jha S, Batut P, Chaisson M, Gingeras TR (2013) STAR: ultrafast universal RNA-seq aligner. Bioinf (Oxf) 29(1):15–21

Dubcovsky J, Dvorak J (2007) Genome plasticity a key factor in the success of polyploid wheat under domestication. Science (New York, NY) 316(5833):1862–1866

Duan J, Xia C, Zhao G. et al. (2012) Optimizing de novo common wheat transcriptome assembly using short-read RNA-Seq data. BMC Genomics 13, 392. https://doi.org/10.1186/1471-2164-13-392

Duchemin W, Dupont P-Y, Campbell MA, Ganley AR, Cox MP (2015) HyLiTE: accurate and flexible analysis of gene expression in hybrid and allopolyploid species. BMC Bioinformatics 16(1):8

Fileccia V, Bertolini E, Ruisi P, Giambalvo D, Frenda AS, Cannarozzi G, Tadele Z, Crosatti C, Martinelli F (2017) Identification and characterization of durum wheat microRNAs in leaf and root tissues. Funct Integr Genomics 17(5):583–598

Foroud NA, Ouellet T, Laroche A, Oosterveen B, Jordan MC, Ellis BE, Eudes F (2012) Differential transcriptome analyses of three wheat genotypes reveal different host response pathways associated with Fusarium head blight and trichothecene resistance. Plant Pathol 61(2):296–314

Friedländer MR, Mackowiak SD, Li N, Chen W, Rajewsky N (2012) miRDeep2 accurately identifies known and hundreds of novel microRNA genes in seven animal clades. Nucleic Acids Res 40(1):37–52

Gao P, Quilichini TD, Zhai C, Qin L, Nilsen KT, Li Q, Sharpe AG, Kochian LV, Zou J, Reddy ASN, Wei Y, Pozniak C, Patterson N, Gillmor CS, Datla R, Xiang D (2021) Alternative splicing dynamics and evolutionary divergence during embryogenesis in wheat species. Plant Biotechnol J 19(8):1624–1643

Garrison E, Marth G (2012) Haplotype-based variant detection from short-read sequencing. arXiv:1207.3907 [q-bio] [online]. Available from: http://arxiv.org/abs/1207.3907. Accessed 15 Jan 2022

Giusti L, Mica E, Bertolini E, De Leonardis AM, Faccioli P, Cattivelli L, Crosatti C (2017) microRNAs differentially modulated in response to heat and drought stress in durum wheat cultivars with contrasting water use efficiency. Funct Integr Genomics 17(2):293–309

Gong F, Huang L, Qi T, Tang G, Liu J, Xiang L, He J, Zheng Y, Liu D, Wu B (2021) Comparative analysis of developing grain transcriptome reveals candidate genes and pathways improving GPC in wheat lines derived from wild emmer. J Appl Genet 62(1):17–25

Gordon A, McCartney C, Knox RE, Ereful N, Hiebert CW, Konkin DJ, Hsueh Y-C, Bhadauria V, Sgroi M, O'Sullivan DM, Hadley C, Boyd LA, Menzies JG (2020) Genetic and transcriptional dissection of resistance to *Claviceps purpurea* in the durum wheat cultivar Greenshank. Theor Appl Genet 133(6):1873–1886

Griffiths-Jones S, Grocock RJ, van Dongen S, Bateman A, Enright AJ (2006) miRBase: microRNA sequences, targets and gene nomenclature. Nucleic Acids Res 34(Database issue):D140–D144

Groli EL, Frascaroli E, Maccaferri M, Ammar K, Tuberosa R (2024) Dissecting the effect of heat stress on durum wheat under field conditions. Front Plant Sci 15:1393349. https://doi.org/10.3389/fpls.2024.1393349

Gu Y, Han S, Chen L, Mu J, Duan L, Li Y, Yan Y, Li X (2021) Expression and regulation of genes involved in the reserve starch biosynthesis pathway in hexaploid wheat (*Triticum aestivum* L.). Crop J 9(2):440–455

Habash DZ, Kehel Z, Nachit M (2009) Genomic approaches for designing durum wheat ready for climate change with a focus on drought. J Exp Bot 60(10):2805–2815

Habash DZ, Baudo M, Hindle M, Powers SJ, Defoin-Platel M, Mitchell R, Saqi M, Rawlings C, Latiri K,

Araus JL, Abdulkader A, Tuberosa R, Lawlor DW, Nachit MM (2014) Systems responses to progressive water stress in durum wheat. PLoS One 9(9):e108431

Hackenberg M, Sturm M, Langenberger D, Falcón-Pérez JM, Aransay AM (2009) miRanalyzer: a microRNA detection and analysis tool for next-generation sequencing experiments. Nucleic Acids Res 37(Web Server issue):W68–W76

Harper AL, Trick M, He Z, Clissold L, Fellgett A, Griffiths S, Bancroft I (2016) Genome distribution of differential homoeologue contributions to leaf gene expression in bread wheat. Plant Biotechnol J 14(5):1207–1214

He F, Wang W, Rutter WB, Jordan KW, Ren J, Taagen E, DeWitt N, Sehgal D, Sukumaran S, Dreisigacker S, Reynolds M, Halder J, Sehgal SK, Liu S, Chen J, Fritz A, Cook J, Brown-Guedira G, Pumphrey M, Carter A, Sorrells M, Dubcovsky J, Hayden MJ, Akhunova A, Morrell PL, Szabo L, Rouse M, Akhunov E (2022) Genomic variants affecting homoeologous gene expression dosage contribute to agronomic trait variation in allopolyploid wheat. Nat Commun 13(1):826

Huang D, Zheng Q, Melchkart T, Bekkaoui Y, Konkin DJF, Kagale S, Martucci M, You FM, Clarke M, Adamski NM, Chinoy C, Steed A, McCartney CA, Cutler AJ, Nicholson P, Feurtado JA (2020) Dominant inhibition of awn development by a putative zinc-finger transcriptional repressor expressed at the *B1* locus in wheat. New Phytol 225(1):340–355

Ison J, Rapacki K, Ménager H, Kalaš M, Rydza E, Chmura P, Anthon C, Beard N, Berka K, Bolser D, Booth T, Bretaudeau A, Brezovsky J, Casadio R, Cesareni G, Coppens F, Cornell M, Cuccuru G, Davidsen K, Vedova GD, Dogan T, Doppelt-Azeroual O, Emery L, Gasteiger E, Gatter T, Goldberg T, Grosjean M, Grüning B, Helmer-Citterich M, Ienasescu H, Ioannidis V, Jespersen MC, Jimenez R, Juty N, Juvan P, Koch M, Laibe C, Li J-W, Licata L, Mareuil F, Mičetić I, Friborg RM, Moretti S, Morris C, Möller S, Nenadic A, Peterson H, Profiti G, Rice P, Romano P, Roncaglia P, Saidi R, Schafferhans A, Schwämmle V, Smith C, Sperotto MM, Stockinger H, Vařeková RS, Tosatto SCE, de la Torre V, Uva P, Via A, Yachdav G, Zambelli F, Vriend G, Rost B, Parkinson H, Løngreen P, Brunak S (2016) Tools and data services registry: a community effort to document bioinformatics resources. Nucleic Acids Res 44(D1):D38–D47

IWGS Consortium, Krasileva KV, Buffalo V, Bailey P, Pearce S, Ayling S, Tabbita F, Soria M, Wang S, Akhunov E, Uauy C, Dubcovsky J (2013) Separating homeologs by phasing in the tetraploid wheat transcriptome. Genome Biol 14(6):R66

Kaushik M, Rai S, Venkadesan S, Sinha SK, Mohan S, Mandal PK (2020) Transcriptome analysis reveals important candidate genes related to nutrient reservoir, carbohydrate metabolism, and defence proteins during grain development of hexaploid bread wheat and its diploid progenitors. Genes 11(5):509

Kim D, Langmead B, Salzberg SL (2015) HISAT: a fast spliced aligner with low memory requirements. Nat Methods 12(4):357–360

Kim SH, Kim DY, Yacoubi I, Seo YW (2021) Development of single-nucleotide polymorphism markers of salinity tolerance for Tunisian durum wheat using RNA sequencing. Acta Agric Scand Sect B Soil Plant Sci 71(1):28–44

Kippes N, VanGessel C, Hamilton J, Akpinar A, Budak H, Dubcovsky J, Pearce S (2020) Effect of phyB and phyC loss-of-function mutations on the wheat transcriptome under short and long day photoperiods. BMC Plant Biol 20(1):297

Klymiuk V, Fatiukha A, Huang L, Wei Z, Kis-Papo T, Saranga Y, Krugman T, Fahima T (2019) Chapter 10 – Durum wheat as a bridge between wild emmer wheat genetic resources and bread wheat. In: Miedaner T, Korzun V (eds) Applications of genetic and genomic research in cereals [online]. Woodhead Publishing, pp 201–230. Available from: https://www.sciencedirect.com/science/article/pii/B9780081021637000107. Accessed 15 Nov 2023

Krasileva KV, Buffalo V, Bailey P, Pearce S, Ayling S, Tabbita F, Soria M, Wang S, Akhunov E, Uauy C, Dubcovsky J, IWGS Consortium (2013) Separating homeologs by phasing in the tetraploid wheat transcriptome. Genome Biol 14(6):R66

Kuang Z, Wang Y, Li L, Yang X (2019) miRDeep-P2: accurate and fast analysis of the microRNA transcriptome in plants. Bioinf (Oxf) 35(14):2521–2522

Kumar J, Rai KM, Pirseyedi S, Elias EM, Xu S, Dill-Macky R, Kianian SF (2020) Epigenetic regulation of gene expression improves Fusarium head blight resistance in durum wheat. Sci Rep 10(1):17610

Kuo T, Frith MC, Sese J, Horton P (2018) EAGLE: explicit alternative genome likelihood evaluator. BMC Med Genet 11(2):28

Law RD, Crafts-Brandner SJ (2001) High temperature stress increases the expression of wheat leaf ribulose-1,5-bisphosphate carboxylase/oxygenase activase protein. Arch Biochem Biophys 386(2):261–267

Leach LJ, Belfield EJ, Jiang C, Brown C, Mithani A, Harberd NP (2014) Patterns of homoeologous gene expression shown by RNA sequencing in hexaploid bread wheat. BMC Genomics 15(1):276

Lei J, Sun Y (2014) miR-PREFeR: an accurate, fast and easy-to-use plant miRNA prediction tool using small RNA-Seq data. Bioinf (Oxf) 30(19):2837–2839

Li A, Liu D, Wu J, Zhao X, Hao M, Geng S, Yan J, Jiang X, Zhang L, Wu J, Yin L, Zhang R, Wu L, Zheng Y, Mao L (2014) mRNA and small RNA transcriptomes reveal insights into dynamic homoeolog regulation of allopolyploid heterosis in nascent hexaploid wheat. Plant Cell 26(5):1878–1900

Li K, Debernardi JM, Li C, Lin H, Zhang C, Jernstedt J, von Korff M, Zhong J, Dubcovsky J (2021a) Interactions between SQUAMOSA and SHORT VEGETATIVE PHASE MADS-box proteins regulate meristem transitions during wheat spike development. Plant Cell 33(12):3621–3644

Li Q, Liu G, Bao Y, Wu Y, You Q (2021b) Evaluation and application of tools for the identification of known microRNAs in plants. Appl Plant Sci 9(3):e11414

Liao Y, Smyth GK, Shi W (2014) featureCounts: an efficient general purpose program for assigning sequence reads to genomic features. Bioinf (Oxf) 30(7):923–930

Liu H, Searle I, Mather D, Able A, Able J (2015a) Morphological, physiological and yield responses of durum wheat to pre-anthesis water deficit stress are genotype-dependent. Crop Pasture Sci 66(10):1024–1038

Liu H, Searle IR, Watson-Haigh NS, Baumann U, Mather DE, Able AJ, Able JA (2015b) Genome-wide identification of microRNAs in leaves and the developing head of four durum genotypes during water deficit stress. PLoS One 10(11):e0142799

Liu B, Asseng S, Müller C, Ewert F, Elliott J, Lobell DB, Martre P, Ruane AC, Wallach D, Jones JW, Rosenzweig C, Aggarwal PK, Alderman PD, Anothai J, Basso B, Biernath C, Cammarano D, Challinor A, Deryng D, Sanctis GD, Doltra J, Fereres E, Folberth C, Garcia-Vila M, Gayler S, Hoogenboom G, Hunt LA, Izaurralde RC, Jabloun M, Jones CD, Kersebaum KC, Kimball BA, Koehler A-K, Kumar SN, Nendel C, O'Leary GJ, Olesen JE, Ottman MJ, Palosuo T, Prasad PVV, Priesack E, Pugh TAM, Reynolds M, Rezaei EE, Rötter RP, Schmid E, Semenov MA, Shcherbak I, Stehfest E, Stöckle CO, Stratonovitch P, Streck T, Supit I, Tao F, Thorburn P, Waha K, Wall GW, Wang E, White JW, Wolf J, Zhao Z, Zhu Y (2016) Similar estimates of temperature impacts on global wheat yield by three independent methods. Nat Clim Chang 6(12):1130–1136

Liu H, Able AJ, Able JA (2019) Genotypic performance of Australian durum under single and combined water-deficit and heat stress during reproduction. Sci Rep 9(1):14986

Liu H, Able AJ, Able JA (2020a) Multi-omics analysis of small RNA, transcriptome, and degradome in T. turgidum—regulatory networks of grain development and abiotic stress response. Int J Mol Sci 21(20):7772

Liu H, Able AJ, Able JA (2020b) Integrated analysis of small RNA, transcriptome, and degradome sequencing reveals the water-deficit and heat stress response network in durum wheat. Int J Mol Sci 21(17):6017

Liu H, Able AJ, Able JA (2020c) Transgenerational effects of water-deficit and heat stress on germination and seedling vigour—new insights from durum wheat microRNAs. Plants 9(2):189

Liu H, Able AJ, Able JA (2021a) Nitrogen starvation-responsive microRNAs are affected by transgenerational stress in durum wheat seedlings. Plants 10(5):826

Liu H, Able AJ, Able JA (2021b) Small RNA, transcriptome and degradome analysis of the transgenerational heat stress response network in durum wheat. Int J Mol Sci 22(11):5532

Langridge P, Alaux M, Almeida NF, Ammar K, Baum M, Bekkaoui F, Bentley AR, Beres BL, Berger B, Braun H.-J, Brown-Guedira G, Burt CJ, Caccamo MJ, Cattivelli L, Charmet G, Civáň P, Cloutier S, Cohan JP, Devaux PJ, Zhang X (2022). Meeting the Challenges Facing Wheat Production: The Strategic Research Agenda of the Global Wheat Initiative. Agronomy, 12(11), 2767. https://doi.org/10.3390/agronomy12112767

Lotti C, Salvi S, Pasqualone A, Tuberosa R, Blanco A (2000) Integration of AFLP markers into an RFLP-based map of durum wheat. Plant Breed 119:393–401

Lowe R, Shirley N, Bleackley M, Dolan S, Shafee T (2017) Transcriptomics technologies. PLoS Comput Biol 13(5):e1005457

Lukasik A, Wójcikowski M, Zielenkiewicz P (2016) Tools4miRs – one place to gather all the tools for miRNA analysis. Bioinformatics 32(17):2722–2724

Ma S, Wang M, Wu J, Guo W, Chen Y, Li G, Wang Y, Shi W, Xia G, Fu D, Kang Z, Ni F (2021) WheatOmics: a platform combining multiple omics data to accelerate functional genomics studies in wheat. Mol Plant 14(12):1965–1968

Maccaferri M, Ricci A, Salvi S, Milner SG, Noli E, Martelli PL, Casadio R, Akhunov E, Scalabrin S, Vendramin V, Ammar K, Blanco A, Desiderio F, Distelfeld A, Dubcovsky J, Fahima T, Faris J, Korol A, Massi A, Mastrangelo AM, Morgante M, Pozniak C, N'Diaye A, Xu S, Tuberosa R (2015) A high-density, SNP-based consensus map of tetraploid wheat as a bridge to integrate durum and bread wheat genomics and breeding. Plant Biotechnol J 13(5):648–663

Maccaferri M, Harris NS, Twardziok SO, Pasam RK, Gundlach H, Spannagl M, Ormanbekova D, Lux T, Prade VM, Milner SG, Himmelbach A, Mascher M, Bagnaresi P, Faccioli P, Cozzi P, Lauria M, Lazzari B, Stella A, Manconi A, Gnocchi M, Moscatelli M, Avni R, Deek J, Biyiklioglu S, Frascaroli E, Corneti S, Salvi S, Sonnante G, Desiderio F, Marè C, Crosatti C, Mica E, Özkan H, Kilian B, De Vita P, Marone D, Joukhadar R, Mazzucotelli E, Nigro D, Gadaleta A, Chao S, Faris JD, Melo ATO, Pumphrey M, Pecchioni N, Milanesi L, Wiebe K, Ens J, MacLachlan RP, Clarke JM, Sharpe AG, Koh CS, Liang KYH, Taylor GJ, Knox R, Budak H, Mastrangelo AM, Xu SS, Stein N, Hale I, Distelfeld A, Hayden MJ, Tuberosa R, Walkowiak S, Mayer KFX, Ceriotti A, Pozniak CJ, Cattivelli L (2019) Durum wheat genome highlights past domestication signatures and future improvement targets. Nat Genet 51(5):885–895

Maccaferri M, Bruschi M, Tuberosa R (2022) Sequence-based marker assisted selection in wheat. In: Reynolds MP, Braun HJ (eds) Wheat improvement: food security in a changing climate. Springer, Cham, pp 513–538

Mandal PK, Rai S, Kaushik M, Sinha SK, Gupta RK, Mahendru A (2019) Transcriptome data of cultivated tetraploid and hexaploid wheat variety during grain development. Data Brief 22:551–556

Mantovani P, Maccaferri M, Sanguineti MC, Tuberosa R, Catizone I, Wenzl P, Thomson B, Carling J, Huttner E, DeAmbrogio E, Kilian A (2008) An integrated DArT-SSR linkage map of durum wheat. Mol Breed 22:629–648

Matsuoka Y (2011) Evolution of polyploid triticum wheats under cultivation: the role of domestication, natural hybridization and allopolyploid speciation in their diversification. Plant Cell Physiol 52(5):750–764

Mazzucotelli E, Sciara G, Mastrangelo AM, Desiderio F, Xu SS, Faris J, Hayden MJ, Tricker PJ, Ozkan H, Echenique V, Steffenson BJ, Knox R, Niane AA, Udupa SM, Longin FCH, Marone D, Petruzzino G, Corneti S, Ormanbekova D, Pozniak C, Roncallo PF, Mather D, Able JA, Amri A, Braun H, Ammar K, Baum M, Cattivelli L, Maccaferri M, Tuberosa R, Bassi FM (2020) The Global Durum Wheat Panel (GDP): an international platform to identify and exchange beneficial alleles. Front Plant Sci [online] 11. Available from: https://www.frontiersin.org/articles/10.3389/fpls.2020.569905. Accessed 15 Nov 2023

Mazzucotelli E, Mastrangelo AM, Desiderio F, Barabaschi D, Maccaferri M, Tuberosa R, Cattivelli L (2024) Gene flow between tetraploid and hexaploid wheat for breeding innovation. In: The wheat genome, pp 135–163. https://doi.org/10.1007/978-3-031-38294-9_8

McCombie WR, McPherson JD, Mardis ER (2019) Next-generation sequencing technologies. Cold Spring Harb Perspect Med 9(11):a036798

McIntosh S, Watson L, Bundock P, Crawford A, White J, Cordeiro G, Barbary D, Rooke L, Henry R (2007) SAGE of the developing wheat caryopsis. Plant Biotechnol J 5(1):69–83

Miao L, Xu W, Liu Y, Huang X, Chen Z, Wang H, Wang Z, Chen Y, Song Q, Zhang J, Han F, Peng H, Yao Y, Xin M, Hu Z, Ni Z, Sun Q, Xing J, Guo W (2024) Reshaped DNA methylation cooperating with homoeolog-divergent expression promotes improved root traits in synthesized tetraploid wheat. New Phytol 242(2):507–523

Michikawa A, Yoshida K, Okada M, Sato K, Takumi S (2019) Genome-wide polymorphisms from RNA sequencing assembly of leaf transcripts facilitate phylogenetic analysis and molecular marker development in wild einkorn wheat. Mol Gen Genomics 294(5):1327–1341

Ni Z, Li H, Zhao Y, Peng H, Hu Z, Xin M, Sun Q (2018) Genetic improvement of heat tolerance in wheat: recent progress in understanding the underlying molecular mechanisms. Crop J 6(1):32–41

Nilsen KT, Walkowiak S, Xiang D, Gao P, Quilichini TD, Willick IR, Byrns B, N'Diaye A, Ens J, Wiebe K, Ruan Y, Cuthbert RD, Craze M, Wallington EJ, Simmonds J, Uauy C, Datla R, Pozniak CJ (2020) Copy number variation of TdDof controls solid-stemmed architecture in wheat. Proc Natl Acad Sci 117(46):28708–28718

Page JT, Udall JA (2015) Methods for mapping and categorization of DNA sequence reads from allopolyploid organisms. BMC Genet 16(Suppl 2):S4

Page JT, Gingle AR, Udall JA (2013) PolyCat: a resource for genome categorization of sequencing reads from allopolyploid organisms. G3 (Bethesda, Md) 3(3):517–525

Patro R, Duggal G, Love MI, Irizarry RA, Kingsford C (2017) Salmon provides fast and bias-aware quantification of transcript expression. Nat Methods 14(4):417–419

Pearce S, Tabbita F, Cantu D, Buffalo V, Avni R, Vazquez-Gross H, Zhao R, Conley CJ, Distelfeld A, Dubcovksy J (2014) Regulation of Zn and Fe transporters by the GPC1gene during early wheat monocarpic senescence. BMC Plant Biol 14(1):368

Pearce S, Vazquez-Gross H, Herin SY, Hane D, Wang Y, Gu YQ, Dubcovsky J (2015) WheatExp: an RNA-seq expression database for polyploid wheat. BMC Plant Biol 15(1):299

Pearce S, Kippes N, Chen A, Debernardi JM, Dubcovsky J (2016) RNA-seq studies using wheat PHYTOCHROME B and PHYTOCHROME C mutants reveal shared and specific functions in the regulation of flowering and shade-avoidance pathways. BMC Plant Biol 16(1):141

Pilcher W, Zandkarimi H, Arceneaux K, Harrison S, Baisakh N (2018) Correction: genome-wide microarray analysis leads to identification of genes in response to herbicide, metribuzin in wheat leaves. PLoS One 13(6):e0199564

Poole RL, Barker GL, Werner K, Biggi GF, Coghill J, Gibbings JG, Berry S, Dunwell JM, Edwards KJ (2008) Analysis of wheat SAGE tags reveals evidence for widespread antisense transcription. BMC Genomics 9(1):475

Puccio G, Ingraffia R, Mercati F, Amato G, Giambalvo D, Martinelli F, Sunseri F, Frenda AS (2023) Transcriptome changes induced by Arbuscular mycorrhizal symbiosis in leaves of durum wheat (*Triticum durum* Desf.) promote higher salt tolerance. Sci Rep 13(1):116

Quint M, Delker C, Franklin KA, Wigge PA, Halliday KJ, van Zanten M (2016) Molecular and genetic control of plant thermomorphogenesis. Nat Plants 2:15190

Ramírez-González RH, Borrill P, Lang D, Harrington SA, Brinton J, Venturini L, Davey M, Jacobs J, van Ex F, Pasha A, Khedikar Y, Robinson SJ, Cory AT, Florio T, Concia L, Juery C, Schoonbeek H, Steuernagel B, Xiang D, Ridout CJ, Chalhoub B, Mayer KFX, Benhamed M, Latrasse D, Bendahmane A, Wulff BBH, Appels R, Tiwari V, Datla R, Choulet F, Pozniak CJ, Provart NJ, Sharpe AG, Paux E, Spannagl M, Bräutigam A, Uauy C, International Wheat Genome Sequencing Consortium (2018) The transcriptional landscape of polyploid wheat. Science 361(6403):eaar6089

Rangan P, Furtado A, Henry R (2020) Transcriptome profiling of wheat genotypes under heat stress during grain-filling. J Cereal Sci 91:102895

Rasool F, Khan MR, Schneider M, Uzair M, Aqeel M, Ajmal W, Léon J, Naz AA (2022) Transcriptome unveiled the gene expression patterns of root architecture in drought-tolerant and sensitive wheat genotypes. Plant Physiol Biochem 178:20–30

Roy B, Sagan V, Haireti A, Newcomb M, Tuberosa R, LeBauer D, Shakoor N (2023) Early detection of drought stress in durum wheat using hyperspectral imaging and photosystem sensing. Remote Sens 16(1):155

Salvi S, Tuberosa R (2015) The crop QTLome comes of age. Curr Opin Biotechnol 32:179–185
Sari E, Cabral AL, Polley B, Tan Y, Hsueh E, Konkin DJ, Knox RE, Ruan Y, Fobert PR (2019) Weighted gene co-expression network analysis unveils gene networks associated with the Fusarium head blight resistance in tetraploid wheat. BMC Genomics 20(1):925
Schreiber AW, Hayden MJ, Forrest KL et al. (2012) Transcriptome-scale homoeolog-specific transcript assemblies of bread wheat. BMC Genomics 13, 492. https://doi.org/10.1186/1471-2164-13-492
Sehgal D, Dhakate P, Ambreen H, Shaik KHB, Rathan ND, Anusha NM, Deshmukh R, Vikram P (2023) Wheat omics: advancements and opportunities. Plants 12(3):426
Seni S, Kaur S, Malik P, Yadav IS, Sirohi P, Chauhan H, Kaur A, Chhuneja P (2021) Transcriptome based identification and validation of heat stress transcription factors in wheat progenitor species Aegilops speltoides. Sci Rep 11(1):22049
Shavit R, Batyrshina ZS, Yaakov B, Florean M, Köllner TG, Tzin V (2022) The wheat dioxygenase BX6 is involved in the formation of benzoxazinoids *in planta* and contributes to plant defense against insect herbivores. Plant Sci 316:111171
Shewry PR (2009) Wheat. J Exp Bot 60(6):1537–1553
Soresi D, Díaz M, Bagnaresi P, Gallo C, Carrera A (2023) Gene and lncRNA expression patterns associated with Qfhs.ndsu-3AS Fusarium head blight resistance QTL in durum wheat. Can J Plant Pathol 45(1):41–56
Stark R, Grzelak M, Hadfield J (2019) RNA sequencing: the teenage years. Nat Rev Genet 20(11):631–656
Tabbita F, Ammar K, Ibba MI, Andrade F, Maccaferri M, Sanguineti MC, Tuberosa R, Guzmán C (2024) Response to heat stress and glutenins allelic variation effects on quality traits in durum wheat. J Agron Crop Sci 210(1):e12686
The International Wheat Genome Sequencing Consortium (IWGSC), Appels R, Eversole K, Stein N, Feuillet C, Keller B, Rogers J, Pozniak CJ, Choulet F, Distelfeld A, Poland J, Ronen G, Sharpe AG, Barad O, Baruch K, Keeble-Gagnère G, Mascher M, Ben-Zvi G, Josselin A-A, Himmelbach A, Balfourier F, Gutierrez-Gonzalez J, Hayden M, Koh C, Muehlbauer G, Pasam RK, Paux E, Rigault P, Tibbits J, Tiwari V, Spannagl M, Lang D, Gundlach H, Haberer G, Mayer KFX, Ormanbekova D, Prade V, Šimková H, Wicker T, Swarbreck D, Rimbert H, Felder M, Guilhot N, Kaithakottil G, Keilwagen J, Leroy P, Lux T, Twardziok S, Venturini L, Juhász A, Abrouk M, Fischer I, Uauy C, Borrill P, Ramirez-Gonzalez RH, Arnaud D, Chalabi S, Chalhoub B, Cory A, Datla R, Davey MW, Jacobs J, Robinson SJ, Steuernagel B, van Ex F, Wulff BBH, Benhamed M, Bendahmane A, Concia L, Latrasse D, Bartoš J, Bellec A, Berges H, Doležel J, Frenkel Z, Gill B, Korol A, Letellier T, Olsen O-A, Singh K, Valárik M, van der Vossen E, Vautrin S, Weining S, Fahima T, Glikson V, Raats D, Číhalíková J, Toegelová H, Vrána J, Sourdille P, Darrier B, Barabaschi D, Cattivelli L, Hernandez P, Galvez S, Budak H, Jones JDG, Witek K, Yu G, Small I, Melonek J, Zhou R, Belova T, Kanyuka K, King R, Nilsen K, Walkowiak S, Cuthbert R, Knox R, Wiebe K, Xiang D, Rohde A, Golds T, Čížková J, Akpinar BA, Biyiklioglu S, Gao L, N'Daiye A, Kubaláková M, Šafář J, Alfama F, Adam-Blondon A-F, Flores R, Guerche C, Loaec M, Quesneville H, Condie J, Ens J, Maclachlan R, Tan Y, Alberti A, Aury J-M, Barbe V, Couloux A, Cruaud C, Labadie K, Mangenot S, Wincker P, Kaur G, Luo M, Sehgal S, Chhuneja P, Gupta OP, Jindal S, Kaur P, Malik P, Sharma P, Yadav B, Singh NK, Khurana JP, Chaudhary C, Khurana P, Kumar V, Mahato A, Mathur S, Sevanthi A, Sharma N, Tomar RS, Holušová K, Plíhal O, Clark MD, Heavens D, Kettleborough G, Wright J, Balcárková B, Hu Y, Salina E, Ravin N, Skryabin K, Beletsky A, Kadnikov V, Mardanov A, Nesterov M, Rakitin A, Sergeeva E, Handa H, Kanamori H, Katagiri S, Kobayashi F, Nasuda S, Tanaka T, Wu J, Cattonaro F, Jiumeng M, Kugler K, Pfeifer M, Sandve S, Xun X, Zhan B, Batley J, Bayer PE, Edwards D, Hayashi S, Tulpová Z, Visendi P, Cui L, Du X, Feng K, Nie X, Tong W, Wang L (2018) Shifting the limits in wheat research and breeding using a fully annotated reference genome. Science 361(6403):eaar7191
Trapnell C, Pachter L, Salzberg SL (2009) TopHat: discovering splice junctions with RNA-Seq. Bioinformatics 25(9):1105–1111
Trick M, Adamski NM, Mugford SG, Jiang C-C, Febrer M, Uauy C (2012) Combining SNP discovery from next-generation sequencing data with bulked segregant analysis (BSA) to fine-map genes in polyploid wheat. BMC Plant Biol 12(1):14
VanGessel C, Hamilton J, Tabbita F, Dubcovsky J, Pearce S (2022) Transcriptional signatures of wheat inflorescence development. Sci Rep 12(1):17224
Vasudevan A, Lévesque-Lemay M, Edwards T, Cloutier S (2023) Global transcriptome analysis of allopolyploidization reveals large-scale repression of the D-subgenome in synthetic hexaploid wheat. Commun Biol 6(1):1–14
Vendramin V, Ormanbekova D, Scalabrin S, Scaglione D, Maccaferri M, Martelli P, Salvi S, Jurman I, Casadio R, Cattonaro F, Tuberosa R, Massi A, Morgante M (2019) Genomic tools for durum wheat breeding: de novo assembly of Svevo transcriptome and SNP discovery in elite germplasm. BMC Genomics 20(1):278
Vergine M, Aprile A, Sabella E, Genga A, Siciliano M, Rampino P, Lenucci MS, Luvisi A, Bellis LD (2017) Cadmium concentration in grains of durum wheat (Triticum turgidum L. subsp. durum). J Agric Food Chem 65(30):6240–6246
Vicente R, Bolger AM, Martínez-Carrasco R, Pérez P, Gutiérrez E, Usadel B, Morcuende R (2019) De novo transcriptome analysis of durum wheat flag leaves provides new insights into the regulatory response to elevated CO_2 and high temperature. Front Plant Sci [online] 10. Available from: https://www.frontiersin.org/journals/plant-science/articles/10.3389/fpls.2019.01605/full. Accessed 10 July 2024

Wan Y, Poole RL, Huttly AK, Toscano-Underwood C, Feeney K, Welham S, Gooding MJ, Mills C, Edwards KJ, Shewry PR, Mitchell RA (2008) Transcriptome analysis of grain development in hexaploid wheat. BMC Genomics 9(1):121

Wang W-C, Lin F-M, Chang W-C, Lin K-Y, Huang H-D, Lin N-S (2009) miRExpress: analyzing high-throughput sequencing data for profiling microRNA expression. BMC Bioinformatics 10(1):328

Wang S, Wong D, Forrest K, Allen A, Chao S, Huang BE, Maccaferri M, Salvi S, Milner SG, Cattivelli L, Mastrangelo AM, Whan A, Stephen S, Barker G, Wieseke R, Plieske J, Lillemo M, Mather D, Appels R, Dolferus R, Brown-Guedira G, Korol A, Akhunova AR, Feuillet C, Salse J, Morgante M, Pozniak C, Luo M, Dvorak J, Morell M, Dubcovsky J, Ganal M, Tuberosa R, Lawley C, Mikoulitch I, Cavanagh C, Edwards KJ, Hayden M, Akhunov E, International Wheat Genome Sequencing Consortium (2014) Characterization of polyploid wheat genomic diversity using a high-density 90 000 single nucleotide polymorphism array. Plant Biotechnol J 12(6):787–796

Wang X, Zhang H, Li Y, Zhang Z, Li L, Liu B (2016) Transcriptome asymmetry in synthetic and natural allotetraploid wheats, revealed by RNA-sequencing. New Phytol 209(3):1264–1277

Wang J, Tian W, Tao F, Wang J, Shang H, Chen X, Xu X, Hu X (2020) TaRPM1 positively regulates wheat high-temperature seedling-plant resistance to Puccinia striiformis f. sp. tritici. Front Plant Sci [online] 10. Available from: https://www.frontiersin.org/articles/10.3389/fpls.2019.01679. Accessed 10 Feb 2023

Xiang D, Quilichini TD, Liu Z, Gao P, Pan Y, Li Q, Nilsen KT, Venglat P, Esteban E, Pasha A, Wang Y, Wen R, Zhang Z, Hao Z, Wang E, Wei Y, Cuthbert R, Kochian LV, Sharpe A, Provart N, Weijers D, Gillmor CS, Pozniak C, Datla R (2019) The transcriptional landscape of polyploid wheats and their diploid ancestors during embryogenesis and grain development. Plant Cell 31(12):2888–2911

Yang G, Liu Z, Gao L, Yu K, Feng M, Yao Y, Peng H, Hu Z, Sun Q, Ni Z, Xin M (2018) Genomic imprinting was evolutionarily conserved during wheat polyploidization. Plant Cell 30(1):37–47

Yousfi S, Kellas N, Saidi L, Benlakehal Z, Chaou L, Siad D, Herda F, Karrou M, Vergara O, Gracia A, Araus JL, Serret MD (2016) Comparative performance of remote sensing methods in assessing wheat performance under Mediterranean conditions. Agric Water Manag 164:137–147

Zou H, Tzarfati R, Hübner S, Krugman T, Fahima T, Abbo S, Saranga Y, Korol AB (2015) Transcriptome profiling of wheat glumes in wild emmer, hulled landraces and modern cultivars. BMC Genomics 16(1):777

Zuluaga DL, Paola DD, Janni M, Curci PL, Sonnante G (2017) Durum wheat miRNAs in response to nitrogen starvation at the grain filling stage. PLoS One 12(8):e0183253

Zuluaga DL, Liuzzi V, Curci PL, Sonnante G (2018) MicroRNAs in durum wheat seedlings under chronic and short-term nitrogen stress. Funct Integr Genomics 18(6):645–657

Zuluaga DL, Blanco E, Mangini G, Sonnante G, Curci PL (2023) A survey of the transcriptomic resources in durum wheat: stress responses, data integration and exploitation. Plants 12(6):1267

The Durum Wheat miRNAome

Erica Mica, Primetta Faccioli, and Cristina Crosatti

Abstract

Small non-coding RNAs are RNA molecules that do not code for a protein, but still retain a fundamental biological role as negative regulators of gene expression; among them the most studied are the miRNAs, 21- and 22-nt long molecules that interfere in *trans* with a target mRNA. Notably, miRNAs are rarely responsible for an on/off regulation, but more often fine tune gene expression involved in several relevant biological processes such as plant development and organ morphogenesis, biotic and abiotic stress response, nutrient homeostasis, flowering time, and secondary metabolism. About 30 publications describing miRNAs in durum wheat are available through the literature. After downloading all miRNA sequences from these publications, a total of 7544 different precursors and 7267 miRNAs have been counted. This chapter reviews the durum wheat miRNAome and focuses on the most relevant miRNA/target-RNA modules as well as their role in plant development and biotic/abiotic stress tolerance.

1 Introduction

Non-coding RNAs include several classes of RNA molecules, transcribed by different RNA polymerase, that do not code for a protein, here including ribosomal RNAs, small nuclear RNAs (snRNAs), small nucleolar RNAs (snoRNAs), transfer RNAs (tRNAs), long non-coding RNAs, circular RNAs, and small non-coding RNAs (ncRNAs). The percentage of genomic sequences belonging to this fraction is consistent, particularly in complex plant genomes. Among small nc-RNAs there is a large plethora of 18- to 25-nt long molecules distinguished for their biogenesis and mechanism of action, that have been discovered 20 years ago in plants (Reinhart et al. 2002). All the known classes of small ncRNAs have the ability of binding, in a sequence-specific manner, a target sequence to promote gene silencing, either at the transcriptional or at the post-transcriptional level. In addition, they all originate from a typical double-strand RNA (dsRNA) precursor, processed by DICER like proteins and they act through a ribonucleoprotein complex known as RNA Inducing Silencing Complex (RISC) that is guided by the small RNA towards the target transcript.

E. Mica
CREA Research Centre for Genomics and Bioinformatics, Fiorenzuola d'Arda, Italy

Department of Sviluppo Sostenibile e la Transizione Ecologica, University of Piemonte Orientale, Vercelli, Italy
e-mail: erica.mica@crea.gov.it

P. Faccioli · C. Crosatti (✉)
CREA Research Centre for Genomics and Bioinformatics, Fiorenzuola d'Arda, Italy
e-mail: primetta.faccioli@crea.gov.it; cristina.crosatti@crea.gov.it

R. Tuberosa et al. (eds.), *The Durum Wheat Genome*, Compendium of Plant Genomes,
https://doi.org/10.1007/978-3-031-84689-2_5

In plants, we can distinguish four main classes of small RNAs: miRNAs, typically 21–22-nt long, trans-acting small interfering RNAs (ta-siRNA) originating as miRNAs from endogenous genes, phased siRNA, and other small RNAs typically 24-nt long, mostly produced by or associated with repetitive elements, introns, natural antisense RNAs, viral genes, and transgenes. The miRNAs come from endogenous genes whose transcripts can fold in a typical stem-loop structure and act *in trans* interfering with target mRNA molecules.

Multiple miRNA coding genes can be grouped in one gene family if they produce a similar mature miRNA. Some miRNA families are highly conserved among plant species, while other are specific to a group of species (i.e., grasses) or even species-specific. Mature miRNAs are produced from longer primary transcripts (pri-miRNA), shaped as a hairpin, which are recognized by DICER and cut in a two-step process, originating a 19–24-nt long dsRNA molecules, with a 2-nt overhang at the 3′ end, called miRNA duplex. Mature miRNAs coupled with *AGO1* (ARGONAUTE 1) in the RISC complex act via post-transcriptional gene silencing, causing mRNA degradation or blocking the translation process of target genes. miRNAs are known to regulate several distinct biological processes, depending on the function of the target gene, such as plant development and organ morphogenesis, biotic and abiotic stress response, nutrient homeostasis, flowering time, and secondary metabolism. miRNAs are rarely responsible for an on/off regulation, but more often fine tune gene expression. This feature, together with their small size, is one of the reasons why small non-coding portion of the genome has been for long time discarded from the analyses and disregarded, being considered not informative nor fundamental. Even though miRNAs and other classes of small RNAs have been largely described in plants, the complexity, and the huge dimension of some plant genomes, has delayed their identification and characterization. This is also the case of durum wheat for which a comprehensive description of miRNAs and other classes of small RNAs is still lacking.

Deep-sequencing technologies have had a big impact on miRNAs and small RNAs identification and characterization since they enable the discovery of an unprecedented number of molecules revealing the complexity of the small RNAs. For example, though deep sequencing became clear that multiple duplexes can originate from the same stem-loop and both strands of the miRNA duplex may be functional, since both can be abundantly represented in the same or different tissues, targeting different genes. The first rules defined first by Ambros et al. (2003) and then by Meyers et al. (2008) to annotate a miRNA, though still valid, needed to be revised (Axtell and Meyers 2018), reducing the stringency on the presence of bulges and mismatches, and requiring sequencing evidence in replicated samples. In parallel with the advancement of next-generation sequencing, increasingly refined and complex bioinformatic pipelines and software have been developed to identify both novel and conserved miRNAs and small RNAs considering many different structural and energetic parameters. It is not always easy to choose the pipeline that better fits to a specific experiment, and not all software are continuously updated and curated from the developers as should be. Conversely, for putative target identification the number of available pipelines is limited and the most commonly used are TargetFinder (Allen et al. 2005; Fahlgren et al. 2007), a perl script that need to be installed locally, psRNATarget (Dai and Zhao 2011), available online with many pre-loaded plant annotated genomes, and Tapir (Bonnet et al. 2010), available online and with the possibility to search also for target mimicry. When target identification is driven by degradome sequencing data (German et al. 2009), then the most widespread tools are CleaveLand4 (Addo-Quaye et al. 2009), PAREsnip (Folkes et al. 2012), sPARTA (Kakrana et al. 2014), and PAREameters (Thody et al. 2020) that facilitate the reduction of false positive targets, identified with a pure in silico search. Nevertheless, all the methods available for target identification are still prone to a high level of false positive, so that it is always necessary to validate the results in vitro, or in vivo with functional and/or transcriptomics-based analyses.

The present chapter is intended to provide the scientific community working on small RNAs with a complete picture of the research efforts done so far to unveil the role of miRNAs in development, stress response, and evolution in durum wheat.

2 Analysis and Classification of the miRNAs Published in Durum and Wild Emmer Wheat

We have considered nearly 30 papers (Table 1), published in a period of 7 years (2015–2022) having as main subject miRNAs identification and analysis in durum wheat (*Triticum turgidum* L. ssp. *durum*) or in its wild relative *T. turgidum* ssp. *dicoccoides* (wild emmer wheat). Most of the considered publications reported the identification of both novel and conserved miRNAs, by means of Illumina deep-sequencing technology. We downloaded, when possible, all the sequences published as pre-miRNA or mature miRNA, for a total of more than 27,000 sequences (16,026 mature miRNAs and 11,061 precursors). Not every publication reported both mature (5′ and 3′) and precursor sequences, for each miRNA gene identified; few papers did not report at all the identified sequences and were discarded by our analysis.

After downloading all the sequences, we divided them in novel and conserved both for durum and wild emmer wheat-derived sequences and removed identical sequences, ending up with 7544 different precursors and 7267 miRNAs that could be from both 5′ and 3′ arm of the precursors. Obviously different precursors could give rise to the same mature miRNAs.

Keeping separated the miRNAs from durum and wild emmer wheat; on the whole there are 4271 diverse and unique miRNA precursors annotated as conserved and 2537 annotated as novel in durum wheat, producing a total of 3906 unique mature sequences. As for wild emmer wheat, we retrieved 892 diverse and unique miRNA precursors annotated as conserved, producing a total of 1190 mature miRNAs, while only the work of Feng et al. (2017) reported 168 novel miRNAs for wild emmer wheat, without reporting precursors sequences.

As expected, most of the precursors, not only for the known miRNAs but also for the novel ones, have been identified in different works and this represents a confirmation of the existence of these molecules. However, lacking an official updated miRNA database, or even a cereal-dedicated one, it is quite common to have mis-annotated novel miRNAs that are already known miRNAs, or highly similar/identical miRNA precursors that are annotated as belonging to different miRNA families. For example, miR1128 is one of the families most often mis-annotated probably because some of the precursors of this family are similar to miR9668 or miR1127 precursors.

It should be noted that works dedicated to long non-coding RNA (lncRNA) identification, quite often include miRNA precursors in this category, since their genomic features match the requisites of the in silico pipelines used to annotate lncRNAs: they are quite long RNA molecules (>200 nt), often mapping to intergenic regions, without any ORFs longer than 300 nt in their sequence. It is controversial if these are bona fide lncRNAs or if they should be better excluded from this class. The fact that most authors identify miRNA genes in their works prompted us to also consider this kind of publications in order to not exclude useful miRNA-related information.

2.1 Analysis of Mature miRNA Sequences

We downloaded and classified all mature miRNAs described in the publications on durum and wild emmer wheat (Table 1), from both 5′ or 3′ arm of the hairpin precursors. Their size is ranging from 17 to 30 nucleotides, being 21 the most common length, while, as expected, molecules shorter than 18 nt or longer of 24 nt are very rare. Small RNAs are typically 21- or 24-nt long, but the miRNAs are usually representing the class of molecules 21 nt long (Fig. 1). Considering the

Table 1 List of papers reporting the characterization or the identification of miRNAs and/or lnc-RNAs in durum wheat and related species, used in the present work

First author	Journal	Year	Species	Variety	Tissues	Housekeeping gene used in qRT-PCR
Fileccia	Funct Integr Genomics	2017	*T. turgidum* ssp. *durum*	Simeto	Leaf/root	–
Dryanova	Genome	2008	Triticeae	–	–	–
Huang	Proc Natl Acad Sci USA	2017	*T. turgidum* ssp. *durum*	NILs	Leaf/flag leaf	–
De Paola	BMC Genomics	2016	*T. turgidum* ssp. *durum*	Ciccio/Svevo	Leaf/root/flag Leaf/spike	*Triticum aestivum* miR159
Debernardi	Development	2017	*T. turgidum* ssp. *durum*	Kronos	Spike	snoR101 (small nucleolar)
Diaz	Mol Biol Rep	2019	*T. turgidum* ssp. *durum*	CBW0101	Flag leaf	*Hordeum vulgare* β-actin
Feng	Genes	2017	*T. turgidum* ssp. *dicoccoides*	B5	Whole plantlet	18S rRNA
Muslu	Int J Mol Sci	2021	*T. turgidum* ssp. *durum and T. turgidum* ssp. *dicoccoides*	–	–	–
Liu	Int J Mol Sci	2020b, c	*T. turgidum* ssp. *durum*	DBA Aurora/L6	Grain	*Triticum turgidum* GAPDH
Cagirici	Sci Rep	2017	*T. turgidum* ssp. *durum and T. turgidum* ssp. *dicoccoides*	Kiziltan/TR39477/TTD	Root	*Triticum aestivum* rRNA26
Akpinar	Sci Rep	2015b	*T. turgidum* ssp. *dicoccoides*	26676G23	–	–
Akpinar	Plant Biotechnol J	2018	*T. turgidum* ssp. *dicoccoides*	TR191/MvGB436	–	–
Matsuda	PLoS One	2017	*T. turgidum* ssp. *durum* and *A. tauschii*	Langdon × *T. tauschii*	Crown tissue	18S rRNA
Kenan Eichler	Genetics	2011	*T. turgidum* ssp. durum and *A. tauschii*	Langdon × *T. tauschii*	Whole plantlet	–
Liu	Sci Rep	2021a	*T. turgidum* ssp. *durum*	DBA Aurora/DBA Artemis	Flag leaf/grain	*Triticum turgidum* GAPDH
Liu	Plants	2020a	*T. turgidum* ssp. *durum*	DBA Aurora/L6	Seedling	–
Liu	Int J Mol Sci	2021b	*T. turgidum* ssp. *durum*	DBA Aurora/DBA Artemis	Flag leaf/grain	*Triticum turgidum* GAPDH
Liu	Int J Mol Sci	2020b, c	*T. turgidum* ssp. *durum*	DBA Aurora/L6	Grain	*Triticum turgidum* GAPDH
Liu	Funct Plant Biol	2017	*T. turgidum* ssp. *durum*	Tamaroi/Yawa/EGA Bellaroy/Tjilkuri	Flag leaf/head	*Triticum turgidum* GAPDH
Liu	Funct Integr Genomics	2016	*T. turgidum* ssp. *durum*	Tamaroi/Yawa/EGA Bellaroy/Tjilkuri	Flag leaf/head	*Triticum turgidum* GAPDH
Liu	PLoS One	2015	*T. turgidum* ssp. *durum*	Tamaroi/Yawa/EGA Bellaroy/Tjilkuri	Flag leaf/head	*Triticum turgidum* GAPDH

Liu	Sci Rep	2019	*T. turgidum* ssp. *durum*	L1-L6/DBA Aurora/WID802/DBA SPES/EGA Bellaroi	Flag leaf	*Triticum turgidum* GAPDH
Giusti	Funct Integr Genomics	2017	*T. turgidum* ssp. *durum*	Ofanto/Cappelli	Leaf	*Triticum aestivum* snoR10 (small nucleolar)
Liu	Plants	2021c	*T. turgidum* ssp. *durum*	DBA Aurora/L6	Seedling	*Triticum turgidum* GAPDH
Zuluaga	Funct Integr Genomics	2018	*T. turgidum* ssp. *durum*	Ciccio/Svevo	Leaf/root	*Triticum turgidum* miR168
Zuluaga	PLoS One	2017	*T. turgidum* ssp. *durum*	Ciccio/Svevo	Leaf/root/flag Leaf/spike	*Triticum turgidum* miR168
Kantar	Planta	2011	*T. turgidum* ssp. *dicoccoides*	Tr39477/Tr38828	Leaf	–
Akpinar	Funct Integr Genomics	2015a	*T. turgidum* ssp. *durum* and *T. turgidum* ssp. *dicoccoides*	Kiziltan/TR39477/TTD-22	Root	–
Zhang	Int J Mol Sci	2019	Triticeae	–	–	–
Maccaferri	Nat Genet	2019	*T. turgidum* ssp. *durum*	Svevo	Multiple tissues	–
Filecci a	Mol Biol Rep	2019	*T. turgidum* ssp. *durum*	Simeto	Leaf/root	U6 snRNA (small nuclear)

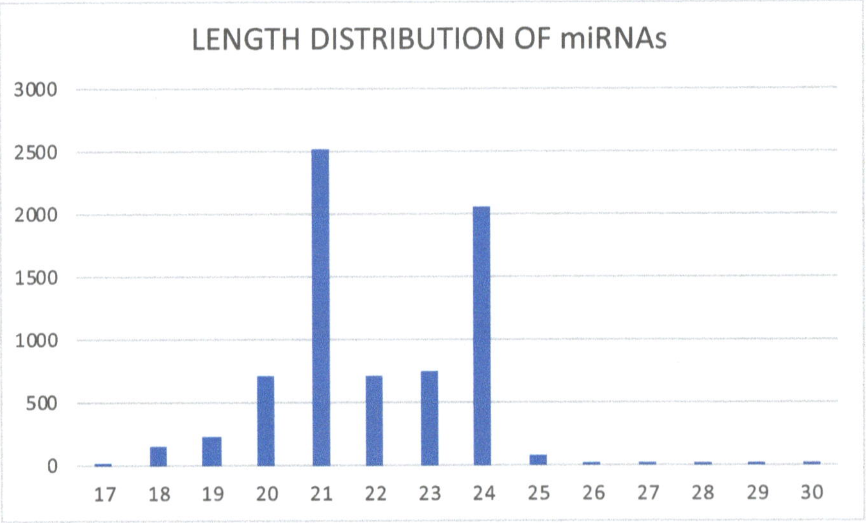

Fig. 1 Length distribution (nucleotides) of all miRNAs published in durum wheat and wild emmer and downloaded from the publications reported in Table 1

sequences of the mature unique miRNAs retrieved, the most common first base at 5 prime end is U (40% of unique sequences).

The number of miRNA-conserved families identified can be deduced considering the miRNAs annotated as "conserved," comparing the data retrieved for durum and wild emmer wheat, with data available online for other plant species. In total, 213 miRNA families span from the most known and conserved ones to the rarest. Among the most known and characterized plant miRNA families, it is worth noting that miR162 is missing in both durum and wild emmer wheat. miR162 was demonstrated to target DCL1 in *Arabidopsis*, tomato and rice (Xie et al. 2003; Zhou et al. 2010; Li et al. 2023) and to be involved as well in stress response (Li et al. 2020, 2023). We did not find any paper reporting miR162 identification in Triticeae except for the work of Fontana et al. (2020) reporting a qRT-PCR on miR162 bread wheat roots under K deficiency and a couple of publications (Akpinar et al. 2015a; Kurtoglu et al. 2014) reporting the presence of miR162 in wheat root miRNAome without releasing the full sequences of the identified miRNAs. This suggests that miR162 family probably exists in *Triticum*, putatively mainly expressed in the root, but we are lacking complete information on precursors and mature miRNA sequences.

Beside miR162, also miR529 and miR535 are missing from our report, even though are well conserved with other plant species. miR535 is similar, in sequence, to miR156 family; hence, it could be possible that it has been misannotated in some papers. Other known families missing from this report are less characterized and conserved.

2.2 Detailed Analysis of miRNA 396 Family

As an example, here we considered miRNA 396 family, a well-known conserved family coding for miRNAs targeting the transcription growth regulating factors (GRFs). In rice and maize, this family includes a total of eight miRNA genes, while in *Arabidopsis*, grapevine, and poplar, there are two, four, and seven genes, respectively. In our database, this family includes 56 unique precursors. Discarding those precursors shorter than 45 nt a total of 47 unique bona fide precursors can be retained. Since the pre-miR396 considered are from different publications and detected using diverse pipelines, we expect that some of these 47 sequences might cluster together, originating from the same locus. Differences might also be due to SNPs arising from sequencing of different accessions. To this end, we aligned the sequences to build a phylogenetic tree with MEGA7 (Fig. 2 and details reported in the legend) that suggests the clear presence of at least eight miR396 genes. Some of the sequences aligned are too divergent, or too short to be clearly assigned to a distinct miRNA gene.

The phylogenetic tree was inferred by using the Maximum Likelihood method based on the Kimura 2-parameter model (Kimura 1980). The bootstrap consensus tree was inferred from 500 replicates (Felsenstein 1985). Branches corresponding to partitions reproduced in less than 50% bootstrap replicates are collapsed. The percentage of replicate trees in which the associated taxa clustered together in the bootstrap test (500 replicates) are shown next to the branches. Initial tree(s) for the heuristic search were obtained automatically by applying Neighbor-Join and BioNJ algorithms to a matrix of pairwise distances estimated using the Maximum Composite Likelihood (MCL) approach, and then selecting the topology with superior log likelihood value. All positions with less than 95% site coverage were eliminated. There was a total of 15 positions in the final dataset. Evolutionary analyses were conducted in MEGA7 (Kumar et al. 2016). Eight clusters are highlighted in different colors, with a bootstrap value >50, that may suggest the presence of right main miRNA genes coding for miR396.

In total, 68 molecules represent mature miR396 with 47 of which are from the 5′ arm corresponding to the most classic mature miR396. Looking at the alignment of these 47 mature miR396, it is evident that in durum wheat mainly two miR396 sequences are present (miR396-1 and miR396-2 in Fig. 3), with a low level of variability at the 5′ end, where iso-miRs present either 0, 1, or 2 T, and a more variable 3′ end. This is a clear example of how different miRNA precursors from different durum wheat varieties and/or expressed in different tissues and treatments might produce different iso-miR. Clearly, some of the variability observed could also be produced by minor sequencing errors.

Published data clearly indicate that for each single miRNA family there is an enormous amount of overlapping hairpins producing almost identical mature miRNAs, plus some other hairpins identified only once in single papers, and it would be necessary to reanalyze all the published raw data for durum wheat and wild emmer, using new and more accurate genomic sequence in order to identify unambiguously all putative miRNAs. This would undoubtedly be of great value to the entire community working on Triticeae representing a comprehensive and complete database of miRNAs, and a first necessary step toward a deeper characterization of these small RNA molecules.

3 Analysis of the miRNAs/mRNA Modules Described in Durum and Wild Emmer Wheat

3.1 miRNAs and Other Small RNAs Involved in Genome Hybridization and Polyploidization

The evolutionary history of the Triticeae genome provides the opportunity to study the role of miRNAs and other small RNAs in the context of hybridization and polyploidization, suggesting that miRNAs and other small RNAs played a key role in shaping genome organization during allopolyploids evolution.

Two publications have investigated the involvement of miRNAs in the genomic events following the interspecific cross between *A. tauschii* and *T. turgidum* ssp. *durum* and the subsequent polyploidization. The triploid hybrid produced, that lately will give rise to a synthetic hexaploidy similar to *T. aestivum*, does rarely produce seeds, and in most cases shows growth abnormalities and necrosis. Kenan-Eichler et al. (2011) considered the role of small ncRNAs in the case of a successful hybridization leading first to triploids (ABD) and then to hexaploid plants (AABBDD) by comparing small RNA and miRNA abundances

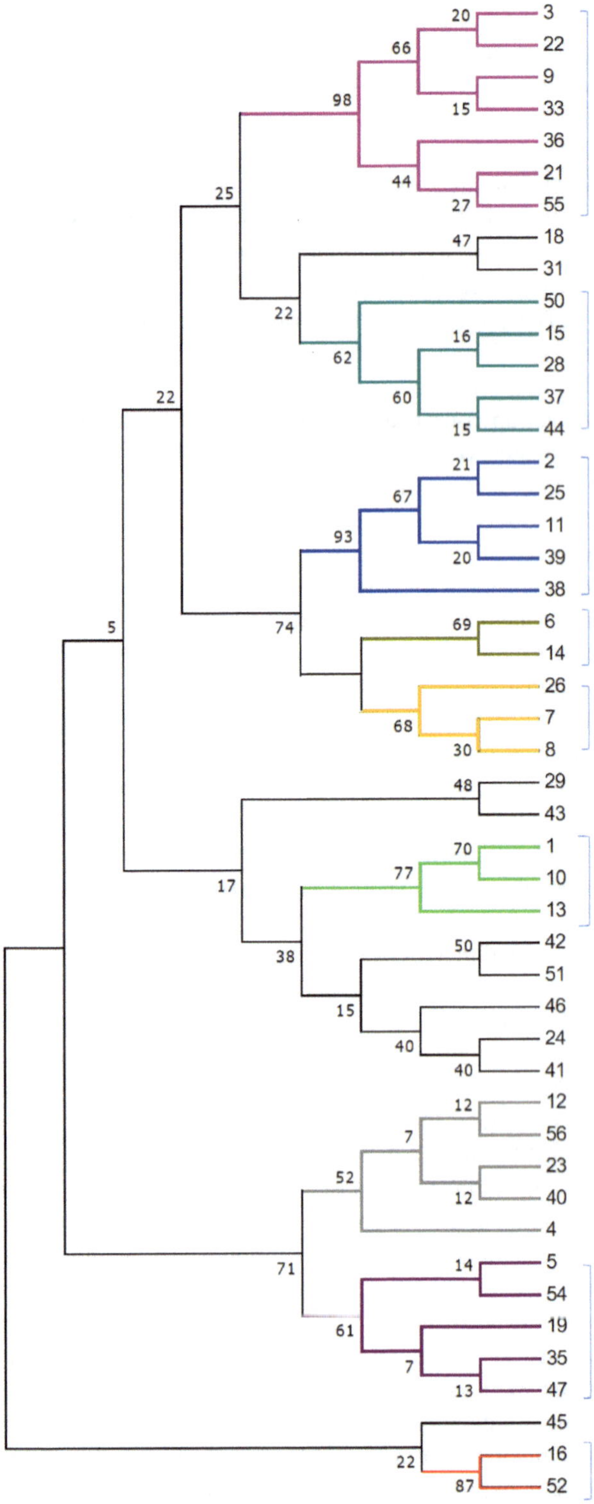

Fig. 2 Phylogenetic analysis of miR396 precursors identified in durum wheat

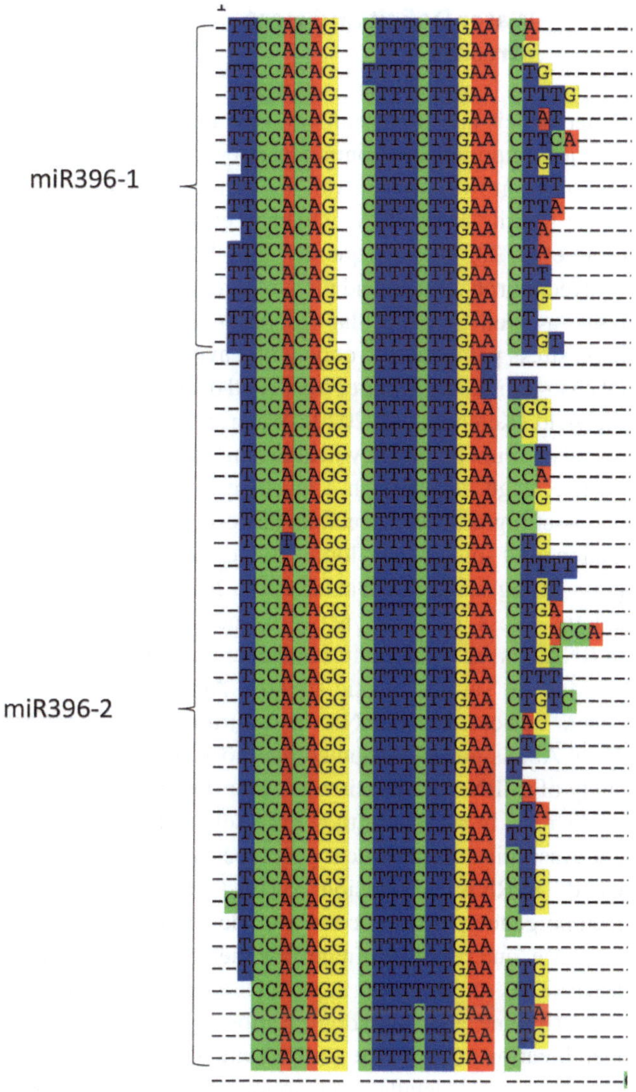

Fig. 3 Alignment of mature miR396 coming from the 5′ arm of the analyzed precursors. Alignment was obtained using MUSCLE, manually curated and visualized using Seaview version 4.7

among diploid, triploid, tetraploid, and hexaploid wheats. Notably, small RNAs decrease with the increase of ploidy level, while the percentage of miRNAs on the total of small RNA molecules increases, from 20% to nearly 45%. The lower level of small RNAs and the concomitant decrease in CpG methylation suggest a dysregulation that eventually activates transposable elements, usually kept silent, because of the genomic shock consequent to polyploidization. The observed increased level of some miRNAs (e.g., miR168) in the hexaploids, on the contrary, could be linked to a genome-wide effect on small RNAs, since this miRNA targets AGO1 and could potentially imbalance small RNA activity and production. Other miRNAs are strongly regulated with the ploidy level and could be linked to an increased vigor of polyploids plants.

Furthermore, in synthetic triploid plants, the repression of miR156 leads to late flowering and increased tillering since miR156 is involved in the regulation of *SQUAMOSA PROMOTER BINDING LIKE* (*SPL*) genes responsible for tillering and determination of the vegetative phase (Matsuda et al. 2017).

3.2 Water Deficit and Heat Stresses

Water deficit and high temperatures are among the most important environmental stresses affecting plant growth in general, including durum wheat (Aprile et al. 2013; Varshney et al. 2018; Groli et al. 2024); consequently, the plants have evolved morphological and physiological adaptations, as well as molecular responses activated upon stress perception to cope with environmental changes retarding plant growth and decreasing the yield (Kercher et al. 2020). Water deficit frequently occurs in durum wheat during the reproductive stage, resulting in a negative impact on grain-setting processes, such as flowering, fertilization, and spikelet development and during the grain filling stage (De Santis et al. 2021). The abiotic stress response network mediated by miRNAs is an important component of plant response to various abiotic stresses. Among the papers dedicated to miRNAs and durum wheat, Jason Able's group conducted the most in-depth study of miRNAs in relation to water deficit and heat stress. In summary, they have selected several Australian durum wheat genotypes and compared miRNA expression against multiple factors: genotype, stress condition, developmental phase, and tissue (grain development and flag leaf). The multi-omics approach (sRNAome, transcriptome and degradome) allowed for the identification of putative miRNA/mRNA modules depicting a detailed description of the complex regulatory network based on miRNAs.

In the first publication (Liu et al. 2015), the Jason Able's group has identified conserved and novel durum miRNAs responsive to pre-anthesis water deficit in a genotypic-dependent manner in drought-tolerant (Tamaroi and Yawa) and drought-sensitive (EGA Bellaroy and Tjilkuri) Australian varieties. Out of 110 conserved and 159 novel miRNAs, 66 conserved and four novel miRNAs, respectively, were found to be responsive to water deficit. Some of them (i.e., miR156, miR159, miR160, miR167, miR319, miR393, miR398, and miR408) were associated with abiotic stress response also in other studies (Zhang et al. 2019), while others were described as responsive to water stress for the first time (e.g., miR1136, miR1432, miR5048, miR5054, miR5071, miR5200, and miR6300), suggesting a possible involvement in species-specific response pathways. The expression analysis highlighted multiple miRNA regulation patterns which might contribute and partly explain the distinct water stress sensitivities of the tested varieties with some miRNAs showing a downregulation in tolerant genotypes versus an upregulation in sensitive ones (Liu et al. 2015).

ARF8 (Auxin Response Factors) and *ARF18*, two key regulators of many auxin-mediated processes in plant development and stress adaptation, have been identified as targets of miR160a in durum wheat and their interaction was experimentally validated by 5′ RLM-RACE (5' RNA Ligase Mediated-Rapid Amplification of c-DNA Ends) (Liu et al. 2016). The expression profiles of the miR160/ARFs module in tolerant and sensitive genotypes have been related to genotypic differences in some physiological traits (e.g., chlorophyll content, leaf relative water content and stomatal conductance) affecting yield components under drought stress conditions. In flag leaves of stress-tolerant genotypes, the upregulation of *ARF8* and *ARF18* matched a corresponding downregulation of miR160a, and this module could in part contribute to coordinate auxin signaling and auxin homeostasis in response to water deficit. The authors proposed a possible use of this module for stress tolerance screening and crop improvement in durum breeding programs (Liu et al. 2017).

When the role of the module miR160/ARFs was tested in ten Australian durum wheat genotypes during single and combined pre-anthesis water deficit and post-anthesis heat stress, the module miR160/ARFs was predominantly activated at early reproductive stages rather than later during grain filling and when heat stress was present. These results are consistent with studies in bread wheat and barley, where miR160 was found to be induced by heat contributing to heat tolerance through auxin signaling and homeostasis (Xin et al. 2010; Kruszka et al. 2014).

The analysis of miRNAs in response to heat stress in durum wheat during the reproductive phase also highlighted the role of the miR396-*HSP* (Heat Shock Protein gene) module. An upregulation of miR396 was found in the tolerant genotype DBA Aurora 5 days after pollination (DAP), while the same response occurred only 15 DAP in sensitive genotype L6. In parallel, two HSP90s (involved in protecting cellular protein stability) were significantly induced at early reproductive stages in both genotypes when heat stress was present. Liu et al. (2019) speculated that the stress-induced HSPs in durum wheat could assist in protecting cellular protein stability against heat.

In two subsequent publications, the same authors performed two multi-omics analysis of the durum small RNAome, transcriptome, and degradome on two Australian durum wheat genotypes with contrasting response to water and heat stress. Single and combined water deficit and heat stress were applied to study the durum wheat response at multiple time-points (5, 15, 25, 35 and 40 DPA) in flag leaves and in developing grains. In flag leaves, the analysis of 120 sRNA libraries has led to the identification of 523 miRNAs (including 55 novel miRNAs), 49,436 differentially expressed genes, and over 100,000 transcript target sites (Liu et al. 2020b). This integrated analysis allowed the identification of key miRNA/mRNA modules exhibiting contrasting patterns between stress-tolerant and -sensitive durum wheat genotypes as well as developmental-dependent patterns reaffirming the complex and coordinated stress-responsive networks mediated by miRNA. The results identified 24 miRNAs expressed exclusively in DBA Aurora and 59 in L6 only; these findings could partly explain the differences in the behavior of the two contrasting genotypes. For instance, the module miR167c-*3p/SCAB1* (Stomatal Closure-related Actin-Binding protein 1), whose miRNA was found expressed only in DBA Aurora, could play a role in coordinating stomatal closure under heat stress in the tolerant genotype. The authors also found developmental stage-specific miRNAs that might play a role in regulating specific processes from flowering to grain maturity. For example, miR9666a-3p/*PLATZ* (PLant AT-rich sequence and Zinc-binding protein) was modulated only at five DAP and affects grain length and number during early grain development in rice (Wang et al. 2019).

In Liu et al. (2020c), the attention was focused on the systematic miRNAome profiling in developing grains at different reproductive stages. Contrasting trends were observed in grain miRNA populations in DBA Aurora and L6 according to the different roles they play in seed development. For instance, miR167c, only detected from 5 to 25 DPA in both genotypes, could play a role in the seed formation through kinesin like-protein, one of its targets.

An interesting investigation carried out applying the stress on parental plants and evaluating the effects on their progeny has shed light on the role of miRNA in the adaptation of durum wheat to drought-prone environments suggesting an implication of miRNA in the transgenerational impact of water and heat stress (Liu et al. 2020a, 2021a, b). It is known that stress-related chromatin modification plays a critical role in transgenerational memory in plants (Lamke and Baurle 2017). The opposite regulation of miR1847-5p/*CBPs* (CREB-Binding Protein genes) in the two different genotypes (DBA Aurora and L6) could ultimately promote histone acetyltransferase-regulated transduction pathways involved in chromatin remodeling. In addition, conserved miRNAs such as miR160, miR444 and miR1847-5p, all downregulated in the tolerant genotype only, could promote the expression of MAPKs (Mitogen-Activated Protein Kinase genes) to facilitate signal transduction. Another research in bread wheat has also demonstrated the involvement of MAPKs in transgenerational tolerance to heat stress (Wang et al. 2016). Heat stress treatment applied to the parents positively affected the progenies by increasing chlorophyll content, grain weight, grain number, and grain total starch content. In this experiment, miR528-5p/F-box was found modulated in DBA Aurora progeny derived from heat-stressed parents, and it has been suggested that this miRNA/mRNA module could contribute to improve antioxidative capability (Liu et al. 2021b) as seen in bread wheat reducing cellular damage in the photosynthetic tissues as shown by the increased chlorophyll content (Zhou et al. 2015; Li et al. 2018). Similarly, two other miRNA/mRNA modules, miR398/*Cu-Zn SOD* and miR5054/*catalase*, could also contribute to an improved heat stress tolerance in the progeny by promoting antioxidant capacity and ROS scavenging under the effects of transgenerational stresses (Liu et al. 2021a, b).

Analysis of small RNA expression in response to drought and heat stress in durum wheat has also been undertaken by other research groups. Giusti et al. (2017) have analyzed fully expanded mature third leaf miRNAome of two durum wheat genotypes (Cappelli and Ofanto) characterized by a contrasting water use efficiency (WUE) largely due to the different stomatal conductance, in response to heat and drought stress. Interestingly, most in silico predicted targets of the miRNAs coming from the differentially expressed precursors have been experimentally linked in other species to mechanisms controlling stomatal movement, a finding in agreement with previous results showing that Cappelli has a lower stomatal conductance than Ofanto (Rizza et al. 2012). Some miRNA/mRNA modules were associated with the different behavior of the two contrasting genotypes. Among them, the miR169/GAD (Glutamate Decarboxylase gene) module is characterized by the downregulation of GAD in Cappelli and is involved in stomatal aperture as demonstrated in *gad1/2* mutant of *Arabidopsis* (Mekonnen et al. 2016). Similarly, the miR5200/*FT*, miR1432/*CALMODULIN LIKE-43*, miR167/*ATPASE FAMILY GENE 2* differentially modulated in Ofanto and Cappelli in response to drought stress are known to be involved in stomatal regulation (Kinoshita et al. 2011; Wu et al. 2013).

The interaction between water stress and the application of mycorrhizae has also an impact on miRNA expression (Fileccia et al. 2019). In the durum wheat cultivar Simeto, mycorrhizae repressed three miRNAs in roots (miR394, miR9666, and miR167-3p) with impact on ABA sensitivity, while in leaves, mycorrhizae modulated miR5384-3p and miR156e-3p affecting trafficking and cell redox homeostasis.

The effect of drought stress on the miRNA expression was investigated also in wild emmer wheat. Kantar et al. (2011) have identified miRNAs in two wild emmer genotypes (Tr39477 and Tr38828T) under dehydration shock using a plant miRNA microarray platform. They found 13 miRNAs differentially regulated in response to drought including miR1435, miR5024, and miR7714, mostly root specific, in the drought-tolerant genotype (Tr39477) only. A similar work was carried out by Akpinar et al. (2015b) comparing the miRNA populations extracted from root tissues of durum wheat cultivar Kızıltan and of two wild emmer accessions (TR39477 and TTD-22 tolerant and sensitive, respectively), exposed to drought stress. Among the miRNAs identified in this work, miR1435 and miR7714 have been validated in rice (Sunkar et al. 2008), and *Brachypodium distachyon* (Bertolini et al. 2013), while for miR5024 experimental evidence exists in *Arabidopsis* (Breakfield et al. 2012), all suggesting that these are miRNAs with potential roles on drought tolerance.

3.3 Other Abiotic Stresses

A single work investigated the role of miRNAs in salinity tolerance in wild emmer (Feng et al. 2017). Through deep-sequencing analysis, a total of 88 conserved and 124 novel miRNAs were identified, and 50 were proven to be salinity-responsive, with 32 significantly upregulated and 18 downregulated. By KEGG analysis, some targets of salinity responsive miRNAs were mapped to salt stress-related pathways such as plant hormone signal transduction, flavonoid biosynthesis, ubiquitin mediated proteolysis, apoptosis, ABC transporter, peroxisome, and DNA repair. Two miRNAs, miR393a and miR393b, showed higher expression under salinity stress. KEGG analysis showed that one of the targets of miR393a is the receptor protein gene *TIR1*, which regulates the hormone signal transduction pathway. GO analysis indicated that the targets of miR393a and miR393b were involved in protein and ATP binding and enzyme activity of transferase or, protein kinase which affected many molecular functions in agreement with previous finding (Gupta et al. 2014).

Eventually, in a study focused on the transcriptome profiling of flag leaves of durum wheat exposed to low temperatures, a list of up- and downregulated lncRNAs were also characterized as additional possible mechanisms of gene expression control (Díaz et al. 2019).

3.4 Nitrogen Stress

Durum wheat productivity largely depends on the availability of nitrogen (N) due to its essential role as a macronutrient for plant growth and development; nevertheless N is also responsible for water pollution; hence, selecting varieties with higher Nitrogen Use Efficiency (NUE) is a major breeding target (Blacklow and Incoll 1981; Xu et al. 2012). Response to variation in N availability is, to some extent, controlled by miRNA and several conserved and novel miRNAs have been found differentially expressed following N depletion in two durum wheat genotypes: Svevo and Ciccio characterized by high and low NUE, respectively (Zuluaga et al. 2017, 2018). In Svevo roots, miR164/NAC (NAM/ATAF/CUC) and miR167/ARF6/ARF8 modules showed the same expression profiles described in *Arabidopsis* and involved in the adaptation of root architecture to reduced N availability (Guo et al. 2005; Liang et al. 2012). Conversely, these miRNAs are characterized by a different expression profile in Ciccio, a cultivar characterized by a lower NUE. The miR827/*NLA* (Nitrogen Limitation Adaptation) module was regulated in the roots of Ciccio only. In *Arabidopsis* and maize, miR827/*NLA* is reported as associated with phosphate absorption and it has been proposed to be necessary to prevent an excessive uptake of phosphate (Zhao et al. 2012; Liang et al. 2012). In addition, Zuluaga et al. (2017) reported a negative relation between the expression of miR399 and its target gene *PHO2* (PHOSPHATE 2) in Svevo roots supporting the hypothesis of the involvement of *PHO2*, *NLA*, miR399, and miR827 in the crosstalk between N and P metabolism and suggested an important role of this modules in the adaptation of durum wheat to a low N during grain filling. Furthermore, a strong downregulation of the novel-61, which does not belong to any previously known miRNA family, was observed in both genotypes during N-stress and associated with N response adaptation in the long-term stress through its target CCAAT-TF (CCAAT-box transcription factor) (Zuluaga et al. 2017).

The miR399/*PHO2* module showed the same expression trend in response to long-term N depletion suggesting that short- and long-term responses are based on common miRNA/mRNA modules (Zuluaga et al. 2018). In the same work, the newly identified miRNA novel-106 showed an immediate strongly downregulation after short-term N depletion negatively correlated with the expression of its target *MYB-A*. The significant and differential expression of novel 61/*CCAT-TF* was, instead, observed in Svevo after long-term N depletion only (Zuluaga et al. 2018).

3.5 Genome-Wide Association Studies for miRNAs Identification

A paragraph apart should be dedicated to the work from Maccaferri et al. (2019) who released, along with the durum wheat genome, also the annotation of 774 miRNA precursors, identified from NGS data of pooled small RNA libraries, representing a total of eight tissues in different developmental, physiological, and stress conditions for a total of 57 tissue/treatment combinations. Among them, it is possible to find biotic stresses (Fusarium head blight and Septoria), nutrient deficiency status, and different spike development stages (including different floral organs): all conditions that have never been considered in other publications about durum wheat nor wild emmer, its progenitor. The rich variety of biological samples analyzed gave the possibility to unveil still untapped resources in the identification of wheat miRNAs, although this discovery work was based on pooled RNAs samples without replicates, so these data cannot be used to infer miRNA expression modulation. Overall, the high proportion of precursors and mature miRNAs uniquely identified in the genome-wide analysis is striking. Out of a total of 829 mature miRNA sequences, 541 are unique and differ from previously annotated sequences; among them, 479 belong to non-conserved, that is, novel, miRNA genes. Even though it is feasible that many new miRNAs are identified when sequencing libraries derived from stress and biological conditions previously uncharacterized, it should be remembered that false positive or misannotation are quite common. For this reason, miRNAs identified in only one research paper should be considered with caution, especially if not validated with other wet experiments.

4 Functional Characterization of miRNAs in Durum Wheat

Since the discovery of the first miRNAs in the plant genomes (Reinhart et al. 2002), the scientific community was urged by the need of understanding the function of these small ncRNAs. The identification and characterization in silico or through wet lab procedures of the target genes of each miRNA represents a clear indication of its biological role but it is not sufficient to describe and define its function. Even if the advent of next-generation sequencing paired with the completion and full annotation of the genomes of many plant species has greatly improved our understanding of the regulatory network governed by miRNAs, we are still lagging behind the functional characterization of these molecules. Some fundamental research has been conducted for the most conserved miRNA families, in model organisms such as *Arabidopsis* and *Oryza sativa*, but concerning the Triticeae clade, and durum wheat in particular, there are very few examples of functional studies. The size of the polyploid genome, the difficulty of stably transforming durum wheat, and the scarce genomic knowledges available for this species, also regarding small RNAs, are huge obstacles towards the success of relevant strategies. Given this constraints, it is worth citing two examples of functional analyses that have been performed in durum wheat.

Debernardi et al. (2017) have proved the molecular role of miR172 in defining spike morphology, threshability, and flowering induction. The miR172 belongs to a miRNA family conserved among most plant species and it has been previously characterized to be involved in juvenile-adult phase transition and hence flowering time definition (Zhu and Helliwell 2011; Wu et al. 2009). The main target of miR172 are the A-class homeotic genes *APETALA2-LIKE* transcription factors, and, in durum wheat, it has been demonstrated that miR172 targets the domestication related Q gene, an AP2-L gene mapping on chr. 5A (Faris et al. 2006; Simons et al. 2006) is responsible for the free-threshing habit of domesticated wheats. The two alleles, q recessive in wild species and Q dominant in most domesticated wheats, present a different affinity for miR172, because of a SNP in the target site, and, consequently, are differently expressed in the developing spikes. The weaker regulatory action of miR172 on Q in domesticated wheats is responsible for homeotic transitions of the external organs of the spikelet (glumes and lemmas) that, in turn, result in a much better threshability of the spike. Moreover, the spike is more compact, and each spikelet shows a higher number of florets, because of a reduced indeterminacy of the spikelet meristem. The same authors confirmed (Debernardi et al. 2020) that miR172 has a redundant regulatory action on the Q paralog gene *AP2L* on chr. 2A and they also showed that miR172/AP2L regulatory module acts on the florigen *FT-1* in leaves, modulating flowering time. In spring varieties, the action of another miRNA, miR156, known to negatively regulate miR172, decreases with plant age, allowing flowering, while in the winter varieties miR156 and miR172 are decoupled and the induction of *VERNALIZATION1* gene is required to induce flowering (Debernardi et al. 2022).

The second example of a functional characterization of a durum wheat miRNA is about a newly discovered miRNA that regulates β-Diketone epicuticular waxes deposition pathway, in particular targeting the Wax1-

CARBOXYLESTERASE (*W1-COE*) gene. This gene belongs to a group of genes mapped on chr. 2BS, defined *W1* locus, that is negatively regulated by *Iw1* (Inhibitor of *W1*) that maps very close to *W1*. Huang et al. (2017) have demonstrated that *Iw1* encodes a recently evolved miRNA gene that encodes a miRNA targeting *W1-COE*. Authors suggest that *Iw1* is a long-non-coding RNAs, that is, a recently evolved long miRNA gene derived by target gene inverted duplication, a quite common mechanism for miRNA gene evolution. The long hairpin structure is, indeed, not a canonical one, but they demonstrated that the most abundant miRNA (miRW1) produced is able to elicit *W1-COE* transcript cleavage. Furthermore, the action of this miRNA induces the production of a multitude of secondary small RNAs that should be responsible for an amplification of the negative regulation of the gene. *W1-COE* gene is responsible for wax deposition and for the consecutive development of glaucousness on mature leaves and spikes, and this miRNA is able to revert a glaucous bread wheat in a non-glaucous one. In wheat, a glaucous appearance is associated with a more stabilized yield in heat-stress prone environments and has been selected during breeding. For this reason, this miRNA gene is almost absent from the most common glaucous modern varieties sequenced so far. These data also confirm how important is the availability of the genomic sequence of different varieties of wheat that might retain useful genomic loci to deepen our understanding on different biological pathways and related functions.

5 Durum Wheat qPCR Housekeeping

Gene expression studies, both small RNA-seq and mRNA-seq, are normally validated by real-time quantitative (reverse transcription) polymerase chain reaction (qRT-PCR) given its accuracy, high sensitivity, and reproducibility. In qRT-PCR analysis, the choice of the appropriate housekeeping gene as an internal control for normalization is crucial. The most frequently used reference genes in plant qRT-PCR analysis are protein-coding genes, such as actin (*ACT*), alpha-Tubulin (*α-TUB*), cyclophilin, gliceraldehyde-3-phosphate dehydrogenase (*GAPDH*), ADP-ribosylation factor1-like protein (*ADP*), and ribosomal RNAs, although their expression has not always been shown to be stable in all tissues and treatments (Ferdous et al. 2015). The use of short length sequence like miRNAs and small nucleolar RNAs (snoRNAs) as reference genes has been proposed for better normalization in miRNAs-qPCR expression analysis in both plants and animals.

There is currently no dedicated study to the identification of reference genes for quantitative expression analysis of miRNAs in durum wheat under various stress conditions, in different tissue or different developmental stages. The many publications dedicated to the analysis of miRNA in durum wheat (Table 1) have used different references such as housekeeping genes (*GADPH* and *ACT*), miRNAs (miR168 and miR159), ribosomal RNA (rRNA 26), and snoRNA (snoRNA10 and snoRNA101). The J. Able groups selected *GADPH* as reference for miRNA analysis across abiotic stress (water deficit and heat stress), tissues (leaves, roots and kernels), and developmental stage (from seedling to mature plants) demonstrating its expression stability in different context. A standardized and reliable qRT-PCR method based on the stem-loop primers protocol (Varkonyi-Gasic et al. 2007) was followed by Giusti et al. (2017) to validate Illumina results on two durum wheat genotypes characterized by a contrasting WUE. In this condition, snoRNA10 showed an expression level similar to mid-expressed miRNA and a better stability than polyUbiquitin and *OEP16* genes (Table 1). Ferdous et al. (2015) evaluated ten candidate reference genes in different tissues, genotypes, and treatments for their ability to normalize miRNA qRT-PCR data in barley. Since most of the genes tested as housekeeping were characterized by some degrees of instability, the authors recommend using a combination of one gene plus one snoRNA as potential references.

6 Conclusions

The works on durum wheat miRNAome here presented clearly highlight the crucial role of miRNAs, particularly in plant development and biotic/abiotic stress response. However, despite the body of evidence there are still few examples in general and none in durum wheat, where allelic variation in miRNA genes or miRNA target sites can be efficiently used to improve plant adaptation to the environment and/or yield. Considering that the activity of miRNAs is depending on their interaction with short binding site in the target genes, any modification in the target site can result in a modulation of the affinity between the two sequences and, consequently, in the level of targeted gene expression. For instance, a mutation in the miRNA binding site of *HvAP2* (APETALA 2) homeotic gene suppresses the cleavage of AP2 resulting in the cleistogamous flower (Nair et al. 2010). Genome editing technology can easily modify targeted region in the genome and has been used in recent year to alter the activity of miRNA leading to new phenotypes (Ferreira and Reis 2023). Although examples are still lacking in durum wheat, it is undoubtful that genome editing is a perfect tool to modulate the miRNA activity in durum as well as in other crops. Future studies will translate the knowledge on durum wheat miRNAome into a real breeding tool.

References

Addo-Quaye C, Miller W, Axtell MJ (2009) CleaveLand: a pipeline for using degradome data to find cleaved small RNA targets. Bioinformatics 25(1):130–131. https://doi.org/10.1093/bioinformatics/btn604. Epub 2008 Nov 18. PMID: 19017659; PMCID: PMC3202307

Akpinar BA, Kantar M, Budak H (2015a) Root precursors of microRNAs in wild emmer and modern wheats show major differences in response to drought stress. Funct Integr Genomics 15(5):587–598. https://doi.org/10.1007/s10142-015-0453-0

Akpinar BA, Yuce M, Lucas S, Vrána J, Burešová V, Doležel J, Budak H (2015b) Molecular organization and comparative analysis of chromosome 5B of the wild wheat ancestor *Triticum dicoccoides*. Sci Rep 5:10763. https://doi.org/10.1038/srep10763

Allen E, Xie Z, Gustafson AM, Carrington JC (2005) microRNA-directed phasing during trans-acting siRNA biogenesis in plants. Cell 121:207–221. https://doi.org/10.1016/j.cell.2005.04.004

Akpinar et al (2018) Chromosome-based survey sequencing reveals the genome organization of wild wheat progenitor Triticum dicoccoides Plant Biotechnology Journal, https://doi.org/10.1111/pbi.12940

Ambros V, Bartel B, Bartel DP, Burge CB, Carrington JC, Chen X, Dreyfuss G, Eddy SR, Griffiths-Jones S, Marshall M, Matzke M, Ruvkun G, Tuschl T (2003) A uniform system for microRNA annotation. RNA 9(3):277-279. https://doi.org/10.1261/rna.2183803

Aprile A, Havlickova L, Panna R et al (2013) Different stress responsive strategies to drought and heat in two durum wheat cultivars with contrasting water use efficiency. BMC Genomics 14:821. https://doi.org/10.1186/1471-2164-14-821

Axtell MJ, Meyers BC (2018) Revisiting criteria for plant microRNA annotation in the era of big data. Plant Cell 30:272–284

Bertolini E, Verelst W, Horner DS, Gianfranceschi L, Piccolo V, Inzè D, Pè ME, Mica E (2013) Addressing the role of microRNAs in reprogramming leaf growth during drought stress in Brachypodium distachyon Molecular Plant 6: 423–443

Blacklow WM, Incoll LD (1981) Nitrogen stress of winter wheat changed the determinants of yield and the distribution of nitrogen and total dry matter during grain filling. Aust J Plant Physiol 8:191–200

Bonnet E, He Y, Billiau K, Van de Peer Y (2010) TAPIR, a web server for the prediction of plant microRNA targets, including target mimics. Bioinformatics 26(12):1566–1568. https://doi.org/10.1093/bioinformatics/btq233

Breakfield NW et al (2012) High-resolution experimental and computational profiling of tissue-specific known and novel miRNAs in Arabidopsis Genome Research 22(1):163–176

Cagirici HB, Alptekin B, Budak H (2017) RNA sequencing and co-expressed long non-coding RNA in modern and wild wheats. Sci Rep 7(1):10670. https://doi.org/10.1038/s41598-017-11170-8

Dai X, Zhao PX (2011) psRNATarget: a plant small RNA target analysis server. Nucleic Acids Res 39:W155–W159. https://doi.org/10.1093/nar/gkr319

De Paola D, Zuluaga DL, Sonnante G (2016) The miRNAome of durum wheat: isolation and characterization of conserved and novel microRNAs and their target genes. BMC Genomics 17:505. https://doi.org/10.1186/s12864-016-2838-4

De Santis MA, Soccio M, Laus MN, Flagella Z et al (2021) Influence of drought and salt stress on durum wheat grain quality and composition: a review. Plants 10(12):2599. https://doi.org/10.3390/plants10122599

Debernardi JM, Lin H, Chuck G, Faris JD, Dubcovsky J (2017) microRNA172 plays a crucial role in wheat spike morphogenesis and grain treshability. Development 144(11):1966–1975. https://doi.org/10.1242/dev.146399

Debernardi JM, Greenwood JR, Jean Finnegan E, Jernstedt J, Dubcovsky J (2020) *APETALA* 2-like genes *AP2L2* and Q specify lemma identity and axillary floral meristem development in wheat. Plant J 101(1):171–187. https://doi.org/10.1111/tpj.14528

Debernardi JM, Woods DP, Li K, Li C, Dubcovsky J (2022) MiR172-APETALA2-like genes integrate vernalization and plant age to control flowering time in wheat. PLoS Genet 18(4):e1010157. https://doi.org/10.1371/journal.pgen.1010157

Díaz ML, Soresi DS, Basualdo J, Cuppari SJ, Carrera A (2019) Transcriptomic response of durum wheat to cold stress at reproductive stage. Mol Biol Rep 46(2):2427–2445. https://doi.org/10.1007/s11033-019-04704-y

Dryanova A, Zakharov A, Gulick PJ (2008) Data mining for miRNAs and their targets in the Triticeae. Genome 51(6):433–443. https://doi.org/10.1139/G08-025

Fahlgren N, Howell MD, Kasschau KD, Chapman EJ, Sullivan CM, Cumbie JS, Givan SA, Law TF, Grant SR, Dangl JL, Carrington JC (2007) High-throughput sequencing of *Arabidopsis* microRNAs: evidence for frequent birth and death of MIRNA genes. PLoS One 2:e219. https://doi.org/10.1371/journal.pone.0000219

Faris J, Simons K, Zhang Z, Gill B (2006) The wheat super domestication gene *Q*. Front Biosci 100:129–148

Felsenstein J (1985) Confidence limits on phylogenies: an approach using the bootstrap. Evolution 39:783–791

Feng K, Nie X, Cui L, Deng P, Wang M, Song W (2017) Genome-wide identification and characterization of salinity stress-responsive miRNAs in wild emmer wheat (*Triticum turgidum* ssp. *dicoccoides*). Genes (Basel) 8(6):156. https://doi.org/10.3390/genes8060156. PMID: 28587281; PMCID: PMC5485520

Ferdous J, Li Y, Reid N, Langridge P, Shi B-J, Tricker PJ (2015) Identification of reference genes for quantitative expression analysis of microRNAs in barley under various stress conditions. PLoS One. https://doi.org/10.1371/journal.pone.0118503

Ferreira SS, Reis RS (2023) Using CRISPR/Cas to enhance gene expression for crop trait improvement by editing miRNA targets. J Exp Bot 74(7):2208–2212. https://doi.org/10.1093/jxb/erad003

Fileccia V, Bertolini E, Ruisi P, Giambalvo D, Frenda AS, Cannarozzi G, Tadele Z, Crosatti C, Martinelli F (2017) Identification and characterization of durum wheat microRNAs in leaf and root tissues. Funct Integr Genomics 17(5):583–598. https://doi.org/10.1007/s10142-017-0551-2

Fileccia V, Ingraffia R, Amato G et al (2019) Identification of microRNAS differentially regulated by water deficit in relation to mycorrhizal treatment in wheat. Mol Biol Rep 46:5163–5174. https://doi.org/10.1007/s11033-019-04974-6

Folkes L, Moxon S, Woolfenden HC, Stocks MB, Szittya G, Dalmay T, Moulton V (2012) PAREsnip: a tool for rapid genome-wide discovery of small RNA/target interactions evidenced through degradome sequencing. Nucleic Acids Res 40(13):e103. https://doi.org/10.1093/nar/gks277

Fontana JE, Wang G, Sun R, Xue H, Li Q, Liu J, Davis KE, Thornburg TE, Zhang B, Zhang Z, Pan X (2020) Impact of potassium deficiency on cotton growth, development and potential microRNA-mediated mechanism. Plant Physiol Biochem 153:72–80. https://doi.org/10.1016/j.plaphy.2020.05.006

German MA, Luo S, Schroth G, Meyers BC, Green PJ (2009) Construction of Parallel Analysis of RNA Ends (PARE) libraries for the study of cleaved miRNA targets and the RNA degradome. Nat Protoc 4(3):356–362. https://doi.org/10.1038/nprot.2009.8

Giusti L, Mica E, Bertolini E, De Leonardis AM, Faccioli P, Cattivelli L, Crosatti C (2017) microRNAs differentially modulated in response to heat and drought stress in durum wheat cultivars with contrasting water use efficiency. Funct Integr Genomics 17:293–309

Groli EL, Frascaroli E, Maccaferri M, Ammar K, Tuberosa R (2024) Dissecting the effect of heat stress on durum wheat under field conditions. Front Plant Sci 15:1393349. https://doi.org/10.3389/fpls.2024.1393349

Guo H, Xie Q, Fei J, Chua N (2005) MicroRNA directs mRNA cleavage of the transcription factor *NAC1* to down-regulate auxin signals for *Arabidopsis* lateral root development. Plant Cell 17:1376–1386. https://doi.org/10.1105/tpc.105.030841. PMID: 15829603

Gupta OP, Meena NL, Sharma I, Sharma P (2014) Differential regulation of microRNAs in response to osmotic, salt and cold stresses in wheat. Mol Biol Rep 41:4623–4629

Huang D, Feurtado JA, Smith MA, Flatman LK, Koh C, Cutler AJ (2017) Long noncoding miRNA gene represses wheat β-diketone waxes. Proc Natl Acad Sci USA 114(15):E3149–E3158. https://doi.org/10.1073/pnas.1617483114

Kakrana A, Hammond R, Patel P, Nakano M, Meyers BC (2014) sPARTA: a parallelized pipeline for integrated analysis of plant miRNA and cleaved mRNA data sets, including new miRNA target-identification software. Nucleic Acids Res 42(18):e139. https://doi.org/10.1093/nar/gku693

Kantar M, Lucas SJ, Budak H (2011) miRNA expression patterns of *Triticum dicoccoides* in response to shock drought stress. Planta 233:471–484. https://doi.org/10.1007/s00425-010-1309-4. PMID: 21069383

Kenan-Eichler M, Leshkowitz D, Tal L, Noor E, Melamed-Bessudo C, Feldman M, Levy AA (2011) Wheat hybridization and polyploidization results in deregulation of small RNAs. Genetics 188(2):263–272. https://doi.org/10.1534/genetics.111.128348

Kercher P, Van Der Meer T, Sujeeth N, Verlee A, Stevens CV, Van Breusegem F et al (2020) Molecular priming as an approach to induce tolerance against abiotic and oxidative stresses in crop plants. Biotechnol Adv 40:107503. https://doi.org/10.1016/j.biotechadv.2019.107503

Kimura M (1980) A simple method for estimating evolutionary rate of base substitutions through comparative studies of nucleotide sequences. J Mol Evol 16:111–120

Kinoshita T, Ono N, Hayashi Y, Morimoto S, Nakamura S, Soda M, Kato Y, Ohnishi M, Nakano T, Inoue S, Shimazaki K (2011) *FLOWERING LOCUS T* regulates stomatal opening. Curr Biol 21:1232–1238

Kruszka K et al (2014) Transcriptionally and post-transcriptionally regulated microRNAs in heat stress response in barley. J Exp Bot 65:6123–6135. https://doi.org/10.1093/jxb/eru353

Kumar S, Stecher G, Tamura K (2016) MEGA7: Molecular Evolutionary Genetics Analysis version 7.0 for bigger datasets. Mol Biol Evol 33:1870–1874

Kurtoglu KY, Kantar M, Budak H (2014) New wheat microRNA using whole-genome sequence. Funct Integr Genomics 14(2):363–379. https://doi.org/10.1007/s10142-013-0357-9

Lamke J, Baurle I (2017) Epigenetic and chromatin-based mechanisms in environmental stress adaptation and stress memory in plants. Genome Biol 18(1):124. https://doi.org/10.1186/s13059-017-1263-6

Li Q, Wang W, Wang W, Zhang G, Liu Y, Wang Y, Wang W (2018) Wheat F-box protein gene *TaFBA1* is involved in plant tolerance to heat stress. Front Plant Sci 9:521

Li XP, Ma XC, Wang H, Zhu Y, Liu XX, Li TT, Zheng YP, Zhao JQ, Zhang JW, Huang YY, Pu M, Feng H, Fan J, Li Y, Wang WM (2020) Osa-miR162a fine-tunes rice resistance to *Magnaporthe oryzae* and yield. Rice (N Y) 13(1):38. https://doi.org/10.1186/s12284-020-00396-2

Li Y, Liu Y, Gao Z, Wang F, Xu T, Qi M, Liu Y, Li T (2023) MicroRNA162 regulates stomatal conductance in response to low night temperature stress via abscisic acid signalling pathway in tomato. Front Plant Sci 14:1045112. https://doi.org/10.3389/fpls.2023.1045112

Liang G, He H, Yu D (2012) Identification of nitrogen starvation-responsive microRNAs in *Arabidopsis thaliana*. PLoS One 7:e48951. https://doi.org/10.1371/journal.pone.0048951. PMID: 23155433

Liu H, Searle IR, Watson-Haigh NS, Baumann U, Mather DE, Able AJ, Able JA (2015) Genome-wide identification of microRNAs in leaves and the developing head of four durum genotypes during water deficit stress. PLoS One. https://doi.org/10.1371/journal.pone.0142799

Liu H, Able AJ, Able JA (2016) Water-deficit stress responsive microRNAs and their targets in four durum wheat genotypes. Funct Integr Genomics. https://doi.org/10.1007/s10142-016-0515-y

Liu H, Able AJ, Able JA (2017) Genotypic water-deficit stress responses in durum wheat: association between physiological traits, microRNA regulatory modules and yield components. Funct Plant Biol 44:538–551

Liu H, Able AJ, Able JA (2019) Genotypic performance of Australian durum under single and combined water-deficit and heat stress during reproduction. Sci Rep 9:14986. https://doi.org/10.1038/s41598-019-49871

Liu H, Able AJ, Able JA (2020a) Transgenerational effects of water-deficit and heat stress on germination and seedling vigour – new insights from durum wheat microRNAs. Plants 9:189

Liu H, Able AJ, Able JA (2020b) Integrated analysis of small RNA, transcriptome, and degradome sequencing reveals the water-deficit and heat stress response network in durum wheat. Int J Mol Sci 21(17):6017. https://doi.org/10.3390/ijms21176017. PMID: 32825615; PMCID: PMC7504575

Liu H, Able AJ, Able JA (2020c) Multi-omics analysis of small RNA, transcriptome, and degradome in *T. turgidum*-regulatory networks of grain development and abiotic stress response. Int J Mol Sci 21(20):7772. https://doi.org/10.3390/ijms21207772. PMID: 33096606; PMCID: PMC7589925

Liu H, Able AJ, Able JA (2021a) Small RNAs and their targets are associated with the transgenerational effects of water-deficit stress in durum wheat. Sci Rep 11(1):3613. https://doi.org/10.1038/s41598-021-83074-7. PMID: 33574419; PMCID: PMC7878867

Liu H, Able AJ, Able JA (2021b) Small RNA, transcriptome and degradome analysis of the transgenerational heat stress response network in durum wheat. Int J Mol Sci 22(11):5532. https://doi.org/10.3390/ijms22115532. PMID: 34073862; PMCID: PMC8197280

Liu H, Able AJ, Able JA (2021c) Nitrogen starvation-responsive microRNAs are affected by transgenerational stress in durum wheat seedlings. Plants 10(5):826. https://doi.org/10.3390/plants10050826

Maccaferri M, Harris NS, Twardziok SO et al (2019) Durum wheat genome highlights past domestication signatures and future improvement targets. Nat Genet 51:885–895. https://doi.org/10.1038/s41588-019-0381-3

Matsuda R, Iehisa JCM, Sakaguchi K, Ohno R, Yoshida K, Takumi S (2017) Global gene expression profiling related to temperature-sensitive growth abnormalities in interspecific crosses between tetraploid wheat and *Aegilops tauschii*. PLoS One 12(5):e0176497. https://doi.org/10.1371/journal.pone.0176497

Mekonnen DV, Flugge U-I, Ludewig F (2016) Gamma-aminobutyric acid depletion affects stomata closure and drought tolerance of Arabidopsis thaliana Plant Science 25–34. https://doi.org/10.1016/j.plantsci.2016.01.005

Meyers BC, Axtell MJ, Bartel B, Bartel DP, Baulcombe D, Bowman JL, Cao X, Carrington JC, Chen X, Green PJ, Griffiths-Jones S, Jacobsen SE, Mallory AC, Martienssen RA, Poethig RS, Qi Y, Vaucheret H, Voinnet O, Watanabe Y, Weigel D, Zhu JK (2008) Criteria for annotation of plant MicroRNAs. Plant Cell 20:3186–3190

Muslu T, Akpinar BA, Biyiklioglu-Kaya S, Yuce M, Budak H (2021) Comparative analysis of coding and non-coding features within insect tolerance loci in wheat with their homologs in cereal genomes. Int J Mol Sci 22(22):12349. https://doi.org/10.3390/ijms222212349

Nair SK et al (2010) Cleistogamous flowering in barley arises from the suppression of microRNA-guided HvAP2 mRNA cleavage. Proc Natl Acad Sci USA 107(1). www.pnas.org/cgi/doi/10.1073/pnas.0909097107

Reinhart BJ, Weinstein EG, Rhoades MW, Bartel B, Bartel DP (2002) MicroRNAs in plants. Genes Dev 16(13):1616–1626. https://doi.org/10.1101/gad.1004402. Erratum in: Genes Dev Sep 1;16(17):2313

Rizza F, Ghashghaie J, Meyer S, Matteu L, Mastrangelo AM, Badeck FW (2012) Constitutive differences in water use efficiency between two durum wheat cultivars. Field Crop Res 125:49–60

Simons KJ, Fellers JP, Trick HN, Zhang Z, Tai YS, Gill BS, Faris JD (2006) Molecular characterization of the major wheat domestication gene *Q*. Genetics 172(1):547–555. https://doi.org/10.1534/genetics.105.044727

Sunkar R, Zhou X, Zheng Y, Zhang W, Zhu J-K (2008) Identification of novel and candidate miRNAs in rice by high throughput sequencing BMC Plant Biology 8:25. https://doi.org/10.1186/1471-2229-8-25

Thody J, Moulton V, Mohorianu I (2020) PAREameters: a tool for computational inference of plant miRNA-mRNA targeting rules using small RNA and degradome sequencing data. Nucleic Acids Res 48(5):2258–2270. https://doi.org/10.1093/nar/gkz1234

Varkonyi-Gasic E, Wu R, Wood M, Walton EF, Hellens RP (2007) Protocol: a highly sensitive RT-PCR method for detection and quantification of microRNAs. Plant Methods 3:12

Varshney RK, Tuberosa R, Tardieu F (2018) Progress in understanding drought tolerance: from alleles to cropping systems. J Exp Bot 69(13):3175–3179

Wang X, Xin C, Cai J, Zhou Q, Dai T, Cao W, Jiang D (2016) Heat priming induces trans-generational tolerance to high temperature stress in wheat. Front Plant Sci 7:501

Wang A, Hou Q, Si L, Huang X, Luo J, Lu D, Zhu J, Shangguan Y, Miao J, Xie Y (2019) The PLATZ transcription factor GL6 affects grain length and number rice. Plant Physiol 180:2077–2090

Wu G, Park MY, Conway SR, Wang JW, Weigel D, Poethig RS (2009) The sequential action of miR156 and miR172 regulates developmental timing in *Arabidopsis*. Cell 138(4):750–759. https://doi.org/10.1016/j.cell.2009.06.031

Wu L, Liu D, Wu J, Zhang R, Qin Z, Liu D, Li A, Fu D, Zhai W, Mao L (2013) Regulation of *FLOWERING LOCUS T* by a microRNA in Brachypodium dystachion. Plant Cell 25:4363–4377

Xie Z, Kasschau KD, Carrington JC (2003) Negative feedback regulation of Dicer-Like1 in *Arabidopsis* by microRNA-guided mRNA degradation. Curr Biol 13(9):784–789. https://doi.org/10.1016/s0960-9822(03)00281-1

Xin M, Wang Y, Yao Y et al (2010) Diverse set of microRNAs are responsive to powdery mildew infection and heat stress in wheat (*Triticum aestivum* L.). BMC Plant Biol 10:123. https://doi.org/10.1186/1471-2229-10-123

Xu G, Fan X, Miller AJ (2012) Plant nitrogen assimilation and use efficiency. Plant Biol 63:153–182

Zhang R, Zhang S, Hao W, Song G, Li Y, Li W, Gao J, Zheng Y, Li G (2019) Lineage-specific evolved microRNAs regulating NB-LRR defense genes in Triticeae. Int J Mol Sci 20(13):3128. https://doi.org/10.3390/ijms20133128

Zhao M, Tai H, Sun S, Zhang F, Xu Y, Li WX (2012) Cloning and characterization of maize miRNA involved in response to nitrogen deficiency. PLoS One 7:e29669. https://doi.org/10.1371/journal.pone.0029669. PMID: 22235323

Zhou M, Gu LF, Li PC, Song XW, Wei LY, Chen ZY, Cao XF (2010) Degradome sequencing reveals endogenous small RNAtargets in rice (*Oryza sativa* L. ssp. *indica*). Front Biol 5:67–90

Zhou SM, Kong XZ, Kang HH, Sun XD, Wang W (2015) The involvement of wheat F-box protein gene *TaFBA1* in the oxidative stress tolerance of plants. PLoS One 10:e0122117

Zhu QH, Helliwell CA (2011) Regulation of flowering time and floral patterning by miR172. J Exp Bot 62(2):487–495. https://doi.org/10.1093/jxb/erq295

Zuluaga DL, De Paola D, Janni M, Curci PL, Sonnante G (2017) Durum wheat miRNAs in response to nitrogen starvation at the grain filling stage. PLoS One 12(8):e0183253. https://doi.org/10.1371/journal.pone.0183253. PMID: 28813501; PMCID: PMC5558935

Zuluaga DL, Liuzzi V, Curci PL, Sonnante G (2018) MicroRNAs in durum wheat seedlings under chronic and short-term nitrogen stress. Funct Integr Genomics 18(6):645–657. https://doi.org/10.1007/s10142-018-0619-7

Application of Hexaploid Bread Wheat SNP Arrays for the Genotyping of Tetraploid Durum Wheat

Martin W. Ganal, Jörg Plieske, Thomas Gross, Eva Grafahrend-Belau, Andreas Polley, Dagmar Kulosa, Naser Poursarebani, Marion S. Röder, and Heike Gnad

Abstract

Genotyping of tetraploid durum wheat (*Triticum turgidum* L. ssp. *durum*, genomes A, B) is currently performed mainly by using single nucleotide polymorphism (SNP) markers. For large-scale parallel genotyping of numerous SNP markers during research and breeding applications, the predominant technology is the use of arrays. Because of the relatively lower significance and smaller-scale breeding efforts for tetraploid durum wheat compared to hexaploid wheat, no comprehensive SNP array development has been performed, specifically for tetraploid wheat. Rather, many different genotyping arrays developed for hexaploid wheat (*Triticum aestivum* L., genomes A, B, D) have also been used for durum wheat. In this review, we describe some important issues in using hexaploid wheat genotyping arrays for the analysis of durum wheat. Specifically, we describe the differences in marker characterization and scoring, effects regarding the analysis of D genome markers with respect to functionality and polymorphism, and provide examples regarding the use of hexaploid wheat arrays in actual durum wheat breeding and research. Finally, we provide an outlook on possible future developments in SNP genotyping for tetraploid durum wheat based on DNA sequencing technologies.

M. W. Ganal (✉) · J. Plieske · T. Gross
E. Grafahrend-Belau · A. Polley · D. Kulosa
N. Poursarebani · H. Gnad
SGS Institut Fresenius GmbH TraitGenetics Section, Gatersleben, Germany
e-mail: martin.ganal@sgs.com

M. S. Röder
Leibniz-Institut für Pflanzengenetik und Kulturpflanzenforschung (IPK), Gatersleben, Germany

1 Introduction

While microsatellite or simple sequence repeat (SSR) markers have been the main tool for hexaploid and tetraploid wheat genotyping (Röder et al. 1998) for more than a decade (mainly 2000–2015), genetic analyses using molecular markers are nowadays predominantly conducted using single nucleotide polymorphism (SNP) markers. SNP markers are defined as differences at individual nucleotides, and they have the advantage of being ubiquitous in a given plant genome and are essentially unlimited in number since they constitute the smallest unit of sequence variation in a genome (Ganal et al. 2009; Thomson 2014). Furthermore, SNPs are often

R. Tuberosa et al. (eds.), *The Durum Wheat Genome*, Compendium of Plant Genomes,
https://doi.org/10.1007/978-3-031-84689-2_6

located directly in genes, while microsatellite markers mostly reside in other genomic regions.

SNPs can easily be analyzed as single markers using various PCR techniques, such as TaqMan or KASP (Ayalew et al. 2019). Moreover, SNPs can be analyzed in multiplex fashion using genotyping arrays. Such arrays are usually defined as microchips on which a large number of SNP markers (thousands to hundred-thousands) can be simultaneously analyzed via the extension of primers that are specific for each individual interrogated SNP. This feature of genotyping arrays makes them highly suitable for the comprehensive, genome-wide analysis of plant material within plant breeding processes or research activities (Rasheed et al. 2017). Furthermore, the cost of genotyping samples with arrays has declined in the last few years due to improved technologies, which enable simultaneous analysis of more samples (Illumina Infinium XT 96-sample arrays and Thermo Fisher Axiom 384-sample arrays).

These high-throughput microarray genotyping technologies are now routinely used in hexaploid and durum wheat for genotyping and have reduced the importance of essentially all other molecular marker technologies including various forms of non-targeted genotyping-by-sequencing technologies (Ganal et al. 2019). Genotyping arrays are now widely used for applications such as marker-assisted backcrossing, genetic mapping of segregating populations, the detection and analysis of genomic locations of marker-trait associations using whole genome scanning, and the precise molecular characterization of wheat lines, varieties, landraces, and wild material (Salvi and Tuberosa 2015; Varshney et al. 2021). Review articles for specific traits have been recently published summarizing many of these results (Colasuonno et al. 2021; Hussain et al. 2021; Maccaferri et al. 2022).

The demand for high-throughput genotyping, in combination with significantly reduced analysis costs, has increased significantly through the use of genomic selection (GS) in breeding (Meuwissen et al. 2001; Hamblin et al. 2011). Through GS, it is now possible to reasonably predict the genetic potential of individual breeding lines and thus eliminate a considerable proportion of low-performing lines before they need to be evaluated extensively in the field (Crossa et al. 2017). It is now not uncommon that thousands of breeding lines are being genotyped each season in a hexaploid wheat breeding program. For durum wheat, with typically much smaller breeding programs, the use of GS has only started in recent years (Juliana et al. 2018; Gesesse et al. 2023).

The economic importance of breeding or research activities in durum wheat is typically one order of magnitude smaller than for hexaploid wheat. Nevertheless, in recent years, significant progress has been made in the molecular analysis of durum wheat for traits of agronomic importance (Milner et al. 2016; Alemu et al. 2020, 2021; Colasuonno et al. 2021; Dagnaw et al. 2022). Other recent highlights are the availability of full genome sequences for durum wheat (cultivar Svevo) and emmer (cultivar Zavitan) or the development of a global tetraploid wheat genetic resources panel (Avni et al. 2017; Maccaferri et al. 2019; Mazzucotelli et al. 2020).

The purpose of this summary is to provide an overview of array-based and other upcoming multiplex genotyping technologies for use in tetraploid durum wheat breeding and research. This is based on available tools in hexaploid wheat and their potential transferability to durum wheat.

2 Hexaploid Wheat Genotyping Arrays for Durum Wheat Genotyping

Due to the much lower importance of durum wheat worldwide compared to hexaploid wheat, no specific genotyping array has been developed for durum wheat. Rather, essentially all genotyping of tetraploid durum wheat (containing the A and B genome) has been performed using genotyping arrays developed for hexaploid wheat that contain markers for the A, B, and D genomes.

The first significant published SNP genotyping array for hexaploid wheat was a 9K Illumina Infinium array (Cavanagh et al. 2013). This array was quickly followed by an improved version (Wheat 90K Array) that contained 90,000 mark-

Table 1 List of published SNP genotyping arrays for hexaploid wheat

	References	Size	Platform	Generation (see text)	Chip sample size	Remarks
1	Cavanagh et al. (2013)	9K	Illumina	First	24	First wheat genotyping array
2	Wang et al. (2014)	90K	Illumina	First	24	First large wheat genotyping array, widely used, contains more than 8000 durum markers
3	Winfield et al. (2016)	820K	Axiom	First	96	Two chips per sample need to be analyzed
4	Allen et al. (2017)	35K	Axiom	Second	384	Derived from 3
5	Soleimani et al. (2020)	15K	Illumina	Second	96	Derived from 2 and other markers
6	Arif et al. (2021)	20K	Illumina	Second	96	Improvement of 5 with markers from 4
7	Gogna et al. (2022)	25K	Illumina	Second	96	Improvement of 6 with markers from 12
8	Gogna et al. (2022)	8K	Illumina	Second	96	Subset of markers from 7
9	Rimbert et al. (2018)	280K	Axiom	First	96	
10	Paux et al. (2022)	35K	Axiom	Second	384	Subset of markers from 9
11	Keeble-Gagnère et al. (2021)	25K	Illumina	Second	96	
12	Gogna et al. (2022)	135K	Axiom	First	96	Markers mainly derived from sequence capture
13	Sun et al. (2020)	660K	Axiom	First	96	
14	Li et al. (2020)	15K	Illumina	Second	n/a	
15	Fan et al. (2022)	55K	Axiom	Second	384	

ers (Wang et al. 2014). This array has been widely used in the wheat community (cited more than 1700 times). For durum wheat, it is important to highlight that the Wheat 90K Array contains a considerable number (>8000) of markers that have been directly identified in durum wheat. This particular feature makes this array somewhat unique and desirable as compared to other wheat genotyping arrays. Through the availability of genome and other sequencing data for a wide range of wheat germplasm, a large number of SNP arrays have now been published for hexaploid wheat (Table 1), and these markers provide an essentially unlimited source of SNP markers for further array development.

Regarding wheat genotyping arrays, two types of arrays could be discriminated (Ganal et al. 2019). The first group are the so-called "first generation" arrays. On such arrays, a large numbers of random SNPs can be found. Typically, these SNPs have been selected without any knowledge about their functionality, precise genome position, and underlying haplotype structure. A typical example for such an array is the Wheat 820K array where many markers from different sources (varieties, landraces, wild wheat ancestral species, etc.) have been included. This results in a relatively limited count of markers useful for genotyping elite material (Winfield et al. 2016).

Over time, so-called "second generation" or optimized arrays have been developed for hexaploid wheat. Such arrays usually contain only SNP markers that have been defined as functional and polymorphic in wheat. Furthermore, one or more additional selection criteria, such as genetic and/or physical map position, haplotype structure, and genome specificity, have been applied. Examples of such optimized arrays include the 35K Wheat Axiom array variants (Allen et al. 2017; Paux et al. 2022) and the 15K/20K/25K Illumina Infinium array series (Soleimani et al. 2020; Arif et al. 2021; Gogna et al. 2022). These optimized arrays allow for the analysis of a large number of samples at reasonable cost (down to 10–20 €/$ per sample including the array, the actual genotyping lab work, and data analysis).

3 Marker Assessment and Quality

In general, the analysis of allopolyploid species is complex (Ganal et al. 2012, 2014) because many SNP marker assays are not genome specific (Fig. 1a) but will fit to a specific locus in more than one of the homeologous genomes. In fact, it is often the case that more than 50% of the SNP assays on a genotyping array detect more than one genome. Furthermore, from those markers, only markers that are polymorphic in only one of the genomes can be used and scored accurately since otherwise they would generate more than three clusters. SNP markers that detect all three genomes in hexaploid wheat but are polymorphic in only one genome produce condensed clusters in allopolyploid species on the intensity scale since the scored allelic situation would actually be AAAAAA, AAAAAB, and AAAABB, and for the tetraploid durum genome, the scored situation would actually be AAAA, AAAB, and AABB (Fig. 1c). Markers that detect only two genomes in hexaploid wheat, of which one is the D genome, are in fact "converted" to genome-specific markers in the durum genome (Fig. 1b).

In summary, compared to the genotyping of hexaploid wheat with highly condensed clusters due to its three genomes, tetraploid durum wheat has the advantage that more markers are genome specific or only one other genome is detected in the background. This ensures that the clusters are less condensed and often easier to score than in hexaploid wheat.

4 Haplotype Structure and Marker Distribution Along Genetic and Physical Maps

A haplotype is represented by a number of polymorphic positions over a certain distance on the DNA strand that are in perfect linkage disequilibrium (LD). When comparing haplotypes in genomic regions over a set of lines, it is typically observed that the physical length of a conserved haplotype is correlated with the amount of genetic recombination (Bhat et al. 2021; Gardiner et al. 2019). Therefore, in centromeric regions, where little recombination is observed, conserved haplotypes can extend over very long regions (over many tens of kilobases or even megabases with hundreds to thousands of markers being in perfect LD). In genetic maps with randomly selected markers, this is reflected by a clustering of genetic markers in the centromeric regions (Jordan et al. 2015, 2022). On the other side, recombination is much more prevalent in regions toward both ends of chromosomes. This results in haplotype blocks that are very short (extending over a few to tens of kilobases). Regions with high levels of recombination are often associated with genic regions.

When physically long haplotype blocks are conserved over a wide range of material (e.g., including landraces and more exotic material), the informativeness of individual markers in perfect linkage disequilibrium is limited. Thus, selecting many markers in the same LD block for an array increases the size of the array, while the amount of real information gain is close to zero unless a very rare recombination event occurs in such a haplotype block. Detecting such rare recombination events does require the use of only a few markers (e.g., three markers extending over the entire haplotype block). SNP markers selected based on physical distance (such as marker per megabase) result in many markers, especially in the centromeric regions, that are in perfect LD and thus are not much informative while increasing the size and costs of an array. On the other side, an array selected based on haplotype structure will typically result in a clustering of the markers toward the ends of the chromosomes with few or very few markers in the centromeric regions.

Due to its longer evolutionary history and the fact that durum wheat varieties contain a higher proportion of the allelic variation from wild material, the extent of haplotype blocks is rather shorter than longer than in hexaploid wheat (Jordan et al. 2015; Sansaloni et al. 2020), and thus, markers of "second generation" arrays are

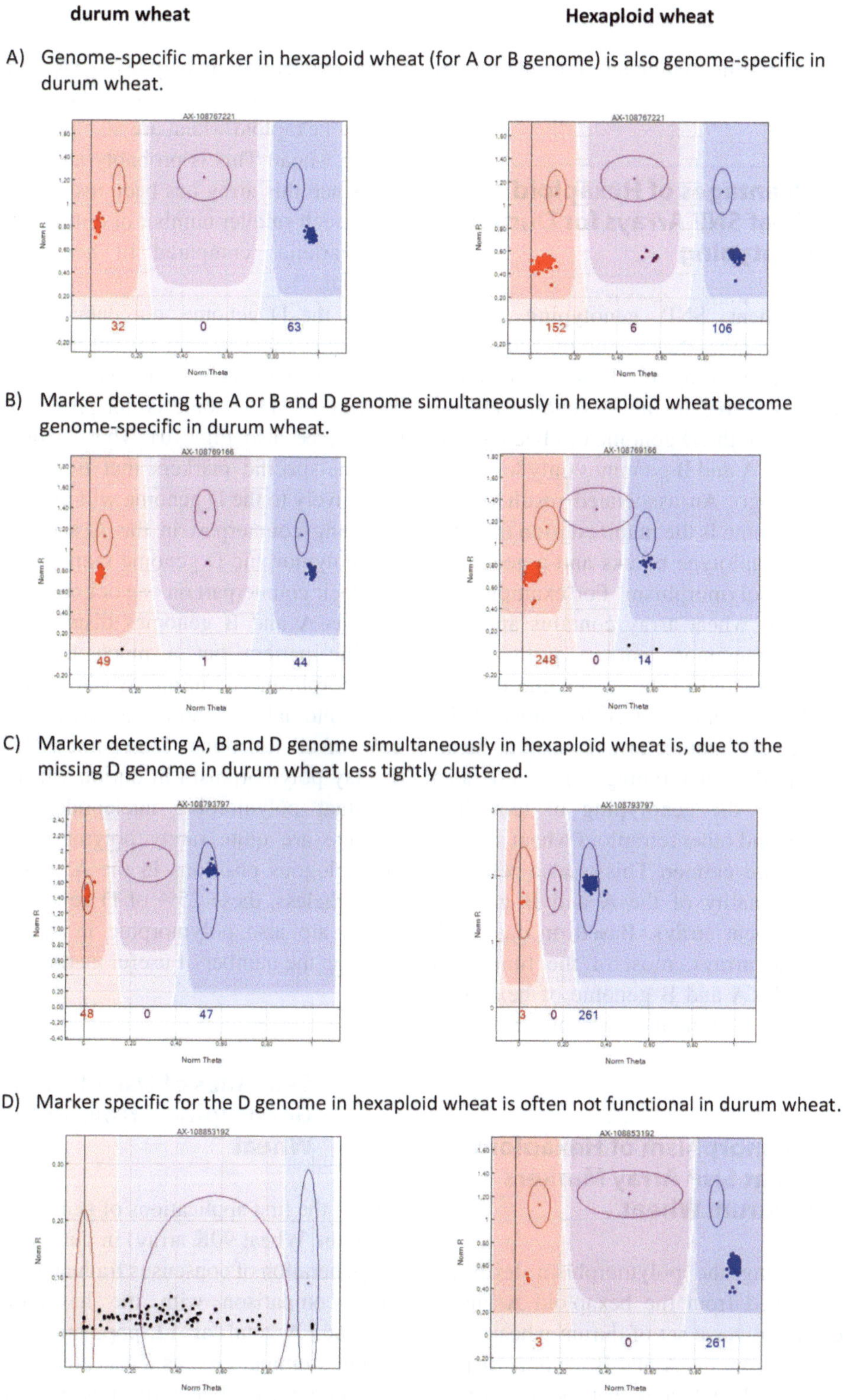

Fig. 1 Cluster distribution of various SNP marker types in durum wheat and hexaploid wheat. (**a**) A genome specific marker in hexaploid wheat (for A or B genome) is also genome specific in durum wheat. (**b**) A marker detecting the A or B and D genome simultaneously in hexaploid wheat becomes genome specific in durum wheat. (**c**) A marker detecting A, B, and D genome simultaneously in hexaploid wheat is, due to the missing D genome in durum wheat, less tightly clustered. (**d**) A marker specific for the D genome in hexaploid wheat is often not functional in durum wheat

also providing little redundant information in durum wheat.

5 Advantages of Hexaploid Wheat SNP Arrays for Durum Genotyping

Hexaploid wheat SNP genotyping arrays, regardless of whether they are optimized or not, usually have many more markers on the A and B genome compared to the D genome (Wang et al. 2014) since the D genome has been added to the ancestral A and B genomes only less than 10,000 years ago. An associated result of the younger D genome is the manifestation of notably lengthier haplotype blocks and a generally lower rate of polymorphism. For example, the optimized 25K wheat array contains approximately three times more markers on the A and B genome compared to the D genome (Gogna et al. 2022). The same is true for most of the other wheat SNP marker arrays, so that hexaploid wheat SNP genotyping arrays are also well-suited for the genotyping of tetraploid durum wheat and other tetraploid wheat such as emmer and wild emmer. This is also reflected by the functionality of the A and B genome markers on wheat arrays. Based on data with various wheat arrays, most of the functional markers on the A and B genome of hexaploid wheat are also useful and scorable in durum wheat.

6 Polymorphism of Hexaploid Wheat SNP Array Markers in Durum Wheat

When analyzing the polymorphism level for markers derived from the hexaploid A and B genomes on a diverse set of durum wheat, most SNP markers from the A and B genome display polymorphism. It indicates that there is a relatively minor ascertainment bias in the A and B genome and most polymorphisms are conserved between tetraploid and hexaploid wheat. Taking the 25K wheat array as reference, internal data show that approximately 57% (A genome) to 63% (B genome) SNP markers that are polymorphic in hexaploid wheat are also polymorphic in durum wheat. This is probably an underestimation since this array has been used to characterize a much smaller number of durum wheat lines and varieties compared to hexaploid wheat material.

For the D genome, our data from the 25K array show that less than 75% of the polymorphic D genome markers are functional in durum wheat because the D genome is not present in durum wheat (see also Fig. 1d). This is logical since genome-specific markers that fit perfectly and exclusively to the D genome will not find a well-matching counterpart in the A and B genome. The polymorphic D genome markers that detect also their counterpart on one or both of the homeologous A and B genomes display a genome-specific pattern but a much lower level of polymorphism in durum wheat. For the 25K array, internal data show that only approximately 25% of all functional D genome SNP markers display polymorphism in durum wheat, indicating that polymorphic nucleotides on the D genome are quite rarely polymorphic at their homeologous positions in the A or B genome. Nevertheless, these 25% of D genome markers, which are also polymorphic in durum wheat, increase the number of useful markers for durum wheat.

7 Examples of Use of Various Genotyping Arrays in Durum Wheat

One of the first applications of genotyping arrays (here the Wheat 90K array) in durum wheat was the generation of consensus framework maps and their comparison with the hexaploid wheat genome. A total of 13 populations including durum x emmer crosses (Maccaferri et al. 2015) with combined 1928 progenies have been analyzed. In total, 26,626 SNP markers could be analyzed and the polymorphic markers were integrated in a consensus map and compared to the hexaploid wheat genome genetic map, thus linking the A and B genomes of tetraploid and hexaploid wheat.

The 15K array (Soleimani et al. 2020) has been used by Taranto et al. (2020) for the analysis of molecular signatures for important traits in durum wheat during historic breeding in Italy or the analysis of grain protein content and kernel weight in a durum x emmer cross (Fatiukha et al. 2020).

The use of the wheat 35K Axiom array has been published for the assessment of a worldwide durum wheat collection with respect to genetic diversity, population structure, and linkage disequilibrium by Roncallo et al. (2021), and for Mediterranean and European durum lines by Ganugi et al. (2021).

The use of the wheat 25K array (Gogna et al. 2022) has recently been described for the characterization and relationship analysis from a Balkan durum wheat collection by Velimirović et al. (2023) or the Ethiopian durum wheat gene pool (Mulugeta et al. 2023a).

Recently, in the publication of Kidane et al. (2019), a total of 1200 lines, including a Nested Association Mapping (NAM) population, were genotyped with the Wheat 90K array and used for the identification of marker-trait associations for important traits and to develop a model for genomic selection including farmers' traditional breeding knowledge that showcases the use of the array in durum landraces (Gesesse et al. 2023; Mulugeta et al. 2023b).

For further information, detailed and comprehensive reviews of the use of predominantly hexaploid wheat genotyping arrays for various traits, including quality, biotic, and abiotic stresses, have recently been provided by Colasuonno et al. (2021), Marcotuli et al. (2022), and Taranto et al. (2023), to which we refer here for these individual traits.

8 Emerging Technologies for Durum Wheat Genotyping

Array-based technologies are by now widely used for durum wheat genotyping. However, the so-called next-generation sequencing (NGS) technologies have made large-scale DNA sequencing extremely cheap, so that it can potentially be used for routine genotyping in durum wheat breeding. Sequence-based genotyping has been made easier through the full genome sequencing of a durum reference (cultivar Svevo) genome (Maccaferri et al. 2019). Considering this advancement, some of the emerging technologies might complement or replace the array-based genotyping technologies. The NGS technologies can roughly be divided into three categories:

The first category is genome sequencing. The sequencing of full durum genomes for the comprehensive SNP identification and genotyping of lines with a high sequence coverage (e.g., > 25X coverage) is now well-established using the Svevo reference sequence (Maccaferri et al. 2019). Nevertheless, the costs of using this technology for routine analysis are still prohibitive (>1000 €/$ per line solely for the sequencing costs and excluding DNA extraction, library preparation, and bioinformatic analysis) based on a genome size of more than 10 gigabases. Whole genome sequencing at high coverage is thus only used for the characterization of important varieties or parental lines such as those used for MAGIC populations (Huang et al. 2015; Hao et al. 2020). Recently, whole genome sequencings has also been used for genotyping by means of SKIM sequencing (Kumar et al. 2021). In SKIM sequencing, individual lines undergo sequencing at reduced coverage (usually 0.5-3X, sometimes up to 5-10X) and the subsequent imputation of missing data relies mainly on the full reference genome sequence data obtained from the parental lines of the respective cross. Alternatively, through a technology called sequence capture, it is now possible to selectively enrich and sequence only the exonic regions of the wheat genome (Winfield et al. 2012; Wang et al. 2023). With that, full costs (from DNA isolation to sequencing) are reduced to a few hundred €/$ per sample, which makes the use of SKIM sequencing at least feasible for research projects where a very high marker density is required (Adhikari et al. 2022). For routine use in durum breeding, SKIM sequencing or Exome capture is still at least 10 times too expensive.

The second category is untargeted genotyping-by-sequencing (GBS). This technology is based on reduced representation sequencing, which means that by various technologies only a certain proportion of the genome (e.g., based on a prior digestion of genomic DNA with restriction enzymes) will be sequenced. The subsequent SNPs' detection will be then restricted solely to the portion of the genome that has been sequenced (Van Orsouw et al. 2007; Elshire et al. 2011; Scheben et al. 2017; Robbana et al. 2019; Miki et al. 2020). Thus, the pure sequencing costs decline to approximately 10 €/$ due to the low amount of sequence information needed (≤1% of the genome needs to be sequenced) and the possibility to multiplex several individual plants in a single sequencing run. The disadvantages of these non-targeted GBS methods are that not in all lines the same fraction of genomic fragments is being sequenced. Usually, only around 10% of the sequenced fragments are detected in the majority (>80%–90%) of the sequenced lines. Furthermore, such GBS data cannot be easily compared between laboratories since the sequenced fragments are highly dependent on the genome-complexity reduction technology used (e.g., the type of restriction enzyme used in fragmentation, the sequencing method, and the sequencing equipment). Finally, many untargeted GBS technologies require a specific license for the process, which increases the total costs per sample to often around 25–50 €/$. This is too high for routine applications in durum breeding. Currently, especially in durum breeding, non-targeted GBS methods are infrequently used.

The most promising emerging category is the Targeted Genotyping-By-Sequencing (tGBS). tGBS is available in two different options, which include amplicon sequencing using (i) two specific amplification primers or (ii) only one specific primer. Usually, a few tens to a few thousand regions in the genome, normally defined by certain candidate genes and/or the presence of one or several known SNPs are amplified simultaneously in each line (Kilian et al. 2012; Bernardo et al. 2015; Campbell et al. 2015). The main advantages of tGBS is that in all investigated lines essentially the same loci are being enriched or amplified. Additional SNPs can be identified in each sequenced amplicon of a given line, which permits the identification of haplotypes and that tGBS can be performed with high multiplexing in terms of a plurality of lines (Ruff et al. 2020). With that and despite the relatively high set-up costs for the amplicon multiplex, the costs for the analysis of diploid samples can be ultimately in the area of around 10 €/$ or less per sample. For tetraploid durum wheat, the costs per sample are somewhat higher since the sequence coverage needs to be increased, especially when both the A and B genomes are amplified simultaneously. The sequence coverage has to be increased further in heterozygous samples since each line needs to be determined reliably for all three allelic states (AA, AB, BB), which requires a certain amount of reads for each locus. Altogether, this means that for the same accuracy as for an inbred diploid individual, the sequence coverage of a heterozygous durum line has to be increased at least by a factor of 4, which causes significantly increasing sequencing costs. So far, the tGBS analysis of durum lines has been published only in a few experimental studies (Holtz et al. 2016).

9 Conclusions

Genotyping of durum wheat has been successfully piggybacking on the development of technologies for the economically much more important hexaploid wheat (*Triticum aestivum* L.). Currently, most of the large-scale durum genotyping is performed using hexaploid wheat SNP genotyping arrays. Most of this activity is now being realized using optimized "second generation" wheat SNP arrays that provide a maximum of information with a set of SNP markers that have been selected based on marker quality, genome distribution, and haplotype structure. Using these arrays, the costs for genotyping have been reduced. This has transformed SNP genotyping into a routine activity for applications such as genetic mapping or association studies, as well as in genomic selection schemes during durum breeding at costs that are lower than the costs of a single small trial plot (approximately 10–20 €/$). In the future, array-based genotyping technologies might still be the method of choice for durum genotyping if the costs per sample can be further reduced through technological developments (e.g., through the simultaneous genotyping of more samples with the same handling costs). Alternatively, durum wheat genotyping might move to sequence-based technologies such as SKIM Sequencing or targeted GBS if accuracy can be improved to the level of array-based genotyping and costs decrease faster than the costs for array-based genotyping. In any case, for durum wheat, the prospects of genotyping large numbers of samples for routine applications in breeding are now very affordable.

References

Adhikari L, Shrestha S, Wu S et al (2022) A high-throughput skim-sequencing approach for genotyping, dosage estimation and identifying translocations. Sci Rep 12(1):17583. https://doi.org/10.1038/s41598-022-19858-2

Alemu A, Feyissa T, Tuberosa R et al (2020) Genome-wide association mapping for grain shape and color traits in Ethiopian durum wheat (*Triticum turgidum* ssp. *durum*). Crop J 8(5):757–768. https://doi.org/10.1016/j.cj.2020.01.001

Alemu A, Feyissa T, Maccaferri M et al (2021) Genome-wide association analysis unveils novel QTLs for seminal root system architecture traits in Ethiopian durum wheat. BMC Genomics 22(1):20. https://doi.org/10.1186/s12864-020-07320-4

Allen AM, Winfield MO, Burridge AJ et al (2017) Characterization of a wheat breeders' array suitable for high-throughput SNP genotyping of global accessions of hexaploid bread wheat (*Triticum aestivum*). Plant Biotechnol J 15(3):390–401. https://doi.org/10.1111/pbi.12635

Arif MAR, Shokat S, Plieske J et al (2021) A SNP-based genetic dissection of versatile traits in bread wheat (*Triticum aestivum* L.). Plant J 108(4):960–976. https://doi.org/10.1111/tpj.15407

Avni R, Nave M, Barad O et al (2017) Wild emmer genome architecture and diversity elucidate wheat evolution and domestication. Science 357(6346):93–97. https://doi.org/10.1126/science.aan0032

Ayalew H, Tsang PW, Chu C et al (2019) Comparison of TaqMan, KASP and rhAmp SNP genotyping platforms in hexaploid wheat. PLoS One 14(5):e0217222. https://doi.org/10.1371/journal.pone.0217222

Bernardo A, Wang S, St Amand P et al (2015) Using next generation sequencing for multiplexed trait-linked markers in wheat. PLoS One 10(12):e0143890. https://doi.org/10.1371/journal.pone.0143890

Bhat JA, Yu D, Bohra A et al (2021) Features and applications of haplotypes in crop breeding. Commun Biol 4(1):1266. https://doi.org/10.1038/s42003-021-02782-y

Campbell NR, Harmon SA, Narum SR (2015) Genotyping-in-Thousands by sequencing (GT-seq): a cost effective SNP genotyping method based on custom amplicon sequencing. Mol Ecol Resour 15(4):855–867. https://doi.org/10.1111/1755-0998.12357

Cavanagh CR, Chao S, Wang S et al (2013) Genome-wide comparative diversity uncovers multiple targets of selection for improvement in hexaploid wheat landraces and cultivars. Proc Natl Acad Sci USA 110(20):8057–8062. https://doi.org/10.1073/pnas.1217133110

Colasuonno P, Marcotuli I, Gadaleta A et al (2021) From genetic maps to QTL cloning: an overview for durum wheat. Plants 10(2):315. https://doi.org/10.3390/plants10020315

Crossa J, Pérez-Rodríguez P, Cuevas J et al (2017) Genomic selection in plant breeding: methods, models, and perspectives. Trends Plant Sci 22(11):961–975. https://doi.org/10.1016/j.tplants.2017.08.011

Dagnaw T, Mulugeta B, Haileselassie T et al (2022) Phenotypic variability, heritability and associations of agronomic and quality traits in cultivated Ethiopian durum wheat (*Triticum turgidum* L. ssp. *Durum*, Desf). Agronomy 12(7):1714. https://doi.org/10.3390/agronomy12071714

Elshire RJ, Glaubitz JC, Sun Q et al (2011) A robust, simple genotyping-by-sequencing (GBS) approach for high diversity species. PLoS One 6(5):e19379. https://doi.org/10.1371/journal.pone.0019379

Fan X, Liu X, Feng B et al (2022) Construction of a novel Wheat 55 K SNP array-derived genetic map and its utilization in QTL mapping for grain yield and quality related traits. Front Genet 13:978880. https://doi.org/10.3389/fgene.2022.978880

Fatiukha A, Filler N, Lupo I et al (2020) Grain protein content and thousand kernel weight QTLs identified in a durum × wild emmer wheat mapping population tested in five environments. Theor Appl Genet 133(1):119–131. https://doi.org/10.1007/s00122-019-03444-8

Ganal MW, Altmann T, Röder MS (2009) SNP identification in crop plants. Curr Opin Plant Biol 12(2):211–217. https://doi.org/10.1016/j.pbi.2008.12.009

Ganal MW, Polley A, Graner EM et al (2012) Large SNP arrays for genotyping in crop plants. J Biosci 37(5):821–828. https://doi.org/10.1007/s12038-012-9225-3

Ganal MW, Wieseke R, Luerssen H et al (2014) High-throughput profiling of genetic resources in crop plants using genotyping arrays. In: Tuberosa R, Graner A, Frison E (eds) Genomics of plant genetic resources. Springer, New York, pp 113–130

Ganal MW, Plieske J, Hohmeyer A et al (2019) High-throughput genotyping for cereal research and breeding. In: Miedaner T, Korzun V (eds) Applications of genetic and genome research in cereals. Woodhead Publishing, Duxford, pp 3–17
Ganugi P, Palchetti E, Gori M et al (2021) Molecular diversity within a Mediterranean and European panel of tetraploid wheat (*T. turgidum* subsp.) landraces and modern germplasm inferred using a high-density SNP array. Agronomy 11(3):414. https://doi.org/10.3390/agronomy11030414
Gardiner LJ, Wingen LU, Bailey P et al (2019) Analysis of the recombination landscape of hexaploid bread wheat reveals genes controlling recombination and gene conversion frequency. Genome Biol 20(1):69. https://doi.org/10.1186/s13059-019-1675-6
Gesesse CA, Nigir B, de Sousa K et al (2023) Genomics-driven breeding for local adaptation of durum wheat is enhanced by farmers' traditional knowledge. Proc Natl Acad Sci USA 120(14):e2205774119. https://doi.org/10.1073/pnas.2205774119
Gogna A, Schulthess AW, Röder MS et al (2022) Gabi wheat a panel of European elite lines as central stock for wheat genetic research. Sci Data 9(1):538. https://doi.org/10.1038/s41597-022-01651-5
Hamblin MT, Buckler ES, Jannink JL (2011) Population genetics of genomics-based crop improvement methods. Trends Genet 27(3):98–106. https://doi.org/10.1016/j.tig.2010.12.003
Hao C, Jiao C, Hou J et al (2020) Resequencing of 145 landmark cultivars reveals asymmetric sub-genome selection and strong founder genotype effects on wheat breeding in China. Mol Plant 13(12):1733–1751. https://doi.org/10.1016/j.molp.2020.09.001
Holtz Y, Ardisson M, Ranwez V et al (2016) Genotyping by sequencing using specific allelic capture to build a high-density genetic map of durum wheat. PLoS One 11(5):e0154609. https://doi.org/10.1371/journal.pone.0154609
Huang BE, Verbyla KL, Verbyla AP et al (2015) MAGIC populations in crops: current status and future prospects. Theor Appl Genet 128(6):999–1017. https://doi.org/10.1007/s00122-015-2506-0
Hussain B, Akpinar A, Alaux M et al (2021) Wheat genomics and breeding: bridging the gap. AgriRxiv:1–57. https://doi.org/10.31220/agriRxiv.2021.00039
Jordan KW, Wang S, Lun Y et al (2015) A haplotype map of allohexaploid wheat reveals distinct patterns of selection on homoeologous genomes. Genome Biol 16(1):48. https://doi.org/10.1186/s13059-015-0606-4
Jordan KW, Bradbury PJ, Miller ZR et al (2022) Development of the Wheat Practical Haplotype Graph database as a resource for genotyping data storage and genotype imputation. G3 (Bethesda) 12(2):jkab390. https://doi.org/10.1093/g3journal/jkab390
Juliana P, Montesinos-López OA, Crossa J et al (2018) Integrating genomic-enabled prediction and high-throughput phenotyping in breeding for climate-resilient bread wheat. Theor Appl Genet 132(1):177–194. https://doi.org/10.1007/s00122-018-3206-3
Keeble-Gagnère G, Pasam R, Forrest KL et al (2021) Novel design of imputation-enabled SNP arrays for breeding and research applications supporting multi-species hybridization. Front Plant Sci 12:756877. https://doi.org/10.3389/fpls.2021.756877
Kidane YG, Gesesse CA, Hailemariam BN et al (2019) A large nested association mapping population for breeding and quantitative trait locus mapping in Ethiopian durum wheat. Plant Biotechnol J 17(7):1380–1393. https://doi.org/10.1111/pbi.13062
Kilian A, Wenzl P, Huttner E et al (2012) Diversity arrays technology: a generic genome profiling technology on open platforms. Methods Mol Biol 888:67–89. https://doi.org/10.1007/978-1-61779-870-2_5
Kumar P, Choudhary M, Jat BS et al (2021) Skim sequencing: an advanced NGS technology for crop improvement. J Genet 100:38. https://doi.org/10.1007/s12041-021-01285-3
Li J, Zhao L, Cheng X et al (2020) Molecular cytogenetic characterization of a novel wheat-Psathyrostachys huashanica Keng T3DS-5NsL•5NsS and T5DL-3DS•3DL dual translocation line with powdery mildew resistance. BMC Plant Biol 20(1):163. https://doi.org/10.1186/s12870-020-02366-8
Maccaferri M, Ricci A, Salvi S et al (2015) A high-density, SNP-based consensus map of tetraploid wheat as a bridge to integrate durum and bread wheat genomics and breeding. Plant Biotechnol J 13(5):648–663. https://doi.org/10.1111/pbi.12288
Maccaferri M, Harris NS, Twardziok SO et al (2019) Durum wheat genome highlights past domestication signatures and future improvement targets. Nat Genet 51(5):885–895. https://doi.org/10.1038/s41588-019-0381-3
Maccaferri M, Bruschi M, Tuberosa R (2022) Sequence-based marker assisted selection in wheat. In: Wheat improvement: food security in a changing climate. Springer International Publishing, Cham, pp 513–538. ISBN: 978-3-030-90675-7
Marcotuli I, Soriano JM, Gadaleta A (2022) A consensus map for quality traits in durum wheat based on genome-wide association studies and detection of ortho-meta QTL across cereal species. Front Genet 13:982418. https://doi.org/10.3389/fgene.2022.982418
Mazzucotelli E, Sciara G, Mastrangelo AM et al (2020) The Global Durum Wheat Panel (GDP): an international platform to identify and exchange beneficial alleles. Front Plant Sci 11:569905. https://doi.org/10.3389/fpls.2020.569905
Meuwissen TH, Hayes BJ, Goddard ME (2001) Prediction of total genetic value using genome-wide dense marker maps. Genetics 157(4):1819–1829. https://doi.org/10.1093/genetics/157.4.1819
Miki Y, Yoshida K, Enoki H et al (2020) GRAS-Di system facilitates high-density genetic map construction and QTL identification in recombinant inbred lines of the wheat progenitor Aegilops taus-

chii. Sci Rep 10(1):21455. https://doi.org/10.1038/s41598-020-78589-4

Milner SG, Maccaferri M, Huang BE et al (2016) A multiparental cross population for mapping QTL for agronomic traits in durum wheat (*Triticum turgidum* ssp. durum). Plant Biotechnol J 14(2):735–748. https://doi.org/10.1111/pbi.12424

Mulugeta B, Ortiz R, Geleta M et al (2023a) Harnessing genome-wide genetic diversity, population structure and linkage disequilibrium in Ethiopian durum wheat gene pool. Front Plant Sci 14:1192356. https://doi.org/10.3389/fpls.2023.1192356

Mulugeta B, Tesfaye K, Ortiz R et al (2023b) Marker-trait association analyses revealed major novel QTLs for grain yield and related traits in durum wheat. Front Plant Sci 13:1009244. https://doi.org/10.3389/fpls.2022.1009244

Paux E, Lafarge S, Balfourier F et al (2022) Breeding for economically and environmentally sustainable wheat varieties: an integrated approach from genomics to selection. Biology 11(1):149. https://doi.org/10.3390/biology11010149

Rasheed A, Hao Y, Xia X et al (2017) Crop breeding chips and genotyping platforms: progress, challenges, and perspectives. Mol Plant 10:1047–1064. https://doi.org/10.1016/j.molp.2017.06.008

Rimbert H, Darrier B, Navarro J et al (2018) High throughput SNP discovery and genotyping in hexaploid wheat. PLoS One 13(1):e0186329. https://doi.org/10.1371/journal.pone.0186329

Robbana C, Kehel Z, Ben Naceur M et al (2019) Genome-wide genetic diversity and population structure of Tunisian durum wheat landraces based on DArTseq technology. Int J Mol Sci 20(6):1352. https://doi.org/10.3390/ijms20061352

Röder MS, Korzun V, Wendehake K et al (1998) A microsatellite map of wheat. Genetics 149(4):2007–2023. https://doi.org/10.1093/genetics/149.4.2007

Roncallo PF, Larsen AO, Achilli AL et al (2021) Linkage disequilibrium patterns, population structure and diversity analysis in a worldwide durum wheat collection including Argentinian genotypes. BMC Genomics 22(1):233. https://doi.org/10.1186/s12864-021-07519-z

Ruff TM, Marston EJ, Eagle JD et al (2020) Genotyping by multiplexed sequencing (GMS): a customizable platform for genomic selection. PLoS One 15(5):e0229207. https://doi.org/10.1371/journal.pone.0229207

Salvi S, Tuberosa R (2015) The crop QTLome comes of age. Curr Opin Biotechnol 32:179–185. https://doi.org/10.1016/j.copbio.2015.01.001

Sansaloni C, Franco J, Santos B et al (2020) Diversity analysis of 80,000 wheat accessions reveals consequences and opportunities of selection footprints. Nat Commun 11(1):4572. https://doi.org/10.1038/s41467-020-18404-w

Scheben A, Batley J, Edwards D (2017) Genotyping-by-sequencing approaches to characterize crop genomes: choosing the right tool for the right application. Plant Biotechnol J 15(2):149–161. https://doi.org/10.1111/pbi.12645

Soleimani B, Lehnert H, Keilwagen J et al (2020) Comparison between core set selection methods using different Illumina marker platforms: a case study of assessment of diversity in wheat. Front Plant Sci 11:1040. https://doi.org/10.3389/fpls.2020.01040

Sun C, Dong Z, Zhao L et al (2020) The Wheat 660K SNP array demonstrates great potential for marker-assisted selection in polyploid wheat. Plant Biotechnol J 18(6):1354–1360. https://doi.org/10.1111/pbi.13361

Taranto F, D'Agostino N, Rodriguez M et al (2020) Whole genome scan reveals molecular signatures of divergence and selection related to important traits in durum wheat germplasm. Front Genet 11:217. https://doi.org/10.3389/fgene.2020.00217

Taranto F, Esposito S, De Vita P (2023) Genomics for yield and yield components in durum wheat. Plants 12(13):2571. https://doi.org/10.3390/plants12132571

Thomson M (2014) High-throughput SNP genotyping to accelerate crop improvement. Plant Breed Biotechnol 2(3):195–212. https://doi.org/10.9787/pbb.2014.2.3.195

van Orsouw NJ, Hogers RC, Janssen A et al (2007) Complexity reduction of polymorphic sequences (CRoPS): a novel approach for large-scale polymorphism discovery in complex genomes. PLoS One 2(11):e1172. https://doi.org/10.1371/journal.pone.0001172

Varshney RK, Bohra A, Yu J et al (2021) Designing future crops: genomics-assisted breeding comes of age. Trends Plant Sci 26(6):631–649. https://doi.org/10.1016/j.tplants.2021.03.010

Velimirović A, Jovović Z, Perović D et al (2023) SNP diversity and genetic structure of "Rogosija", an old Western Balkan durum wheat collection. Plants 12(5):1157. https://doi.org/10.3390/plants12051157

Wang S, Wong D, Forrest K et al (2014) Characterization of polyploid wheat genomic diversity using a high-density 90,000 single nucleotide polymorphism array. Plant Biotechnol J 12(6):787–796. https://doi.org/10.1111/pbi.12183

Wang H, Bernardo A, St Amand P et al (2023) Skim exome capture genotyping in wheat. Plant Genome:e20381. https://doi.org/10.1002/tpg2.20381

Winfield MO, Wilkinson PA, Allen AM et al (2012) Targeted re-sequencing of the allohexaploid wheat exome. Plant Biotechnol J 10(6):733–742. https://doi.org/10.1111/j.1467-7652.2012.00713.x

Winfield MO, Allen AM, Burridge AJ et al (2016) High-density SNP genotyping array for hexaploid wheat and its secondary and tertiary gene pool. Plant Biotechnol J 4(5):1195–1206. https://doi.org/10.1111/pbi.12485

Genetic Resources for Trait Mapping in Durum Wheat

Mercy Wairimu Macharia, Matteo Dell'Acqua, and Mario Enrico Pè

Abstract

Durum wheat (*Triticum durum* Desf.) is a staple grain crop for millions of people globally. It is also a genetic tool used by scientists and breeders to identify the genetic basis of complex traits in wheat, and a bridge to facilitate transfer of beneficial alleles across the wheat species complex. Many advances have been achieved in durum wheat genetics thanks to the development of a rich collection of genetic resources, spanning from recombinant lines in bi-parental and multiparental crosses, to diversity collections representing durum wheat landraces cultivated worldwide. Here, we discuss the many genetic resources available in durum wheat and their use for improvement and mapping of quantitative trait loci (QTL) for wheat productivity, nutritional quality, and resilience against biotic and abiotic factors.

Keywords

Durum wheat · RILs · Mapping · QTL · Genetics

M. W. Macharia · M. Dell'Acqua · M. E. Pè (✉)
Institute of Plant Sciences,
Scuola Superiore Sant'Anna, Italy
e-mail: m.pe@sssup.it

1 Durum Wheat: Staple Crop and Genetic Tool

The cultural and agricultural importance of durum wheat (*Triticum durum* Desf.) is rooted in the Neolithic revolution, when it was domesticated from wild emmer in the Fertile Crescent (Civáň et al. 2013). Ever since, durum wheat has been at the basis of the Old-World diets and is now cultivated throughout the world. It is much appreciated as it performs well in semiarid regions and is adapted to a range of climatic conditions (Sall et al. 2019), generally fetching prices higher than other cereals, including bread wheat (*Triticum aestivum*) (EU Commodity Price Dashboard 2022). Among the world's commonly cultivated cereals, durum wheat ranks tenth with an average yearly production of about 40 million tons, with less than one-tenth of it being used for feed (https://www.igc.int/en/default.aspx).

Today, durum wheat's traditional and modern varieties are cultivated throughout North America, Southern Europe, North Africa, Middle East and Central Asia, Ethiopia, Australia, and India (Tuberosa et al. 2021; Peters Haugrud et al. 2024). In many of these regions, durum wheat faces contrasting environmental constraints that are worsened by the climate crisis. Erratic and dwindling rainfall, heat and drought, and increased incidence of pests and diseases increasingly curtail production and call for the development and cultivation of climate-resilient wheat

R. Tuberosa et al. (eds.), *The Durum Wheat Genome*, Compendium of Plant Genomes,
https://doi.org/10.1007/978-3-031-84689-2_7

varieties (Kahiluoto et al. 2019), including durum cultivars. Continued improvement in durum wheat cultivars is therefore paramount to ensure sustained production and mitigate both biotic and abiotic constraints, considering the success of synthetic wheat, particularly in China (Li et al. 2024).

The quest for sustainable intensification of agriculture under climate change calls for a better use and understanding of natural allelic variation (Langridge et al. 2022). In this sense, the re-introduction of favorable alleles that were negatively modified or even eliminated during domestication and breeding offers a successful pathway for major breakthroughs in wheat breeding (Rahman et al. 2020). In this context, the interest in durum wheat also depends on its potential as a *genetic bridge* to transfer desirable allelic variants (haplotypes) from wild forms and/landraces to cultivated wheat as advocated by Maccaferri et al. (2015) and Klymiuk et al. (2019). Durum wheat is an allotetraploid species (2n = 4x = 28, genomes AABB), the last of a series evolved from *Triticum urartu* (2n = 2x = 14, genomes AA) and an unknown donor of the BB genome, to wild emmer (*Triticum turgidum* ssp. *dicoccoides*), to domesticated emmer (*Triticum turgidum* ssp. *dicoccum*), and finally from cultivated naked forms of emmer to durum (Jorgensen et al. 2017). Wild emmer is a progenitor of all forms of cultivated wheat including bread wheat (*Triticum aestivum*, $2n = 6x = 42$, genomes AABBDD). This places durum wheat in an intermediate position between wild and cultivated forms, so that research on durum wheat connects across the wheat species complex supporting breeding of both tetraploid and hexaploid wheat (Chatzav et al. 2010; Li et al. 2020).

Genetics research may further disclose durum wheat's potential by describing the genetic basis of its traits of agronomic relevance. Conventional plant breeding exploits intra-species genetic variation for phenotypic selection of interesting traits (Cui et al. 2015). However, a considerable amount of genetic variation remains masked at the phenotypic level and uncharacterized at the molecular level. DNA markers and genomic technologies revolutionized the analysis of complex agronomic traits in crops, including durum wheat (Collard and Mackill 2007; Tuberosa et al. 2021). Modern breeding based on marker-assisted selection (MAS) and biotechnological approaches, including new breeding technologies (NBT), require the identification of the molecular targets underlying quantitative trait loci (QTL) and contributing to the expression of the target traits (Maccaferri et al. 2022). This information comes from the QTLome dissection through QTL mapping, a statistical endeavor aimed at linking phenotypic data to genotypic data to explain the molecular mechanisms at the basis of complex traits, ultimately localizing the chromosomal regions of interest and describing genes and genetic mechanisms underlining trait variation (Salvi and Tuberosa 2015).

QTL mapping may dramatically speed up the development of new crop varieties, and its effectiveness depends on the availability of genetic tools to properly describe the allelic variation available to mapping. Molecular markers including RFLP (Blanco et al. 1998), DArT-SSR (Mantovani et al. 2008; Maccaferri et al. 2015), AFLPs (Nachit et al. 2001), and DaRT (Peleg et al. 2008) have been widely used in the past to construct linkage maps and support QTL mapping in durum (Colasuonno et al. 2021). Currently, the complete sequence of the durum wheat genome (Maccaferri et al. 2019), along with that of bread wheat (Appels et al. 2018) and of wild emmer (Avni et al. 2017), allows for an improved understanding of the genetic basis of the genomic landscape in which QTL are located, including a better identification and annotation of candidate genes (Soriano et al. 2021). In this genomics era, single nucleotide polymorphism (SNP) markers are widely used to construct durum wheat linkage maps (Maccaferri et al. 2015; Marcotuli et al. 2017), describe its diversity (Kabbaj et al. 2017), and conduct genome-wide association studies (GWAS) (Mengistu et al. 2016; Kidane et al. 2017a, b) at a definition never attained before. This wealth of information enables improved gene annotation and localization, and eventually QTL cloning and gene validation (Maccaferri et al. 2022).

Here, we describe recent and notable examples of trait mapping in durum wheat, with a special focus on genetic resources developed by the community and made available to scientists and breeders. We start discussing durum wheat as a bridge across the wheat species complex, and we then move on to describe segregant populations and their use in QTLome dissection and QTL mapping (Salvi and Tuberosa 2015). We then describe collections of durum wheat natural allelic variation, and finally mutant populations developed to facilitate and validate gene/QTL mapping and cloning.

2 Durum Wheat for Gene Transfer Across the Wheat Species Complex

Wheat's wild relatives are a reservoir of alleles with great potential for crop improvement (Rong et al. 2000; Zhang et al. 2016). A large diversity exists untapped in more distant wild relatives in regard to environmental adaptation (Brunazzi et al. 2018) and quality traits (Talini et al. 2020). Crosses between wild and cultivated genotypes, however, are marred with problems related to sexual incompatibility, lack of hybrid viability, and sterility (Zamir 2001; Dempewolf et al. 2017). Even when successful, wild crosses typically have much lower quality and yield due to linkage drag (Brown and Marshall 1986; Salamini et al. 2002). To improve crosses fertility and facilitate the reduction of linkage drag, durum wheat has been used as an intermediate step to shuttle genes from wild relatives to bread wheat (Klymiuk et al. 2019). Success stories mostly refer to oligogenic traits, most notably disease resistance. The *Yr* genes, responsible for stripe rust resistance, have been transferred from wild emmer to bread wheat through durum (Xu et al. 2013; Yaniv et al. 2015). Similarly, the *Gpc-B1* gene for high grain protein and mineral contents has been introgressed into elite wheat varieties using a durum chromosome substitution line (6BS) (Distelfeld and Fahima 2007; Tabbita et al. 2017). This is also the case of fusarium head blight resistance genes (Buerstmayr et al. 2013) and powdery mildew *Bgt* resistance genes (Li et al. 2009; Ben-David et al. 2010). Also with abiotic stress resistance, valuable traits can be found in wild relatives: a wild emmer QTL for drought resistance has been introgressed in bread wheat through durum (Merchuk-Ovnat et al. 2016).

3 Development and Use of Segregant Populations in Durum Wheat

QTL mapping relies on the identification of coinheritance patterns between genetic diversity and phenotypic diversity in segregating populations. Once QTL are identified, markers such as Kompetitive allele-specific PCR (KASP) may be developed to assist MAS, as in the case of disease resistance (Maccaferri et al. 2008a; Kidane et al. 2017a; Maccaferri et al. 2022). Throughout the years, a handful of mapping populations have been developed in durum wheat to capture the great diversity in this species. The simplest type of mapping population can be developed crossing any two parental lines, chosen because of the difference in their trait values (Collard et al. 2005). Crossing such lines produces an F_1 offspring whose genomes are a uniform representation of the parental genome. At that point, F_1 lines can be advanced to the next generation by selfing, commencing the segregation of allelic variation inherited from the parental lines. Different mapping strategies based on molecular markers can then be used to trace the co-inheritance of alleles at specific markers and trait values. Backcrosses can also be used to break linkage drag and make uniform the genetic background in which QTL haplotypes have been inserted by recombination, as in the case of the fine mapping for the QTL for fusarium head blight resistance (Buerstmayr et al. 2013) and introgressions carrying the *Sr47* gene for stem rust resistance (Klindworth et al. 2023).

By self-crossing of F_1 lines, it is possible to produce segregating F_2, the simplest segregating mapping population attainable. For decades, durum wheat F_2 populations have been produced to select and map various traits including grain yield (Liu et al. 1994; Maccaferri et al. 2008b) and grain quality (Clarke et al. 1997). F_2 populations are flexible genetic tools and recently have been used to validate multi-trait phenotyping methods aimed at speeding up the breeding process in durum wheat (Alahmad et al. 2018). Indeed, these populations are handy to characterize contrasting phenotypes in parental pairs, providing the means to track beneficial traits quickly and efficiently. However, F_2 populations still segregate extensively and will not maintain their allelic combinations when propagated to further generations. In order to create an immortalized set of lines, offsprings of segregating populations must be selfed for several generations creating recombinant inbred lines (RILs). RIL genomes are highly homozygous mosaics of the diversity available in founder lines with minimal segregation left. Once RILs are established as pure line population, they can be used repeatedly to assess QTL for different phenotypes under different environments and thus can be shared among research groups (Maccaferri et al. 2011; Fatiukha et al. 2020). In durum wheat, interspecific RILs have been used to construct genetic linkage maps in the range of 2000–3000 cM (Peleg et al. 2008; Avni et al. 2014; Zaïm et al. 2020). Several genetic maps have been developed in durum wheat RILs and combined in a consensus map (Maccaferri et al. 2015), eventually supporting the high-quality genome assembly of this species (Maccaferri et al. 2019).

Durum wheat RILs have been used to map QTL for several traits of agronomic relevance. Particularly successful are the examples coming from oligogenic disease resistance traits. Durum wheat RIL populations were used to identify and describe Fusarium head blight resistance QTL, later introduced in durum wheat cultivars (Prat et al. 2017; Zhao et al. 2018), as well as adult plant stem rust resistance (Mago et al. 2022). Several candidate genes responsible for leaf rust resistance have also been identified (Maccaferri et al. 2008a; Aoun et al. 2017; Kthiri et al. 2019). Durum wheat RILs have been successful in mapping phenology and yield components (Maccaferri et al. 2008a; Graziani et al. 2014), grain mineral content (Peleg et al. 2009a), and grain quality (Yan et al. 2018). Environmental stress QTL can also be targeted using RILs, as in the case of drought resistance (Peleg et al. 2009b). A set of RIL populations with partially overlapping pedigrees combining landraces and ICARDA improved lines have been recently used to map QTL for yield components in drought conditions (Zaïm et al. 2020). Grain quality is a trait of prime importance in durum wheat, and RILs developed based on pasta quality have been used to identify QTL for grain color (Zhang et al. 2008). The advantage of RILs is that they can be indefinitely reproduced and screened in different experiments, validating QTL in different environments, as in the case of quality traits (Roncallo et al. 2012). Once QTL mapping studies are collated, meta-QTL analysis can be used to combine their findings to consolidate and prioritize genomic loci for breeding (Iannucci et al. 2017). QTL meta-analysis involves the construction of a consensus map of different types of markers or estimating consensus regions where QTL or meta-QTL are concentrated. A similar approach allowed to prioritize candidate regions for abiotic, biotic stress and quality traits QTL in durum wheat (Soriano et al. 2021; Bruschi et al. 2024; Cabas-Lühmann et al. 2024; Mohammadi and Mohammadi 2024).

In recent years, multiparental populations (MPPs) were established as advanced mapping and pre-breeding tools to better leverage genetic diversity available in crop genetic resources (Scott et al. 2020). MPPs are based on the intercrossing of more than two parental lines, resulting in higher diversity and denser recombination than classic biparental crosses (Huang et al. 2015). Multiparent advanced generation intercrosses (MAGIC) and nested association mapping (NAM) MPPs have been developed in durum wheat. MAGIC populations will usually have several founder lines (4, 8, or 16) that are intercrossed over multiple generations in a balanced scheme before MAGIC RILs are generated. A durum wheat MAGIC population was developed crossing four elite durum wheat cultivars which differ in geographic origin and in traits of agronomic interest, including plant height, grain yield, heading date, and maturity date (Milner et al. 2016). NAM populations involve crosses between a common founder line and various other lines to produce interconnected biparental NAM populations. The NAM scheme is an open crossing scheme that can involve any number of lines. In durum wheat, three RIL populations were developed by crossing a leaf rust susceptible line to three resistant lines, eventually mapping six genomic regions involved in leaf rust resistance (Kthiri et al. 2019). A durum wheat NAM was developed in Ethiopia, establishing a mapping population and a pre-breeding resource seeking the combination of an elite variety with international background with 50 different local landraces. A preliminary characterization of the Ethiopian NAM showed high diversity segregating in the population, identifying loci for phenology and plant structure (Kidane et al. 2019). The population was also used to test participatory methods accelerating genetic gain for local adaptation (Gessesse et al. 2023).

Regardless of the crossing scheme, the development of RIL panels is indeed resource intensive, as it requires several generations of inbreeding. On the other hand, double haploids (DH) can be used to achieve complete homozygosity within one generation, consequently speeding up gene mapping (Niu et al. 2014). Durum wheat DH have been developed and used to map QTL for a number of traits, including grain protein concentration (Suprayogi et al. 2009; Ruan et al. 2021), gluten content (Ruan et al. 2020a), and yield components (Liu et al. 2020).

4 Diversity Collections and GWAS

In the advent of high-throughput genotyping technologies enabled by the genomics revolution, GWAS emerged as a powerful tool for detecting genomic regions associated with natural variation (Mackay 2001). Durum wheat germplasm is highly diverse, and collections of landraces may provide a good representation of alleles relevant to breeding (Royo et al. 2010; Mengistu et al. 2016; Kabbaj et al. 2017). GWAS does not require the production of segregant populations as it relies on historic recombination events accumulated in any collection of germplasm. The characterization of diversity collections with GWAS does not require or allow for the production of linkage maps, and rather relies on physical maps derived from genome sequencing (Maccaferri et al. 2019).

GWAS potential has been largely demonstrated in durum wheat. A collection of elite durum wheat varieties were used to identify genetic loci for drought-adaptive phenological traits and grain yield after phenotyping across 15 environments highly variable for water availability (Maccaferri et al. 2011). Likewise, a GWAS study of elite durum wheat varieties sourced from CIMMYT identified QTL for agronomic traits under different environments including drought stress and heat stress (Sukumaran et al. 2018). Association mapping has also been used to dissect durum grain and flour quality in commonly used durum wheat breeding lines (Fiedler et al. 2017).

Diversity panel approaches have been used to assemble and study previously unexplored Ethiopian landrace germplasm, showing its potential in contributing to breeding loci for production traits and phenology traits (Mengistu et al. 2016). The uniqueness of Ethiopian durum wheat germplasm has been confirmed by several studies employing GWAS to map the loci of breeding relevance, including resistance to Septoria (Kidane et al. 2017a), grain shape and color (Alemu et al. 2020), and root system architecture (Alemu et al. 2021). Ethiopian durum wheat landraces have also been used as a benchmark to demonstrate the quantitative nature of smallholder farmers' preference traits, mapping QTL for farmers' traditional knowledge along QTL for agronomic traits (Kidane et al. 2017b). Studies on landraces enabled by genomics and GWAS on durum wheat unveiled a large reservoir of variation and potential of traditional knowledge in breeding programs (Mancini et al. 2017). Durum wheat landraces sourced from the Mediterranean region were also evaluated for the genetic basis of yield, phenology, and biomass using GWAS, revealing associations with high breeding relevance (Soriano et al. 2017). With a similar approach, landraces and cultivars sourced from environments spread across Morocco, Lebanon, Mauritania, and Senegal were evaluated by GWAS and used to identify QTL for flowering time (Gupta et al. 2020). Diversity panel-based methods also allow to assemble tetraploid wheats beyond *Triticum durum*, further expanding the allele pool to search for associations with traits of agronomic relevance (Mangini et al. 2018). In one such panel containing seven subspecies of tetraploid wheat, several QTL for grain protein content were identified, some of which showed cosegregation with nitrogen-related genes (Nigro et al. 2019). Another approach to enlarge the allelic diversity used in GWAS is to collate larger world durum wheat collections, which can efficiently identify the genetic basis of commonly targeted agronomic traits (Wang et al. 2019).

5 Mutant Populations

When not to be found in segregant populations or diversity panels, allelic variation can be created with artificial mutagenesis. Artificial mutations add to those that occur spontaneously in nature, creating new variations that may be harnessed up by breeders or used to further understand molecular processes. Although mutations leading to changes in chromosome structures often result in negative effects, breeders and researchers may leverage them to increase the number of recombination events and break disadvantageous linkages in low-recombination regions of the genome. To support crop breeding, mutations can be induced in crops by non-targeted methods such as physical irradiation (e.g., X-rays or UV) (Beyaz and Yildiz 2017), chemical mutagenesis (Jankowicz-Cieslak and Till 2016), and mutagenesis mediated by endogenous mechanisms such as T-DNA insertion and transposons (Krysan et al. 2002; Bennetzen and Wang 2014). Beneficial alleles can be further accessed and manipulated with methods such as TILLING (Targeting Induced Local Lesions in Genomes) (McCallum et al. 2000) and gene editing (Wolter et al. 2019; Nigro et al. 2023).

Chemical-induced mutations using Ethyl methanesulfonate (EMS) have been successfully utilized to obtain high-yielding varieties in durum wheat (Mehmet and Ahmet 2004). EMS durum wheat lines were recently used to select for increased waxiness (Li et al. 2020). In the Mediterranean area, mutant varieties of durum wheat that were developed by irradiation have been in cultivation for decades, showing the remarkable potential for mutagenesis breeding (Scarascia-Mugnozza et al. 1993).

At the nexus of mutagenesis in plant breeding and "-omics" is gene function analysis, which determines the basis of traits from varying and/or mutant alleles. Mutagen libraries for various crops exist, including wheat (Lyu 2017). These libraries are a resource to the genetics community as they can be interrogated and used to source specific mutants complementing gene mapping and accelerating candidate gene validation. The annotation of the durum wheat reference genome supports high-precision identification of candidate genes, providing an ideal complement to mutant libraries (Maccaferri et al. 2019). TILLING populations have been successfully developed in wheat, including durum wheat cultivar *Kronos* (Uauy et al. 2009) and *Svevo* (Bovina et al. 2014). TILLING populations are available to the community and contribute substantially to the validation of candidate genes deriving from QTL mapping (http://www.wheat-tilling.com/). For example, TILLING lines have been used to identify alleles with potential for high temperature stress exploitation, so that diagnostic markers could be designed for testing and validation of these mutant alleles in breeding material (Comastri et al. 2018). *Kronos* TILLING lines were also used for provitamin biofortification of durum, when a null mutant for lycopene epsilon cyclase, a gene in the carotenoid pathway, supported the accumulation of β-carotene (Sestili et al. 2019). A bioinformatic tool to facilitate selection of mutations discovered by TILLING is also available, facilitating the use of this resource (Fruzangohar et al. 2019).

Genome editing, CRISPR/Cas system, has revolutionized genetic engineering and has immense potential in crop improvement due to its precision and versatility (Nigro et al. 2023; Fu et al. 2024). However, it is still limited by technical design difficulties, off-target effects, and homologous recombination (Zhang et al. 2018). In wheat, genome editing is challenging as large chromosomal regions are hardly amenable to meiotic recombination and often, success is dependent on genotype (Koeppel et al. 2019). In a proof-of-concept study, genome editing, coupled with haploid induction, was utilized in bread wheat and durum wheat to generate new allelic variants for plant height (Budhagatapalli et al. 2020). Currently, genome editing in durum wheat is demonstrated in several studies. A CRISPR-Cas9 multiplexing strategy was used to transform cv *Svevo* by knocking off immunogenic associated proteins and thus reduced the amount of potential allergens associated with adverse body reactions (Camerlengo et al. 2020).

6 Beyond Quantitative Trait Locus Mapping

Although QTL mapping and reverse genetics based on mutagenesis provide the means to unravel the genetic basis of some durum wheat traits, much of the agronomic performance of the crop remains under complex and polygenic control. These traits include kernel-related traits (Sun et al. 2020), grain protein content (Ruan et al. 2021), Fusarium head blight resistance (Ruan et al. 2020b), carotenoid content (Colasuonno et al. 2019), root system architecture (Kirschner et al. 2021; Ober et al. 2021; Maccaferri et al. 2022), and grain yield (Maccaferri et al. 2008b).

In addition, traits have complex interactions so that QTL for agronomic performance are often pleiotropic. In durum wheat, QTL for root system architecture have been shown to overlap with grain yield (Canè et al. 2014). Phenological traits have common QTL with Fusarium head blight resistance (Ruan et al. 2020b). The expression of QTL also varies with interactions across genes and environment, hampering their fruitful utilization for genetic improvement (Collins et al. 2008), particularly in terms of impact of climate resilience (Snowdon et al. 2019). This limitation can be addressed, among other methods, by capturing constitutive QTL with major effects through extensive multi-environment phenotyping and able to highlight the components of the QTLome (Salvi and Tuberosa 2015; Tuberosa et al. 2021). Constitutive QTL for plant height and grain yield have been detected in durum wheat across environments with a broad range of water availability (Blum and Tuberosa 2018), and reasonably stable QTL have also been identified in wheat under drought and heat stress (Pinto et al. 2010).

The precise mapping of a QTL is crucial in that it affects its application in map-based cloning and marker-assisted breeding (Mackay 2016). Indeed, detecting a QTL is less difficult than identifying its precise position. The key factors influencing the accuracy of QTL mapping include marker density, heritability of the trait (Tuberosa 2012), QTL position on chromosome, and the mapping procedure employed (Su et al. 2013). Not all these parameters are under experimental control. Although trait heritability has a large effect on QTL mapping, it is intrinsic to the trait itself, as well as the genetic background, and thus cannot be manipulated (Su et al. 2015). However, marker density and sample size are important factors in increasing QTL mapping precision and efficiency, and both are steadily increasing in durum wheat studies (Tuberosa and Pozniak 2014; Tuberosa et al. 2021). Also due to the complexity of molecular mechanisms underlying the target traits, out of the high numbers of QTL identified in durum wheat, few have been associated with functional pathways and/or genes (Habash et al. 2014; Mangini et al. 2021; Colasuonno et al. 2021; Condorelli et al. 2022).

Bulked segregant analysis (BSA) is another technique that aims to identify genomic loci involved in traits of interest. BSA involves pooling samples of segregating individuals from a single cross (Michelmore et al. 1991). This method is especially powerful with oligogenic traits with high heritability. BSA has been used to identify new sources of resistance to leaf rust and stem rust in durum wheat (Aoun et al. 2017). A radiation hybrid population of durum wheat was analyzed using sequencing of BSA to refine the species cytoplasm specific locus (Bassi et al. 2016).

QTL mapping and GWAS can also be complemented by transcriptome analysis. Typically, individuals with contrasting traits are analyzed for the identification of candidate genes through differential expressions. Transcriptomic analysis of durum wheat under different biological conditions has been conducted targeting many conditions including cold stress (Díaz et al. 2019), nitrogen starvation (Curci et al. 2017), and heat stress (Vicente et al. 2019). The transcriptome reference assembly for durum wheat cv *Svevo* has been recently assembled (Vendramin et al. 2019), supporting these types of approaches.

Thanks to the accumulation of data and germplasm resources in durum wheat (Maccaferri et al. 2019; Mazzucotelli et al. 2020), genomics selection (GS) can be used to further accelerate breeding decisions. GS involves the selection of marker/genotypic profile at a genome-wide scale to select desirable genotypes (Hickey et al. 2017). While GS has revolutionized plant breeding, including durum wheat, it is still confounded by the complexity of quantitative traits, particularly their influence on the environment (Montesinos-López et al. 2019). Breeding programs targeting these traits carry out large-scale trials in different locations to capture G × E interaction, bringing up the cost of GS (Bhat et al. 2016). These large breeding trials may also not be representative of varieties ideal for cultivation in marginal and heterogenous environments. Thus, participatory and decentralized breeding may be used to complement these large-scale trials (van Etten et al. 2019; van Etten et al. 2023; Gessesse et al. 2023). Notably, such decentralized methods have been pioneered on durum wheat landraces in Ethiopia. There, the triadic comparison of technologies (tricot) was used to engage thousands of farmers, combining durum wheat genomics, farmers' knowledge and preferences, and environmental data to achieve a more accurate selection of locally adapted varieties (de Sousa et al. 2021).

7 Conclusion

Durum wheat is an important cereal that provides food to millions of people worldwide. Beyond its agronomic potential, durum wheat is an important genetic tool as it acts as a bridge across the wheat species, thus holding immense potential in bread wheat improvement. Over the years, durum wheat breeding has benefited from the development of molecular markers, quantitative trait loci mapping, gene cloning, new gene editing tools, availability of a draft genome, and decreasing cost of sequencing technologies, which have eventually allowed for the exploitation of genetic resources and development of improved varieties (Maccaferri et al. 2022; Peters Haugrud et al. 2024). The reduction in trait mapping time supported by genomics (Watson et al. 2018), speed breeding (Alahmad et al. 2018), and innovative genomic selection approaches as 3D breeding (de Sousa et al. 2021) and editing (Fu et al. 2024) can support further durum wheat breeding and better understanding of the complex nature of its traits, particularly those encoded by QTL.

References

Alahmad S, Dinglasan E, Leung KM et al (2018) Speed breeding for multiple quantitative traits in durum wheat. Plant Methods 141(14):1–15

Alemu A, Feyissa T, Tuberosa R et al (2020) Genome-wide association mapping for grain shape and color traits in Ethiopian durum wheat (*Triticum turgidum* ssp. *durum*). Crop J 8:757–768

Alemu A, Feyissa T, Maccaferri M et al (2021) Genome-wide association analysis unveils novel QTLs for seminal root system architecture traits in Ethiopian durum wheat. BMC Genomics 22:1–16

Aoun M, Kolmer JA, Rouse MN et al (2017) Inheritance and bulked Segregant analysis of leaf rust and stem rust resistance in durum wheat genotypes. Phytopathology 107:1496–1506

Appels R, Eversole K, Feuillet C et al (2018) Shifting the limits in wheat research and breeding using a fully annotated reference genome. Science 361:6403

Avni R, Nave M, Eilam T et al (2014) Ultra-dense genetic map of durum wheat × wild emmer wheat developed using the 90K iSelect SNP genotyping assay. Mol Breed 34:1549–1562

Avni R, Nave M, Barad O et al (2017) Wild emmer genome architecture and diversity elucidate wheat evolution and domestication. Science 357:6346

Bassi FM, Ghavami F, Hayden MJ et al (2016) Fast-forward genetics by radiation hybrids to saturate the locus regulating nuclear–cytoplasmic compatibility in *Triticum*. Plant Biotechnol J 14:1716

Ben-David R, Xie W, Peleg Z et al (2010) Identification and mapping of *PmG16*, a powdery mildew resistance gene derived from wild emmer wheat. Theor Appl Genet 121:499–510

Bennetzen JL, Wang H (2014) The contributions of transposable elements to the structure, function, and evolution of plant genomes. Annu Rev Plant Biol 65:505–530

Beyaz R, Yildiz M (2017) The use of gamma irradiation in plant mutation breeding. Plant Eng 17:33–46

Bhat JA, Ali S, Salgotra RK et al (2016) Genomic selection in the era of next generation sequencing for complex traits in plant breeding. Front Genet 7:221

Blanco A, Bellomo MP, Cenci A et al (1998) A genetic linkage map of durum wheat. Theor Appl Genet 97:721–728

Blum A, Tuberosa R (2018) Dehydration survival of crop plants and its measurement. J Exp Bot 69:975–981

Bovina R, Brunazzi A, Gasparini G et al (2014) Development of a TILLING resource in durum wheat for reverse- and forward-genetic analyses. Crop Pasture Sci 65:112–124

Brown AHD, Marshall DR (1986) Wild species as genetic resources for plant breeding. In: Proceedings of the plant breeding symposium DSIR. Agronomy Society of New Zealand, Lincoln, pp 9–15

Brunazzi A, Scaglione D, Talini RF et al (2018) Molecular diversity and landscape genomics of the crop wild relative *Triticum urartu* across the Fertile Crescent. Plant J 94(4):670–684

Bruschi M, Bozzoli M, Ratti C et al (2024) Dissecting the genetic basis of resistance to soil-borne cereal mosaic virus (SBCMV) in durum wheat by bi-parental mapping and GWAS. Theor Appl Genet 137(9):213

Budhagatapalli N, Halbach T, Hiekel S et al (2020) Site-directed mutagenesis in bread and durum wheat via pollination by cas9/guide RNA-transgenic maize used as haploidy inducer. Plant Biotechnol J 18:2376–2378

Buerstmayr M, Alimari A, Stener B, Buerstmayr H (2013) Genetic mapping of QTL for resistance to Fusarium head blight spread (type 2 resistance) in a *Triticum dicoccoides* × *Triticum durum* backcross-derived population. Theor Appl Genet 126:2825–2834

Cabas-Lühmann P, Schwember AR, Arriagada O et al (2024) Meta-QTL analysis and candidate genes for quality traits, mineral content, and abiotic-related traits in wild emmer. Front Plant Sci 15:1305196

Camerlengo F, Frittelli A, Sparks C et al (2020) CRISPR-Cas9 multiplex editing of the α-amylase/trypsin inhibitor genes to reduce allergen proteins in durum wheat. Front Sustain Food Syst 0:104

Canè MA, Maccaferri M, Nazemi G et al (2014) Association mapping for root architectural traits in durum wheat seedlings as related to agronomic performance. Mol Breed 344(34):1629–1645

Chatzav M, Peleg Z, Ozturk L et al (2010) Genetic diversity for grain nutrients in wild emmer wheat: potential for wheat improvement. Ann Bot 105:1211–1220

Civáň P, Ivaničová Z, Brown TA (2013) Reticulated origin of domesticated emmer wheat supports a dynamic model for the emergence of agriculture in the fertile crescent. PLoS One 8:e81955

Clarke J, Leisle D, Kopytko GL (1997) Inheritance of cadmium concentration in five durum wheat crosses. Crop Breed Gen Cytol 37:1722–1726

Colasuonno P, Marcotuli I, Blanco A et al (2019) Carotenoid pigment content in durum wheat (*Triticum turgidum* L. var durum): an overview of quantitative trait loci and candidate genes. Front Plant Sci 10:1347

Colasuonno P, Marcotuli I, Gadaleta A, Soriano JM (2021) From genetic maps to QTL cloning: an overview for durum wheat. Plants 10(2):315

Collard BC, Mackill DJ (2007) Marker-assisted selection: an approach for precision plant breeding in the twenty-first century. Philos Trans R Soc B Biol Sci 363:557–572

Collard BCY, Jahufer MZZ, Brouwer JB, Pang ECK (2005) An introduction to markers, quantitative trait loci (QTL) mapping and marker-assisted selection for crop improvement: the basic concepts. Euphytica 142:169–196

Collins NC, Tardieu F, Tuberosa R (2008) Quantitative trait loci and crop performance under abiotic stress: where do we stand? Plant Physiol 147:469

Comastri A, Janni M, Simmonds J et al (2018) Heat in wheat: exploit reverse genetic techniques to discover new alleles within the *Triticum durum* sHsp26 family. Front Plant Sci 9:1337

Condorelli GE, Newcomb M, Groli EL et al (2022) Genome wide association study uncovers the QTLome for osmotic adjustment and related drought adaptive traits in durum wheat. Genes 13(2):293

Cui Y, Zhang F, Xu J et al (2015) Mapping quantitative trait loci in selected breeding populations: a segregation distortion approach. Heredity 115:538–546

Curci PL, Aiese Cigliano R, Zuluaga DL et al (2017) Transcriptomic response of durum wheat to nitrogen starvation. Sci Rep 71(7):1–14

de Sousa K, van Etten J, Poland J et al (2021) Data-driven decentralized breeding increases prediction accuracy in a challenging crop production environment. Commun Biol 41(4):1–9

Dempewolf H, Baute G, Anderson J et al (2017) Past and future use of wild relatives in crop breeding. Crop Sci 57:1070–1082

Díaz ML, Soresi DS, Basualdo J et al (2019) Transcriptomic response of durum wheat to cold stress at reproductive stage. Mol Biol Rep 46:2427–2445

Distelfeld A, Fahima T (2007) Wild emmer wheat as a source for high-grain-protein genes: map-based cloning of Gpc-B1. Isr J Plant Sci 55:297–306

EU Commodity Price Dashboard (2022) Available online at: https://agriculture.ec.europa.eu/system/files/2022-04/commodity-price-dashboard_2022-04_en_0.pdf. Last Access September 18th 2023

Fatiukha A, Filler N, Lupo I et al (2020) Grain protein content and thousand kernel weight QTLs identified in a durum × wild emmer wheat mapping population tested in five environments. Theor Appl Genet 133:119–131

Fiedler JD, Salsman E, Liu Y et al (2017) Genome-wide association and prediction of grain and semolina quality traits in durum wheat breeding populations. Plant. Genome 10:3 plantgenome2017-05

Fruzangohar M, Kalashyan E, Kalambettu P et al (2019) Novel Informatic tools to support functional annotation of the durum wheat genome. Front Plant Sci 10:1244

Fu M, Liu S, Che Y et al (2024) Genome-editing of a circadian clock gene *TaPRR95* facilitates wheat peduncle growth and heading date. J Genet Genomics 51. https://doi.org/10.1016/j.jgg.2024.05.011

Gessesse CA, Nigir B, de Sousa K et al (2023) Genomics-driven breeding for local adaptation of durum wheat is enhanced by farmers' traditional knowledge. Proc Natl Acad Sci USA 120(14):e2205774119

Graziani M, Maccaferri M, Royo C et al (2014) QTL dissection of yield components and morpho-physiological traits in a durum wheat elite population tested in contrasting thermo-pluviometric conditions. Crop Pasture Sci 65(1):80–95

Gupta P, Kabbaj H, El Hassouni K et al (2020) Genomic regions associated with the control of flowering time in durum wheat. Plants 9(12):1628

Habash DZ, Baudo M, Hindle M et al (2014) Systems responses to progressive water stress in durum wheat. PLoS One 9:108431

Hickey JM, Chiurugwi T, Mackay I, Powell W (2017) Genomic prediction unifies animal and plant breeding programs to form platforms for biological discovery. Nat Genet 49:1297–1303

Huang BE, Verbyla KL, Verbyla AP et al (2015) MAGIC populations in crops: current status and future prospects. Theor Appl Genet 1286(128):999–1017

Iannucci A, Marone D, Russo MA et al (2017) Mapping QTL for root and shoot morphological traits in a durum wheat × *T. dicoccum* segregating population at seedling stage. Int J Genomics 2017:6876393

Jankowicz-Cieslak J, Till B (2016) Chemical mutagenesis of seed and Vegetatively propagated plants using EMS. Curr Protoc plant Biol 1:617–635

Jorgensen C, Luo M-C, Ramasamy R et al (2017) A high-density genetic map of wild emmer wheat from the Karaca Dağ region provides new evidence on the structure and evolution of wheat chromosomes. Front Plant Sci 0:1798

Kabbaj H, Sall AT, Al-Abdallat A et al (2017) Genetic diversity within a global panel of durum wheat(Triticum durum)landraces and modern germplasm reveals the history of alleles exchange. Front Plant Sci 8:1277

Kahiluoto H, Kaseva J, Balek J et al (2019) Decline in climate resilience of European wheat. Proc Natl Acad Sci USA 116(1):123–128

Kidane YG, Hailemariam BN, Mengistu DK et al (2017a) Genome-wide association study of Septoria tritici blotch resistance in Ethiopian durum wheat landraces. Front Plant Sci 8:1586

Kidane YG, Mancini C, Mengistu DK et al (2017b) Genome wide association study to identify the genetic base of smallholder farmer preferences of durum wheat traits. Front Plant Sci 8:1230

Kidane YG, Gesesse CA, Hailemariam BN et al (2019) A large nested association mapping population for breeding and quantitative trait locus mapping in Ethiopian durum wheat. Plant Biotechnol J 17:1380

Klindworth DL, Salsman E, Underdahl J et al (2023) Linkage drag analysis in three Aegilops speltoides introgressions carrying Sr47 in modern durum and hard red spring wheat germplasm. Theor Appl Genet 136:168. https://link.springer.com/article/10.1007/s00122-023-04409-8

Kirschner GK, Rosignoli S, Guo L et al (2021) *ENHANCED GRAVITROPISM 2* encodes a STERILE ALPHA MOTIF-containing protein that controls root growth angle in barley and wheat. Proc Natl Acad Sci USA 118(35):e2101526118

Klymiuk V, Fatiukha A, Huang L et al (2019) Durum wheat as a bridge between wild emmer wheat genetic resources and bread wheat. Appl Genet Genomic Res Cereal:201–230

Koeppel I, Hertig C, Hoffie R, Kumlehn J (2019) Cas endonuclease technology-a quantum leap in the advancement of barley and wheat genetic engineering. Int J Mol Sci 11:2647

Krysan P, Young JC, Jester PJ et al (2002) Characterization of T-DNA insertion sites in *Arabidopsis thaliana* and the implications for saturation mutagenesis. OMICS 6:163–174

Kthiri D, Loladze A, N'Diaye A et al (2019) Mapping of genetic loci conferring resistance to leaf rust from three globally resistant durum wheat sources. Front Plant Sci 1247

Langridge P, Alaux M, Almeida NF et al (2022) Meeting the challenges facing wheat production: the strategic research agenda of the global wheat initiative. Agron 12:2767

Li G, Fang T et al (2009) Molecular identification of a new powdery mildew resistance gene Pm41 on chromosome 3BL derived from wild emmer (*Triticum turgidum* var. *dicoccoides*). Theor Appl Genet 119:531–539

Li S, Zhong X, Zhang X et al (2020) Production of waxy tetraploid wheat (*Triticum turgidum* sp. *durum* L.) by EMS mutagenesis. Genet Resour Crop Evol 67:433–443

Li H, Zhu L, Fan R et al (2024) A platform for whole-genome speed introgression from *Aegilops tauschii* to wheat for breeding future crops. Nat Protoc 19(2):281–312

Liu CY, Shepherf KW, Gras PW (1994) Grain yield and quality characteristics grain yield and quality characteristics of chromosome ID and IB substitution lines

in durum wheat and their F2-derived progeny lines. I. Comparisons among the tetraploid phenotypes. J Cereal Sci 0(1):23–32
Liu T, Wu L, Gan X et al (2020) Mapping quantitative trait loci for 1000-grain weight in a double haploid population of common wheat. Int J Mol Sci 21(11):3960
Lyu J (2017) Reverse genetics: wheat solution. Nat Plants 32(3):1–1
Maccaferri M, Mantovani P, Tuberosa R et al (2008a) A major QTL for durable leaf rust resistance widely exploited in durum wheat breeding programs maps on the distal region of chromosome arm 7BL. Theor Appl Genet 117:1225–1240
Maccaferri M, Sanguineti MC, Corneti S et al (2008b) Quantitative trait loci for grain yield and adaptation of durum wheat (*Triticum durum* Desf.) across a wide range of water availability. Genetics 178:489–511
Maccaferri M, Sanguineti MC, Demontis A et al (2011) Association mapping in durum wheat grown across a broad range of water regimes. J Exp Bot 62:409–438
Maccaferri M, Ricci A, Salvi S et al (2015) A high-density, SNP-based consensus map of tetraploid wheat as a bridge to integrate durum and bread wheat genomics and breeding. Plant Biotechnol J 13:648–663
Maccaferri M, Harris NS, Twardziok SO et al (2019) Durum wheat genome highlights past domestication signatures and future improvement targets. Nat Gen 51(5):885–895
Maccaferri M, Bruschi M, Tuberosa R (2022) Sequence-based marker assisted selection in wheat. In: Wheat improvement: food security in a changing climate. Springer International Publishing, Cham, pp 513–538
Mackay TFC (2001) The genetic architecture of quantitative traits. Annu Rev Genet 35:303–339
Mackay TFC (2016) Quantitative genetic variation. Encycl Evol Biol:372–376
Mago R, Chen C, Xia X et al (2022) Adult plant stem rust resistance in durum wheat Glossy Huguenot: mapping, marker development and validation. Theor Appl Genet 135:1541–1550
Mancini C, Kidane YGYG, Mengistu DKDK et al (2017) Joining smallholder farmers' traditional knowledge with metric traits to select better varieties of Ethiopian wheat. Sci Rep 7:13076
Mangini G, Gadaleta A, Colasuonno P et al (2018) Genetic dissection of the relationships between grain yield components by genome-wide association mapping in a collection of tetraploid wheats. PLoS One 13:e0190162
Mangini G, Blanco A, Nigro D et al (2021) Candidate genes and quantitative trait loci for grain yield and seed size in durum wheat. Plants 10:312
Mantovani P, Maccaferri M, Sanguineti MC et al (2008) An integrated DArT-SSR linkage map of durum wheat. Mol Breed 22:629–648
Marcotuli I, Gadaleta A, Mangini G et al (2017) Development of a high-density SNP-based linkage map and detection of QTL for β-Glucans, protein content, grain yield per spike and heading time in durum wheat. Int J Mol Sci 18(6):1329
Mazzucotelli E, Sciara G, Mastrangelo AM et al (2020) The global durum wheat panel (GDP): an international platform to identify and exchange beneficial alleles. Front Plant Sci 0:2036
McCallum CM, Comai L, Greene EA, Henikoff S (2000) Targeting induced local lesions IN genomes (TILLING) for plant functional genomics. Plant Physiol 123:439–442
Mehmet AS, Ahmet Y (2004) Induced mutations for yield and its components in durum wheat (*Triticum durum* Desf.). Food Agric Environ 2:285–290
Mengistu D et al (2016) High-density molecular characterization and association mapping in Ethiopian durum wheat landraces reveals high diversity and potential for wheat breeding. Plant Biotechnol J 14:1800–1812
Merchuk-Ovnat L, Barak V, Fahima T et al (2016) Ancestral QTL alleles from wild emmer wheat improve drought resistance and productivity in modern wheat cultivars. Front Plant Sci 7:452
Michelmore RW, Paran I, Kesseli RV (1991) Identification of markers linked to disease-resistance genes by bulked segregant analysis: a rapid method to detect markers in specific genomic regions by using segregating populations. Proc Natl Acad Sci USA 88:9828–9832
Milner SG, Maccaferri M, Huang BE et al (2016) A multiparental cross population for mapping QTL for agronomic traits in durum wheat (*Triticum turgidum* ssp. durum). Plant Biotechnol J 14:735–748
Mohammadi M, Mohammadi R (2024) Potential of tetraploid wheats in plant breeding: a review. Plant Sci 112155
Montesinos-López OA, Montesinos-López A, Tuberosa R et al (2019) Multi-trait, multi-environment genomic prediction of durum wheat with genomic best linear unbiased predictor and deep learning methods. Front Plant Sci 10:479328
Nachit MM, Elouafi I, Pagnotta A et al (2001) Molecular linkage map for an intraspecific recombinant inbred population of durum wheat (*Triticum turgidum* L. var. *durum*). Theor Appl Genet 102:177–186
Nigro D, Gadaleta A, Mangini G et al (2019) Candidate genes and genome-wide association study of grain protein content and protein deviation in durum wheat. Planta 249:1157–1175
Nigro D, Smedley MA, Camerlengo F, Hayta S (2023) Using gene editing strategies for wheat improvement. In: A roadmap for plant genome editing. Cham, Springer Nature Switzerland, pp 183–201
Niu Z, Jiang A, Hammad WA et al (2014) Review of doubled haploid production in durum and common wheat through wheat × maize hybridization. Plant Breed 133:313–320
Ober ES, Alahmad S, Cockram J et al (2021) Wheat root systems as a breeding target for climate resilience. Theor Appl Genet 134:1645–1662
Peleg Z, Saranga Y, Suprunova T et al (2008) High-density genetic map of durum wheat × wild emmer wheat based on SSR and DArT markers. Theor Appl Genet 117:103–115

Peleg Z, Cakmak I, Levent AE et al (2009a) Quantitative trait loci conferring grain mineral nutrient concentrations in durum wheat 3 wild emmer wheat RIL population. Theor Appl Genet 119:353–369

Peleg Z, Fahima T, Krugman T et al (2009b) Genomic dissection of drought resistance in durum wheat × wild emmer wheat recombinant inbreed line population. Plant Cell Environ 32:758–779

Peters Haugrud AR, Achilli AL, Martínez-Peña R, Klymiuk V (2024) Future of durum wheat research and breeding: insights from early career researchers. The Plant Genome 2024:e20453

Pinto R, Reynolds MP, Mathews KL et al (2010) Heat and drought adaptive QTL in a wheat population designed to minimize confounding agronomic effects. Theor Appl Genet 121:1001–1021

Prat N, Guilbert C, Prah U et al (2017) QTL mapping of Fusarium head blight resistance in three related durum wheat populations. Theor Appl Genet 130:13–27

Rahman S, Islam S, Yu Z et al (2020) Current Progress in understanding and recovering the wheat genes lost in evolution and domestication. Int J Mol Sci 21:1–19

Roncallo PF, Cervigni GL, Jensen C et al (2012) QTL analysis of main and epistatic effects for flour color traits in durum wheat. Euphytica 185:77–92

Rong JK, Millet E, Manisterski J, Feldman M (2000) A new powdery mildew resistance gene: introgression from wild emmer into common wheat and RFLP-based mapping. Euphytica 115:121–126

Royo C, Maccaferri M, Álvaro F et al (2010) Understanding the relationships between genetic and phenotypic structures of a collection of elite durum wheat accessions. F Crop Res 119:91–105

Ruan Y, Yu B, Knox RE et al (2020a) High density mapping of quantitative trait loci conferring gluten strength in Canadian durum wheat. Front Plant Sci 0:170

Ruan Y, Zhang W, Knox RE et al (2020b) Characterization of the genetic architecture for Fusarium head blight resistance in durum wheat: the complex Association of Resistance, flowering time, and height genes. Front Plant Sci 0:2071

Ruan Y, Yu B, Knox RE et al (2021) Conditional mapping identified quantitative trait loci for grain protein concentration expressing independently of grain yield in Canadian durum wheat. Front Plant Sci 0:447

Salamini F, Özkan H, Brandolini A et al (2002) Genetics and geography of wild cereal domestication in the near east. Nat Rev Genet 3:429–441

Sall AT, Chiari T, Legesse W et al (2019) Durum wheat (*Triticum durum* Desf.) origin, cultivation, and potential expansion in sub-Saharan Africa. Agronomy 9(5):263

Salvi S, Tuberosa R (2015) The crop QTLome comes of age. Curr Opin Biotechnol 32:179–185

Scarascia-Mugnozza GT, D'Amato F, Avanzi S (1993) Mutation breeding for durum wheat (*Triticum turgidum* ssp. *durum* Desf.) improvement in Italy. In: Plant mutation breeding for crop improvement

Scott MF, Ladejobi O, Amer S et al (2020) Multi-parent populations in crops: a toolbox integrating genomics and genetic mapping with breeding. Hered 1256(125):396–416

Sestili F, Garcia-Molina MD, Gambacorta G et al (2019) Provitamin a biofortification of durum wheat through a TILLING approach. Int J Mol Sci 20:5703

Snowdon RJ, Stahl A, Wittkop B et al (2019) Reduced response diversity does not negatively impact wheat climate resilience. Proc Natl Acad Sci USA 116:10623

Soriano JM, Malosetti M, Roselló M et al (2017) Dissecting the old Mediterranean durum wheat genetic architecture for phenology, biomass and yield formation by association mapping and QTL meta-analysis. PLoS One 12:e0178290

Soriano JM, Colasuonno P, Marcotuli I, Gadaleta A (2021) Meta-QTL analysis and identification of candidate genes for quality, abiotic and biotic stress in durum wheat. Sci Rep 111, 11:1–15

Su C, Qiu X, Ji Z (2013) Study of strategies for selecting quantitative trait locus mapping procedures by computer simulation. Mol Breed 314(31):947–956

Su C, Liu ZB, Li YG (2015) Influence of various quantitative trait loci (QTL) mapping methods on the mapping accuracy under varying heritability levels. Genet Mol Res 14:13003–13012

Sukumaran S, Reynolds MP, Sansaloni C (2018) Genome-wide association analyses identify QTL hotspots for yield and component traits in durum wheat grown under yield potential, drought, and heat stress environments. Front Plant Sci 0:81

Sun L, Huang S, Sun G et al (2020) SNP-based association study of kernel architecture in a worldwide collection of durum wheat germplasm. PLoS One 15:e0229159

Suprayogi Y, Pozniak C, Clarke F et al (2009) Identification and validation of quantitative trait loci for grain protein concentration in adapted Canadian durum wheat populations. Theor Appl Genet 119:437–448

Tabbita F, Pearce S, Barneix AJ (2017) Breeding for increased grain protein and micronutrient content in wheat: ten years of the GPC-B1 gene. J Cereal Sci 73:183–191

Talini RF, Brandolini A, Miculan M et al (2020) Genome-wide association study of agronomic and quality traits in a world collection of the wild wheat relative *Triticum urartu*. Plant J 102(3):555–568

Tuberosa R (2012) Phenotyping for drought tolerance of crops in the genomics era. Front Phys 3:347

Tuberosa R, Pozniak C (2014) Durum wheat genomics comes of age: introduction to the special issue on durum wheat genomics. Mol Breed 34:1527–1530

Tuberosa R, Cattivelli L, Ceriotti A et al (2021) Editorial: proceedings of FSTP3 congress–a sustainable durum wheat chain for food security and healthy lives. Front Plant Sci 12:675510

Uauy C, Paraiso F, Colasuonno P et al (2009) A modified TILLING approach to detect induced mutations in tetraploid and hexaploid wheat. BMC Plant Biol 9:1–14

van Etten J, de Sousa K, Aguilar A et al (2019) Crop variety management for climate adaptation supported by citizen science. Proc Natl Acad Sci USA 116:4194–4199

van Etten J, de Sousa K, Cairns J et al (2023) Data-driven approaches can harness crop diversity to address heterogeneous needs for breeding products. Proc Natl Acad Sci USA 120(14):e2205771120

Vendramin V, Ormanbekova D, Scalabrin S et al (2019) Genomic tools for durum wheat breeding: De novo assembly of *Svevo* transcriptome and SNP discovery in elite germplasm. BMC Genomics 20:1–16

Vicente R, Bolger AM, Martínez-Carrasco R et al (2019) De novo transcriptome analysis of durum wheat flag leaves provides new insights into the regulatory response to elevated CO2 and high temperature. Front Plant Sci 10:1605

Wang S, Xu S, Chao S et al (2019) (2019) a genome-wide association study of highly heritable agronomic traits in durum wheat. Front Plant Sci 10:919

Watson A, Ghosh S, Williams MJ et al (2018) Speed breeding is a powerful tool to accelerate crop research and breeding. Nat Plants 41(4):23–29

Wolter F, Schindele P, Puchta H (2019) Plant breeding at the speed of light: the power of CRISPR/Cas to generate directed genetic diversity at multiple sites. BMC Plant Biol 191(19):1–8

Xu L, Wang M, Cheng P et al (2013) Molecular mapping of Yr53, a new gene for stripe rust resistance in durum wheat accession PI 480148 and its transfer to common wheat. Theor Appl Genet 126:523–533

Yan J, Xue WT, Yang RZ et al (2018) Quantitative trait loci conferring grain selenium nutrient in durum wheat × wild emmer wheat RIL population. Czech J Genet Plant Breed 54(2018):52–58

Yaniv E, Raats D, Ronin Y et al (2015) Evaluation of marker-assisted selection for the stripe rust resistance gene *Yr15*, introgressed from wild emmer wheat. Mol Breed 35:1–12

Zaïm M, Kabbaj H, Kehel Z et al (2020) Combining QTL analysis and genomic predictions for four durum wheat populations under drought conditions. Front Genet 11:316

Zamir D (2001) Improving plant breeding with exotic genetic libraries. Nat Rev Genet 212(2):983–989

Zhang W et al (2008) QTL analysis of pasta quality using a composite microsatellite and SNP map of durum wheat. Theor Appl Genet 117:1361–1377

Zhang H, Zhang L, Wang C et al (2016) Molecular mapping and marker development for the Triticum dicoccoides-derived stripe rust resistance gene *YrSM139-1B* in bread wheat cv. Shaanmai 139. Theor Appl Genet 129:369–376

Zhang Y, Massel K, Godwin ID, Gao C (2018) Applications and potential of genome editing in crop improvement. Genome Biol 191(19):1–11

Zhao M, Leng Y, Chao S et al (2018) Molecular mapping of QTL for Fusarium head blight resistance introgressed into durum wheat. Theor Appl Genet 1319(131):1939–1951

Evolution of TILLING as a Useful Tool for Durum Wheat Functional Studies and Breeding Programs

Ermelinda Botticella, Francesco Camerlengo, Francesco Sestili, and Samuela Palombieri

Abstract

The development of improved crop varieties remains a significant challenge in agriculture, especially in light of climate change, increasing world population, and regional conflicts. To address these challenges, it is crucial to understand the genetic basis of desirable traits and harness genetic variability through breeding programs. Various strategies, including traditional breeding methods and modern biotechnological approaches, have been employed to generate genetic diversity in crops. Among these approaches, Targeting Induced Local Lesions IN Genomes (TILLING) is a reverse genetic tool that combines the use of mutagenesis and molecular techniques for the identification of point mutations in target genes. TILLING has been successfully applied to various plant species, facilitating the creation of novel allelic series and the study of gene function.

In durum wheat, TILLING has been widely used to improve yield and nutritional value and give biotic and abiotic stress resilience. In addition, the TILLING platform has recently been employed to investigate promoter mutations and their effects on gene expression and phenotype modulation. Methodological advances, such as next-generation sequencing (NGS), have improved mutation discovery and reduced costs, making this approach accessible and efficient for durum wheat breeding programs.

In conclusion, TILLING is a valuable tool for durum wheat improvement, offering insights into gene function, generating genetic diversity, and addressing challenges in agricultural sustainability. Continued research and technological innovations in TILLING hold promise for further enhancing crop resilience, nutritional quality, and yield in durum wheat and other major crops.

Keywords

TILLING · Durum wheat · Nutritional value · Yield · Biotic and abiotic stress

E. Botticella
Institute of Sciences of Food Production (ISPA), National Research Council (CNR), Lecce, Italy

F. Camerlengo
Department of Agricultural and Food Sciences, University of Bologna, Bologna, Italy

F. Sestili · S. Palombieri (✉)
Department of Agriculture & Forest Sciences, University of Tuscia, Viterbo, Italy
e-mail: palombieri@unitus.it

R. Tuberosa et al. (eds.), *The Durum Wheat Genome*, Compendium of Plant Genomes,
https://doi.org/10.1007/978-3-031-84689-2_8

1 Introduction

The development of improved crop varieties has been and still is one of the major challenges in agriculture. Today, climate change, increasing world population, and regional conflicts call for targeted breeding strategies to mitigate their negative effects on crop production and quality. To modify the traits of interest, it is necessary to know the genes that are responsible and understand their impact on the phenotype. The most important components of the success of breeding programs are the genetic variability and inheritance of traits.

Various strategies and techniques can be employed to generate new genetic variability, which is crucial for both functional studies and breeding programs. Different approaches can be used to increase or utilize genetic diversity, including traditional breeding (crosses), mutagenesis, transgenesis, cisgenesis, and genome editing. In this context, chemical mutagenesis is the most used tool to introduce functional polymorphism in the genome of agriculturally relevant species. This approach offers several advantages, including its applicability to all plant species, low costs, and the generation of small mutations (often single nucleotide polymorphisms) capable of producing allelic series (Sikora et al. 2011).

Two main strategies can be followed to study mutagenized plants, known as forward and reverse genetics (Szurman-Zubrzycka et al. 2023). The forward strategy begins with the observation of an interesting phenotype and then attempts to determine the mutated gene responsible for this phenotype. The numerous sequencing programs of plant genomes increased the availability of gene sequences and changed the approach for the studies of gene function. This approach is known as reverse genetics and starts with the detection of a severe mutation within the gene of interest and then the observation of the phenotype caused by the mutation (Griffiths et al. 2004).

Since 2000, the development of a new reverse genetics approach defined as Targeting Induced Local Lesions IN Genomes (TILLING) has facilitated the study of gene function and the creation of new allelic series through the identification of point variants in mutagenized populations (McCallum et al. 2000a; McCallum et al. 2000b). TILLING consists of a platform useful for the screening of induced mutants in target genes and combines the use of traditional chemical mutagenesis with molecular tools to select the induced mutations. In contrast to transgenic methods, one of the most important advantages is that the plants obtained by TILLING are not considered genetically modified (Okabe et al. 2011).

Initially tested on *Arabidopsis thaliana*, the application of TILLING was later extended to numerous plant species including rice, wheat, barley, maize, tomato, tobacco, and lotus (Till et al. 2004; Wu et al. 2005; Slade et al. 2005; Caldwell et al. 2004; Menda et al. 2004; Reddy et al. 2012; Horst et al. 2007; Talamè et al. 2008). Regarding durum wheat, several TILLING mutants were obtained, addressing different traits such as qualitative and agronomical aspects controlled by QTLs (Salvi and Tuberosa 2015) and making this approach suitable to create new genetic variability useful for breeding and functional studies (Table 1).

2 Tilling Methodology

The steps to create a TILLING platform are as follows: (1) mutagen treatment of seeds or pollen; (2) genomic DNA extraction and pooling; (3) screening to select mutations in genes of interest (choice of screening method).

Different chemical mutagen agents can be applied, but the most used ones are ethyl methane sulfonate (EMS) and sodium azide (NaN_3) responsible for random transitions from G/C to A/T in plant genomes. Their application on seeds or pollen permits obtaining large mutant populations, characterized by an elevated density of induced point mutations spread throughout the genome (Sikora et al. 2011). The density of induced mutations in the genome depends on the target tissue, mutagen concentration, and exposure time. The optimal conditions of mutagen application must be determined experimentally

Table 1 Overview of different TILLING approaches in durum wheat

Topic	Variety	Treatment	Strategy/tools	Trait	Gene	Reference
Nutritional quality	Kronos	EMS	CEL1 Li-Cor technique	Starch modification	*GBSSI*	Slade et al. (2005)
	Kronos	EMS	CEL1 Li-Cor technique	Starch modification	*SBEIIa*	Slade et al. (2012)
	Kronos	EMS	CJE and on-denaturing polyacrylamide detection method	Starch modification	*SBEIIa/b*	Hazard et al. (2012)
	Svevo	EMS	HRM	Starch modification	*SBEIIa*	Sestili et al. (2015)
	Marco Aurelio	EMS	DHPLC, (CelI)/agarose gel	Carotenoid pathway	*β-LCY-6; ε-LCY-3*	Colasuonno et al. (2016)
	Kronos	EMS	In silico exome capture (Krasileva et al. 2017)	Carotenoid Biosynthesis	*LCYE*	Richaud et al. (2018)
	Kronos	EMS	In silico exome capture (Krasileva et al. 2017)	Carotenoid	*LCYE*	Sestili et al. (2019)
	Kronos	EMS	In silico exome capture (Krasileva et al. 2017)	Carotenoid Biosynthesis	*βHYD1*	Garcia Molina et al. (2021)
	Kronos	EMS	In silico exome capture (Krasileva et al. 2017)	Carotenoid Biosynthesis	*LCYE/ HYD2*	Yu et al. (2022)
	Kronos	EMS	In silico exome capture (Krasileva et al. 2017)	Mineral content	*MRP3*	Frittelli et al. (2023a)
	Kronos	EMS	In silico exome capture (Krasileva et al. 2017)	Mineral content	*IPK1*	Frittelli et al. (2023b)
	Kronos	EMS	dsDNA cleavage enzyme, fragment Analyzer (AATI)	Immunogenicity	*DME*	Wen et al. (2022)
	Kronos	EMS	In silico exome capture (Krasileva et al. 2017)	Immunogenicity	*DRE2*	Wen et al. (2022)
Yield	Kronos	EMS	CJE and on-denaturing polyacrylamide detection method	Seed size	*GW2*	Simmonds et al. (2016)
	Kronos	EMS	In silico exome capture (Krasileva et al. 2017)	Tiller number	*TB1*	Volkman et al. (2022)
	Kronos	EMS	Mapping strategy in silico exome capture (Krasileva et al. 2017)	Plant height	*RhtA1/B1*	Mo et al. (2018)
	Divide	EMS	Forward screen, sequence	Plant height	*RhtA1*	Brown et al. (2022)
	Kronos	EMS	In silico exome capture (Krasileva et al. 2017)	Plant height	*PLATZ1*	Zhang et al. (2023a, b)
	Kronos	EMS	In silico exome capture (Krasileva et al. 2017)	Amyloplast size	*PARC6*	Esch et al. (2023)
Abiotic and biotic stress	CHAM-1	EMS	Forward screening	Drought stress	–	Kahama et al. (2019)
	Kronos CHAM-1	EMS	In silico exome capture (Krasileva et al. 2017)	Heat stress	*Hsp26*	Comastri et al. (2018)
	Kronos	EMS	In silico exome capture (Krasileva et al. 2017)	Flowering time	*VRN1*	Chen and Dubcovsky (2012)
	Kronos	EMS	In silico exome capture in promotor region	Flowering time	*VRN1-A*	Zhang et al. (2023a)
	Kronos	EMS	Forward screening, mapping-by-sequencing approach	Chlorosis	*YES-1*	Harrington et al. (2019)
	Kronos	EMS	In silico exome capture (Krasileva et al. 2017)	Carotenoid biosynthesis	*βHYD1/2*	Bekkering et al. (2023)
	Kronos	EMS	In silico exome capture (Krasileva et al. 2017)	Resistance to powdery mildew	*Mlo*	Ingvardsen et al. (2019)

and depend on species, genotype, and ploidy level (Parry et al. 2009). In general, the optimal dose for creating a TILLING population produces approximately 50% lethality in the treated seeds (Talebi et al. 2012) and produces high density point mutations sufficient to generate a series of allelic mutations in all genes.

To avoid chimeric plants, genomic DNA is extracted from leaves of the M_2 generation if the treatment has been carried out on the seeds, or M_1 generation if the pollen has been mutagenized. After purification and accurate quantitation, the genomic DNA is organized into pools (from 2 to 10 samples) to reduce the screening time and expenses (Gilchrist et al. 2006). In addition, the presence of SNPs in the DNA pools aids in the formation of heteroduplex molecules, which are easy to detect using different analytical techniques. Among these, the first technique used by the inventors of TILLING was high-performance liquid denaturing chromatography (dHPLC), a chromatographic technique that detects base pair changes using heteroduplex analysis (McCallum et al. 2000b). This method has been successfully used on a mutagenized *Arabidopsis* population, but it has a limitation: the low number of samples that can be analyzed simultaneously.

The same research team has developed a screening technique based on the use of an S1 nuclease, CEL1 (Colbert et al. 2001). This endonuclease was suitable for detecting heterozygous polymorphisms in DNA pools because it cleaves mismatches of all types into a double-stranded DNA molecule. The amplicons are obtained using two gene-specific primers carrying different dye labels, and after incubation with CEL1, the amplicons are separated by denaturing gel electrophoresis and detected in two separate channels by LI-COR scanners.

This strategy provides the size of the fragments produced by the cut, giving an approximate location of the mutation within the amplicon with a sensitivity to detect the cleavage product generated by CEL1 in an eightfold pool. The technique was the most common method of mutation discovery for the first two decades of TILLING research. To reduce the time and cost of mutation screening, another more sensitive and faster method based on high-resolution melting (HRM) was developed (Wittwer et al. 2003).

This technique is based on the study of melting profiles produced by the loss of fluorescent from the intercalating dyes of a PCR product. After amplification of a specific target, followed by denaturation and re-annealing of the double-stranded DNA (dsDNA) product, the melting curves are generated by monitoring the fluorescence of a saturation dye that does not inhibit PCR. This analysis produces distinctive DNA melting curve profiles that allow for the differentiation of allelic variants also through subtle and small sequence variations (Reed et al. 2007). HRM analysis can be performed immediately after PCR in the same reaction tube using highly accurate optical detection systems, such as the Rotor-Gene (Qiagen, Hilden, Germany) or the Lightscanner (Idaho Technology Inc., Salt Lake City, USA), which allow for detection of single base mismatches in amplicons up to 500 bp. This method does not require the use of the electrophoresis gel, reducing the detection time, and increasing the resolution (Bovina et al. 2014; Sestili et al. 2015).

After the identification of the sample with the mutation, amplicon sequencing is needed to identify sequence variants because the number and position of mutations within the amplicon cannot be identified using this method alone.

Next-generation sequencing (NGS) platforms have been used in mutation detection (Rigola et al. 2009) and to improve the workflow of both TILLING and EcoTILLING (Tsai et al. 2011; Irshad et al. 2019). This strategy was first developed by the Comai lab and allows to directly determine the sequence of each mutation and its effect on the function of the protein without the use of marked primers, expensive fluorophores, or restriction enzymes, reducing the possibility of false positives (Taheri et al. 2017). TILLING by sequencing can be applied directly to small genomes, through various sample pooling strategies (Krothapalli et al., 2013), or to large genomes via genome-wide resequencing of coding regions (exome capture), enabling efficient SNP discovery and facilitating comprehensive reverse genetic screening (King et al. 2015; Krasileva et al. 2017; Mo et al. 2018). The limitation of this technique is that a genomic reference with more complete coverage of homologues is required for accurate mutation calling (Taheri et al. 2017). Nevertheless, the exome capture technique has been successfully applied by the group of Dubcovsky (Krasileva et al. 2017) turning TILLING into an in silico approach, cheap and easily accessible by all laboratories. The information obtained from this work is publicly available (see paragraph 1.3), allowing the scientific community to identify a mutation in a target gene and to obtain the specific mutant. Several groups have used this tool to identify mutants of interest and to develop durum wheat lines with better nutritional characteristics, higher yield, and resistance to different biotic and abiotic stresses (Table 1).

Over the years, further improvements to the TILLING strategy have been made by individual laboratories to address many potential bottlenecks and to use the best strategy for the species of interest by developing new methodologies such us iTILLING, DEco-TILLING, Self-EcoTILLING, NEATTILL (Bush and Krysan 2010; Garvin and Gharrett 2007; Wang et al. 2008; Sreelakshmi et al. 2010). In *Arabidopsis*, iTILLING aims to reduce the efforts needed to manage the mutant lines by collecting M_2 seeds in bulk from the mutagenized population of plants. All the steps (seedling growth, tissue collection, and DNA extraction) are performed in 96-well format and the mutations are then identified using HRM-curve analysis of gene-specific polymerase chain reaction products. Only plants carrying mutations of interest are then transferred to soil to complete the mutant selection in less than four months (Bush and Krysan 2010).

DEco-TILLING (double stranded Eco-TILLING) is a low-cost system that can be applied for SNP discovery by reducing ascertainment bias (Garvin and Gharrett 2007). The innovation of this method is to involve complete digests of both strands using CEL1 at all mispaired sites in a DNA heteroduplex. The TaqMan, TM-Shift, or an equivalent assay are used to conduct subsequent survey of variability, which is both more rapid and cost-effective. In durum wheat, Denaturing High Performance Liquid Chromatography (dHPLC) technology in melting temperature conditions was used to detect the heteroduplex with more sensitivity than cleavage assay based on agarose gel (Colasuonno et al. 2016).

3 Tilling Platforms in Cereal Species

Since its inception as an alternative to insertional mutagenesis in *Arabidopsis thaliana* (McCallum et al. 2000b), the TILLING approach has gained widespread popularity in plant research for generating new genetic diversity. To date, TILLING populations and associated platforms provide valuable resources for researchers and breeders aimed at improving cereal characteristics through targeted mutagenesis. Mutant populations of cereal crops are essential for comprehensive phenotypic characterization, crucial for both forward and reverse genetic studies, as well as for breeding purposes. It is worth noting that making a TILLING platform publicly accessible necessitates the creation of a comprehensive database where all data regarding phenotypes, identified mutations, and seed availability are meticulously documented. However, TILLING is not without its challenges. The process involves random mutagenesis, which can lead to background mutations, many of which potentially deleterious or severely affecting the germination rate. Additionally, a high-throughput screening method is required for mutation discovery across large populations. The advent of NGS technologies, coupled with various sample pooling strategies, has significantly enhanced mutation discovery, leading to the expansion of TILLING by sequencing.

Several TILLING projects have been specifically developed for cereals, including rice, maize, barley, durum, and bread wheat. Rice, being not only a vital food source but also a model system in functional genomics, witnessed the creation of its first TILLING platform in the indica variety IR64, which is widely cultivated in Southeast Asia (Wu et al. 2005). To facilitate access to mutant lines, a comprehensive database was established within the International Rice Information System (IRIS; http://www.iris.irri.org), documenting the phenotypes and gene functions uncovered by researchers.

The efficacy of the mutagenesis process was explored in two mutagenized rice populations generated using chemical mutagens EMS and a combination of sodium azide plus methyl-nitrosourea (Az-MNU). The mutation density was estimated to be one mutation per 300 kb for the EMS population and one mutation per 200 kb for the Az-MNU population, a density higher than previously reported in rice populations (Till et al. 2007). Serrat et al. (2014) introduced a novel TILLING method for rice, employing EMS mutagenesis in mature seed-derived calli, significantly accelerating the process of obtaining mutant populations for genetic analysis.

The researchers achieved a high mutagenesis rate, with one mutation every 451 kb, by screening plants for two senescence-related genes in 2400 individuals from a mutant population of 6912. In 2013, mutagenesis was used to generate the first rice TILLING platform of a European temperate japonica accession composed of 1860 M_2 EMS-mutagenized lines from the cv. Volano. The mutation density was one mutation every 373 kb, demonstrating the effectiveness of this approach for targeted improvement of European temperate rice germplasm (Casella et al. 2013).

The diploid and smaller genome of barley compared with other cereal species genomes makes it a good model for genetic studies in cereal crops. TILLING populations and platforms were also developed in barley in the last 20 years (Caldwell et al. 2004). A sodium azide mutagenized barley population of cv. Morex, named TILLMore, was developed by Talamè et al. (2008), with a mutation density of approximately one mutation every 374 kb. Another barley TILLING population, created using NaN_3 in the Danish variety Lux, comprises approximately 10,000 lines, with the authors estimating 2000 mutations per barley genome (Lababidi et al. 2009). The use of EMS to induce mutations and augment genetic diversity led to the development of a TILLING population from the two-rowed malting cultivar "Barke," consisting of 10,279 M_2 mutants, offering a high-throughput gene discovery platform. The average mutation frequency was found to be approximately one mutation every 0.5 Mb on average (Gottwald et al. 2009).

The HorTILLUS population, derived from the spring barley cv. Sebastian, was developed using two chemical mutagens, NaN_3 and MNU, encompassing over 9600 M_2 plants. The average mutation density in the HorTILLUS population was estimated at one mutation every 477 kb, making it the largest and best-characterized TILLING population reported for barley to date (Szurman-Zubrzycka et al. 2018). In 2019, Schreiber et al. presented a highly mutagenized barley TILLING population of the cv. Golden Promise, significant for barley transformation and functional genomics. The combination of a mutagenized barley population with EMS and a high-throughput screening approach (whole exome capture sequencing and pooled amplicon sequencing) led to the discovery of a high mutation density, with one mutation every 154 kb (Schreiber et al. 2019).

The establishment of high-throughput TILLING platforms for mutant identification relies not only on the mutagen treatment but also on the mutation screening method used for genotype analysis. A population of EMS-induced mutations, consisting of 8525 M_3 lines in the barley landrace Hatiexi, was developed. This was complemented by an amplicon-seq pipeline that utilized a three-dimensional DNA pooling strategy along with high-throughput sequencing of the pooled amplicons. This approach allowed to obtain a cost-effective TILLING platform, enabling the identification of mutations in an efficient and high-throughput manner. The mutation frequency in the population varied from 1.51 to 4.09 mutations per megabase (Jiang et al. 2022).

FIND-IT (Fast Identification of Nucleotide variants by droplet DigITal PCR) is a novel tool facilitating the rapid identification and isolation of specific genetic variants using large-scale sample pooling of mutagenized individuals combined with droplet digital PCR (ddPCR) (Knudsen et al. 2022). The effectiveness of this method has been validated by screening variant libraries comprising 500,000 barley individuals and isolating more than 125 targeted barley gene knockout lines, which could expedite functional gene analysis.

The method was validated with barley, but it is applicable to any other organism. A barley variant, identified using FIND-IT, has been registered as a new cultivar and is in commercial pilot trials within less than four years of variant identification (Knudsen et al. 2022).

Mutation density varies among different crops and is influenced by several factors, including the type of mutagen used, the specific screening method employed for mutation discovery, and the genome complexity of the species. It is noteworthy that higher mutation densities are not always desirable, as they can increase the likelihood of deleterious mutations. The large genome size and high repetitive DNA content in wheat pose challenges for gene function studies and breeding programs, making them more time-consuming.

Despite these challenges, significant genomic resources have been developed, facilitating the cloning of important genes. The first TILLING populations published in wheat were created by Slade et al. (2005) from both the hexaploid wheat variety Express and the tetraploid wheat variety Kronos, with mutation frequencies of one mutation every 24 kb and one mutation every 40 kb, respectively. Subsequently, Uauy et al. (2009) developed TILLING populations in both tetraploid and hexaploid wheat using EMS as a chemical mutagen. The tetraploid population was developed from the durum variety Kronos, while the hexaploid population from the Hard Red Spring common wheat breeding line UC1041 + Gpc-B1/Yr36. The mutation densities reported were high in both the hexaploid (~1/38 kb) and tetraploid (~1/51 kb) populations (Uauy et al. 2009). A TILLING population from the spring wheat cv. Cadenza was produced using mutagenesis at Rothamsted Research in 2005 and the agronomic and morphological properties of the mutant population were examined over a three-year field experiment, observing traits like ear emergence, plant height, and grain color (Rakszegi et al. 2010). The researchers also evaluated the breadmaking quality parameters of selected mutant lines. These studies aimed to explore the range of phenotypic diversity achievable by mutagenesis in hexaploid wheat, highlighting the high tolerance of polyploid species to mutations (Rakszegi et al. 2010). Species with high ploidy levels, like wheat, can largely tolerate mutations because essential genes are complemented by copies from other sets of chromosomes. However, this genetic buffering can also limit the variety and intensity of phenotypic traits observed in a mutant population. In 2015, the TILLING population of cv. Cadenza was used to demonstrate the feasibility of next-generation sequencing through exon capture for genome resequencing as a mutation detection method in polyploid wheat. A mutation rate of approximately 35 SNPs per Mb was detected, similar to the rate estimated by PCR-based TILLING (King et al. 2015).

Another TILLING population of 2610 M_2 mutants was developed using the common wheat cultivar Jinmai 47, showing diverse phenotypic mutations; various phenotypes with altered morphological and agronomic traits were observed, indicating a valuable resource for wheat research. High mutation densities were found, with RAPD and intron-spliced junction (ISJ) analysis showing 1/34 kb and TILLING revealing 1/47 kb, leading to the identification of 31 novel alleles in three targeted genes (Chen et al. 2012).

An in silico wheat TILLING resource database was developed by Krasileva et al. (2017). They catalogued more than 10 million mutations in the protein-coding regions of 2735 mutant lines of tetraploid and hexaploid wheat: 1535 lines of the tetraploid durum wheat variety Kronos and 1200 lines of the hexaploid bread wheat variety Cadenza. Over 90% of wheat genes per population had at least one truncation or deleterious missense mutation. The process involved exome capture and Illumina sequencing for 2735 mutant lines.

The raw sequencing data was aligned to the RefSeqv1.0 genome of bread wheat Chinese Spring, facilitating the identification of mutations and the prediction of their effects according to the gene models. Approximately 90% of the captured genes exhibited alleles predicted to be harmful, such as premature stop codons, splice site mutations, and non-synonymous amino acid changes. The mutation database created is accessible on Ensembl Plants and on www.wheat--tilling.com, with seed stocks available as resources for gene function studies and wheat variety improvement. Recently, Zhang et al. (2023a) used the same EMS-Kronos lines to remap exome capture reads to Chinese Spring RefSeqv1.1. In comparison with the original study, they found 59% more nonsynonymous substitutions and 49% more truncation mutations. Furthermore, the researchers were the first to employ the TILLING strategy to sequence regulatory regions in addition to coding regions. They conducted a promoter capture assay to sequence 2-kb regions upstream of all high-confidence annotated genes and identified two mutations in a regulatory region of *VRN-A1* gene to showcase the potential of this type of mutation. The promoter mutation dataset is an important resource for inducing fine-tuned regulation of gene expression that produces a phenotypic series. Such mutations are particularly useful for genes whose complete silencing causes severe or lethal plant phenotypes. By manipulating the expression of these genes, it is possible to study their function or generate intermediate phenotypes valuable for breeding programs.

Another valuable genetic resource for functional genomics, applicable to tetraploid and hexaploid wheat, was generated by the mutagenesis (EMS) of the diploid wheat *Triticum monococcum* producing 1532 M_2 families with a mutation frequency of 1/90 kb (Rawat et al. 2012).

The mutation frequency is higher compared to other diploid TILLING populations from grass species. Due to the presence of only one homoeologous locus for any gene, diploid wheats facilitate the validation of gene function and structure. In contrast to polyploid wheat, most diploid wheat mutants allow for straightforward genotype-to-phenotype relationships and can be used for both forward- and reverse-genetic studies, making diploid wheat a valuable model for investigating gene function in polyploid wheat.

TILLING populations and platforms have been established in other cereal species such as maize (Till et al. 2004), sorghum (Xin et al. 2008), and oat (Chawade et al. 2010), significantly contributing to functional genomics studies and breeding programs. Despite the emergence of new targeted genome editing strategies, including CRISPR/Cas9-based methods, TILLING continues to play a crucial role in plant research as a valuable tool for generating genetic variation for genomics and breeding projects. Even with the advent of new genomics methods and increased data storage capacity, TILLING may be utilized even more efficiently to discover new genes and enhance variability.

4 Durum Wheat Genotypes with Improved Nutritional Value

4.1 Manipulation of Starch Composition

The starch composition has been the first trait to be targeted through TILLING in durum wheat. Starch constitutes more than 60% of the semolina derived from durum wheat and significantly influences the nutritional value of durum wheat-derived foods. Natural mutations in key genes involved in starch synthesis were found to be rare, compounded by limitations in detection methods, thus impeding the identification of valuable mutant genotypes. TILLING has proven to be highly effective in generating novel variability in starch phenotypes within wheat. Extensive screening among natural mutant collections often failed in detecting alleles of interest. SDS-PAGE-based screening of starch enzymes was cumbersome and yielded poor results due to the low resolution of the electrophoretic profile of starch granule proteins (SGP). Various enzymes pivotal to starch synthesis have been targeted using classical and biotechnological methods, elucidating their functions and providing novel variability in starch phenotypes in wheat (Botticella et al. 2021). The application of TILLING in targeting starch biosynthesis has led to the development of innovative genotypes suitable for food production.

Granule-bound starch Synthase (*GBSSI*) and *Starch Synthase* (*SSIIa*) null genotypes were engineered in durum wheat through a genetics approach employing both natural and chemical-induced genetic variability (Lafiandra et al. 2010).

The first gene screened by TILLING in the Kronos EMS population was *GBSSI*, encoding a Granule Bound Starch Synthase responsible for amylose chain elongation in plants. Slade and co-workers (2005) reported the identification of a nonsense mutation in the A genome homoeologous and predicted deleterious polymorphisms in the B genome homeoallele.

Following the elucidation of the Starch Branching Enzyme IIa (SBEIIa) functionality in bread and durum wheat (Regina et al. 2006; Sestili et al. 2010), TILLING has been applied by various researchers in breeding programs aiming to produce durum wheat lines with a higher resistant starch (RS) content. The term "resistant starch" refers to a starch fraction that behaves functionally like dietary fiber, conferring significant health benefits. RS aids in lowering the glycemic index of wheat-derived foods (Botticella et al. 2021) and promoting gut health. One way to increase RS in food concerns breeding for high amylose (apparent amylose) genotypes. Amylose consists of linear α 1–4 linked glucan chains that cluster together through H-bonds in supramolecular structures, impeding degradation by amylases in the stomach.

RS is assimilated in the large bowel, where gut-microbiome ferments it to produce Short Chains Fatty acids (SCFA), crucial for colonic health. The first high amylose (HA) durum wheat was created by pyramiding the mutations *SBEIIa-A* carrying a premature stop codon, and *SBEIIa-B* affecting a splice junction in the genetic background of cv. Kronos (Slade et al. 2012).

The resulting double null homozygous genotype exhibited an amylose content of 47.4% of total starch compared to 24% in the control line, with RS increasing to 4.7% compared to 0.8% in the control (Slade et al. 2012). Concurrently, in the same cv. Kronos, Hazard et al. (2012) combined two different mutations in the homoeoalleles *SBEIIa-A* and *SBEIIa-B*, resulting in a modest increase in amylose content and resistant starch.

Although the two SBEIIa null genotypes were produced in the same cv. Kronos, differences in the type and location of polymorphism affected gene expression and functionality to varying extents. In subsequent work, Hazard et al. (2012) demonstrated that combining the SBEIIa and SBEIIb mutations substantially increased the amylose and RS content, leading to the registration of the HA new breeding line SBEIIa/b-AB (Hazard et al. 2014). Notably, in the SBEIIa/b null line, the increase in RS did not fully correlate with the change in amylose; it can be speculated that SBEIIb could affect starch structure differently from the SBEIIa isoform or there may be limitations in the RS determination method. In both the studies described above, the CEL1 Li-Cor technique was used to detect mutations in twofold pools of genomic DNA. An HA line was developed in cv. Svevo by combining mutations in both homoalleles of *SBEIIa*, resulting in an increase in the amylose level from 27 to 52% of total starch, with RS rising from 0.8 to 6% (Sestili et al. 2015). The four high-amylose lines described here exhibited a modest to slight decrease in seed yield and total starch content, possibly due to the interference in starch biosynthesis machinery. HA durum wheat has been the subject of numerous investigations aimed at exploring its potential to enhance the nutritional value of durum-wheat-derived food. Preliminary in vivo studies suggested the potential of HA Svevo semolina to reduce glycemic index in spaghetti, with a suggested minimum RS content of ~7%, corresponding to an amylose content of ~58% required to lower GI (Sissons et al. 2020). HA pasta provided higher dietary fiber and resistant starch compared to the control, with minimal impacts on pasta's technological properties.

4.2 Micronutrient Biofortification

Reducing micronutrient malnutrition, or hidden hunger, is one of the primary challenges outlined in the 2030 Agenda for Sustainable Development.

This deficiency often stems from inadequate diets characterized by low intakes of essential vitamins and minerals (Von Grebmer et al. 2014). The Food and Agriculture Organization (FAO) estimates that 17% of the population suffers from micronutrient deficiencies (WHO 2019).

Among the most prevalent micronutrient deficiencies are iron (Fe), zinc (Zn), iodine (I), and various vitamins, disproportionately affecting women and children. This deficiency weakens the immune system, leading to infections and other issues such as diarrhea, stunting, blindness, anemia, fetal malformations, and developmental difficulties (Frittelli et al. 2023a).

Addressing essential nutrient deficiencies in diets can be achieved through the biofortification of staple crops like wheat. Carotenoids, crucial nutrients in the human diet, serve as significant sources of β-carotene, a precursor to provitamin A.

The absence of carotenoids in the diet is linked to severe diseases prevalent in low-income countries, where staple crops lack essential micronutrients. Given its widespread consumption, biofortifying durum wheat emerges as a valuable strategy to enhance β-carotene levels in diets. TILLING has been employed to engineer carotenoid biosynthesis in durum wheat, selecting mutations in key biosynthetic genes in the EMS-Kronos (Richaud et al. 2018; Sestili et al. 2019; Garcia Molina et al. 2021; Yu et al. 2022; Bekkering et al. 2023) and EMS-Marco Aurelio TILLING platforms (Colasuonno et al. 2016).

Researchers focused on enzymes regulating the competitive branch of the pathway that directs lycopene toward α-carotene synthesis and enzymes involved in β-carotene turnover.

Several studies investigated the potential of loss of function mutations in key genes, paving the way for the development of durum wheat enriched in provitamin A. Two enzymes, lycopene-ε-cyclase (LCYE) and β-hydroxylase (β-HYD), have been selected based on results obtained in other cereals. LCYE enzyme has been identified as a target to be abolished to enhance β-carotene content in plants from a blockchain perspective. Inhibition of LCYE activity redirects the substrate, lycopene, toward the β-carotene biosynthetic branch, effectively hindering α-carotene synthesis in various crops. Differently, the abolishment of β-HYD advantages the accumulation of β-carotene versus the synthesis of the downstream xanthophylls.

A nonsense mutation in *LCYE -A* homeoallele was associated with a 70% increase of β-carotene in leaves, with no significant change in seeds, as reported by two independent groups in two different cultivars, EMS-Kronos and Svevo-EMS (Richaud et al. 2018; Sestili et al. 2019). Similar results were observed in a line carrying a mutation affecting the splice junction in the *LCYE-B* homeoallele.

Combining the two mutations proved effective, resulting in a substantial 75% increase in mature seed β-carotene levels, while simultaneously reducing α-carotene by 99% (Sestili et al. 2019). Building on this success, the same research group conducted a breeding program to pyramid null mutations in the two homoeoalleles responsible for β-HYD1 (Garcia Molina et al. 2021). The resulting double null line exhibited a 70% increase in β-carotene content in seeds compared to controls. Recently, in EMS-Kronos, the combination of loss of function mutations in *LCYE* and a *βHYD-2*, a paralogue of *βHYD-1*, enhanced β-carotene by 6-fold in flour; however, the total carotenoid content was reduced to 35% of the control (Yu et al. 2022). By contrast, the simultaneous silencing of *LCYE* and *βHYD-1* led to a 3.5-fold increase in β-carotene content accompanied by a 30% reduction in total carotenoids. (Palombieri et al. 2025). Hence, further research is needed to identify key actors in carotenoid metabolism to significantly enhance provitamin A content in durum wheat.

Minerals are essential micronutrients provided through diet. A diet poor in micronutrients is associated with the onset of serious diseases but can also lead to less severe consequences with a slight impact on general well-being.

One strategy for the biofortification of crops in micronutrients has been to lower the content of phytic acid (Myo-inositol-1,2,3,4,5,6-hexakisphosphate (InsP6)). Due to its high load in phosphates, phytic acid (PA) sequesters important cations in the form of phytate salts, which are indigestible by monogastric animals lacking phytases in the digestive tract. Lowering of PA in durum wheat has been achieved by targeting key enzymes involved in PA transport to the vacuole where it is stored in the seed. Loss of function mutants in the homeologs coding for the MRP3 transporter has been identified in the EMS-Kronos population in durum wheat. The complete null genotype showed a significant reduction in PA (−84.5%) and increases of 38.4% for Mg, 32.6% for S, 186.3% for Fe, 78.4% for Zn, and 36.3% for Mn (Frittelli et al. 2023a). However, phytic acid is a major phosphorous storage sink in the plant and plays pivotal roles in plant development. Several low PA (lpa) mutations are lethal for plants, such as the lpa2 mutation targeting the IPK1 enzyme, which catalyzes the last step of PA biosynthesis. In EMS-Kronos, the complete abolishment of enzyme functionality resulted in the loss of seed germinability (Frittelli et al. 2023b).

Another application of TILLING was related to a study focused on reducing immunogenicity in durum wheat semolina (Wen et al. 2022). In this regard, wheat is known to trigger important immunogenic reactions such as celiac disease, allergy, and gluten sensitivity. Toxicity is primarily triggered by protein fractions belonging to the gluten polymer, specifically low molecular weight subunits (LMW-GSs) and gliadins. DEMETER (DME) and Derepressed for Ribosomal protein S14 Expression (DRE2) enzymes play roles in the demethylation of genes encoding gliadins and LMW-GSs in wheat, essential for their transcriptional activation. Loss-of-function mutants were selected in the EMS-Kronos population for both genes and fully characterized for protein content and composition. The suppression of *DME* and *DRE2* genes reduces gluten content without affecting cooking properties. However, this approach is not entirely viable due to the role of these genes in pollen vitality and germination. Future research must focus on developing durum wheat with lowered immunogenicity, possibly by identifying mutations limiting disruption of gene expression in the endosperm compartment.

5 Durum Wheat Genotypes with Improved Yields

Despite the increasing human population and escalating demand for wheat, the low rates of genetic progress in wheat yields are a critical concern for global food security. Grain yield conventionally comprises numerous sub-traits known as "yield components," regulated by polygenic factors alongside significant environmental influences. Key genetically controlled traits influencing yield include spikelets per head, seed size, and productive tiller number. Recently, TILLING has made significant contributions to the identification of new alleles that can consistently enhance yield potential in durum wheat. Here, we report a few examples of yield-related traits targeted by TILLING in durum wheat.

The *Grain Weight 2* (*GW2*) gene encodes a RING-type E3 ubiquitin ligase and functions as a negative regulator of grain width and weight in rice (Song et al. 2007). An EMS-Kronos splice acceptor site mutation in *TdGW2-A1* increased thousand grain weight in a series of BC-derived lines, providing further support to the role of GW2 as a negative regulator of grain size (Simmonds et al. 2016). The mutant line had significantly higher TGW (+10.6%) and larger grain area (+6.8%) than lines with the Kronos wild-type allele. It was shown that GW2 acts on maternal tissue before anthesis to restrict seed size in wheat (Simmonds et al. 2016). This is one of the few examples in which mutation in only one homoeologous affected the phenotype in durum wheat.

Plant yield can be raised by increasing the productive tiller number. One such gene that affects tillering is *Teosinte Branched1* (*TB1*), first identified in maize (*Zea mays* L.) (Doebley et al. 1997). *TB1* is a member of the TCP family of transcriptional regulators, known as negative regulators, that suppress lateral branching, axillary buds growth, and branch internode elongation (Cubas et al. 1999) and it has conserved function in maintaining this regulatory role. Although TB1 *null* mutations increased the tiller number, grain yield was not improved in durum wheat (Volkman et al. 2022). It is presumed that an increase in grain yield can be achieved by combining the TB1 *null* mutations with alleles of genes known to increase seeds per head and seed weight.

Incorporation of semi-dwarfing alleles of the *Reduced height* (*Rht*) gene into hexaploid wheat (*Triticum aestivum* L.) varieties increased yield during the middle of the twentieth century (Pearce et al. 2011). Similarly, the introgression of the semi-dwarf allele *Rht-B1b* into durum wheat improved grain yield by ~9% compared with the wild-type allele, but it significantly decreased height and coleoptile length (Pandey et al. 2015). Reduction in height and coleoptile length raises concerns related to mechanical harvest and poor seed emergence in deep sowing, respectively (Addisu et al. 2010; Pandey et al. 2015). In this context, the creation of novel allelic variability for the *Rht* gene can provide a novel opportunity to reduce its negative side effects. Several yield-focused research programs generated novel allelic variability in the two homeologs, *Rht-A1* and *Rht-B1*, which affected both plant height and coleoptile length in durum wheat.

TILLING by sequencing was used to detect genetic loci that are putatively associated with agronomically important traits. In detail, Mo and colleagues' study (Mo et al. 2018) combined the use of a mapping strategy and exome sequencing in an EMS-Kronos population to identify genetic variants in a tall phenotype mutant. A missense mutation bearing the semidwarf *Rht-1Bb* allele partially suppressed its semidwarf phenotype, increasing both plant height and coleoptile length. The *Rht-B1b-E529K* allele conferred plant height and coleoptile length intermediate between *Rht-B1b* and *Rht-B1a*. In a further work, a novel allele of *RhtA1* homoeologous identified by screening the EMS population cv. Divide produced a line with intermediate height and coleoptile length when combined with wild-type *Rht-B1* (Brown et al. 2022).

Unlike the *Rht* genes of the Green Revolution, the gene *PLATZ1* underlying the *Rht25* locus causes the GA-sensitive dwarfing phenotype. This gene was isolated and fully characterized using previously developed segregating populations and loss-of-function TILLING mutants (Zhang et al. 2023b). *PLATZ1* encodes for a zinc-dependent DNA-binding protein with two conserved noncanonical zinc finger domains and interacts with *Rht1*, modulating the effect of *DELLA* on wheat plant height. Natural and EMS-loss-of-function mutations in this gene reduced plant height, allowing to select alternative dwarf alleles in modern wheat germplasm that have positive effects on wheat productivity. As starch is the main component of wheat flour, increasing starch content in the seed is an interesting approach to improving durum wheat yield. The *PARC6* gene is known to alter amyloplast size in model species (Glynn et al. 2008). Recently, null mutations in *PARC6* were found to affect amyloplast size in the durum wheat cv. Kronos with no effect on total starch content, indicating that amyloplast size is not a limiting factor for starch yield (Esch et al. 2023).

6 Durum Wheat Genotypes Resilient to Biotic and Abiotic Stresses

Global climate change exacerbates the adverse effects of abiotic and biotic stresses in crops, leading to increased yield losses. In this context, enhancing durum wheat resilience to multiple stress conditions is a crucial prerequisite for a more sustainable agriculture. TILLING can be utilized to decipher the function of genes or discover new alleles for resistance/tolerance to various stresses.

In Morocco, where durum wheat is predominantly cultivated in arid and semiarid regions and exposed to water deficit, an EMS-durum wheat population (cv. CHAM-1) was phenotypically selected based on plant vigor, yield, and disease resistance. The screening led to the identification of 262 M_9 lines. The most productive genotypes (19 mutagenized lines) were tested for four seasons in the experimental farm at the INRA station (Kahama et al. 2019). The authors demonstrated that the EMS treatment applied to the CHAM-1 durum wheat cv. increased genetic variability, particularly noticeable under drought conditions, with high variation coefficients obtained for grain number per spike (34.5%), plant height (10.5%), and wall-bound peroxidases (46.2%). This work showed how mutagenesis facilitated the introduction of new variability to a commercial variety, correlating the decrease in ionically wall-bound peroxidases with drought adaptation.

Another prevalent issue affecting the Mediterranean Basin, where much of durum wheat production is concentrated, is high temperature stress. Comastri et al. (2018) utilized two different TILLING populations (EMS-Kronos and EMS-CHAM1) to identify new alleles in four members of the chloroplast-localized small heat shock proteins (sHSP) encoded by the *Hsp26* gene family. These proteins play a crucial role in safeguarding the photosynthetic machinery during high-temperature stress episodes, and the newly identified alleles hold promise for developing thermotolerant durum wheat genotypes. In this work, 50 TILLING mutant lines were selected and specific KASP (Kompetitive Allele Specific PCR) markers were designed to track SNPs in future breeding programs.

A sustainable strategy to limit yield losses caused by multiple stresses involves the identification of new alleles in genes that regulate productivity. Understanding processes that control flowering time, a key component for adaptation and yield in many species, is valuable for designing crops that respond optimally to extreme climatic conditions. Chen and Dubcovsky (2012) used TILLING mutants of durum wheat (cv. Kronos) to demonstrate that the vernalization gene *VRN1* is not essential for flowering and regulates the *VRN2* flowering repressor in the leaves. This work also highlighted the presence of redundant genes for wheat flowering, offering new targets to modify wheat varieties for better adaptation to environmental changes.

Zhang et al. (2023a) targeted the same gene, *VRN1-A*, to find mutations in the promoter region. By sequencing 2-kb regions upstream of all high-confidence annotated genes from 1513 EMS-Kronos lines using a promoter capture assay, they found mutations in the transcription factor binding sites of *VRN1-A* gene that significantly altered its expression, including one mutation that reduced the number of spikelets per spike. These findings underscored the biological significance of mutations in the promoter region, which can modulate gene expression to produce intermediate phenotypes suitable for both basic research and commercial applications.

Studying phenotypes common to different stresses, such as diseases and nutrient deficiencies, helps elucidate the complex mechanisms regulating disease resistance and nutrient homoeostasis under different environmental conditions. Harrington et al. (2019) identified a dominant chlorosis phenotype using a forward screen strategy in a TILLING durum wheat population.

This genotype showed chlorophyll pigment degradation in leaves of wheat grown under different stress conditions. The authors used a mapping-by-sequencing (GBS) approach to define the locus *Yellow Early Senescence-1* and fine-mapped it to a region of 4.3 Mb, containing 59 high-confidence genes.

Crop yield and stress resilience can also be enhanced by modifying biochemical traits critical for other physiological traits, including photosynthesis and growth regulation. Carotenoids, crucial metabolites for major plant physiological processes, including photosynthesis and stress response, all important for improving plant performance (Condorelli et al. 2022). Additionally, some carotenoids serve as precursors of vitamin A in the diet. Bekkering et al. (2023) characterized loss-of-function mutants (EMS-Kronos) for *HYD1* and *HYD2* genes to understand their roles in carotenoid biosynthesis and their effects on plant vigor. These paralogous genes, involved in the hydroxylation reaction converting carotenes to xanthophylls, exhibit functional redundancy, with *HYD1* mainly involved in carotenoid metabolism in leaves and *HYD2* in mature grain tissues. Further studies are needed to understand the phenotypic plasticity and environmental responses of combinatorial HYD mutants (Bekkering et al. 2023).

Crop diseases also compromise yield and food safety, and identifying resistant genotypes is crucial for reducing yield losses (Bigini et al. 2021). The recessive mutation in the *Mlo* gene in barley provides effective and lasting resistance against powdery mildew, a serious cereal disease (Büschges et al. 1997). Ingvardsen et al. (2019) used captured exome information from the EMS-Kronos population to identify knock-out mutations in *Mlo* homeoalleles. By using TILLING mutants, the authors demonstrated the potential to introduce durable protection in tetraploid wheat crops against powdery mildew attacks by combining mutations in *Mlo-A1* and *Mlo-B1* homoeoalleles.

7 Conclusion

Since its introduction in 2000, TILLING has been widely used in plant research to study gene function and generate new allelic diversity useful for breeding programs. This strategy can be applied to most organisms, proving extremely useful in polyploid species due to the redundant genes that allow tolerance of very high mutation densities.

In durum wheat, several traits have been studied, and new allelic variants have been identified. This genetic variability is exploitable to improve nutritional and agronomic quality, as well as resistance/resilience to different stresses.

The main bottlenecks of this technique, its time-consuming nature and excessive cost in detecting mutations of interest in a large population, have been addressed through methodological developments: initially, high-throughput molecular techniques were employed, and more recently, next-generation sequencing approaches were used. Several mutation discovery methods have been implemented, reducing equipment costs and increasing efficiency and accuracy, making TILLING accessible to all research groups. Nowadays, TILLING has become an in silico procedure, allowing the proliferation of new gene and genome targets, saving time and money.

A recent implementation of the TILLING by sequencing technique has been the discovery of mutations in promoter regions, which can be used to modulate spatial or temporal expression of target genes. Mutations in regulatory regions can result in subtle phenotypic variations differently from mutations in coding regions, which often produce extreme phenotypes.

Differently from new genome editing systems, TILLING-derived genotypes can be used for commercial purposes. However, the major drawbacks include the potential for random mutagenesis to cause negative pleiotropic effects due to the introduction of unwanted mutations, which must be eliminated through genetic backcrosses, often laborious and time-consuming.

Additionally, in polyploid species like durum wheat, obtaining the desired phenotype requires detecting mutations in both homoeoalleles and combining them by developing double mutants through crosses. These limitations are overcome in modern genome editing methods, which are specific to the target sequence (avoiding off-target mutations), and the design of optimized RNA guides can facilitate the simultaneous suppression of both homeoalleles in durum wheat. On the other hand, TILLING does not require an efficient transformation system or expensive laboratory equipment and can be successfully used for improving polyploid species.

All things considered, thanks to improvements over the past 20 years and the availability of publicly sequenced wheat mutant lines, TILLING remains an easy and economically convenient strategy to obtain an allelic series in a gene of interest, with applications in both basic research and breeding purposes.

References

Addisu M, Snape JW, Simmonds JR, Gooding MJ (2010) Effects of reduced height (*Rht*) and photoperiod insensitivity (*Ppd*) alleles on yield of wheat in contrasting production systems. Euphytica 172:169–181

Bekkering CS, Yu S, Isaka NN, Sproul BW, Dubcovsky J, Tian L (2023) Genetic dissection of the roles of β-hydroxylases in carotenoid metabolism, photosynthesis, and plant growth in tetraploid wheat (*Triticum turgidum* L.). Theor Appl Genet 136:8. https://doi.org/10.1007/s00122-023-04276-3

Bigini V, Camerlengo F, Botticella E, Sestili F, Savatin DV (2021) Biotechnological resources to increase disease-resistance by improving plant immunity: a sustainable approach to save cereal crop production. Plants 10:1146. https://doi.org/10.3390/plants10061146

Botticella E, Savatin DV, Sestili F (2021) The triple jags of dietary fibers in cereals: how biotechnology is longing for high fiber grains. Front Plant Sci 14(12):745579. https://doi.org/10.3389/fpls.2021.745579

Bovina R, Brunazzi A, Gasparini G, Sestili F, Palombieri S, Botticella E, Lafiandra D, Mantovani P, Massi A (2014) Development of a TILLING resource in durum wheat for reverse-and forward-genetic analyses. Crop Pasture Sci 65:112–124

Brown MM, Martin JM, Jobson EM, Hogg AC, Carr PM, Giroux MJ (2022) Evaluating the impact of *Rht* hypomorphic mutations in durum wheat. Crop Sci 62:247–258

Büschges R, Hollricher K, Panstruga R et al (1997) The barley *Mlo* gene: a novel control element of plant pathogen resistance. Cell 7(88):695–705. https://doi.org/10.1016/s0092-8674(00)81912-1

Bush SM, Krysan PJ (2010) iTILLING: a personalized approach to the identification of induced mutations in Arabidopsis. Plant Physiol 154:25–35. https://doi.org/10.1104/pp.110.159897

Caldwell DG, McCallum N, Shaw P, Muehlbauer GJ, Marshall DF, Waugh R (2004) A structured mutant population for forward and reverse genetics in Barley (*Hordeum vulgare* L.). Plant J. 40(1):143–150. https://doi.org/10.1111/j.1365-313X.2004.02190.x

Casella L, Greco R, Bruschi G et al (2013) TILLING in European rice: hunting mutations for crop improvement. Crop Sci 53:2550–2562. https://doi.org/10.2135/cropsci2012.12.0693

Chawade A, Sikora P, Bräutigam M et al (2010) Development and characterization of an oat TILLING-population and identification of mutations in lignin and β-glucan biosynthesis genes. BMC Plant Biol 10:1–13

Chen A, Dubcovsky J (2012) Wheat TILLING mutants show that the vernalization gene *VRN1* down-regulates the flowering repressor VRN2 in leaves but is not essential for flowering. PLoS Genet 8(12):e1003134. https://doi.org/10.1371/journal.pgen.1003134

Chen L, Huang L, Min D, Phillips A, Wang S, Madgwick PJ, Parry MA, Hu YG (2012) Development and characterization of a new TILLING population of common bread wheat (*Triticum aestivum* L.). PLoS One 7:e41570. https://doi.org/10.1371/journal.pone.0041570

Colasuonno P, Incerti O, Lozito ML, Simeone R, Gadaleta A, Blanco A (2016) DHPLC technology for high-throughput detection of mutations in a durum wheat TILLING population. BMC Genet 17:17–43

Colbert T, Till BJ, Tompa R, Reynolds S, Steine MN, Yeung AT, McCallum CM, Comai L, Henikoff S (2001) High-throughput screening for induced point mutations. Plant Physiol. 126:480–484. https://doi.org/10.1104/pp.126.2.480

Comastri A, Janni M, Simmonds J, Uauy C, Pignone D, Nguyen HT, Marmiroli N (2018) Heat in wheat: exploit reverse genetic techniques to discover new alleles within the *Triticum durum* sHsp26 family. Front Plant Sci 19(9):1337. https://doi.org/10.3389/fpls.2018.01337

Condorelli GE et al (2022) Genome wide association study uncovers the QTLome for osmotic adjustment and related drought adaptive traits in durum wheat. Genes 13:293. https://doi.org/10.3390/genes13020293

Cubas P, Lauter N, Doebley J, Coen E (1999) The TCP domain: a motif found in proteins regulating plant growth and development. Plant J. 18(2):215–222. https://doi.org/10.1046/j.1365-313x.1999.00444.x

Doebley J, Stec A, Hubbard L (1997) The evolution of apical dominance in maize. Nature 3(386):485–488. https://doi.org/10.1038/386485a0

Esch L, Ngai QY, Barclay JE, McNelly R, Hayta S, Smedley MA, Smith AM, Seung D (2023) Increasing amyloplast size in wheat endosperm through mutation of PARC6 affects starch granule morphology. New Phytol 240:224–241. https://doi.org/10.1111/nph.19118

Frittelli A, Botticella E, Palombieri S, Masci S, Celletti S, Fontanella MC, Astolfi S, De Vita P, Volpato M, Sestili F (2023a) The suppression of TdMRP3 genes reduces the phytic acid and increases the nutrient accumulation in durum wheat grain. Front Plant Sci 18(14):1079559. https://doi.org/10.3389/fpls.2023.1079559

Frittelli A, Palombieri S, Quagliata G et al (2023b) The silencing of TdIPK1 genes enhances micronutrient concentration in durum wheat grain. Current Plant Biol 35:100309

Garcia Molina MD, Botticella E, Beleggia R, Palombieri S, De Vita P, Masci S, Lafiandra D, Sestili F (2021) Enrichment of provitamin A content in durum wheat grain by suppressing β-carotene hydroxylase 1 genes with a TILLING approach. Theor Appl Genet 134:4013–4024. https://doi.org/10.1007/s00122-021-03944-6

Garvin MR, Gharrett AJ (2007) DEco-TILLING: an inexpensive method for single nucleotide polymorphism discovery that reduces ascertainment bias. Mol Ecol Notes 7:735–746. https://doi.org/10.1111/j.1471-8286.2007.01767.x

Gilchrist EJ, O'Neil NJ, Rose AM, Zetka MC, Haughn GW (2006) TILLING is an effective reverse genetics technique for Caenorhabditis elegans. BMC Genom 7:1–16

Glynn JM, Froehlich JE, Osteryoung KW (2008) *Arabidopsis* ARC6 coordinates the division machineries of the inner and outer chloroplast membranes through interaction with PDV2 in the intermembrane space. Plant Cell 20:2460–2470. https://doi.org/10.1105/tpc.108.061440

Gottwald S, Bauer P, Komatsuda T, Lundqvist U, Stein N (2009) TILLING in the two-rowed barley cultivar "Barke" reveals preferred sites of functional diversity in the gene HvHox1. BMC Res Notes 2:258. https://doi.org/10.1186/1756-0500-2-258

Griffiths A, Wessler S, Lewontin R, Gelbart W, Suzuki D, Miller J (2004) Dissection of gene function. In: Freeman (ed) An introduction to genetic analysis, 8th edn. WH New York

Harrington SA, Cobo N, Karafiátová M, Doležel J, Borrill P, Uauy C (2019) Identification of a dominant chlorosis phenotype through a forward screen of the Triticum turgidum cv. Kronos TILLING population. Front Plant Sci 24(10):963. https://doi.org/10.3389/fpls.2019.00963

Hazard B, Zhang X, Colasuonno P, Uauy C, Beckles DM, Dubcovsky J (2012) Induced mutations in the starch branching enzyme II (*SBEII*) genes increase amylose and resistant starch content in durum wheat. Crop Sci 52:1754–1766. https://doi.org/10.2135/cropsci2012.02.0126

Hazard B, Zhang X, Naemeh M, Dubcovsky J (2014) Registration of durum wheat germplasm lines with combined mutations in *SBEIIa* and *SBEIIb* genes conferring increased amylose and resistant starch. J Plant Regist 8:334–338. https://doi.org/10.3198/jpr2014.02.0007crg

Horst I, Welham T, Kelly S, Kaneko T, Sato S, Tabata S, Parniske M, Wang TL (2007) TILLING mutants of *Lotus japonicus* reveal that nitrogen assimilation and fixation can occur in the absence of nodule-enhanced sucrose synthase. Plant Physiol. 44:806–820. https://doi.org/10.1104/pp.107.097063

Ingvardsen CR, Massange-Sánchez JA, Borum F, Uauy C, Gregersen PL (2019) Development of *mlo*-based resistance in tetraploid wheat against wheat powdery mildew. Theor Appl Genet 132(11):3009–3022. https://doi.org/10.1007/s00122-019-03402-4

Irshad A, Guo H, Zhang S, Gu J, Zhao L, Xie Y, Xiong H, Zhao S, Ding Y, Ma Y, Liu L (2019) EcoTILLING reveals natural allelic variations in starch synthesis key gene *TaSSIV* and its haplotypes associated with higher thousand grain weight. Genes 10:307. https://doi.org/10.3390/genes10040307

Jiang C, Lei M, Guo Y et al (2022) A reference-guided TILLING by amplicon-sequencing platform supports forward and reverse genetics in barley. Plant Commun 11:100317. https://doi.org/10.1016/j.xplc.2022.100317

Kahama I, Kohaich K, Baaziz M, Labhilili M (2019) Quantitative and qualitative aspects of soluble and ionically wall-bound peroxidases of EMS-mutagenized durum wheat lines (*Triticum durum*), as tested for yield at drought conditions. Agric Res 8:148–157

King R, Bird N, Ramirez-Gonzalez R, Coghill JA, Patil A, Hassani-Pak K, Uauy C, Phillips AL (2015) Mutation scanning in wheat by exon capture and next-generation sequencing. PLoS One 10:e0137549

Knudsen S, Wendt T, Dockter C et al (2022) FIND-IT: accelerated trait development for a green evolution. Sci Adv 8(34):eabq2266. https://doi.org/10.1126/sciadv.abq2266

Krasileva KV, Vasquez-Gross HA, Howell T, Bailey P, Paraiso F, Clissold L et al (2017) Uncovering hidden variation in polyploid wheat. Proc Natl Acad Sci USA. 114:913–921. https://doi.org/10.1073/pnas.1619268114

Krothapalli K, Buescher EM, Li X, Brown E, Chapple C, Dilkes BP, Tuinstra MR (2013) Forward genetics by genome sequencing reveals that rapid cyanide release deters insect herbivory of Sorghum bicolor. Genetics 195:309–318

Lababidi S, Mejlhede N, Rasmussen SK, Backes G, Al-Said W, Baum M, Jahoor A (2009) Identification of barley mutants in the cultivar 'Lux' at the Dhn loci through TILLING. Plant breeding 128:332–336. https://doi.org/10.1111/j.1439-0523.2009.01640.x

Lafiandra D, Sestili F, D'Ovidio R, Janni M, Botticella E, Ferrazzano G, Silvestri M, Ranieri R, DeAmbrogio E (2010) Approaches for the modification of starch composition in durum wheat. Cereal Chemistry 87(1):28–34

McCallum CM, Comai L, Greene EA, Henikoff S (2000a) Targeting induced local lesions in genomes (TILLING) for plant functional genomics. Plant Physiol 123:439–442

McCallum CM, Comai L, Greene EA, Henikoff S (2000b) Targeted screening for induced mutations. Nat Biotechnol 18:455–457

Menda N, Semel Y, Peled D, Eshed Y, Zamir D (2004) In silico screening of a saturated mutation library of tomato. Plant J 38:861–872. https://doi.org/10.1111/j.1365-313X.2004.02088.x

Mo Y, Howell T, Vasquez-Gross H, de Haro LA, Dubcovsky J, Pearce S (2018) Mapping causal mutations by exome sequencing in a wheat TILLING population: a tall mutant case study. Mol Genet Genomics 293:463–477. https://doi.org/10.1007/s00438-017-1401-6

Okabe Y, Asamizu E, Saito T, Matsukura C, Ariizumi T, Brès C, Rothan C, Mizoguchi T, Ezura H (2011) Tomato TILLING technology: development of a reverse genetics tool for the efficient isolation of mutants from micro-Tom mutant libraries. Plant Cell Physiol 52:1994–2005

Pandey M, Singh AK, DePauw RM, Bokore FE, Ellouze W, Knox RE, Cuthbert RD (2015) Coleoptile length, gibberellin sensitivity, and plant height variation of durum wheat in Canada. Can J Plant Sci 95:1259–1264

Palombieri S, Frittelli A, Garcia Molina M et al (2025) Pyramiding of mutations in lycopene ε-cyclase and β-hydroxylase 1 increases β-carotene content and modifies carotenoid metabolism in durum wheat. Plant Physiol Biochem:110007. https://doi.org/10.1016/j.plaphy.2025.110007

Parry MA, Madgwick PJ, Bayon C, Tearall K, Hernandez-Lopez A, Baudo M, Rakszegi M, Hamada W, Al-Yassin A, Ouabbou H (2009) Mutation discovery for crop improvement. J Exp Bot 60:2817–2825

Pearce S, Saville R, Vaughan SP (2011) Molecular characterization of *Rht-1* dwarfing genes in hexaploid wheat. Plant Physiol 157(4):1820–1831. https://doi.org/10.1104/pp.111.183657

Rakszegi M, Kisgyörgy BN, Tearall K, Shewry PR, Láng L, Phillips A, Bedő Z (2010) Diversity of agronomic and morphological traits in a mutant population of bread wheat studied in the Healthgrain program. Euphytica 174:409–421

Rawat N, Sehgal SK, Joshi A et al (2012) A diploid wheat TILLING resource for wheat functional genomics. BMC Plant Biology 12:1–11

Reddy TV, Dwivedi S, Sharma NK (2012) Development of TILLING by sequencing platform towards enhanced leaf yield in tobacco. Ind Crops Prod 40:324–335

Reed GH, Kent JO, Wittwer CT (2007) High-resolution DNA melting analysis for simple and efficient molecular diagnostics. Pharmacogenomics 8:597–608

Regina A, Bird A, Topping D, Bowden S, Freeman J, Barsby T, Kosar-Hashemi B, Li Z, Rahman S, Morell M (2006) High-amylose wheat generated by RNA interference improves indices of large-bowel health in rats. Proc Natl Acad Sci USA 103:3546–3551. https://doi.org/10.1073/pnas.0510737103

Richaud D, Stange C, Gadaleta A, Colasuonno P, Parada R, Schwember AR (2018) Identification of *Lycopene epsilon cyclase* (*LCYE*) gene mutants to potentially increase β-carotene content in durum wheat (*Triticum turgidum* L.ssp. *durum*) through TILLING. PLoS One 13(12):e0208948. https://doi.org/10.1371/journal.pone.0208948

Rigola D, van Oeveren J, Janssen A, Bonne A, Schneiders H, van der Poel HJ, van Orsouw NJ, Hogers RC, de Both MT, van Eijk MJ (2009) High-throughput detection of induced mutations and natural variation using KeyPoint technology. Plos One 4:4761

Salvi S, Tuberosa R (2015) The crop QTLome comes of age. Curr Opin Biotechnol 32:179–185. https://doi.org/10.1016/j.copbio.2015.01.001

Schreiber M, Barakate A, Uzrek N, Macaulay M, Sourdille A, Morris J, Hedley PE, Ramsay L, Waugh R (2019) A highly mutagenised barley (cv. Golden Promise) TILLING population coupled with strategies for screening-by-sequencing. Plant Methods 15:99. https://doi.org/10.1186/s13007-019-0486-9

Serrat X, Esteban R, Guibourt N, Moysset L, Nogués S, Lalanne E (2014) EMS mutagenesis in mature seed-derived rice calli as a new method for rapidly obtaining TILLING mutant populations. Plant Methods 10:1. https://doi.org/10.1186/1746-4811-10-5

Sestili F, Janni M, Doherty A, Botticella E, D'Ovidio R, Masci S, Jones HD, Lafiandra D (2010) Increasing the amylose content of durum wheat through silencing of the *SBEIIa* genes. BMC Plant Biol 14(10):144. https://doi.org/10.1186/1471-2229-10-144

Sestili F, Palombieri S, Botticella E, Mantovani P, Bovina R, Lafiandra D (2015) TILLING mutants of durum wheat result in a high amylose phenotype and provide information on alternative splicing mechanisms. Plant Sci 233:127–133

Sestili F, Garcia-Molina MD, Gambacorta G, Beleggia R, Botticella E, De Vita P, Savatin DV, Masci S, Lafiandra D (2019) Provitamin a biofortification of durum wheat through a TILLING approach. Int J Mol Sci. 14(20):5703. https://doi.org/10.3390/ijms20225703

Sikora P, Chawade A, Larsson M, Olsson J, Olsson O (2011) Mutagenesis as a tool in plant genetics, functional genomics, and breeding. Int J Plant Genomics 314829. https://doi.org/10.1155/2011/314829

Simmonds J, Scott P, Brinton J, Mestre TC, Bush M, Del Blanco A, Dubcovsky J, Uauy C (2016) A splice acceptor site mutation in *TaGW2-A1* increases thousand grain weight in tetraploid and hexaploid wheat through wider and longer grains. Theor Appl Genet 129(6):1099–1112. https://doi.org/10.1007/s00122-016-2686-2

Sissons M, Sestili F, Botticella E, Masci S, Lafiandra D (2020) Can manipulation of durum wheat amylose content reduce the glycaemic index of spaghetti? Foods 9:693. https://doi.org/10.3390/foods9060693

Slade AJ, Fuerstenberg SI, Loeffler D, Steine MN, Facciotti D (2005) A reverse genetic, non-transgenic approach to wheat crop improvement by TILLING. Nat Biotechnol. 23:75–81. https://doi.org/10.1038/nbt1043

Slade AJ, McGuire C, Loeffler D, Mullenberg J, Skinner W, Fazio G, Holm A, Brandt KM, Steine MN, Goodstal JF, Knauf VC (2012) Development of high amylose wheat through TILLING. BMC Plant Biol 14(12):69. https://doi.org/10.1186/1471-2229-12-69

Song XJ, Huang W, Shi M, Zhu MZ, Lin HX (2007) A QTL for rice grain width and weight encodes a previously unknown RING-type E3 ubiquitin ligase. Nat Genet 39:623–630. https://doi.org/10.1038/ng2014

Sreelakshmi Y, Gupta S, Bodanapu R et al (2010) NEATTILL: a simplified procedure for nucleic acid extraction from arrayed tissue for TILLING and other high-throughput reverse genetic applications. Plant Methods 6:3. https://doi.org/10.1186/1746-4811-6-3

Szurman-Zubrzycka ME, Zbieszczyk J, Marzec M et al (2018) HorTILLUS-A rich and renewable source of induced mutations for forward/reverse genetics and pre-breeding programs in Barley (Hordeum vulgare L.). Front Plant Sci 21(9):216. https://doi.org/10.3389/fpls.2018.00216. PMID: 29515615; PMCID: PMC5826354

Szurman-Zubrzycka M, Kurowska M, Till BJ, Szarejko I (2023) Is it the end of TILLING era in plant science? Front Plant Sci 14:1160695. https://doi.org/10.3389/fpls.2023.1160695

Taheri S, Abdullah TL, Jain SM et al (2017) TILLING, high-resolution melting (HRM), and next-generation sequencing (NGS) techniques in plant mutation breeding. Mol Breeding 37:40. https://doi.org/10.1007/s11032-017-0643-7

Talamè V, Bovina R, Sanguineti MC, Tuberosa R, Lundqvist U, Salvi S (2008) TILLMore, a resource for the discovery of chemically induced mutants in barley. Plant Biotechnol J. 6:477–485. https://doi.org/10.1111/j.1467-7652.2008.00341

Talebi AB, Talebi AB, Shahrokhifar B (2012) Ethyl methane sulphonate (EMS) induced mutagenesis in

Malaysian rice (cv. MR219) for lethal dose determination. Am J Plant Sci 3:1661–1665

Till BJ, Reynolds SH, Weil C, Springer N et al (2004) Discovery of induced point mutations in maize genes by TILLING. BMC Plant Biol. 4:12. https://doi.org/10.1186/1471-2229-4-12. PMID: 15282033; PMCID: PMC512284

Till BJ, Cooper J, Tai TH, Colowit P, Greene EA, Henikoff S, Comai L (2007) Discovery of chemically induced mutations in rice by TILLING. BMC Plant Biol 11:719. https://doi.org/10.1186/1471-2229-7-19

Tsai H, Howell TNR, Missirian V et al (2011) Discovery of rare mutations in populations: TILLING by sequencing. Plant Physiol 156:1257–1268

Uauy C, Paraiso F, Colasuonno P, Tran RK, Tsai H, Berardi S, Comai L, Dubcovsky J (2009) A modified TILLING approach to detect induced mutations in tetraploid and hexaploid wheat. BMC Plant Biol 28:9–115. https://doi.org/10.1186/1471-2229-9-115

Volkman MM, Martin JM, Hogg AC, Wright L, Hale C, Carr PM, Giroux MJ (2022) Durum wheat *Teosinte Branched1* null mutations increase tillering. Crop Sci. 62:1522–1530

Von Grebmer K, Saltzman A, Birol E et al (2014) Synopsis: 2014 global hunger index: the challenge of hidden hunger. Intl Food Policy Res Inst 83

Wang GX, Imaizumi T, Li W, Saitoh H, Terauchi R, Ohsako T, Tominaga T (2008) Self-EcoTILLING to identify single-nucleotide mutations in multigene family. Pestic Biochem Phys 92:24–29

Wen N, Osorio CE, Brew-Appiah RAT (2022) Targeting Induced Local Lesions in the wheat *DEMETER* and *DRE2* genes, responsible for transcriptional derepression of wheat gluten proteins in the developing endosperm. Front Nutr 3(9):847635. https://doi.org/10.3389/fnut.2022.847635

Wittwer CT, Reed GH, Gundry CN, Vandersteen JG, Pryor RJ (2003) High-resolution genotyping by amplicon melting analysis using LCGreen. Clin Chem 49:853–860

World Health Organization (2019) The state of food security and nutrition in the world 2019: safeguarding against economic slowdowns and downturns (vol. 2019). Food & Agriculture Org

Wu JL, Wu C, Lei C et al (2005) Chemical- and irradiation-induced mutants of indica rice IR64 for forward and reverse genetics. Plant Mol Biol. 59:85–97. https://doi.org/10.1007/s11103-004-5112-0. PMID: 16217604

Xin Z, Wang ML, Barkley NA, Burow G, Franks C, Pederson G, Burke J (2008) Applying genotyping (TILLING) and phenotyping analyses to elucidate gene function in a chemically induced sorghum mutant population. BMC Plant Biol 14(8):103. https://doi.org/10.1186/1471-2229-8-103

Yu S, Li M, Dubcovsky J, Tian L (2022) Mutant combinations of lycopene ε-cyclase and β-carotene hydroxylase 2 homoeologs increased β-carotene accumulation in endosperm of tetraploid wheat *(Triticum turgidum* L.) grains. Plant Biotechnol J 20:564–576. https://doi.org/10.1111/pbi.13738

Zhang J, Xiong H, Burguener GF et al (2023a) Sequencing 4.3 million mutations in wheat promoters to understand and modify gene expression. Proc Natl Acad Sci USA 120:e2306494120. https://doi.org/10.1073/pnas.2306494120

Zhang J, Li C, Zhang W, Zhang X, Mo Y, Tranquilli GE, Vanzetti LS, Dubcovsky J (2023b) Wheat plant height locus *RHT25* encodes a PLATZ transcription factor that interacts with DELLA (RHT1). Proc Natl Acad Sci USA 120(19):e2300203120. https://doi.org/10.1073/pnas.2300203120

Meta-Analysis of QTL in Durum Wheat: State of the Art and Perspectives in the Post-genome Era

Daniela Marone, Jose Miguel Soriano, and Elisabetta Mazzucotelli

Abstract

Quantitative Trait Loci (QTL) meta-analysis relies on compiling QTL data from multiple independent genetic backgrounds and/or environments to detect stable and reliable meta-QTL (mQTL), and potentially increase their mapping resolution, thus prioritizing genomics regions to implement marker-assisted selection (MAS) and, eventually, positional cloning. This review summarizes insights obtained by meta-analyses based on QTL mapped in durum wheat, considering all relevant agronomic traits from biotic and abiotic resistance, to yield and quality performance. Meta-analysis output may increase selection accuracy and efficiency, thus enhancing genetic gains in durum wheat breeding programs. The availability of the durum wheat reference genome facilitates hypothesizing candidates among genes that underline meta-QTL, thus opening to integration with transcriptome data, haplotype information, and marker–trait associations (MTAs) from genome-wide association studies. Lastly, the inclusion of both bread and durum wheat studies in meta-QTL analyses identifies shared and specific trait variation and thus might suggest loci to transfer across the two species in the breeding programs.

D. Marone
Council for Agricultural Research and Economics (CREA) – Research Centre for Cereal and Industrial Crops, Foggia, Italy
e-mail: daniela.marone@crea.gov.it

J. M. Soriano
Department of Agricultural and Forest Sciences and Engineering, University of Lleida, Lleida, Spain
e-mail: josemiguel.soriano@udl.cat

E. Mazzucotelli (✉)
Council for Agricultural Research and Economics (CREA) – Research Centre for Genomics and Bioinformatics, Fiorenzuola d'Arda (PC), Italy
e-mail: elisabetta.mazzucotelli@crea.gov.it

1 Meta-Analysis of QTL: Definition and Methods

The meta-analysis is a statistical tool that combines quantitative trait loci (QTL) data for the same trait from different sources in a single study to verify if QTL are common to different crosses, although their alleles can be different. Specific features of the bi-parental mapping populations (parents, population type, and size) and environments used for phenotyping strongly influence the results of QTL analysis (Collins et al. 2008). Consequently, the transfer of QTL and associated markers across different breeding programs is risky since the effectiveness of a QTL in another genomic background and in a different environment cannot be ensured. Meta-analysis of QTL represents a powerful approach to circumvent this issue (Salvi and Tuberosa 2015). By compil-

R. Tuberosa et al. (eds.), *The Durum Wheat Genome*, Compendium of Plant Genomes,
https://doi.org/10.1007/978-3-031-84689-2_9

ing QTL data from multiple independent genetic backgrounds and/or environments, the meta-analysis of QTL detects stable and reliable mQTL, while determining their global effect and potentially reducing the confidence interval (CI) (Soriano et al. 2021). An additional advantage is the improvement in mapping resolution for the selection of more informative molecular markers for MAS and for the discovery of relevant candidate genes upon physical mapping of mQTL on the reference genome of the target crop (Soriano et al. 2020). According to Löffler et al. (2009) to consider a mQTL for MAS, three criteria are required: small supporting intervals of the mQTL, high number of initial QTL, and high phenotypic variance explained (PVE) by the initial QTL. Together, mQTL analysis may increase selection accuracy and efficiency, thus enhancing genetic gains in crop breeding programs.

A meta-analysis method suitable for QTL data was developed by Goffinet and Gerber (2000). This method indicates the most likely number of "real" QTL underlying a pool of QTL of independent experiments. It also provides consensus positions for these QTL. The approach involves several steps (Fig. 1), including the construction of a consensus map based on individual linkage maps or, if available, the use of the organism reference map, the projection of QTL identified from each study onto the consensus map, and, finally, the estimation of meta- or consensus-QTL effects. The most specialized and user-friendly software to run this kind of analysis is Biomercator (Arcade et al. 2004), widely contributing to identify mQTL in different crops (reviewed in Mores et al. 2021). It provides a tool to automate map compilation and compute the QTL meta-analysis.

It is implicit that the effectiveness of meta-analysis of QTL and its potential in dissecting the genetic bases of complex quantitative traits is related to the number of studies and QTL combined, and diversity of populations considered. Additionally, mQTL can be eventually corroborated by MTAs reported in genome-wide association studies (GWAS) and ortho-mQTL. The last are the results of comparative genomic approaches which seek mQTL at genomic regions that are co-linear between related crop species, thus allowing transferability of information between different species and facilitating the identification of causal genes among the synthetic conserved genes.

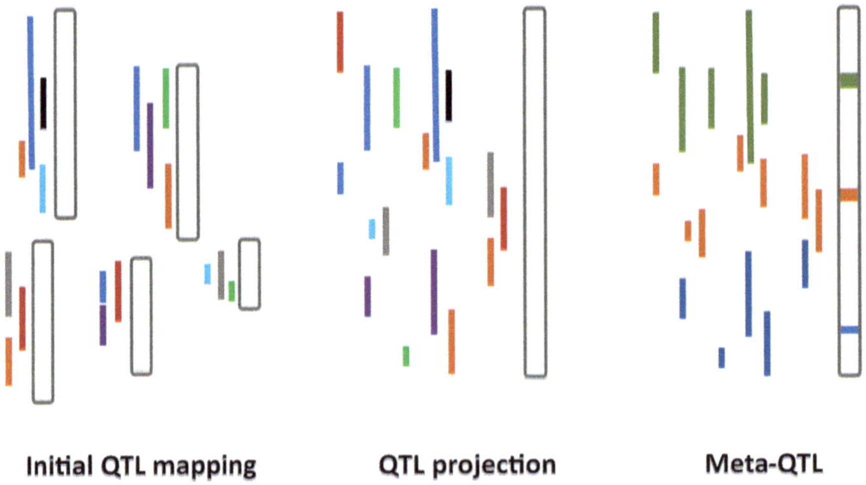

Fig. 1 The meta-QTL approach: from an initial QTL identification in the same chromosome of different mapping populations, where different colors of QTLs represent the different traits, a projection against a reference genome or consensus map (one chromosome is represented) is done. From this projection, the meta-analysis identifies consensus genomic regions with lower confidence interval (with the same colors different QTLs for the same trait belonging to the same meta-QTL)

2 Meta-Analysis of QTL in Wheat

The availability of many genetic resources and corresponding linkage mapping studies for many agronomic traits have provided an opportunity to conduct many meta-analyses of QTL in common wheat. A noteworthy example of a comprehensive and successful meta-analysis of QTL conducted in bread wheat for yield-related traits was recently published by Saini et al. (2022). It is on a collection of 8998 QTL from 230 reports published during 1999–2020 on yield and related traits. The study resulted in the identification of 141 mQTL, with an 8.8 average fold reduction in CI. Confirmation by co-location with MTAs from GWAS was also achieved for half of consensus QTL. Among them 63 mQTL were found stable and robust, because each based on at least ten QTL, while 13 mQTL were considered suitable for breeding application, having small CI, large LOD value, and high level of phenotypic variation explained.

About durum wheat, many QTL studies have been performed to investigate biotic and abiotic stresses, yield, phenology, and quality traits, as also recently reviewed by Colasuonno et al. (2021). However, just a few meta-analyses of QTL have been conducted specifically in this wheat species, although some meta-analyses on common wheat also included studies on tetraploid wheat mapping populations. We have collected all meta-analyses including QTL identified in tetraploid wheat, as reported in Table 1. The deployment of common genotyping technologies (mostly Illumina 90 K Wheat SSNP Beadchip Array) supported these integrative analyses, which can provide useful indications to help breeders for development of new durum cultivars. Combining data from both common and durum wheat can also clarify which QTL are shared or that, being specific of one species, could be advantageously transferred among them.

3 Meta-QTL Analysis in Durum Wheat for Disease Resistance

Many QTL mapping studies have been conducted to identify sources of genetic disease resistance in the tetraploid germplasm. About QTL meta-analysis, leaf and yellow rusts have been targeted in durum wheat so far, but also other diseases like powdery mildew and tan spot, in all cases combined with many more QTL studies reported for bread wheat. In these meta-analyses, both field resistance to a specific crop pathogen and different race/isolate specific resistance for the same pathogen have been considered. Thus, an interesting outcome of these analyses is the opportunity to identify genomic regions conferring resistance to multiple races of the same pathogen.

Among biotic factors, rusts (leaf, stripe and stem) are considered the most damaging pathogens for the durum wheat crop. Soriano and Royo (2015) compared the genomic regions controlling leaf rust (*Puccinia triticina*) resistance across different wheat populations, including three Recombinant Inbred Lines (RIL) populations of tetraploid wheat, in which resistance to both adult and seedling plant stage was evaluated. Through meta-analysis, the 90% of 327 initial QTL for leaf rust resistance clustered in 35 consensus QTL over the genome, thus the number of QTL was narrowed down by 73% and the supporting interval by 50% in comparison with those reported for the initial QTL. Recently, Amo and Soriano et al. (2021) updated this analysis by incorporating mapping populations genotyped with SNP arrays. A total of 393 QTL from 50 studies were collected and grouped in 75 mQTL that were refined to 15 high-confidence mQTL (hcmQTL) based on the length of the CI (10 cM or 20 Mb). The candidate genes discovered within the hcmQTL were checked for differential expression in silico using data from three transcriptome studies, resulting in 92 differentially expressed genes (DEGs). In addition, the co-location with known *Lr* genes was investigated. Eight known leaf rust resistance genes (*Lr13, Lr14a, Lr41, Lr42, Lr46, Lr60, Lr63,* and *Lr68*) were co-localized with mQTL. Notably, authors found co-location of the mQTL on chr. 7B with two leaf rust genes, *Lr68* and *Lr14a*. *Lr68* is known to confer adult plant resistance in bread wheat to the majority of *P. triticina* isolates with low-to-medium infection types and is linked to small but noticeable leaf tip necrosis. *Lr14a*, identified in emmer and durum wheat and transferred to common wheat, confers seedling resis-

Table 1 Meta QTL studies for yield, quality, biotic and abiotic stress, root-related traits from durum wheat and durum/bread wheat QTL studies

Trait	Reference	N studies	N crosses	N Traits	N single QTL	N mQTL	initial QTL CI	MQTL CI	Breeding mQTL
Yield-related	Soriano et al. (2017)	24	23	26	409	77	14.5	5.44	14
	Avni et al. (2018)	9	8	1	53	15	nr	nr	1
	Arriagada et al. (2022)	25	24	19	729	76	24.3	6.8	12
Quality	Roselló et al. (2018)	20	21	5	345	45	18	6.4	3
	Colasuonno et al. (2019)	19	12[a]	2	172	81	8.8	8.8	n.r
	Soriano et al. (2021)	15	12	17	171	85[a]	25.5[a]	4.9[a]	10
Biotic stress	Marone et al. (2013)	23	18	2	101	24	23.4	7.3	?
	Soriano and Royo (2015)	19	20	14	144	35	20.8	8.4	5
	Liu et al. (2020a, b)	15	18	14	104	19	14.8	4.9	?
	Soriano et al. (2021)	24	23	8	71	85[a]	25.5[a]	4.9[a]	10
	Amo and Soriano (2021)	50	57	21	393	75	11.8	5.4	15
Abiotic stress	Soriano et al. (2021)	3	3	8	48	85[a]	25.5[a]	4.9[a]	10
	Acuna Galindo (2012)	30	24	3	858	66	21.6	5.8	?
Root	Soriano et al. (2021)	4	4	1	79	85[a]	25.5[a]	4.9a	10
	Petrarulo et al. (2015)	11	13	30	141	22	nr	9.4	?
	Iannucci et al. (2017)	13	16	36	117	34	12.8	5.6	?

CI mean size of the confidence interval in cM
cM centimorgan
n.r. not reported
[a]Includes data from abiotic and biotic stress resistance, and root-related QTL

tance, and is thought to be associated with the stem rust and powdery mildew resistance genes *Sr17* and *Pm5*. Consequently, this mQTL could provide an interesting region of multiple disease resistance.

QTL for resistance to stripe rust (*Puccinia striiformis* f. sp. *tritici*) from both bread and durum wheat studies have been recently target of meta-analysis (Jan et al. 2021; Kumar et al. 2023). Jan et al. (2021) collected a total of 353 QTL from 2000 to 2020, 184 of which were incorporated into 61 mQTL identifying ten important genomic regions that included six breeders' mQTL (PVE >20%) and four mQTL hotspots were selected to be used in wheat breeding. The search for candidate genes reported 409 transcripts that encoded known R proteins or showed in silico differential expression to stripe rust infection. Kumar et al. (2023) accomplished a more comprehensive QTL meta-analysis on bread wheat, again including some durum wheat mapping populations. They collected 505 QTL for stripe rust resistance from 101 linkage-based interval mapping studies, and the QTL were projected on a new consensus linkage map involving 138,574 markers. Despite the higher number of

QTL collected, 67 mQTL were identified, a number that is close to the mQTL identified in the previous meta-analysis conducted by Jan et al. (2021), suggesting that the variability for yellow rust resistance in wheat has been almost fully explored. Interestingly, 44 mQTL were validated by co-localization with marker–trait associations reported in 20 earlier GWAS conducted for stripe rust resistance in wheat, or with known major resistant genes, as *Yr5, Yr7, Yr16, Yr26, Yr30, Yr43, Yr44, Yr64*, and *YrH52*. Furthermore, transcriptome data about wheat plant reaction to yellow rust infection identified 59 differential expressed genes with defense-related functions which are promising candidates of the hcmQTL. Among them, NBS-LRR domain-containing genes, protein kinase family genes, receptor-like kinases (RLKs), WRKY transcription factors, and genes encoding other zinc finger-containing proteins, all known to have major functions in host–pathogen interactions.

Powdery mildew (*Blumeria graminis* f. sp. *tritici*) represents another damaging disease of wheat. An attempt to study the genetic bases of powdery mildew resistance in wheat through meta-analysis was reported by Marone et al. (2013). This report included both durum and bread wheat and collected QTL data from a total of 23 mapping studies and Pm resistance genes from 58 different references. Twenty-four mQTL were found with a mean 3.2-fold reduction in CIs relative to those of the original QTL. Very interesting is the mQTL located on chromosome 2A that was characterized by a very restricted confidence interval (0.6 cM) and for which the resistance gene *Pm4b* has been found to be located just under, or the two mQTL mapped on chromosome 1A which are in a region involving three *Pm3* genes that might represent an allelic series at the *Pm3* locus. Generally, the very small genetic intervals of the mQTL obtained that include clusters of genes involved in the control of this trait as reported by Marone et al. (2013) represent an important tool that can be used for marker-assisted selection/pyramiding in wheat-breeding programs, or for map-based cloning.

Lastly, the foliar disease tan spot, caused by the fungal pathogen *Pyrenophora tritici-repentis* (Ptr), has been target of a meta-analysis in bread and durum wheat (Liu et al. 2020a). More than 100 QTL linked to resistance to tan spot induced by a variety of *Ptr* races and isolates have been identified in literature. They clustered in 19 meta-QTL and three out of them located on chromosomes 2A, 3B, and 5A showed large genetic effects and have been found to confer resistance to multiple races. The genetic determinant of such mQTL could be single resistant genes with low race -specificity or clusters of resistant genes closely located on wheat genome. Lines resistant to multiple races are very interesting as they could be parents in durum wheat breeding programs.

4 Meta-QTL Analysis in Durum Wheat for Abiotic Stresses and Root-Related Traits

Abiotic stresses, and drought events with heat waves above all, represent a serious problem for durum wheat as they could strongly affect yield and quality and they are of major concern in the frame of current climate changes. Thus, interesting genomic regions involved in the response to a combination of heat and drought stress were identified from an analysis conducted in common wheat, but also including five tetraploid populations (Acuna Galindo 2012). The meta-analysis was launched on a QTL database with 854 QTL from 30 different reports and on 80 different traits related to drought stress, heat stress, and adaptive traits in non-stressed environments. The analysis resulted in 66 mQTL regions distributed throughout the wheat genome, with at least one mQTL detected on each of the 21 wheat chromosomes. A reduction in the average 95% CI from 21.6 cM to 5.8 cM was observed in that work when comparing the CI of individual QTL present within mQTL regions to the CI of the mQTL. Most regions co-localized for both heat and drought stress, although both drought and heat stress specific mQTL regions were also identified.

Root-related traits are considered of great importance for yield performance under drought conditions, particularly under the current climate changes and global warming, even if they remain poorly understood (Ober et al. 2021; Tuberosa et al. 2021; Langridge et al. 2022). A detailed analysis of the genetic basis of morphological root traits under controlled conditions has been carried out in a durum wheat Recombinant Inbred Lines (RIL) population by Petrarulo et al. (2015) where several traits were considered (length, surface area, average diameter, volume, and number of tips). The same authors compared their results with those reported in literature for root architecture, also including mapping studies in bread wheat, by carrying a meta-analysis. From a collection of a total of 141 QTL linked to growth and root apparatus 22 mQTL resulted, most of which involved in the control of more than one root trait. An interesting mQTL was for located on chromosome 2A as it merged an individual QTL for root surface area and two additional QTL, the first of which affected root length, surface area, volume, and number of tips, and the second one affected the number of root forks, root length, surface area, and volume. Analogously, the chromosome regions involved in the control of shoot and root architecture identified in a durum wheat x *T. dicoccum* segregating population have been compared with other available data for the same traits (Iannucci et al. 2017). In particular, the meta-analysis performed by these authors on QTL collected from 12 independent studies resulted in 23 mQTL involved in one or more root-related traits, whereas three regions remained novel after the analysis and could be of interest to validate them as targets for breeding programs for optimization of root function in field conditions.

Finally, Soriano and Alvaro (2019) performed a meta-analysis for root-related traits in a survey of 30 studies, three of them carried out in durum wheat, collected from 2006 to 2017. Authors reduced the 754 initial QTL to 94 mQTL, eight of them considered as breeding mQTL following the Löffler et al. (2009) approach, therefore highly promising to achieve effective root improvement in wheat. Candidate gene analysis found 35 genes controlling root traits and/or drought stress tolerance. Notably, this comprehensive collection of QTL for root traits highlighted the presence of some significantly enriched genomic regions, particularly on chromosomes 1A and 1B, 2A and 2B, 3A and 3B, and 4B.

5 Meta-QTL Analysis for Yield

Yield is the most important trait in plant breeding but at the same time the most complex one involving numerous genes with low effect and highly influenced by the environment. Thus, its genetic dissection becomes difficult. To reduce this complexity, grain yield is divided in different components, mainly grain weight, grain number or spike number, assuming a decrease of the number of genomic regions involved in each trait, besides an increase in trait heritability (Graziani et al. 2014).

Meta-QTL analyses for yield and related traits have been conducted by different authors in wheat (Zhang et al. 2010; Acuña-Galindo et al. 2015; Tyagi et al. 2015; Liu et al. 2020b; Xie and Sparkes 2021; Yang et al. 2021); however, the studies involving durum wheat mapping populations are less represented.

More recently, Soriano et al. (2017) reported the analysis of 477 QTL for yield and phenology from both bread and durum wheat projected onto the comprehensive durum wheat consensus map (Maccaferri et al. 2015). Among them, 409 QTL were integrated in 71 mQTL (Table 1). As reported by the authors, QTL meta-analysis was able to reduce the length of the confidence intervals up to 63%. Among the reported traits, grain weight and number were the most represented in the QTL studies (19 and 16 studies, respectively). The integration with phenological traits allowed the authors to find pleiotropic genome regions involving both type of characters. The study reported by Avni et al. (2018) used only tetraploid wheat populations to dissect the genetic architecture for grain weight in durum wheat. This study combined a QTL analysis based on the Svevo x Zavitan Recombinant Inbred Lines (RIL) population with a QTL meta-analysis laying on other seven mapping populations among tetraploid wheats. Authors found a meta-QTL on linkage group 6A resulting from the introgression from wild wheats that affected positively the grain weight. In both cases, QTL meta-analysis was successfully applied to reduce the number of genomic regions and their size for candidate gene analyses. More recently, Arriagada et al. (2022) carried out a QTL meta-analysis for yield-related traits in durum wheat under different water regimes, as water deficit is one of the abiotic stresses most influencing yield losses. The authors review a total of 724 QTLs for 19 different traits. From them, 468 QTLs were reported under rainfed conditions and 256 under irrigation. From 590 projected QTLs, 421 were grouped into 76 mQTLs in both water conditions, with 12 genomic regions showing stable mQTLs. From these mQTLs authors reported the selection of 15 genes differentially expressed under drought stress. Notably, a mQTL on chromosome 2A (*YIELD_MQTL2A.2*_D) linked to TGW and HI co-localized with a mQTL (*durumMQTL2A.3*) previously described for traits related to abiotic stress (vegetation index and chlorophyll content) in durum wheat (Soriano et al. 2021). Among the candidate genes of this locus, there is a CBL-interacting protein kinase 2-like, whose homologue conferred drought tolerance by regulating stomatal closure upon over-expression in transgenic tobacco plants (Wang et al. 2016). This strong MQTL for stress and for yield under rainfed conditions can be useful in durum wheat breeding programs, in which the objective is to increase grain yield under drought conditions.

6 Meta-QTL Analysis for Quality

During the last century, in many countries wheat breeding efforts were concentrated on yield increase, being grain quality a secondary objective (Sanchez-Garcia et al. 2015). In the last three decades, the breeders' interest for improvement of quality traits greatly increased driven by new requirements of the industry and consumers (Colasuonno et al. 2019; Nigro et al. 2024). For durum wheat, important quality traits are grain protein content (GPC) and quality, which affects gluten strength (GS), grain color as consequence of carotenoid content, test weight (TW), and more recently also content in anti-oxidant compounds mainly as phenolic acids. However, improving the quality of durum wheat presents several limitations due to the quantitative nature of the trait and the costly and time-consuming evaluation of end-products, and for some traits the recurrent trade-off with grain yield. In this sense, the identification of QTL involved in the trait variation and its linked molecular markers will result of special interest for durum wheat quality breeding. The complexity of the QTLome for quality in durum wheat has been reduced by means of QTL meta-analysis as reported by several authors (Roselló et al. 2018; Colasuonno et al. 2019; Soriano et al. 2021). The study reported by Roselló et al. (2018) combined a GWAS in durum wheat Mediterranean landraces with a QTL meta-analysis using 20 published QTL studies for five quality traits: GPC, GS, yellow index (YI), yellow pigment content (YPC), and TW. Both analyses were mapped against the consensus map of durum wheat reported in Maccaferri et al. (2015) to compare the genomic positions for MTAs and mQTL. From a total of 345 QTL, 249 were projected onto the consensus map and 204 of them were grouped in 45 consensus regions with at least two original QTL. These results from landraces could be of special interest to enhance quality traits in breeding programs as protein content has decreased because of past breeding activities, which concentrated on increasing yield potential. Authors found that the DArT marker wPt-2737 located on chromosome 7B showed the greatest variance explained for protein content and was present mainly in east Mediterranean landraces.

The study from Colasuonno et al. (2019) reported the analysis of 172 QTL for carotenoid content from 19 studies, including those resulted from biparental populations (71) and GWAS panels (101). QTL were positioned in the durum wheat consensus map (Maccaferri et al. 2015). However, although authors did not perform a QTL meta-analysis, they identified 81 clusters of different QTL based on the overlapping of the QTL confidence intervals.

7 The Durum QTLome: Collecting QTL for Multi-traits

Of major interest are meta-analysis considering different traits in order to identify regions with major QTL effects and possibly with pleiotropic and/or tight linkage effects, both rather important factors to consider in order to more effectively leverage the QTLome haplotpe diversity (Salvi and Tuberosa 2015). Recently, a meta-QTL meta-analysis for both biotic (including rusts, Fusarium head blight, powdery mildew) and abiotic stresses (drought tolerance) and 17 quality traits has been carried out specifically in durum wheat (Soriano et al. 2021). From 41 independent studies and 34 different mapping populations, a total of 368 single QTL for 36 phenotypic traits were collected, divided in the different trait categories as follows: 171 for quality-related traits, 127 to abiotic stress, and 71 to biotic stress. Meta-analysis resulted in 85 mQTL involving different traits. The number of QTL per mQTL ranged from two to 11, and the number of traits from one to seven, with six mQTL involving five or more different traits. An interesting example was the mQTL on chromosome 2B that included the higher number of single QTL for quality, leaf rust resistance, and abiotic stress. Forty-one mQTL (47%) were derived from the clustering of QTL from three or more different studies on different parental lines; thus, they can be considered stable across environments, while 17 mQTL were considered the most suitable for MAS according to Löffler et al. (2009). Authors also compared the position of mQTL with significant associations from GWAS. Notably, 48 out of 352 MTAs were found within mQTL regions. The traits LR resistance, Normalized Difference Vegetation Index (NDVI) and YPC showed the highest number of MTAs per biotic stress, abiotic stress and quality, respectively. Moreover, known major genes for adaptation were found coincident with mQTL. For example, the mQTL *DurumMQTL5A.5* and *DurumMQTL7B.9* included the vernalization genes *Vrn-A1* and *Vrn-B3*, respectively, being useful for breeders to develop spring habit without cold requirements for flowering, while the *DurumMQTL4B.4* includes the dwarfing. Lastly, a candidate gene approach was also performed through the Svevo genome sequence (Maccaferri et al. 2019) and in silico RNAseq analysis (Ramírez-González et al. 2018) on the most relevant mQTL. Valuable candidate genes within marker peaks have been retrieved, potentially useful, once duly validated, to speed up breeding programs. Among them, authors found an interesting case of study in a mQTL located on chromosome 2B in a region with QTL for root-related traits (Maccaferri et al. 2016) and soil-borne cereal mosaic virus resistance (Maccaferri et al. 2012), where an NBS-LRR gene was overexpressed in both biotic and abiotic stress conditions. Authors suggested that this may indicate a link between the two traits with a pleiotropic effect between root development and pathogen growth.

Statistical approaches for meta-analysis combining QTL from linkage and MTA from association mapping have not been implemented so far. A simple, but still effective approach is collecting and anchoring QTL and MTAs on a common framework (consensus map or reference genome). Although most of meta-analysis advantages are missed, co-location of QTL and MTA allows identification of major genomic regions and prioritization of targets for in-depth genetic analysis and breeding. Thus, QTL and MTAs for major trait categories reported in tetraploid wheat, including domestication, fertility (grain-yield-related traits, GY), heading date (HD), plant height (PH), biomass, grain yellow pigment content (GYPC), grain quality in general, disease resistance, and root architecture-related traits, were considered for the projection onto the assembly of the reference durum wheat genome (Maccaferri et al. 2019). A dataset of a total 2105 QTL signals was compiled based on published literature of QTL (n = 1162) and genome-wide association studies (n = 943) in *T. turgidum* biparental and tetraploid wheat collections, respectively (as of June 30th, 2018). Out of 2105 recorded mQTL, the summary included 47 for domestication, 775 for grain yield grain-yield-related traits, 200 for phenology (mainly heading date), 104 for plant height, 133 for biomass, 407 for disease response, 185 for grain quality in general, 233 for root architecture-related traits, and 21 for other traits. They were projected to DW genome assembly using a high-density reference binned Svevo × Zavitan linkage map as genetic framework anchored to the physical assembly, enriched of additional DArT™, SSR, and EST-SSR/STS markers with matching genetic and physical locations. It was found that near 60% of all QTL (1272 over 2105) located in highly recombinogenic distal regions, with a QTL density of 6.2 mQTL/10 Mb. The most prominent examples were the 46 QTL for shattering (brittle rachis phenotype), threshability, threshing time, threshing efficiency, and tenacious glumes phenotypes. Twenty-one out of 46 were detected within selection cluster intervals, most of them located in the distal, recombining regions. To deepen this analysis, a comparative alignment between QTL and selection signals was conducted to extract indications about relevant traits and corresponding regions relevant to domestication/adaptive evolution across the selection and breeding history of tetraploid wheat. Thus, domestication-related QTL from mapping populations obtained by crossing (wild) emmer with a modern durum (most frequently Langdon or Svevo) were informative for signals detected in the (wild)-emmer to durum transitions. Indeed, among 46 QTL for shattering (brittle rachis phenotype), threshability, threshing time, threshing efficiency and tenacious glumes phenotypes, 21 were detected within selection cluster intervals. Similarly, strong and narrow selection signals in emmer-durum wheat transition were associated with QTL for grain protein content, a trait that underwent a progressive dilution during domestication, due to the continuous unconscious selection for yield capacity mainly driven by starch accumulation and kernel weight. Co-location was also observed for many QTL for yield-related traits and disease response and signals detected in the landrace-cultivar transition highlighting main effects of modern breeding. Finally, it is noteworthy to observe that the QTL peak of heading time on chromosome 2A and 2B obtained from QTL mapping studies based on the Illumina 90 K SNP array was projected with a precision of 1 to 2 Mb maximum close to the *Ppd-A1* and *Ppd-B1* corresponding genes in Svevo, highlighting the importance of relying on a high-quality reference genome as well as good heritability of the target traits.

8 Ortho-Meta-QTLs: Analyzing QTL Across Crop Species

Ortho-meta-QTL analysis is a method used in genetics to study the genetic basis of complex traits across species. It combines two approaches: orthologous comparative mapping and meta-analysis of QTL studies. Wheat is a crop with a large and complex genome, and many important traits, such as yield, disease resistance, and grain quality, are controlled by multiple genes and QTL whose effect may also be due to haplotypic variation in noncoding sequences. Comparative genomics with less complex species (e.g., rice and barley) by ortho-mQTL approaches will allow to gain insights into the genetic control of complex traits in wheat and potentially identify candidate genes or genetic markers that can be targeted for breeding purposes. Ortho-MQTL mining can validate crop specific analyses and suggest key regulatory genes with evolutionary history and conservative function (Khahani et al. 2020). Such knowledge can aid in the development of improved wheat varieties with desirable agronomic traits. Some examples of ortho-meta-QTL analysis are reported. Kumar et al. (2021) identified ortho-mQTLs for thermotolerance in wheat (both common and durum) and rice. Although authors identified 85 wheat mQTLs, only seven of them had their correspondence in the rice genome. Among them, two ortho-mQTLs were considered the most promising for breeding as they revealed gene collinearity in the syntenic region. Saini et al. (2021) identified ortho-MQTLs and candidate genes for nitrogen use efficiency (NUE) and root system architecture (RSA) between wheat, maize, and rice. The syntenic regions of 19 selected wheat mQTLs were compared with previously published maize mQTLs and nine ortho-mQTLs were identified for RSA. No ortho-mQTLs were identified in maize for NUE. To identify candidate genes for NUE between wheat and rice, a set of 60 rice genes were used. Authors found 49 wheat orthologues in mQTL regions for 35 rice genes. From them, 18 genes showed more than one wheat orthologue, whereas 17 genes showed only one orthologue in wheat mQTLs. Another comparative genomic analysis among wheat, rice, barley, and maize performed by Saini et al. (2022) allowed identifying wheat gene homologues under the mQTL and defining 24 ortho-mQTL for grain yield. Singh et al. (2022) identified ortho-meta-QTLs and candidate genes for grain Fe and Zn contents in wheat. From 11 mQTLs, authors found high confidence CGs in four mQTLs that were used for comparative mapping with rice and maize, and finally, ortho-mQTLs in rice and maize were identified for two of the wheat mQTLs. In a recent study, Kumar et al. (2023) used data from experiments under water stress conditions to identify ortho-mQTLs in wheat, maize, and rice for yield components. Using the synteny information among wheat, rice, and maize, 12 ortho-mQTLs were identified. Out of them, eight were conserved across all the three species and four were conserved either among wheat and maize or wheat and rice. Forty-nine collinear genes were found in the 12 ortho-mQTLs between wheat and maize and 28 collinear genes with rice.

9 Limitations and Perspectives of Meta-Analysis of QTL in Durum Wheat

Although QTL meta-analysis has been reported as a powerful approach for reducing confidence intervals of QTL regions and to identify linked markers for specific traits (Martinez et al. 2016), regrettably it has not been widely applied for MAS and/or positional cloning in wheat. The use of mapping populations with different marker coverage can underestimate mQTL positions; therefore, QTL with large confidence intervals should not be considered for meta-analysis. Additionally, an initial step should be performed prior to meta-analysis to identify suggestive QTL with low LOD value to preferably exclude from the analysis. The use of the first linkage maps developed for a species without anchor markers to reference or consensus maps can make difficult QTL projection to these maps. Likewise, the use of some old types of molecular markers (e.g., RFLP and AFLP) without physical position in the genome sequence can restrict their use for candidate gene search. Lastly, meta-analyses lye on the amount of single studies provided by the community of geneticists of the target crop. The small size of the durum wheat community is, therefore, a limitation, that can be overcome only combing durum together with bread wheat mapping studies, as reported in many examples and clearly advocated in Maccaferri et al. (2015). The inclusion of both bread and durum wheat studies in meta-QTL analyses could lead to know which loci are specific to tetraploid and hexaploid wheat species, respectively, and in the meanwhile high and stable loci could be transferred across the two species in the breeding programs. Notably, extensive and advanced resources are now available for durum wheat genomics that can support linkage and association mapping studies and therefore also future meta-analyses. These include large germplasm reference collections representative of the tetraploid genetic diversity (the so-called TGC, Tetraploid wheat Germplasm Collection, Maccaferri et al. 2019, and GDP-Global Durum Panel, Mazzucotelli et al. 2020) and their genotyping data obtained through the common widely used Illumina technology (90 K Infinium iSelect wheat SNP Chip). Currently, the availability of the genomic sequence in durum wheat (Maccaferri et al. 2019) offers new perspectives in meta-analysis as it allows for the projection of meta-QTL directly onto the genome, thus facilitating the search for genes underpinning a certain trait. Haplotype information could be also integrated in this analysis, and much effort should be done to integrate data from QTL mapping and genome-wide association studies, still far from being reached.

Meta-GWAS has shown to be powerful in dissecting complex human diseases, but its application in major crops is limited to a few examples. In meta-GWAS, summary statistics (such as SNP effects and their standard errors) of multiple datasets are used to calculate a global p-value for each SNP considering all environments and traits and thus it should result in higher power and less false positive associations compared to individual GWAS studies (Bolormaa et al. 2014). Although some software for meta-GWAS (i.e., Metal) require complete phenotypic records for all individuals in all experiments, besides comprehensive SNP matrix, the statistical frameworks for unbalanced phenotypes have been provided (Bolormaa et al. 2014; Battelfield et al. 2018). Two applications have been reported so far for common wheat, the first one on CYMMIT germplasm for milling and baking quality-related traits (Battelfield et al. 2018) and a second on germplasm of the Australian breeding program for agronomic and quality traits under a range of normal and heat-stressed environments (Joukhadar et al. 2021). About durum wheat, no meta-GWAS have been conducted so far; however, Marcotuli et al. (2022) identified QTL hotspots for quality traits by integrating independent GWAS studies in a reference map for durum wheat. A total of 10 GWAS studies, comprising 395 marker-trait associations (MTA) on 57 quality traits, with more than 1500 genotypes from nine association panels, were used to investigate consensus QTL hotspots representative of a wide durum wheat genetic variation.

Lastly, genomic data should be deeply integrated and combined with other -omics sciences such as transcriptomics or proteomics for prioritizing candidate genes of meta-QTL/MTA, or for combining QTL from traditional traits with those from molecular traits, the so called genetical genomics experiment.

Acknowledgements/Funding JMS is a Serra-Hunter fellow funded by the Generalitat de Catalunya. Financial support was obtained through the project "InnoWheatRes – Advanced breeding with Innovative strategies to improve biotic stresses resistance in durum wheat" under the National Recovery and Resilience Plan (NRRP), Mission 4, Component 2, Investment 1.1, (Grant Assignment Decree No. 104 adopted on 2 February 2022 by the Italian Ministry of Ministry of University and Research), and by the project "WheatSecurity- Identification and sustainbale deployment of wheat genetic diversity to enhance the resilience and security of the European food supply", under the Call SusCrop Era-net.

References

Acuna Galindo, MA (2012) Meta-analysis of wheat QTL regions associated with heat and drought stress. Theses and Dissertations 645

Acuña-Galindo MA, Mason RE, Subramanian NK, Hays DB (2015) Meta-analysis of wheat QTL regions associated with adaptation to drought and heat stress. Crop Sci 55(2):477–492

Amo A, Soriano JM (2021) Unravelling consensus genomic regions conferring leaf rust resistance in wheat via meta-QTL analysis. Plant Genome 15:e20185. https://doi.org/10.1002/tpg2.20185

Arcade A, Labourdette A, Falque M et al (2004) BioMercator: integrating genetic maps and QTL towards discovery of candidate genes. Bioinformatics 20:2324–2326. https://doi.org/10.1093/bioinformatics/bth230

Arriagada O, Gadaleta A, Marcotuli I, Maccaferri M, Campana M, Reveco S, Alfaro C, Matus I and Schwember AR (2022) A comprehensive meta-QTL analysis for yield-related traits of durum wheat (Triticum turgidum L. var. durum) grown under different water regimes. Front. Plant Sci. 13:984269. https://doi.org/10.3389/fpls.2022.984269

Avni R, Oren L, Shabtay G et al (2018) Genome based meta-QTL analysis of grain weight in tetraploid wheat identifies rare alleles of *GRF4* associated with larger grains. Genes 9:636. https://doi.org/10.3390/genes9120636

Battelfield SD, Sheridan JL, Silva LDCE et al (2018) Breeding-assisted genomics: applying meta-GWAS for milling and baking quality in CIMMYT wheat breeding program. PLoS One 13(11). https://doi.org/10.1371/journal.pone.0204757

Bolormaa S, Pryce JE, Reverter A, Zhang Y, Barendse W, Kemper K, et al (2014) A Multi-Trait, Meta-analysis for Detecting Pleiotropic Polymorphisms for Stature, Fatness and Reproduction in Beef Cattle. PLoS Genet 10(3): e1004198. https://doi.org/10.1371/journal.pgen.1004198

Colasuonno P, Marcotuli I, Blanco A et al (2019) Carotenoid pigment content in durum wheat (*Triticum turgidum* L. var *durum*): an overview of quantitative trait loci and candidate genes. Front Plant Sci 10:1347. https://doi.org/10.3389/fpls.2019.01347

Colasuonno P, Marcotuli I, Gadaleta A et al (2021) From genetic maps to QTL cloning: an overview for durum wheat. Plants 10:315. https://doi.org/10.3390/plants10020315

Collins NC, Tardieu F, Tuberosa R (2008) Quantitative trait loci and crop performance under abiotic stress: where do we stand? Plant Physiol. Jun;147(2):469–86. doi: https://doi.org/10.1104/pp.108.118117

Goffinet B, Gerber S (2000) Quantitative trait loci: a meta-analysis. Genetics 155:463–473. https://doi.org/10.1093/genetics/155.1.463

Graziani M, Maccaferri M, Royo C et al (2014) QTL dissection of yield components and morpho-physiological traits in a durum wheat elite population tested in contrasting thermo-pluviometric conditions. Crop Pasture Sci 65(1):80–95

Iannucci A, Marone D, Russo AM et al (2017) Mapping QTL for root and shoot morphological traits in a durum wheat *(T. dicoccum)* segregating population at seedling stage. Int. J Genomics 2017:6876393. https://doi.org/10.1155/2017/6876393

Jan I, Saripalli G, Kumar K et al (2021) (2021) meta-QTL analysis for stripe rust resistance in wheat. Sci Rep 11:22923. https://doi.org/10.1038/s41598-021-02049-w

Joukhadar R, Thistlethwaite R, Trethowan R et al (2021) Meta-analysis of genome-wide association studies reveal common loci controlling agronomic and quality traits in a wide range of normal and heat stressed environments. Theor Appl Genet 134(7):2113–2127. https://doi.org/10.1007/s00122-021-03809-y

Khahani B, Tavakol E, Shariati V, Fornara F (2020) Genome wide screening and comparative genome analysis for meta-QTLs ortho-MQTLs and candidate genes controlling yield and yield-related traits in rice. BMC Genomics 21:1–24. https://doi.org/10.1186/s12864-020-6702-1

Kumar S, Singh VP, Saini DK et al (2021) Meta-QTLs, ortho-MQTLs, and candidate genes for thermotolerance in wheat (*Triticum aestivum* L.). Mol Breeding 41:69. https://doi.org/10.1007/s11032-021-01264-7

Kumar A, Saini DK, Saripalli G et al (2023) Meta-QTLs, ortho-meta QTLs and related candidate genes for yield and its component traits under water stress in wheat (*Triticum aestivum* L.). Physiol Mol Biol Plants 29:525–542. https://doi.org/10.1007/s12298-023-01301-z

Langridge P, Alaux M, Almeida NF et al (2022) Meeting the challenges facing wheat production: the strategic research agenda of the global wheat initiative

Liu Y, Salsman E, Wang R et al (2020a) Meta-QTL analysis of tan spot resistance in wheat. Theor Appl Genet 133:2363–2375. https://doi.org/10.1007/s00122-020-03604-1

Liu H, Mullan D, Zhang C et al (2020b) Major genomic regions responsible for wheat yield and its components as revealed by meta-QTL and genotype–phenotype association analyses. Planta 252:65. https://doi.org/10.1007/s00425-020-03466-3

Löffler M, Schoen CC, Miedaner T (2009) Revealing the genetic architecture of FHB resistance in hexaploid wheat (*Triticum aestivum* L.) by QTL meta-analysis. Mol Breed 23:473–488. https://doi.org/10.1007/s11032-008-9250-y

Maccaferri M, Francia R, Ratti C et al (2012) Genetic analysis of *soil-borne cereal mosaic virus* response in durum wheat: evidence for the role of the major quantitative trait locus *QSbm.ubo-2BS* and of minor quantitative trait loci. Mol Breeding 29:973–988. https://doi.org/10.1007/s11032-011-9673-8

Maccaferri M, Ricci A, Salvi S et al (2015) A high-density, SNP-based consensus map of tetraploid wheat as a bridge to integrate durum and bread wheat genomics and breeding. Plant Biotechnol J 13(5):648–663. https://doi.org/10.1111/pbi.12288

Maccaferri M, El-Feki W, Nazemi G et al (2016) Prioritizing quantitative trait loci for root system architecture in tetraploid wheat. J Exp Bot 67:1161–1178. https://doi.org/10.1093/jxb/erw039

Maccaferri M, Harris NS, Twardziok SO et al (2019) Durum wheat genome highlights past domestication signatures and future improvement targets. Nat Genet 51:885–895. https://doi.org/10.1038/s41588-019-0381-3

Marcotuli I, Soriano JM, Gadaleta A (2022) A consensus map for quality traits in durum wheat based on genome-wide association studies and detection of ortho-meta QTL across cereal species. Front Genet 13:982418. https://doi.org/10.3389/fgene.2022.982418

Marone D, Russo MA, Laidò G et al (2013) Genetic basis of qualitative and quantitative resistance to powdery mildew in wheat: from consensus regions to candidate genes. BMC Genomics 14:562. https://doi.org/10.1186/1471-2164-14-562

Martinez AK, Soriano IM, Tuberosa R et al (2016) Yield QTLome distribution correlates with gene density in maize. Plant Sci 242:300–309. https://doi.org/10.1016/j.plantsci.2015.09.022

Mazzucotelli E, Sciara G, Mastrangelo AM et al (2020) The global durum wheat panel: an international platform to identify and exchange beneficial alleles. Front Plant Sci 11:569905. https://doi.org/10.3389/fpls.2020.569905

Mores A, Borrelli GM, Laidò G et al (2021) Genomic approaches to identify molecular bases of crop resistance to diseases and to develop future breeding strategies. Int J Mol Sci 22:5423. https://doi.org/10.3390/ijms22115423

Nigro D, Blanco E, Mangini G et al (2024) Identifying QTL for grain protein content independent from grain yield-related traits in durum wheat. J Cereal Sci 117:103894. https://doi.org/10.1016/j.jcs.2024.103894

Ober ES, Alahmad S, Cockram J et al (2021) Wheat root systems as a breeding target for climate resilience. Theor Appl Genet 134(6):1645–1663

Petrarulo M, Marone D, Ferragonio P et al (2015) Genetic analysis of root morphological traits in wheat. Mol Gen Genomics 290:785–806. https://doi.org/10.1007/s00438-014-0957-7

Ramírez-González RH, Borrill P, Lang D et al (2018) The transcriptional landscape of polyploid wheat. Science 361:eaar6089. https://doi.org/10.1126/science.aar6089

Roselló M, Royo C, Álvaro F et al (2018) Pasta-making quality QTLome from Mediterranean durum wheat landraces. Front Plant Sci 9:1512. https://doi.org/10.3389/fpls.2018.01512

Saini DK, Chopra Y, Pal N, Chahal et al (2021) Meta-QTLs, ortho-MQTLs and candidate genes for nitrogen use efficiency and root system architecture in bread wheat (*Triticum aestivum* L.). Physiol Mol

Biol Plants 27:2245–2267. https://doi.org/10.1007/s12298-021-01112-0

Saini DK, Srivastava P, Pal N et al (2022) Meta-QTL, ortho-meta-QTL and candidate genes for grain yield and associated traits in wheat (*Triticum aestivum* L.). Theor Appl Genet 135(3):1049–1081. https://doi.org/10.1007/s00122-021-04018-3

Salvi S, Tuberosa R (2015) The crop QTLome comes of age. Curr Opin Biotechnol 32:179–185. https://doi.org/10.1016/j.copbio.2015.01.001

Sanchez-Garcia M, Alvaro F, Peremartí A et al (2015) Changes in bread-making quality attributes of bread wheat varieties cultivated in Spain during the 20th century. Eur J Agron 63:79–88. https://doi.org/10.1016/j.eja.2014.11.006

Singh R, Saripalli G, Gautam T et al (2022) Meta-QTLs, ortho-MetaQTLs and candidate genes for grain Fe and Zn contents in wheat (*Triticum aestivum* L.). Physiol Mol Biol Plant 28:637–650. https://doi.org/10.1007/s12298-022-01149-9

Soriano JM, Alvaro F (2019) Discovering consensus genomic regions in wheat for root-related traits by QTL meta-analysis. Sci Rep 9(1):10537. https://doi.org/10.1038/s41598-019-47038-2

Soriano JM, Royo C (2015) Dissecting the genetic architecture of leaf rust resistance in wheat by QTL meta-analysis. Phytopathology 105(12):1585–1593. https://doi.org/10.1094/PHYTO-05-15-0130-R

Soriano JM, Malosetti M, Roselló M et al (2017) Dissecting the old Mediterranean durum wheat genetic architecture for phenology, biomass and yield formation by association mapping and QTL meta-analysis. PLoS One 12(5):e0178290. https://doi.org/10.1371/journal.pone.0178290

Soriano JM, Colasuonno P, Marcotuli I et al (2021) Meta-QTL analysis and identification of candidate genes for quality, abiotic and biotic stress in durum wheat. Sci Rep 11:11877. https://doi.org/10.1038/s41598-021-91446-2

Soriano JM (2020) Molecular Marker Technology for Crop Improvement. Agronomy, 10:1462. https://doi.org/10.3390/agronomy10101462

Tuberosa R, Frascaroli E, Maccaferri M, Salvi S (2021) Understanding and exploiting the genetics of plant root traits. In: Understanding and improving crop root function. Burleigh Dodds Science Publishing, pp 395–466

Tyagi S, Mir RR, Balyan HS et al (2015) Interval mapping and meta-QTL analysis of grain traits in common wheat (*Triticum aestivum* L.). Euphytica 201:367–380. https://doi.org/10.1007/s10681-014-1217-y

Wang Y, Sun T, Li T, Wang M, Yang G, He G (2016) A CBL-interacting protein kinase TaCIPK2 confers drought tolerance in transgenic tobacco plants through regulating the stomatal movement. PLoS One 11:e0167962. https://doi.org/10.1371/journal.pone.0167962

Xie Q, Sparkes DL (2021) Dissecting the trade-off of grain number and size in wheat. Planta 254:3. https://doi.org/10.1007/s00425-021-03658-5

Yang Y, Amo A, Wei D et al (2021) Large-scale integration of meta-QTL and genome-wide association study discovers the genomic regions and candidate genes for yield and yield-related traits in bread wheat. Theor Appl Genet 134:3083–3109. https://doi.org/10.1007/s00122-021-03881-4

Zhang LY, Liu DC, Guo XL et al (2010) Genomic distribution of quantitative trait loci for yield and yield-related traits in common wheat. J Int Plant Biol 52:996–1007. https://doi.org/10.1111/j.1744-7909.2010.00967.x

Genomics-Enhanced Pyramiding of Alleles for Abiotic Stress Tolerance in Durum Wheat

Amit Gautam, Yaman Jabbour, Miguel Sanchez-Garcia, Andrea Visioni, Rodomiro Ortiz, and Filippo Maria Bassi

Abstract

Climate change is increasingly damaging the durum wheat crop across most regions of the world. Generating durum wheat genotypes primed with the genetic capacity to withstand drought and heat stress is essential to ensure future harvests. However, the interaction between genotypes and the changing environments generates data that are often complex to interpret. Here, we revise the main literature to define the specific detrimental effects on different traits caused by drought and heat stress occurring at different stages during the season. The best physiological tools and biometric approaches for identifying and selecting these major traits are then presented, in the optic of integrating them into a durum wheat breeding program that specifically targets the genetic improvement in abiotic stress tolerance. Accordingly, some of the main sources of germplasm to help generate useful diversity are also reported. Here are also reported the major loci identified from the literature as linked to adaptation to abiotic stresses, including those that have already been transformed with success via cis- or trans-genesis or editing approaches. In addition, the use of genomic models to integrate also minor alleles for tolerance is described, providing the scheme used by the International Center for Agricultural Research in the Dry Areas (ICARDA) program as an example of integration of the various technologies described here. Overall, this chapter provides an ideal starting point to design or revise breeding approaches targeting the adaptation to the changing climates, exploiting the latest genomic and phenotypic approaches to maximize the rates of genetic gain and ensure that durum wheat remains a crop of choice for future farmers.

Amit Gautam and Yaman Jabbour contributed equally with all other contributors.

A. Gautam
ICARDA Amlaha, Bhopal, India

Y. Jabbour
GCSAR, Aleppo, Syria

M. Sanchez-Garcia · A. Visioni · F. M. Bassi (✉)
ICARDA, Morocco, Rabat, Morocco
e-mail: f.bassi@Consultative Group on International Agricultural Research (CGIAR).org

R. Ortiz
SLU, Alnarp, Sweden

1 Introduction

Durum wheat (*Triticum turgidum* spp. *durum*) is a disomic tetraploid species (genome AABB, $2n = 4x = 28$), which is economically important and widely cultivated. The most important growing areas are in the Mediterranean Basin, North

R. Tuberosa et al. (eds.), *The Durum Wheat Genome*, Compendium of Plant Genomes,
https://doi.org/10.1007/978-3-031-84689-2_10

America, and Southwest Asia (Maccaferri et al. 2014; Sall et al. 2019; Ltaief and Krouma 2023; Saini et al. 2023). Due to global warming, the crops growing in these regions are often exposed to drought and heat stresses both early and late in the season. These in turn not only affect the rate of plant growth and development causing major reductions in productivity (Hu et al. 2014; Barlow et al. 2015; Gautam et al. 2015; He et al. 2017; Ni et al. 2018; Tramblay et al. 2020; Qi et al. 2022; Devate et al. 2023; Kumar et al. 2023; Zaim et al. 2023; Tedone et al. 2023) but also affect seed quality (Tabbita et al. 2024). Sowing stress tolerant cultivars remains the most cost-effective and sustainable practice for climate change adaptation, especially if managed with smart agronomic practices (Dhankher and Foyer 2018).

Genotype x environment interactions (particularly genotype x year interactions) are the most confounding and critical effects that breeders need to face when attempting to derive new durum cultivars well adapted to the erratic climatic conditions (Beres et al. 2020). These interactions are assessed by breeders via the use of multi-location and multi-year trials selecting for highly stable cultivars, with an average grain yield of 4232 kg ha^{-1} (Subira et al. 2015; Bassi and Sanchez-Garcia 2017; Sall et al. 2018). However, the complexity and cost of this endeavor limit its scalability, hence reducing the ability of breeders to achieve fast genetic progress. Therefore, many geneticists have tried to partition these interactions into the individual traits contributing to them, and then worked to unravel the mechanisms of tolerance for each (Ni et al. 2018; El Hassouni et al. 2019).

The analysis of quantitative trait loci (QTL) strategically links phenotypic and genotypic data to understand the genetic control of complex traits (Miles and Wayne 2008). Many traits are associated with the wheat response to abiotic stresses, which are under the control of even more QTL (Sukumaran et al. 2018; Tricker et al. 2018; Hu et al. 2019; Liu et al. 2019; Soriano et al. 2021; Arriagada et al. 2022; Ma et al. 2022; Zhikai et al. 2022; Amanda et al. 2023; Raj et al. 2023). In fact, quantitative traits are mostly controlled by many loci interacting with each other under strong environmental influence (Colasuonno et al. 2020). In that sense, the possibility to tag a given exploitable trait with a reliable marker not affected by the environment would greatly simplify the work of breeders (Colasuonno et al. 2014). However, the high number of putative loci identified to date and the lack of available markers to track these in segregating populations have prevented breeding programs from using them at scale (Bassi et al. 2019).

Dozens of QTL have been identified to date as linked to the response of wheat to drought stress (Collins et al. 2008; Maccaferri et al. 2008; Kadam et al. 2012; Mwadzingeni et al. 2016, 2017; Xu et al. 2017; Soriano et al. 2021; Arriagada et al. 2022; Ma et al. 2022; Zhikai et al. 2022; Amanda et al. 2023; Raj et al. 2023) and extreme heat (Shirdelmoghanloo et al. 2016; Ogbonnaya et al. 2017; El Hassouni et al. 2019). In fact, most of these QTL have been identified when exposing the germplasm to combined drought and heat stress, since they often occur together in natural environments (Pinto et al. 2010; Liu et al. 2019; Arriagada et al. 2020).

The vast number of loci discovered as being associated with the response to the two main abiotic stresses (drought and heat) has challenged the durum wheat research community. The inability to efficiently compare results across trials and germplasm types has prevented the identification of major loci of shared interest. The availability of a high-density tetraploid wheat consensus map which integrates 13 datasets from independent biparental populations of durum wheat cultivars (*Triticum turgidum* ssp. *durum*), cultivated emmer (*T. turgidum* ssp. *dicoccum*) and their ancestor (wild emmer, *T. dicoccoides*) coupled with the release of the durum wheat genome sequence assembly (Maccaferri et al. 2019) finally provides a highly informative scaffold to best align QTL and mutants for target traits and identify the most useful ones. The addition of the Global Durum Panel (Mazzucotelli et al. 2020), a global effort to identify and share most of the available genetic diversity available for durum wheat, has also provided the durum community with a common set to validate the discoveries, and thereafter share them freely. With these three new tools in place, there is great hope that the durum wheat research community will finally be able to translate their discoveries into actual usable tools and resources for breeders to tackle the complexity of abiotic stress tolerance.

In this chapter, we review the complexity of drought and heat stress in durum wheat and define the main tolerance mechanisms available to breeders. Furthermore, we define the physiological and genomic tools that can be used to track these traits in a large-scale breeding program. Ultimately, we merge this knowledge to propose a strategic breeding scheme that takes advantage of all available tools to achieve higher rates of genetic gain despite the strong genotype × environment and genotype × year interactions, which often occur in most of durum wheat growing regions.

2 Interactions Between Durum Wheat and the Environment

Drought stress takes place when soil humidity is low due to a negative balance between the evapotranspiration and amount of moisture reaching the soil (Lipiec et al. 2013). Heat stress occurs due to the rise in air temperature beyond a threshold level for a minimum amount of time. Unfortunately, most of the major durum wheat growing regions experience both of these climatic occurrences in most years. Figure 1 gives as example of the average monthly maximum temperatures and rainfall in four major durum wheat production locations: Adana in Turkey, Foggia in Italy, Jerez de la Frontera in Spain, and Settat in Morocco. It is evident that as the durum wheat crops progress toward the reproductive phase (between April and May) the temperatures rise beyond the stress level and the moisture availability drops (Dreesen et al. 2012). Another important factor is the vast range of season-to-season variation. In fact, while the average climatic data help visualize the overall trend, it is extremely rare to experience an "average season." The second part of Fig. 1 reports the moisture and average maximum temperature recorded at the International Center for Agricultural Research in the Dry Areas (ICARDA) main breeding station of Marchouch, Morocco, during two consecutive seasons. Marchouch station is considered a representative site for the target population of environments (TPE) as it represents very well the growing conditions of many North African production areas. During the first season (2019–2020 in this case), ICARDA was conducting the preliminary Stage 1 yield trials. The best material was then selected to advance to Stage 2 yield trials at the same station in the following season. However, the first season experienced very severe drought with just a total of 178 mm of moisture and average maximum temperatures that exceeded 23 °C. In comparison, the following season (Stage 2) was very favorable, with 422 mm of total rainfall and acceptably low temperatures averaging 20 °C. In this example, the productivity recorded for the second season was more than double that of the first year. Also, the best entries identified in season 1 rarely confirmed their performances in season 2, showing a very strong genotype × year interaction. These types of interactions complicate the selection process, and hence need to be critically addressed by breeders in their strategy to best adapt germplasm to climatic stresses.

3 Effects of Stresses at Different Phases of the Durum Crop

The concurrent occurrence of drought and terminal heat stress is common in most durum growing areas (Fig. 1). However, terminal stress is not the only type of occurrence and probably not the most damaging. Instead, climate change has brought forward an increase in frequency of prolonged warm periods during the fall to winter transition, which is often associated with below average rainfall distribution (Mahato 2014; Lesk et al. 2016; Sabella et al. 2020). Early season stresses can cause even more damage as these prevent the crop from properly establishing, while at the same time it promotes a severe reduction in its natural

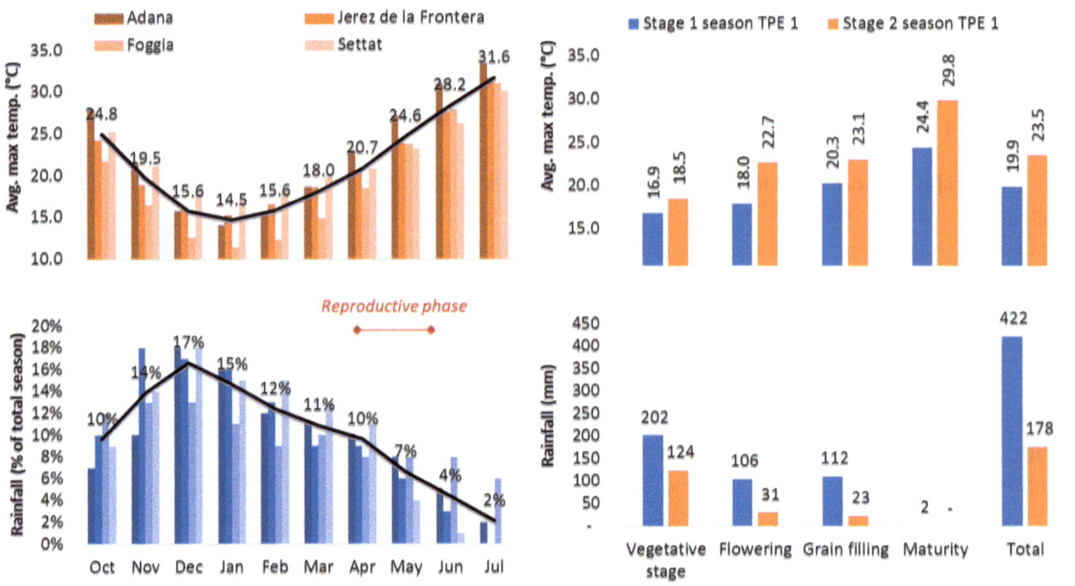

Fig. 1 Average rainfall distribution and temperatures in four major durum cultivating areas of the past decade, and what they entail when breeding at one target population of environment (TPE) with strong seasonal variation

growing cycle. In this section, we revise the major components of grain yield and how these are affected by the occurrence of drought and heat stress at different plant development stages of the cropping season (Chenu et al. 2013; Daryanto et al. 2016; Sarto et al. 2017). Figure 2 shows the phenological stages based on the Freeke's scale (Hayes 2019) of the durum crop and how these contribute to its main grain yield components, as defined by Slafer and Rawson (1994). To the same graph are added four timings of stress occurrence: (i) early in the season as the seed germinate and until the leaf sheet lengthen; (ii) during crop transition to the erect phase, as the nodes are generated and the flag leaf emerges; (iii) occurring during flowering, as the head emerges and pollination and grain setting occurs; (iv) terminal stress that occurs after flowering is completed and grain filling begins until harvest time.

3.1 Effect of Stresses on Phenology

The vast majority of durum wheat cultivation worldwide utilizes true spring types or facultative winter plants to ensure completion of the crop cycle before the beginning of the warmest months (Sall et al. 2018; Gupta et al. 2020). Some pockets of winter-type durum wheat can be found in Central Asia and in some mountain regions in West Asia and East Africa, but these represent a small fraction of global production. The true spring types do not require any vernalization period before transitioning to flowering and therefore are preferred in those areas where the winter temperatures are mild, and the wet spring season is short like Northern Europe, the United States, Asia, and the South Pacific. Hence, the flowering of these plant types is mostly controlled by the accumulation of growing degree days (GDD), since photoperiodism does not change within location despite climate change. The facultative winter types instead require a short period of cool temperatures to transition effectively to flowering, or in absence of these low temperatures, it needs more GDD to flower. These plant types are vastly utilized in Europe and other regions where winters are cool and at least two wet months can be expected in spring like in Germany.

The length of the crop season is a major factor influencing the grain yield of durum wheat. A shorter time to maturity can help avoiding terminal

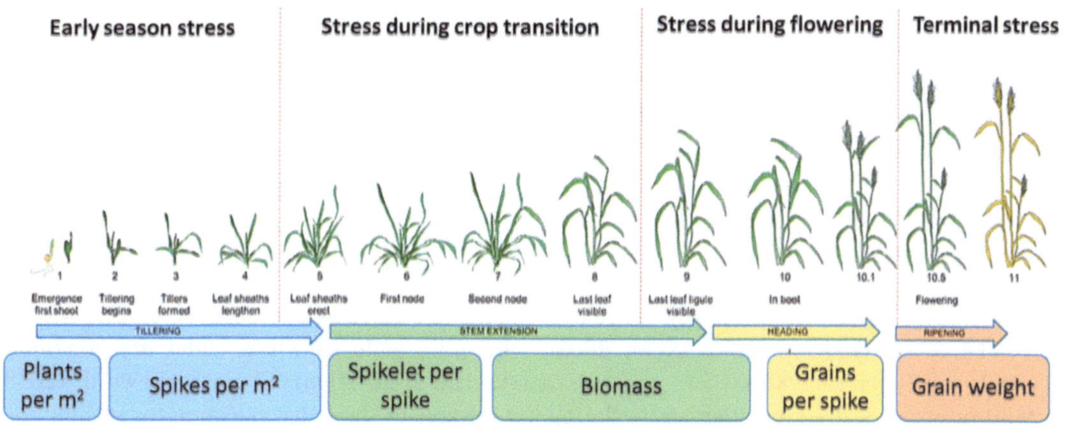

Fig. 2 Yield components affected by stresses during different crop stages. (Modified from Slafer and Rawson (1994) and displayed on Feeke's scale as developed by Hyles et al. (2020))

stresses, but it will also result in lower yield potential. Instead, a longer crop season allows for higher accumulation of biomass, and therefore higher yield potential would expose the plant to higher risks of losses in case of terminal stress. The occurrence of higher temperatures (heat stress) at any of the stages of plant development before flowering will inevitably impact the phenology of the plant. Faster GDD progression will accelerate the flowering of true spring types; for facultative winter types instead, the warmer days might delay flowering by preventing the reaching of the vernalization requirements (Parmesan and Hanley 2015). Combining delayed flowering with warmer and drier seasons would mean that the crop cannot complete properly its grain setting and filling, ultimately resulting in low harvests. Drought during the crop growth cycle may affect grain yield (Baloch et al. 2012; Zhang et al. 2018a; Blum and Tuberosa 2018; Khadka et al. 2020). That is why in regions where warm winters and drought are common, there is a strong preference for true spring types that have the plasticity to adapt to each season based on the temperatures alone.

3.2 Early Season Stress

The occurrence of drought right after planting can deprive the grains of the initial moisture that is necessary to start germination and ensure a good establishment of the seedling (Kaya et al. 2006; Wahid et al. 2007; Farooq et al. 2009; Ayalew et al. 2018). In turn, this reduces total plants per m^2. In dryland agriculture, most authors suggest the planting of 200 kg of seeds per hectare as the ideal rate to face the potential reduction in germination (Weldearegay et al. 2012; Beres et al. 2020; Iqbal et al. 2021).

During the first phases of the durum plant lifecycle, the tillers are formed. This is a major factor influencing grain yield in dry conditions (García del Moral et al. 2003; Moragues et al. 2006; Royo et al. 2009). Moisture stress after germination causes low tiller formation and abortion. A 72% reduction in grain yield was reported when severe drought occurred during tillering stage, which was more damaging than when it occurred at the reproductive stages (Ding et al. 2018) or during grain filling (Keyvan2010; Dolferu 2014). Analogously, other authors confirmed that drought stress occurring during the vegetative stage caused up to 54% reduction in grain yield due to the decrease in the number of spikes per unit area (Saeidi et al. 2015; Gupta et al. 2001; Aslam et al. 2015; Shamuyaria et al. 2019), and wheat biomass (Naeem et al. 2015; Ding et al. 2018).

The tiller number and crop establishment are also major factors influencing the ability of the durum wheat plant to cover the ground. Good early plant vigor ensures shading of the soil and guarantees that moisture does not evaporate before reaching the developing roots (Reynolds et al. 2004; Dodig et al. 2008; Blum 2011; Richards et al. 2011; Tuberosa 2012; Elazab et al. 2015). Vigorous seedlings are best suited to compete against parasitic weeds due to their shading effect and favor higher water and nutrient use efficiency (Zhang et al. 2014). Because of this, early vigor is considered by many a good candidate trait for assessing the ability of wheat genotypes to adapt to drought at the early growth stage (Rebetzke et al. 1999; Reynolds et al. 2005; Reynolds and Tuberosa. 2008).

Mild or occasional temperature stress during the early phases of the cropping season does not seem to have a direct negative effect on the major yield components, except as cause of higher evapotranspiration and therefore moisture stress, or influencing phenology as described previously. However, severe heat stress with temperatures exceeding 30 °C can have a tremendous effect by causing dehydration of the plant, which is typically followed by closure of the stomata and rolling of the leaf blades. As such, the photosynthetic capacity of the plant is drastically reduced (Gautam et al. 2014), thereby preventing the production of the biomass needed for good crop establishment, and occasionally resulting in the complete dehydration of the plant (Ristic et al. 2007; Cossani and Reynolds 2012; Iqbal 2019). Imposing severe heat stress on 14-old day seedlings of wheat was shown to cause a reduction in the total number of leaves produced, lower seedling height, a drop on chlorophyll content, and hence affected the final biomass (Maulana et al. 2020). It was also reported that a large portion of seedling wilted before reaching booting stage, probably due to reactive oxygen species (ROS) that were released from the thylakoid as stress response mechanism (Iurlaro et al. 2016).

3.3 Stress Occurring During the Transition Phase

During this growth phase, the number of spikelets per spike and biomass are determined. A reduction in spikelet formation would ultimately affect yield via reduction in the total number of grains that each spike produces, while a reduction in biomass would affect the potential yield and overall harvest index. Thus, occurrence of climatic stresses during this phase has a detrimental effect on these two traits (Ding et al. 2018). Saeidi et al. (2015) confirmed that drought occurring during stem elongation has a major effect on plant biomass, and ultimately it reduces the availability of stem carbohydrate reserves, which are estimated to contribute up to 40% of the grain yield of wheat under dry conditions (Ehdaie et al. 2008). El Hassouni et al. (2018) were able to show that selection for specific root angles could ultimately result in better drought tolerance for durum wheat. Under dry conditions occurring at the transition phase, the shoot biomass was highly correlated to root biomass and grain yield, and genotypes with narrow root angle capable of reaching the moisture stored in deeper soil layers were able to accumulate higher biomass obtaining 28–42% grain yield advantage.

Hlavacova et al. (2018) showed that the combined effect of drought and heat occurrence in this phase had as major effect the reduction in grain number. In an elegant experiment imposing moderate elevated temperatures (28 °C) on wheat, Balla et al. (2021) also noticed that heat stress at this stage had a major effect on the grain number produced, but that the effect was less marked than when the heat stress occurred at booting stage. Hence, breeders seeking to develop cultivars capable of withstanding stress in this phase will have to focus on selections for higher biomass and grain number in occurrence of the stress.

3.4 Stress Occurring During Flowering

The transition to flowering and completion of the reproductive cycle is deemed by many authors as one of the most sensitive stages to abiotic stresses (Giorno et al. 2013). Within the closed florets, before the spike becomes visible on the stem, meiosis occurs to produce the haploid gametes. This process is extremely sensitive to temperature stress: it was demonstrated that exposing wheat plants to temperatures of 30 °C and 35 °C during meiosis and the premeiotic interphase, 50 and 80% of the assessed meiotic progressions did not advance past the pre-meiosis–leptotene phase, respectively, thus resulting in a corresponding reduction on the number of grains produced (Draeger and Moore 2017). Once meiosis is completed, pollen is produced and deposited onto the awaiting female gamete. During this reproductive phase, high temperatures have been shown to cause up to 40% reduction in pollen viability, which results in a reduction of 15–23% in seed number (Bheemanahalli et al. 2019) and 32–34% in seed weight (Nasehzadeh and Ellis 2017). Similarly, El Hassouni et al. (2019) confirmed that imposing heat stress at the time of flowering reduced grain yield in durum wheat by 54%, and grain number was the main trait affected by this stress. Sall et al. (2018) assessing large collections of durum wheat along the Senegal River, where daily temperatures exceed 31 °C throughout the growing cycle, also confirmed that the maintenance of grain number despite the stress was the main tolerance mechanism used by top yielding genotypes.

Moisture stress during the reproductive stage causes less damage as compared to heat, mostly leading to similar but reduced effects. Still, even short periods of moisture stress during meiosis and gametogenesis can lead to the reduction in the number of grains produced per spike (Barber et al. 2015; Onyemaobi et al. 2017; Dolferus et al. 2013). However, the main effect attributed to moisture stress in this phase is the reduction in the leaf water potential, which leads to reduced transpiration rate and an overall increase in the canopy temperature, bringing forth the same symptoms as heat stress (Turner et al. 2001; Tuberosa 2012: Gautam et al. 2015).

3.5 Stress Occurring After Flowering

After embryos are effectively produced, the grain filling phase starts. During the month(s) following grain setting, the plant starts a complex set of changes to translocate the carbohydrates and nutrients accumulated as biomass to the developing grains (Wei et al. 2017). These elements will form the starch, which represents the edible and commercial part of durum wheat. Durum wheat production vastly occurs in conditions where drought and heat stress often occur during this terminal filling stage, and it is therefore normal that vast part of the research for these stresses focuses on this phase (Martinez et al. 2011; Liu et al. 2020; Olsen 2020).

The occurrence of heat stress during this phase was demonstrated to have a milder effect than when occurring in the previous one (Farooq et al. 2011; Shanmugam et al. 2013; Dwivedi et al. 2017; Bheemanahalli et al. 2019). Its primary effect is the reduction in the length of this phase, promoting faster transaction to dehiscence (Borrill et al. 2015). The grain filling rate, which is the speed at which elements are translocated from the source to the sink, is often affected by heat stress since it promotes faster transpiration rates and hence faster nutrient reallocation to the grains (Baillot et al. 2018; Liu et al. 2020). Hence, the major effect of heat stress at this phase is not a reduction on grain number, but rather of grain size (Nasehzadeh and Ellis 2017; Bheemanahalli et al. 2019). In fact, the accelerated grain filling rate due to heat cannot compensate for the shorter grain filling duration, especially when combined with higher transpiration rate of moisture from the soil (Wahid et al. 2007; Ruggiero et al. 2017; Schachtman and Goodger 2008). A severe drought at this stage causes a high grain yield loss (Cai et al. 2011; Wang et al. 2016), due to the reduction in both duration of the grain filling period (Merah and Monneveux 2015b; Alghabari and Ihsan 2018), and rate of gain filling (Nass and Reiser 1975; Borrill et al. 2015), which led to reduce the weight of kernels (Royo et al. 2000). For this reason, the occurrence of drought stress during this phase is detrimental for durum wheat grain yield. Gevrek and Atasoy (2012) reported that 28% losses in grain yield were due to a reduction in the weight of grain when drought stress occurred during grain filling stage; in addition the drought during this stage affects the quality and size of grain by reduction in the protein and starch accumulation (Wardlaw and Willenbrink 2000; Nagy-Reder et al. 2021).

The occurrence of heat stress during this phase was demonstrated to have a milder effect than when occurring in the previous one (Bheemanahalli et al. 2019; Farooq et al. 2011; Shanmugam et al. 2013; Dwivedi et al. 2017). Its primary effect is the reduction in the length of this phase, promoting faster transaction to dehiscence (Borrill et al. 2015). The grain filling rate, which is the speed at which elements are translocated from the source to the sink, is often affected by high temperature since it promotes faster transpiration rates and hence faster nutrient reallocation to the grains (Baillot et al. 2018; Liu et al. 2020). Hence, the major effect of heat stress at this phase is not a reduction in grain number, but rather in grain size (Nasehzadeh and Ellis 2017; Bheemanahalli et al. 2019). In fact, the advantageous faster grain filling rate due to high temperature is not sufficient to compensate for the negative shortening of the grain filling time, especially when combined with higher transpiration rate of moisture from the soil (Wahid et al. 2007; Ruggiero et al. 2017; Schachtman and Goodger 2008).

The occurrence of drought at this stage results in major yield loss (Cai et al. 2011; Wang et al. 2016). It impacts negatively both the duration of the grain filling period (Alghabari and Ihsan 2018; Merah and Monneveux 2015b) and the rate of grain filling (Nass and Reiser 1975; Borrill et al. 2015), ultimately resulting in smaller and lighter kernels (Royo et al. 2000). Gevrek and Atasoy (2012) reported that 28% losses in yield were due to a reduction in the weight of grain when drought stress occurred during grain filling stage. In addition, the drought stress during this stage affects the quality and size of grain by reduction in the protein and starch accumulation (Wardlaw and Willenbrink 2000). The reallocation of nutrients is mediated by the plant transpiration mechanism that favors their movement upward in the phloem toward the grains (Alghabari and Ihsan 2018; Merah and Monneveux 2015b). Lack of moisture in the soil prevents the normal transpiration rate and causes a reduction in the amount of nutrients reallocated to starch (Farooq et al. 2009; Kilic and Yagbasanlar 2010; Bogale and Tesfaye 2016; Liu et al. 2016). Hence, in durum wheat the selection of genotypes with large kernels is both a strategy to meet the marker demand as well to promote drought tolerance.

Roots are the organs responsible for capturing and providing moisture and nutrients to the plant. Several authors have identified this organ as the most strategic to maintain normal transpiration rate when moisture stress occurs during grain filling (Manschadi et al. 2006; Reynolds and Tuberosa. 2008; Christopher et al. 2008). In particular, the root angle is deemed to be an excellent predictor of the ability of durum roots to penetrate deeper in the soil layers and find residual moisture there (Ryser and Eek 2000; Manschadi et al. 2008; Wasson et al. 2012; Meister et al. 2014; Richard et al. 2015; Maccaferri et al. 2016; Alahmad et al. 2019; Alemu et al. 2021; Ober et al. 2021). Accessing this moisture allows to extend the grain filing period and foster transpiration rates, which ultimately result in better kernel weight and size (Ruiz et al. 2018). El Hassouni et al. (2018) confirmed that genotypes with narrower root angle, and hence deeper root architecture, gained 35% in grain yield and 9% in the weight of 1000 kernels when exposed to drought during the grain filling phase.

4 Scalable Phenotyping Screening Tools for Determining Stress Response

All grain yield components can be negatively affected by abiotic stresses occurring at different periods during the crop season (Groli et al. 2024). The cereal physiologists have developed several tools to measure the impact of the climate on the crop as it occurs. Some of these are complex and require vast time and resources for their application and are hence not suitable for large-scale application in breeding programs. Others instead are extremely scalable and represent a strategic addition to improve the selection precision (Tuberosa 2012) and accuracy of breeders

Table 1 Physiological tools available for deployment by breeding programs to measure specific traits associated with abiotic stress response

Tool	Trait measured	Plots screened per person per day	References
Infrared Thermometer	Plant canopy temperature	1000	Gautam et al. (2015)
CANOPEO	Early ground cover	2000	Patrignani and Ochsner (2015)
Green Seeker (Handhel 505) NDVI	Biomass	2000	Gutierrez-Rodriguez et al. (2004)
Spike Density Red, Green, Blue (Red, Green, Blue (RGB)) software	Spike density	1000	Jin et al. (2019), Madec et al. (2019
Grain scanner	Grain characteristics	500	Wahn et al. (2014)
Pasta strainer method	Root angle	50	El Hassouni et al. (2018)

(Table 1). The key to success is the identification of proxy traits that are genetically associated with yield response under stress but have greater heritability to favor faster genetic gain (Royo et al. 2006, 2007) and should be easier to measure or help reduce the characterization costs.

Durum wheat germplasm with high grain yield in well-watered condition had high chlorophyll content and low canopy temperature. It is considered that wheat germplasm with a low canopy temperature can maintain high transpiration and photosynthetic rate as well as produce a high yield under moisture-stressed condition (Talebi 2011; Sallam et al. 2019; Lepekhov 2022). In addition, there was a significant positive correlation between chlorophyll content and yield in different environments. Significant correlation between canopy temperature and chlorophyll content with grain yield under moisture-stressed conditions and stress susceptibility index value indicated the potential for screening wheat germplasm for drought response. Correlation analysis revealed that grain yield under both rainfed and irrigated conditions was mostly determined by grain number (Pinter et al. 1990; Amani et al. 1996; Bilge et al. 2008; Rebetzke et al. 2013; Groli et al. 2024). Soil and plant analyzer development (SPAD) showed significant and positive correlations with grain yield and number (Ghimire et al. 2015; Kandel 2020). Similarly canopy temperature was reported as significantly and negatively correlating with grain and biomass yield per plant under late and very late conditions and number of grains per spike and 1000 grain weight under drought and heat stress conditions (Rebetzke et al. 2013; Gautam et al. 2015). Hence, this characteristic can be used as an important selection trait in a breeding program when doing field trials. In addition, canopy temperature depression positively correlated to grain yield, spike yield, and grain numbers per spike (Gautam et al. 2015). Research suggests that chlorophyll content (SPAD) and canopy temperature can be used as one of the important criteria for the selection of stable genotypes under drought and heat stressed conditions and can be considered as a cheap method of indirect selection for improving drought and terminal heat stress tolerance in durum wheat, thereby improving its productivity (Bahar et al. 2008; Mehmet et al. 2011; Talebi et al. 2010; Gautam et al. 2015; Jokar et al. 2018; Khanghahi et al. 2021). In addition, this trait is highly heritable, and it has shown positive correlation with grain yield (Kandic et al. 2009). Furthermore, screening for this trait is fast and scalable rather by applying a visual rating score (Mullan and Garcia 2012) or digital imaging using an Red, Green, Blue (RGB) camera (Condorelli et al. 2018; Zhang et al. 2018b; Walter et al. 2019). Hence, the combination of ease of screening, high heritability, and importance for tolerance has made early vigor a trait routinely screened by breeders attempting to deliver better wheat cultivars adapted to early season droughts.

The number of spikes per m^2 and plot biomass are excellent predictors of the plot yield at harvest—especially when stress occurs during the transitional phase—but they are extremely time consuming to measure on large scale. Novel solutions have been proposed to scale the characterizations for these traits. The normalized difference vegetation index (NDVI), derived from near-infrared reflectance sensors, is widely used to assess vegetative greenness and canopy photosynthetic capacity. The resulting index can be reliably converted into estimates of green area index (GAI), leaf area index (LAI), nutrient status, and overall plant biomass (Pietragalla and Vega 2012). Different sizes of NDVI sensors are available to measure this index at different levels: low on the plot, high in the sky, or even from satellites (Pietragalla and Vega 2012). The greenness of entire field crop could be estimated by the NDVI tool "GreenSeeker spectral sensor," which has been considered more integrated than soil and plant analyzer development (SPAD) as it measured the senescence from the whole crop canopy compared to the SPAD that uses only one leaf (Lopes et al. 2012),

While the use of unmanned aerial vehicle (UAV) and satellite images holds the promise of rapid recording of vast areas, to date their application is often compromised by the complexity of handling the resulting images and converting these into plot level information that is why the most practical way to measure NDVI remains at plot level through hand-held or tractor carried sensors (Condorelli et al. 2018). The NDVI approach is, however, unable to count the number of spikes reliably. Instead, object recognition software has been developed to use simple Red, Green, Blue (RGB) cameras to identify and then count the spikes of wheat. A machine learning approach is used in this case, by first teaching the software what spikes look like, and then running this at scale models. Those models are based on the assumptions that at the end of the cropping season tiller number reflects with high accuracy the number of spikes. Jin et al. (2019) reported a very high correlation indicating that the stem density could be a good proxy of the ear density. They also reported that the stem diameters follow a gamma distribution over each microplot with an average diameter close to 2 mm the stem density, and plant height is closely related to the above-ground biomass at harvest with a relative Root Mean Square Error (Root Mean Square Error (RMSE)). A low-cost methodology for ear number assessment is based on zenithal Red, Green, Blue (RGB) color digital images taken from above the crop in natural light conditions and a three-step image processing (Laplacian frequency filter, Median filter, and Find Maxima) was proposed by Fernandez-Gallego et al. (2018). While the previous method estimates ear number after harvest counting cut stems, this method allows ear counting at anthesis. Total above-ground biomass and ear number are two important traits for the selection of genotypes in cereal breeding programs; therefore, the ability of estimating both traits before harvest could facilitate the selection of promising genotypes and reduce trial cost.

Above-ground biomass can be estimated in a destructive way cutting plant manually or in non-destructive way using a Light Detection and Ranging (LiDAR). For instance, Jimemnez-Berni et al. (2018) field-validated non-destructive estimation of canopy height, ground cover, and above-ground biomass using a LiDAR mounted on a semi-automated ground-based phenomobile. They found very strong correlations with manually collected data for canopy height (R^2: 0.99 error: 0.017 m), above-ground cover (R^2: 0.92; slope 1.02), and biomass using three-dimensional (3D) voxel index (R^2: 0.92) and using 3D profile index (R^2: 0.93), respectively. The effective repeatability of LiDAR measurements of both canopy height and above-ground biomass using the previous method was confirmed in a subsequent study, using data collected from different growing stages in a field trial of 98 wheat lines. A similar experiment reported that data for both traits, which were collected with an HTP system, were extremely repeatable, independent of the direction in which they were collected in the field. Furthermore, the same experiment showed that solid data can be collected from 2400 plots per hour with potential to increase the number to 7400 plots per hour (Walter et al. 2019). The 3Dpi algorithm was also used for above-ground biomass estimation using and UAV-LiDAR giving a slightly lower correlation for both traits (Harkel et al. 2020).

Thermal cameras (or thermal long-wave infrared cameras; TIR) are sensors able to record the emitted radiation and providing images representing the temperature values per pixels. TIR offer the advantage of stress detections before the appearance of the symptoms (Chaerle et al. 2007). On the other hand, the measurements or thermograms need to be interpreted considering the external factors that influence the plant canopy temperature such as the environment and plant morphology. As recently reviewed by Pineda et al. (2021), a number of these factors are directly associated with traits related to plant water balance. The most common parameters used are NDVI, crop water status index (CWSI), stomatal conductance, maximum temperature difference (MTD), normalized water canopy temperature (NRCT), and average canopy temperature (Tav). It is well known that increases in canopy temperature in field conditions are often caused by an abiotic stress or in combinations, such as heat and drought. The ability of a plant to successfully complete the reproductive phase and set seeds is a critical factor in determining its ability to withstand heat stress occurring during flowering. Hence, thermography can be used by breeders to predict, for example, grain yield. El Sayed et al. (2017) found that NRCT, relative water content (RWC), and canopy water content (CWC) were closely and significantly associated with grain yield, and the prediction ability substantially increased (R^2: 0.97) combining in a partial least square regression model NRCT, RWC, and CWC. Garcia-Romero et al. (2019) reported that germplasm with the lowest GY presented elevated CT during the anthesis and grain filling periods, and vice versa. Hence, germplasm with a higher resistance to drought and heat can be identified as plants with cooler leaves (Berger et al. 2010). Similar findings were reported also by Crain et al. (2018).

In recent years, the field-based high-throughput phenotyping (HTP) system has become popular; HTP offers the great advantage of reducing the time and the work amount for data collection and at the same time they increase the solidity of the datasets recorded (Moreira et al. 2020). It is noteworthy to say that HTP, despite the substantial progress achieved in the past years, still remains as one of the main bottlenecks in plant breeding (Araus and Cairns 2014). In a recent review, Li et al. (2021) classified HTP platform into three main classes (i) greenhouses and growth chamber under strictly controlled conditions, (ii) ground-based proximal phenotyping, and (iii) aerial large-scale remote sensing. Performing HTP in controlled environment offers the advantage of avoiding the variation due to genotype × environment interaction and it is widely used for study plant response to specific growth conditions (Li et al. 2021). The type of HTP approach depends in part on the type of research being pursued. Institutions that are focusing on breeding often combine the use of ground-based phenotyping (GBP) systems with aerial unmanned vehicles (AUV) systems. Genomic Breeding Value (GBV) system offers a better resolution as reported by several authors (Crain et al. 2018; Garcia-Romero et al. 2019; Li et al. 2021). On the other hand, AUV offer a faster data acquisition, but data processing is usually time consuming. The HTP approach used by the International Center for Agricultural Research in the Dry Areas (ICARDA) is based on a tractor mounted GPS-assisted multi-sensor head including high-definition Red, Green, Blue (RGB) and multi-spectral cameras and LiDAR among others (Rutkoski et al. 2016; Araus et al. 2018). This approach allows for rapid and accurate measuring of phenotypic traits such as green fraction, senescence, vegetation indices, LAI, biomass, spike number, and plant height (York et al. 2018). In order to produce useful and meaningful datasets to be integrated into breeding programs, a complete HTP system also requires a fully developed standardized procedure for data acquisition, storage, and analysis. Often data analysis is provided as paid service from the HTP system manufacturer or in alternative ideally any research institution can develop its own data analysis pipeline. The One Consultative Group on International Agricultural Research (CGIAR) in collaboration with other public institutions, under the umbrella of the Excellence in Breeding Initiative (EiB), developed a freely available standard operating procedures manual for UAV phenotyping (https://

excellenceinbreeding.org/toolbox/tools/standard-operating-procedures-uav-phenotyping).

4.1 In-Field Assessment Tools

The ability of a plant to successfully complete the reproductive phase and set seeds is a critical factor in determining its ability to withstand heat stress occurring during flowering. Therefore, thermography can be used to effectively measure the level of plant stress. It has been shown that wheat genotypes with lower canopy temperature can maintain higher transpiration and photosynthetic rate when moisture-stress occurs after flowering (Talebi 2011; Sallam et al. 2019; Lepekhov 2022). Canopy temperature was also reported as negatively correlated to grain yield, biomass, number of grains per spike, and 1000-grains weight under terminal drought and heat stress conditions (Rebetzke et al. 2013; Gautam et al. 2015). Garcia-Romero et al. (2019) reported that genotypes with the lowest yield presented elevated canopy temperatures during the anthesis and grain filling periods. Hence, genotypes with better tolerance to drought and heat can be identified as plants with cooler leaves (Berger et al. 2010; Crain et al. 2018). This trait can be measured using affordable infrared thermometers and a vast number of plots can be screened in one day. Alternatively, thermal cameras (or thermal long-wave infrared cameras; TIR) are sensors able to record the emitted radiation of plants by generating images representing the temperature values per pixels. Hence, TIR offer the advantage of stress detections before the appearance of the symptoms (Chaerle et al. 2007). Compared to the infrared thermometers, these tools are more precise, and can be used to measure the temperatures of specific tissues, such as the spike during the fertilization phase. However, they are also significantly more expensive and require more complex interpretation of the external factors that influence the plant canopy temperature. Pineda et al. (2021) reviewed a number of factors directly correlated to traits related with plant water balance. Because of its scalability, low cost, ease of use, and strong correlation with yield and its components, the use of thermal infrared sensor shas become a proxy vastly utilized by breeders, while thermal cameras remain a tool of choice for more advanced studies of stress response.

Early plant vigor, also defined as early ground cover, results from the combination of the number of plants germinating and overall leaf biomass produced. As such, it is an ideal proxy to define how well each genotype is adapted to the occurrence of early season stress. Easy-to-use tools have been developed to measure this trait using a visual rating score (Mullan and Garcia 2012) or digital imaging by red, green, and blue (Red, Green, Blue (RGB)) camera (Zhang et al. 2018b; Walter et al. 2019) via automatic pixel classification. The most affordable and practical Red, Green, Blue (RGB) method is CANOPEO (Patrignani and Ovhsner 2015). This app uses the camera of a smart phone to capture plot pictures and return the estimated green area by integrating in an algorithm color ratio of red to green, blue to green, and an excess green index. It has been proven to accurately estimate canopy coverage in soybean, corn, sorghum, and wheat (Jauregui et al. 2015; Patrignani and Ochsner 2015; Chung et al. 2017; Goodwin et al. 2018) and even partially predict grain yield and biomass (Jauregui et al. 2015; Chung et al. 2017; Goodwin et al. 2018). Additionally, CANOPEO can also estimate green area coverage from a continuous video of one or several plots, reducing the time needed to collect data per plot and increasing the accuracy of the measurement (Patrignani and Ochsner 2015). In addition, this trait is highly heritable, and it has shown positive correlation with grain yield (Kandic et al. 2009).

The biomass is an excellent predictor of the final plot yield in wheat, especially when stress occurs during the transitional phase. However, the actual biomass can only be computed after the harvest has been completed, it remains costly to record in large scale, and the data becomes available at the same time of yield anyway. The normalized difference of vegetation index (NDVI) utilizes a near-infrared reflectance sensor to measure vegetative greenness and canopy photosynthetic rate. This index can be further converted to a green area index (GAI), leaf area index (LAI), and the overall plant biomass (Pietragalla and Vega 2012). Different sizes of NDVI sensors are available to measure this index at different levels: low on the plot, high in the sky, or even from satellites (Pietragalla and Vega 2012). The GreenSeeker spectral sensor is one of such tools which uses NDVI to measure the progression to senescence from the whole crop canopy (Lopes et al. 2012).

The light detection and ranging (LiDAR) sensor also provide a non-destructive method to estimate biomass. Jimemnez-Berni et al. (2018) used a LiDAR mounted on a semi-automated ground vehicle to confirm that it predicted with accuracy canopy height (r^2: 0.99), ground cover (r^2: 0.92), and biomass (r^2: 0.93). The main advantage of LiDAR is that it estimates both canopy height and above-ground biomass, which is a more precise estimate of total biomass. Deploying a LiDAR on a high-throughput phenotyping (HTP) system achieved the scoring of 2400 plots per hour (Walter et al. 2019) matching the correlations described above, while the prediction of above-ground biomass using a LiDAR positioned on an unmanned aerial vehicle (UAV) achieved a lower correlation (Harkel et al. 2020).

The number of spikes is also a critical component of yield, which is negatively affected by stress occurring early in the cycle. Its recording is a time consuming and tedious work that requires many man-hours to be completed. An alternative that has been recently proposed is the use of high-resolution Red, Green, Blue (RGB) cameras to capture images of the plots in a cost-effective manner, followed by the deployment of object recognition software capable of identifying and then counting the spikes of wheat. A machine learning approach is used in this case, by first teaching the software what spikes look like, and then running this at scale models. One such method using Red, Green, Blue (RGB) color digital images taken from above the crop in natural light conditions, followed by a three-step image processing method (Laplacian frequency filter, Median filter, and Find Maxima), was proposed by Fernandez-Gallego et al. (2018).

In recent years, the use of field based high-throughput phenotyping (HTP) systems has gained popularity. HTP offers the great advantage of assessing germplasm in near-real conditions, testing a vast number of entries with high repeatability of results (Moreira et al. 2020). Li et al. (2021) classified HTP platform into three main classes (i) greenhouses and growth chamber under strictly controlled conditions, (ii) ground-based proximal phenotyping, and (iii) aerial large-scale remote sensing. Performing HTP in controlled environment offers the advantage of avoiding the variation due to GxE interaction and it is widely used to study plant response to specific growth conditions (Li et al. 2021). The type of HTP approach depends in part on the type of research being pursued. Institutions that are focusing on breeding often combine the use of ground-based phenotyping (GBP) systems with aerial unmanned vehicles (AUV) systems. Genomic Breeding Value (Genomic Breeding Value (GBV)) system offer a better resolution as reported by several authors (Crain et al. 2018; Garcia-Romero et al. 2019; Li et al. 2021), while AUV offer a faster data acquisition, but data processing is usually time consuming.

ICARDA durum program has adopted a GBP approach by deploying a tractor mounted, GPS-assisted multi-sensor adjustable head composed of several high-definition Red, Green, Blue (RGB), multi-spectral cameras, and LiDAR develop in partnership with HiPhen (Fig. 3). This "phenomobile" allows for rapid and accurate measuring of phenotypic traits such as green fraction, senescence, vegetation indices, biomass, spike number, and plant height (York et al. 2018). In order to produce useful and meaningful datasets to be timely integrated into breeding programs, a complete HTP system also requires a fully developed standardized procedure for data acquisition, storage, and analysis. Especially the analysis part can now be offered as paid service, which helps lowering the overall cost of deploying these tools.

4.2 Post-Harvest Assessment Tools

Thousand kernel weight (TKW) is a trait with high heritability (Royo et al. 2002; García del Moral et al. 2003; Rharrabti et al. 2003; Wang et al. 2009; Maccaferri et al. 2011; Graziani et al. 2014; Li et al. 2015; Wurschum et al. 2018),

Fig. 3 ICARDA "Pheno Buggy" taking measurements in a Stage 1 yield trial of the ICARDA durum wheat program

thus being very suitable as target for improvement through crossbreeding (Lopes et al. 2012; Feng et al. 2018) and a critical component of grain yield when terminal drought occurs. The grain size, and consequently its weight, is determined by grain length and width. Interestingly, grain length is under strong genetic control, while its width fluctuates based on the available moisture (Xin et al. 2020; Elhadi et al. 2021). Hence, selection for grain length generates a faster genetic gain and is therefore a major strategy for promoting drought tolerance (Gegas et al. 2010; Wu et al. 2015; Kehel et al. 2020; Khadka et al. 2020). Grain shape is also of great importance since spherical grains tend to produce more flour yield due to the lower surface/volume ratio (Evers et al. 2000). To facilitate the screening for this trait, several applications have been developed to dissect grain volume. One of the most affordable and scalable applications is the Grain Scan, which uses a home-grade scanner to collect images of grains, followed by analysis with a form distinction algorithm that determines all main grain characteristics such as length, width, and perimeter (Whan et al. 2014). As mentioned previously, grain length is highly heritable and presents a great opportunity to make rapid genetic gain for adaptation to drought occurring after flowering (Kehel et al. 2020; Khadka et al. 2020).

4.3 Below-Ground Assessment of Traits

The root system plays a pivotal role in ensuring moisture and nutrients availability at all stages of the crop development (Ober et al. 2021). However, it is still unpractical to conduct direct root assessments in the field. A scalable alternative was presented by Alahmad et al. (2019) by deploying a clear pot method to visualize and score thousands of plants for their seminal root angle. Still, this method does not always predict well the mature root angle (El Hassouni et al. 2018) nor the ultimate field performance. One alternative presented by El Hassouni et al. (2018) was the use of a pasta-trainer method to sample the actual mature roots. However, while this method allows to predict reasonably well the final field performance, it is only scalable to medium level as it remains quite time consuming. A better alternative would then be the use of proxy traits that are easy to measure. The temperature of the canopy is tightly linked to the water status of the plant and therefore represents a good predictor for the ability of the roots to reach moisture in deeper soil layers at the end of the season. In this regard, El Hassouni et al. (2018) found a strong linear relationship between total root number and total tiller number, which suggests this above-ground trait as a valuable proxy to assess the extent of development of the root system.

5 Biometric Approaches to Improve Field Assessments

Climatic stresses are far from linear and their effect on a field trial is often complex to extricate. In fact, the effect of the environment and especially the year/season are often the major sources of variation for grain yield in nearly all multi-locations studies of durum wheat, accounting for more than half of the total effect (Bassi and Sanchez-Garcia 2017; Macro et al. 2017). Breeders have since long learned the importance of blocking this variation using specific field layouts and ensured sufficient replications within and across environments. Controlling the effect of environment and year therein is fundamental for a breeding program that aims at addressing climatic stress tolerance (Garcia et al. 2003; Bassi and Nachit 2019). Still, the effect of seasons and locations is not the only confounding effect to a trial, and in fact a large part of the variation occurs within the environment. Heat stress typically occurs uniformly across a trial, but when this happens around the time of flowering, different genotypes could face it at different phenological stages, therefore being more or less exposed to its damaging effects. That is why the use of simple biometric solutions, such as using flowering time as covariate and/or suitable controls in any heat stress analysis, can prove quite useful to control some of the errors (Gupta et al. 2020). However, high temperatures tend to be quite selective on the germplasm, and normally good trial heritability is obtained (Sall et al. 2018).

In the case of drought, research stations can accumulate large experimental errors due to many factors affecting the unaccounted variability amount of moisture available in the soil. For instance, the slope of the trial can favor moisture accumulation in the lower points, especially when rainfall is limited. Similarly, the diverse depth of the soil layers across a trial means that roots must reach different depths to find residual moisture, and it might also be available in different amounts. Finally, another common cause of trial error in dryland agriculture is the rotation system applied, with some rotating crops consuming more moisture than others.

Altogether, these factors cause severe spatial variation across a trial, resulting in higher error and lower experimental heritability. Blocking these micro-variations by field layout is often too complex to achieve. Instead, biometricians have provided breeders with a simpler solution: the use of row by column designs allows for spatial correction of field trials (Shackley and Anderson 1995; Vetch et al. 2019; Francki et al. 2021). Adopting this field layout has been facilitated by the adoption of the design package CycDesigN 7.0 (https://www.vsni.co.uk/software/cycdesign/, John and Williams 1995, 1998; Piepho et al. 2018).) as described by Williams et al. (2006) and Hartung et al. (2019) provide further details. Figure 4 demonstrates an example of deploying this statistical model for the experimental station of Marchouch using data from the two seasons shown in Fig. 1. The use of corrections boosted the trial heritability in the case of six columns by 24 rows experiment conducted in an alpha lattice design with two replications. It is possible to observe in the graphic representation how entries falling onto rows 9 and 15 had an average yield of 2400 kg ha^{-1}, but once spatial correction was done, the entries in row 10 were corrected to lower yields, while for row 15 the values were confirmed. Given the higher accuracy that this approach ensures and the relative no cost of deploying it, it seems prudent to promote its use across all programs.

6 Sources of Stress Tolerance Available to Improve Durum Wheat Adaptation

To develop durum wheat cultivars well adapted to stress tolerance, it is critical to first identify sources of useful alleles (Cattivelli et al. 2008; Fleury et al. 2010; Hu and Xiong 2014; Mwadzingeni et al. 2016; Mazzucotelli et al. 2020).

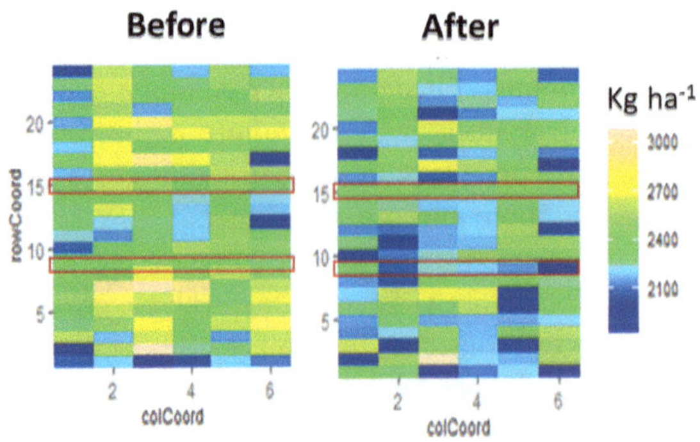

Trait	Spatial adjustment	
	Before	After
2019-20	Trial heritability	
Flowering time	0.13	0.45
1,000-kernels W.	0.10	0.35
Grain yield	0.16	0.76
2020-21		
Flowering time	0.21	0.63
1,000-kernels W.	0.39	0.39
Grain yield	0.58	0.79

Fig. 4 Improving drought-affected trial heritability by spatial adjustment. Examples of 2019–2020 and 2020–2021 seasons at the experimental station of Marchouch (Morocco) in a trial of 6 columns by 24 rows to assess 72 genotypes repeated twice in alpha lattice design. The legend shows the spatial distribution for grain yield expressed in kg ha^{-1}

6.1 Modern and Cultivated Germplasm

Until the early twentieth century, landraces were the only types of durum wheat cultivars being grown. Most of them belonged to the Mediterranean type (Ali et al. 1992) and were tall, susceptible to lodging, and late flowering. Landraces were part of a pastoral farming system aimed at self-supporting and at satisfying the needs of the households in rural areas by means of different animals and crops, in support, within agro-ecosystems characterized by high levels of diversification (Marina et al. 2019; Xynias et al. 2020; Junhua et al. 2021).

The progressive release of more productive cultivars with improved quality occurred in synchrony with developments in agriculture and management practices. It was the synergy between breeding and management that generated the most remarkable results with the introduction of the *Rht-1* dwarfing genes (Swaminathan 2014) responsible for a drastic reduction in plant height and an increase in the harvest index for similar biomass levels (Giunta et al. 2007; Lopes et al. 2015). In the 1960s, the substantial increase in grain yield obtained with the semi-dwarf cultivars generated the so-called Green Revolution in the Global South, particularly in Asia and the Southern Cone of South America (Swaminathan 2014). In 2014, 100% of the total area planted with durum wheat in European Union (EU) and high-income countries was occupied by semi-dwarf cultivars, and 77% of these cultivars traced their lineage to the Consultative Group on International Agricultural Research (Consultative Group on International Agricultural Research (CGIAR)) germplasm (Lantican et al. 2005, 2016).

Despite the general interest in identifying sources from exotic germplasm, the modern gene pool has been developed and selected to adapt to the stress conditions occurring in the field. Assessing the available genetic variation in durum wheat, several authors (Kabbaj et al. 2017; Bassi and Nachit 2019; Mazzucotelli et al. 2020) have realized that the modern germplasm is extremely rich of allelic variation, but that these variations are maintained mostly by separated breeding programs. The exchange of germplasm among breeding programs across countries remains limited, and that has been causing a narrowing of the genetic basis within programs (Mkhabela et al. 2020). The international breeding efforts conducted by the One Consultative Group on International Agricultural Research (Consultative Group on International Agricultural Research (CGIAR)) breeding programs provide strategic solutions to this issue, importing and exporting valuable alleles across borders (Mkhabela et al. 2019, 2020). A more targeted effort was developed by the Durum Wheat Expert

Working Group by developing and distributing the Global Durum Panel (GDP). This panel combines a large portion of the known diversity for durum wheat and provides an ideal support to identify, validate, and exchange alleles across borders. A similar approach but more oriented toward mapping and cloning of novel loci of the QTLome is being pursued on a broader scale of haplotype diversity and lower linkage disequilibrium (LD) with the tetraploid wheat global collection (TGC; Maccaferri et al. 2019). More recently, Illumina 90K data and haplotype structure analysis allowed for the assembly of the tetraploid wheat core collection (TCC; 432 accessions, see GrainGenes) for capturing and cloning more effectively the significant sequences of target QTL from the A and B genomes shared by durum and bread wheat. Notably, the TCC spans a linkage disequilibrium decay rate from 195 kb in wild emmer to 1.5 Mb in domesticated emmer and durum landraces, up to 4.5 Mb in durum cultivars. Therefore, the TCC is an optimal platform for the QTL dissection and allele-mining (Maccaferri et al. 2019). The TCC includes a golden level of 24 accessions (included in an ongoing international project for the durum wheat pangenome sequencing), 250 accessions for the silver level, and up to 432 for the bronze level.

The GDP, TGC, and TCC are detailed as Global Durum Genomic Resource in GrainGenes at https://wheat.pw.usda.gov/GG3/global_durum_genomic_resources. The tetraploid wheat gene pool, including 11 *T. turgidum* subspecies spread across Europe, North Africa, Asia, and Ethiopia is a valuable resource for improving both durum and bread wheat, either by traditional backcrossing or by generating synthetic wheats, or by targeted editing/cisgenesis. Despite the general interest in identifying sources from exotic germplasm, the modern gene pool has been developed and selected to adapt to the stress conditions occurring in the field. Assessing the available genetic variation in durum wheat, several authors (Kabbaj et al. 2017; Bassi and Nachit 2019; Mazzucotelli et al. 2020) have realized that the modern germplasm is extremely rich of allelic variation, but that these variations are maintained mostly by separated breeding programs. The exchange of germplasm among breeding programs across countries remains limited, and that has been causing a narrowing of the genetic basis within programs (Mkhabela et al. 2020). The international breeding efforts conducted by the Consultative Group on International Agricultural Research (Consultative Group on International Agricultural Research (CGIAR)) breeding programs represent strategic solutions to this issue, importing and exporting valuable alleles across borders (Mkhabela et al. 2019, 2020). A more targeted effort was developed by the durum wheat expert working group by developing and distributing the Global Durum Panel (GDP) and the tetraploid wheat core collection (TCC). These panels combine most of the known diversity for durum and other tetraploid wheat and provide an ideal support to identify, validate, clone, and exchange valuable alleles across borders.

6.2 Ancient Germplasm and Landraces

Landraces represent the basis of modern breeding. The most successful durum wheat cultivar of all time "Senatore Cappelli" was bred in 1915 by Nazareno Strampelli starting from a selection of a landrace population probably from North Africa (Di Fonzo et al. 2005; Scarascia Mugnozza 2005). This same principle was applied by most durum wheat breeding programs worldwide, selecting one or more landrace populations as the initial founders. A more recent example is the ICARDA drought tolerant cultivar "Om Rabi" (syn Cham 5), which was derived from the CIMMYT cultivar "Jori 69" and the West Asian landrace "Haurani" (Nachit 1992; Kabbaj et al. 2017). This was first released in 1981 in Algeria (Nachit 1992) followed by other 21 countries that still cultivate it today (Latican et al. 2016), and it was also extensively used as parent for the following generations.

Beside these examples, several screenings of landraces have revealed exploitable traits and alleles for heat and drought tolerance (Distelfeld et al. 2006; Wasaya et al. 2018), for instance, the

West Asian landrace selection and landrace intercrossing, such as "Senatore Cappelli" and "Capeiti 8" (a cross between "Cappelli" and a Syrian landrace selection). The second group included durum landraces from West Asia including "Haurani," well-known as the most widely cultivated landrace population in its area of origin, showing characteristic black spike that has been demonstrated to possess several developmental and morphological traits relevant for adaptation to low water availability and high temperatures, widely exploited and it is then not a surprise that it constitutes one of the founders of the ICARDA durum wheat breeding program since its inception (Elings and Nachit 1991; Pagnotta et al. 2005; Kabbaj et al. 2017). Another example is the "Karak" landrace that showed long shoots and roots (Marina et al. 2019).

Recent studies reported PIC values of 0.26 for durum modern germplasm (Chao et al. 2017), 0.19 for a set of both landraces and modern lines (Ren et al. 2012), and 0.18 in a collection of 168 durum wheat accessions of different origins (Roncallo et al. 2019). Analogously, the analysis of molecular variance (AMOVA) clustering based on geography and breeding program of origin showed that only 13% of the total genetic variance could be captured among groups, while most diversity remained among individuals within clusters. These results concur with those reported by Soriano et al. (2016) with 172 landraces from 21 countries, by Roncallo et al. (2019) with a panel of 168 durum accessions and by N'Diaye et al. (2018) with a panel of Canadian durum cultivars where only 10% of variation was captured among groups. Other research considering similar panels reported capturing over 30% of the total genetic variance by clustering germplasm based on kinship matrix but using relatively higher k values. Breeding programs and landraces were the richest in rare alleles. Furthermore, phylogenetic analysis among landraces indicated available genetic diversity for this crop and shall guide its targeted use by breeders (Kabbaj et al. 2017; Robbana et al. 2019). Similarly, the importance of very old Mexican landraces as potential sources of genetic diversity was highlighted for key quality traits in the development of modern wheat cultivars with different grain textures (Marcela et al. 2013).

An initial study on some of the used landraces showed the tolerance to drought and heat according to the investigated traits. The "Karak" landrace showed long shoots (48% relative to its control), the longest roots (45% relative to its control), and the highest proline content (483% relative to its control) (Marina et al. 2019; Xynias et al. 2020; Junhua et al. 2021) and hence, it might be the most drought tolerant landrace (Al-Khateeb et al. 2017). Landraces such as "Sakha 64" and "Sakha 63" had tolerance selected in Egypt following selection to many stress conditions (Al-Naggar et al. 2020). Current durum wheat selection may lead to yield stability and specific adaptation, which provides the opportunity for this germplasm source to be useful for the genetic improvement in both drought and cold tolerance in durum wheat (Mohammadi et al. 2014).

6.3 Secondary Gene Pool and Wild Relatives

The secondary gene pool and wild relatives (WR) of durum wheat represent a potential and excellent source of useful alleles for climatic stress adaptation (Zaim et al. 2017; Kumar et al. 2018; Vincent et al. 2019; Jovovic et al. 2020; El Haddad et al. 2021). In fact, this type of germplasm survives in marginal lands, exposed to stresses, and survives without any help by farmers (Zaharieva et al. 2001; Nevo et al. 2002; Yesayan et al. 2009; Kiani et al. 2015; Zhang et al. 2016a). However, the use of this type of germplasm comes with its own difficulty and hence breeders rarely turn to it for abiotic stress adaptation. Nonetheless, a shift has been seen in the past decade and more breeders seek to introgress the wild alleles into modern germplasm (Pour-Aboughadareh et al. 2020; Budak et al. 2015). One approach is to develop trait-specific genetic stocks (TGS) that can be used to generate new populations. These trait-specific genetic stocks are the "immortal" germplasm and are ideal for trait mapping, identification of candidate genes and re-sequencing research, as well as for developing a super-pangenome for wheat due to the costs and complexity associated with pre-breeding. Nowadays more than ever the role of public sector research becomes critical to provide freely as parental lines based on specific needs, but it was also demonstrated that the use of WR in three-way crosses can be even more strategic to rapidly deploy these novel alleles. Sall et al. (2019) tested 24 durum wheat lines for tolerance to extreme heat along the Senegal River, which revealed that the top and most stable yielders were derived from interspecific hybridizations with *Triticum dicoccoides* and *Aegilops speltoides*. El Hassouni et al. (2019) exposed a set of durum elites to extreme heat stress by applying plastic tunnels at the time of flowering, reaching peaks of 46 °C, and identified as top tolerant genotypes lines derived from WR. Similarly, Zaim et al. (2017) screened wild relative-derived elites across drought-prone environments in North Africa, to confirm that these out-yielded the best elites and cultivars. In the case of Zaim et al. (2017), the use of *T. dicoccoides* as parent was strategic to gain tolerance to terminal drought stress. Aberkane et al. (2021) evaluated 77 durum wheat lines for drought and heat tolerance to show that three WR derivatives were the best under both stresses. These findings are in good agreement with several authors that also identify the progenitor of durum wheat as an ideal source for drought adaptation (Nevo et al. 1984; Feldman and Millet 1993; Monneveux et al. 2000; Plamenov 2003; Peleg et al. 2005; Nachit et al. 2016; Trethowan 2014; El Haddad et al. 2021).

7 Use of Major Effect Alleles for Abiotic Tolerance

The identification of major alleles contributing to tolerance to heat and drought is a critical step to favor their pyramiding into cultivars ready to face the changing climate (Afzal et al. 2017; Mwadzingeni et al. 2017; Langridge et al. 2022). Tagging these alleles with markers represents a strategic solution to conduct selection without the confounding effect of the environment (Semenov and Halford 2009; Reynolds and Rebetzke 2011; Zarei et al. 2013; Maccaferri et al. 2022). To be exploitable by breeders, these alleles need to have a clear and strong effect on the final phenotype, hence accounting for a large portion of the phenotypic variation, and also be highly heritable to simplify their introgression. In that sense, these are defined here as major effect alleles in what has been defined the QTLome, namely the major effects QTL with additive effects reasonably large and heritable to allow for their mapping and, eventually, cloning (Salvi and Tuberosa 2015; Kong et al. 2023). Extensive research has been conducted accordingly, but here we revise only the major QTL that have been identified by various research groups as associated with stress tolerance and that for this reason are of greater importance for durum wheat breeders (Khan et al. 2011; Ibrahim et al. 2012; Saini et al. 2022).

Several researchers investigated the genetic basis of grain yield and its components (Maccaferri et al. 2008; Laidò et al. 2014; Soriano et al. 2016), crop phenology (Soriano et al. 2017), or biomass (Soriano et al. 2016) in durum wheat. QTL meta-analysis has been achieved in durum wheat for different traits such as grain yield (Wahid et al. 2007 Ibrahim et al. 2012) crop phenology (Hanocq et al. 2007; Wahid et al. 2007), disease resistance (Loffler et al. 2009; Goudemand et al. 2013; Soriano et al. 2016), plant height (Griffiths et al. 2012), grain-related traits (Quraishi et al. 2011; Sheoran et al. 2015), and root-related traits (Darzi-Ramandi et al. 2017; Rosello et al. 2019).

Maccaferri et al. (2008), in a milestone paper for the durum community, indicated that 16 QTL found after screening of 249 recombinant inbred lines (RIL) evaluated across 16 environments (10 rainfed and 6 irrigation) in different Mediterranean locations. Two QTL (*QYld.idw-2B* and *QYld.idw-3B*) were highly significant and stable in eight and seven environments with a coefficient of determination (R^2) of 21.5 and 13.8%, respectively. Zaim et al. (2020) using four RIL populations tested under drought stress, identified six quantitative trait loci (QTL) associated with grain yield, One of these QTL, located on chromosome 2B, explained between 4.2% and 13.1% of the variation in grain yield across the different populations. Roncallo et al. (2017) used a RIL tested across drought-affected environments to further confirm the major locus on chromosome 3B, which in this case explained 23% of the variance for grain yield. These two major QTL on chromosomes 2B and 3B are therefore ideal targets for durum breeders seeking to improve drought adaptation.

Thousand kernel weight is one of the most critical yield components. Xu et al. (2017) reported six QTL for TKW located on chromosomes 1B, 4B, 6A, 6B, and 7A. Similarly, Avni et al. (2018) reported using a RIL population derived from the cross between elite durum wheat and wild emmer tested across four environments, to define a large QTL explaining 20.9% of the variance for TKW located on chromosome 1B. Graziani et al. (2014), Fatiukha et al. (2020), and Zaim et al. (2020) confirmed this finding, identifying smaller effect QTL linked to TKW in the same region. Tura et al. (2020) also identified one QTL (QTgw.aww-1B) for TKW on chromosome 1B using double haploid population evaluated in 32 field trials across three countries under drought and heat stress. Therefore, this region on 1B is certainly of great interest to breeders seeking to improve drought tolerance.

An important study by Alahmad et al. (2019) using nested mapping populations identified a major QTL (*qSRA-6A*) on chromosome 6A responsible for controlling the seminal root angle. This same QTL was also previously reported by Maccaferri et al. (2016) and Rossello et al. (2019), and later by Alemu et al. (2021) and Schmidt et al. (2020), making it another region of choice for integration into breeding programs (Ober et al. 2021).

El Hassouni et al. (2019) imposed on a panel of durum wheat lines severe heat stress at the time of flowering, to reveal three novel loci on chromosomes 1AL, 5BL, and 6BS. These QTL had been previously reported in hexaploid wheat but were otherwise novel for durum wheat. Their effect of different allele combination on grain yield using a separate validation set was significant and its result confirmed that those genotypes with positive allele at all three QTLs shown to increase yield by 8% when field tested under heat stress. These were further converted to Kompetitive Allele-Specific PCR (KASP) and made available for use in breeding programs. Sukumaran et al. (2018) utilized 208 durum wheat lines assessed under heat stress to define two marker–trait association (MTAs) for grain yield on chromosomes 2BS and 4AL. The one located on 2BS was suggested to overlap with the QTL already described previously for drought adaptation (Maccaferri et al. 2008), while the one on 4AL appears to be novel and might warrant additional consideration by breeders. However, it is somewhat surprising that while vast literature exists about genomic dissection of drought tolerance in durum wheat, the two reported above are the only articles we could find on heat tolerance.

7.1 Validation of Markers for Drought and Heat Tolerance

The identification of major alleles is an important step to further our understanding of the genetic mechanisms contributing to stress adaptation. However, this discovery is of limited practical use for breeders, unless it is paired with a step of marker conversion to a technology that is readily exploitable by breeders and then validated for its ability to predict phenotypic changes in elite germplasm. In fact, QTL discovery is typically performed using vast genotyping arrays that remain today too expensive to be used in large scale for selecting among thousands of progenies. The durum community has elected three major sequence-based genotyping technologies as the preferred options for QTL discovery: DArT Seq (Baloch et al. 2017; Robbana et al. 2019, 2021), 35K Axiom Breeder Array (Kabbaj et al. 2017), and Illumina 90K (Maccaferri et al. 2014; Mazzucotelli et al. 2020).

Thanks to the availability of the "Svevo" genome assembly (Maccaferri et al. 2019), it has become relatively easy to align the sequences tagged by each of the three genotyping technologies to the genome scaffold. This alignment allows to compare the position of major loci across different experiments and hence facilitates the identification of novel and useful sources. Still, these platforms are not suitable for screening progenies of breeding programs, and a fourth valid alternative has arisen. The competitive allele-specific polymerase chain reaction (Kompetitive Allele-Specific PCR (KASP)) assay is a PCR-based approach that can reach high level of multiplexing (i.e., 24 to 192 Single Nucleotide Polymorphism (SNP)s) at a very competitive price (Rasheed et al. 2016; Bassi et al. 2016; Anuarbek et al. 2020; Bhavjot et al. 2020; Maccaferri et al. 2022).

To achieve marker conversion, the sequences of the markers tagging a QTL of interest are used to design three Kompetitive Allele-Specific PCR (KASP) primers: one common primer, designed on a sequence that is identical in all allelic versions, and at least two specific primers, each tagging a Single Nucleotide Polymorphism (Single Nucleotide Polymorphism (SNP)) specific to one allele (Maccaferri et al. 2022). These Kompetitive Allele-Specific PCR (Kompetitive Allele-Specific PCR (KASP)) then need to be validated by genotyping some of the lines where the QTL was originally discovered to confirm their allelic call, but also an additional group of elites to confirm its capacity to discriminate phenotypic variation. Good examples of this conversion and validation process can be seen in Bassi et al. (2019) and El Hassouni et al. (2019).

Unfortunately, many authors of durum wheat research fail to complete their studies with the conversion and validation step, leaving resources that would be of great interest for breeders completely under exploited. Table 2 summarizes to the best of our knowledge a list of Kompetitive Allele-Specific PCR (Kompetitive Allele-Specific PCR (KASP)) markers currently available (i.e., validated) for selecting durum wheat progenies for their adaptation to heat and drought stress.

7.2 Fine-Mapping Major Alleles for Abiotic Stress Tolerance

The identification of QTL is the first step, followed by the conversion and validation of markers for assisted selection (MAS). However, the suitability of a marker for MAS is highly depen-

Table 2 Kompetitive Allele-Specific PCR (KASP) markers validated for abiotic tolerance in durum wheat

Marker ID	Stress	Trait	References
AX-95182463	Heat	Yield	El Hassouni et al. (2019)
AX-95260810	Heat	Yield	El Hassouni et al. (2019)
AX-94408589	Heat	Yield	El Hassouni et al. (2019)
AWN_BWc8266_5AL	Drought	Yield	Khalid et al. (2019)
BS0006632_51	Drought	Yield	Mahhdiyeh et al. (2021)
1fehw3	Drought	Yield	Rasheed et al. (2016)
Dreb-B1	Drought	Yield	Rasheed et al. (2016)
Sus1-7B-2932IND	Drought	Superoxide dismutase	Khalid et al. (2019)
AX-108886949	Drought	Grain weight	Lin et al. (2021)
TaSnRK2.3-1A	Drought	Grain weight	Ur Rehman et al. (2021)
TaSnRK2.3-1B	Drought	Grain weight	Ur Rehman et al. (2021)
TaSAP-7B	Drought	Grain weight	Ur Rehman et al. (2021)
TaPARG-2A	Drought	Grain weight	Ur Rehman et al. (2021)
TaPPH	Drought	Grain weight	Ur Rehman et al. (2021)
TaSnRK2.9-5A	Drought	Grain weight	Ur Rehman et al. (2019)
TaMoc-2433	Drought	Grains per spike	Rasheed et al. (2016)
AX-94721936	Drought	Spikelet per spike	Ma et al. (2020)
AX-110129912	Drought	Spikes per m^2	Liu et al. (2020)
AX- 110914105	Drought	Spikelet per spike	Li et al. (2021)
TaDRO-5A_Single Nucleotide Polymorphism (SNP)	Drought	Deep roots	Zhang et al. (2018c)
TaDRO-5B_InDel	Drought	Deep roots	Zhang et al. (2018c)

dent on its proximity to the actual locus responsible for the detected phenotypic variation. Hence, the ideal marker would be the gene (or non-coding sequence; for an example in maize see Salvi et al. 2013) itself and even better the functional Single Nucleotide Polymorphism (Single Nucleotide Polymorphism (SNP)) responsible for the phenotypic variability. To achieve this level of resolution, the initial QTL discovery needs to be followed by a fine mapping procedure, to refine the position of the gene controlling the phenotype, and ultimately achieve the actual cloning of the gene (Edae et al. 2014; Sheoran et al. 2019; Maccaferri et al. 2022).

One method to clone a gene is by forward genetics, starting from a phenotype of interest to then pinpoint the gene controlling it. Huang et al. (2006) followed a classical forward genetic approach to saturate the genomic interval of a QTL linked to salinity tolerance (*Nax1*) on chromosome 2AL of durum wheat and identify one open reading frame as the most likely candidate. Bassi et al. (2016) deployed a "fast forward" genetics approach by using bulk sequencing to fine map the genomic region of a QTL controlling the interaction between the cytoplasm and the nucleus of durum wheat, ultimately identifying one candidate gene and its functional Single Nucleotide Polymorphism (Single Nucleotide Polymorphism (SNP)). Additional examples of gene cloning by forward genetics exist for durum wheat, but these mostly refer to disease resistance loci. As to candidates for drought-related traits in durum wheat, Condorelli et al. (2022) described potential candidates for active osmotic adjustment, a scarcely investigated trait notwithstanding its demonstrated importance (Blum 2017).

An alternative approach to gene cloning is the use of reverse genetics, starting from generating mutants for a given gene and then determining the phenotyping effect. There are two major mutants' populations available to the durum community, both developed via ethyl-methanesulfonate (EMS) mutation "Kronos" (Krasileva et al. 2017). Comastri et al. (2018) conducted a classical reverse genetic study using the Kronos population by first identifying in silico the sequence of the durum heat shock protein 26 (*TdHsp26*), to then seek mutants for it to validate the phenotypic effect of different alleles. Harrington et al. (2019) also exploited the Kronos population via "fast reverse" genetics approach, using bulk sequencing of mutants to identify a small region of 4.3 Mbp in size on chromosome 3A responsible for controlling accelerated senescence in durum wheat. More recently, the Kronos in silico collection allowed the validation of a gene controlling root angle in both barley and durum wheat (Kirschner et al. 2021).

Overall, there are substantially no cloned genes for abiotic stress tolerance of durum wheat to date, so a major effort is needed by the community to appropriately dissect the QTLome (Salvi and Tuberosa 2015; Maccaferri et al. 2022) achieve this for supporting breeders in their attempt to implement gene editing and, hopefully, mitigate the effects of climate change.

8 Transgenic, Cisgenic, and Gene Editing Approaches to Adapt Durum Wheat to Abiotic Stress

Once a strong candidate gene for a trait of interest has been identified via fine mapping, its delivery into superior cultivars can follow two main pipelines: accelerated molecular breeding, using DNA markers and rapid backcrossing schemes, or direct cisgenic or transgenic breeding. Rapid introgression of individual genes via molecular breeding is now routinely achieved in as little as one year using speed breeding (Alahmad et al. 2018), thus making this the approach of choice for durum wheat. Also, the current regulations prevent the cultivation of genetically modified durum wheat in most of its main growing areas. In parallel, the use of genome editing approaches based on CRISPR-Cas9 proteins has great potential for application, but it remains unclear on how these will be regulated in the future (Peters Haugrud et al. 2024.

Despite its limited commercial use, some interesting trans- and cis-genic approaches have been attempted in durum wheat as academic validation studies. A good example of a transgenic approach has been proposed by Djemal and Khoudi (2016), who placed under 35S constitutive promotion into tobacco *TdSHN1*, a durum wheat transcription factor belonging to the APETALA/Ethylene Responsive factor (AP2/ERF) family. The transgenic plants exhibited enhanced drought tolerance. Similarly, Ben Saad et al. (2019) placed under constitutive expression into tobacco plant *TdAnn12*, a gene from durum wheat that belongs to the Annexin family and known for their expression across organs in linkage with stress response (Davies 2014; Ijaz et al. 2017). In this case, the transgenic plants also showed improved tolerance to both drought and salt stress under glasshouse conditions. Ebel et al. (2018) extracted *TdTIFY11a* from the Tunisian durum wheat cultivar "Oum Rabiaa 3." *TdTIFY11a* is a JAZ/TIFY protein regulating several processes of plants including stress responses. Transformation of this gene into *Arabidopsis* under constitutive expression confirmed its role in providing salinity tolerance when grown on 150 mM of NaCl media. MeRed, Green, Blue (RGB)y et al. (2021) isolated *TtNAC2A* gene from durum wheat, a NAC transcription factor that plays diverse roles in plant development and stress responses. Overexpression of this gene into *Arabidopsis* confirmed its role in activating several drought-related genes and promoting overall drought tolerance of transformed plants.

Cisgenesis instead refers specifically to the transformation of a wheat plant using a gene from another wheat plant or a wheat wild relative. Ayadi et al. (2019) used this approach for placing the durum wheat aquaporin *TdPIP2;*1 gene under constitutive expression into the *Tunisian durum* cultivar "Maali." Aquaporins are proteins responsible for the transportation of water and other molecules through the membrane of the roots, and hence are deemed of strategic importance for drought stress tolerance. Assessing of the wild type and transformed version under drought in the glasshouse confirmed clear superiority of the second thanks to a more vigorous germination, higher biomass production, and better nutrients uptake. Yang et al. (2019) placed two drought-responsive element binding (DREB) genes isolated from bread wheat (*TaDREB3* and *TaCBF5L*) under the expression of two durum wheat stress responsive promoters *TdHDZipI-3* and *TdHDZipI-4* into wheat plants. The authors confirmed that the expression could be modulated via the imposition of water stress and that the transformed plants performed better under controlled conditions than the wild type when exposed to this stress in the glasshouse. Zang et al. (2018) transformed *TaPEPKR2*, a phosphoenolpyruvate carboxylase kinase-related kinase gene identified in common wheat as involved with heat stress response, into the wheat cultivar "Liaochun 10" to confirm that its overexpression provided enhanced heat stress tolerance.

The number of experiments using genome editing techniques in durum wheat is still limited but expanding rapidly. Zhang et al. (2016b) targeted *TdGASR7-A1* and *-B1*, two durum wheat homeologs associated with the negative control of grain width and weight. In their experiment, it was possible to confirm that simultaneous mutants for all four allelic states could be obtained in T1, and none carried the CRISPR/Cas9 construct. This was the first example of generating mutated durum lines with larger grain size, without any traceable transgene. Other editing experiments have been conducted for durum wheat to improve its dietary acceptance by reducing its allergenic effect on celiac affected consumers (Camerlengo et al. 2020), but none thus far linked to stress tolerance. In fact, the first step required for transformation or editing is a clear understanding of the genes involved and the functional Single Nucleotide Polymorphism (SNP)s responsible for the phenotypic effect. To date, there are no genes cloned in durum wheat for stress adaptation, and hence no possible targets for editing. However, there are some interesting results obtained by stable transformation (*TdSHN1, TdAnn12, TdTIFY11a, TtNAC2A, TdPIP2;*1, *TaDREB3, TaCBF5,* and *TaPEPKR2*) that could now be investigated for the available diversity, and then target the best alleles by editing. Such an approach could generate non-transgenic superior parental material to be incorporated into breeding programs.

9 Incorporating Minor Effect Alleles for Abiotic Tolerance

The effect of "year" (i.e., different seasons) often explains the largest fraction of phenotypic variation for grain yield under abiotic stresses. As such, it would be critical to control this factor in the breeding programs to achieve faster genetic gain, but this task is far from trivial. While it is true that breeding programs occur in the field and that each season differences are captured when selecting germplasm annually; however, there is no guarantee that this alternation of different seasons truly results in better lines being selected. In the climatic example presented for Marchouch station in Fig. 1, there was substantially no correlation between the grain yield results obtained for stage 1 under severe drought in season 2019–2020 and for stage 2 the following season that was instead favorable (Fig. 5). Both error types occurred, with type I error causing the selection of top performing entries in Stage 1 that resulted as poor performers the following season, and type II error causing poor performing Stage 1 to become outstanding lines in Stage 2 the following season.

The best approach to control the "year" effect would be to use data combined across multiple seasons to define which entries are truly performing well. However, this in turn would require breeding programs to perform selections only after multiple years of data are gathered. Such a solution, while correct, is very complex to apply in a program since most breeders rely on imposing strong selection intensity between Stage 1 and Stage 2 to achieve rapid rates of genetic gain at reasonable cost after just one season of data. A more realistic approach would instead be to use multiple diverse environments to mimic the year-to-year variation and select genotypes that are stable across them (Bassi and Sanchez-Garcia 2017; Watson et al. 2018). Still, doing so at Stage 1 presents an additional array of difficulties. For instance, seed availability is often scarce at this stage of the breeding pipeline. Also, the use of accelerated breeding methods such as speed breeding (Alahmad et al. 2018; Watson et al. 2018; Hickey et al. 2019) or shuttle breeding (Borlaug 1968; Gale and Youssefian 1985; Kronstad and Rajaram 1990) mean short turnaround time between seasons, which can cause strains on the logistics of harvesting and replanting rapidly across locations.

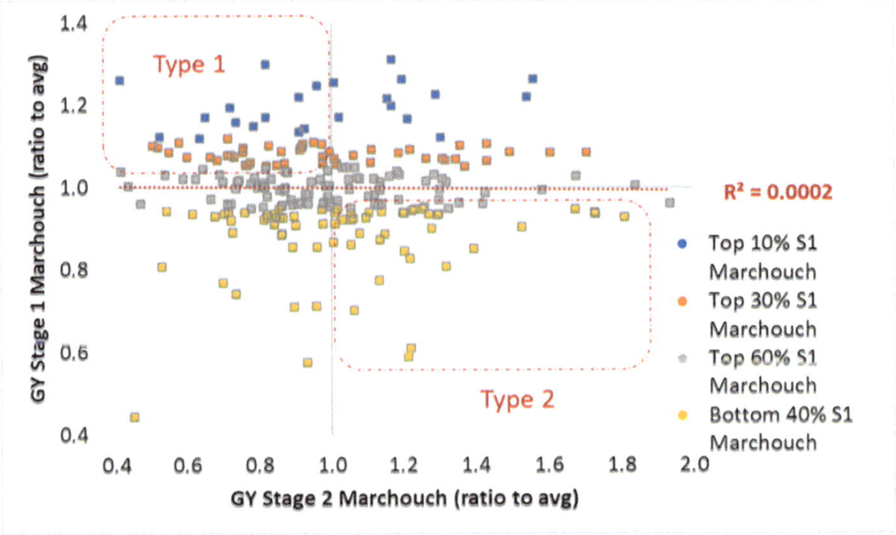

Fig. 5 Effect of two consecutive seasons (see Fig. 1) on the selection of best performing durum wheat entries from Stage 1 conducted in 2019–2020 season and Stage 2 conducted in 2020–2021 season at the station of Marchouch (Morocco)

The use of genomic predictions has been suggested by many authors as a strategic solution to model the genotype by environment and genotype by year effects in the early generations of the breeding cycle. Genomic predictions rely on the use of a training set of germplasm for which phenotypic and genotypic data are available, to model (i.e., predict) the performances of a second set of germplasm for which only genotypic data are known. There are several models to do so, but in general they all rely on the idea of defining an "effect" for each allele tagged by a marker and then define the performance of a line based on the sum of these effects, independently on how small each individual effect is. In that sense, it is often referred to as "minor allele effect" approach to distinguish it from MAS, which relies instead on "major allele effects" that control large portion of the phenotypic variation. Adopting the advocated "iceberg" metaphor of the QTLome (Salvi and Tuberosa 2015; Maccaferri et al. 2022), genomic selection (GS) allows for the full exploitation of the beneficial haplotypes in the submerged portion of the genome iceberg and as such with additive effects too small to be adequately mapped and exploited through MAS and, eventually and most importantly, cloned and edited.

An example of genomic selection is provided in Fig. 6 using data from the ICARDA durum wheat breeding program, where the Stage 1 phenotypic value obtained at one location (defined as TPE 1) were assessed for their ability to predict the performances of Stage 2 trial at the same location but in a different season (TPE1), and at two other locations with different conditions (TPE2 and TPE3). As already shown in Fig. 4, the field data obtained for Stage 1 in one season at one location do not predict well Stage 2 at the same site but in a different season, since climatic conditions were profoundly different (accuracy = −0.02). Also, the ability of predicting Stage 2 trials at different sites is only casually better (accuracy = 0.07 and 0.19). On the other hand, the use of phenotypic Stage 2 data combined across locations predicts with good accuracy the performances of the Stage 2 at all sites (accuracy = 0.45–0.75). Shifting from phenotypic values to genomic estimated breeding value (GEBV) improves mildly the prediction accuracy of Stage 2 performances when one location of Stage 1 is used for training the model (accuracy gain = +0.01 to +0.10), but it becomes very effective when Stage 2 combined data are used for training.

Hence, GEBV can be used with good effect to mimic/predict performances across locations and years to select entries that are more stable and with acceptable yield, even without planting these at multiple sites. Still, for the predictions to be reliable they need to be estimated starting from a training set that is highly related to the population under selection (Bassi et al. 2016;

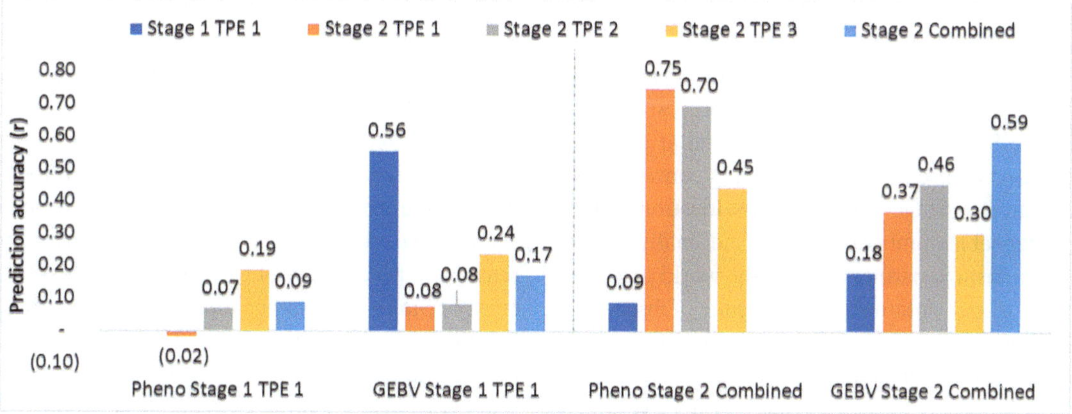

Fig. 6 Prediction accuracy of Stage 1 and Stage 2 yield trials conducted across the target population of environments (TPE) using the phenotypic data (Pheno) as they are or by modeling genomic estimated breeding value (GEBV)

Zaim et al. 2020). This presents a challenge for all durum wheat breeding programs, because the new Stage 1 entries are supposedly different than the entries of the previous year, and hence poor relativeness exist among the set to be predicted and used for its training. An elegant solution has been devised by Cruz et al. (2021) via the use of sparse testing. In this approach, some progenies from the same family are planted at one location and other progenies at other locations to ensure that a highly related training dataset is included for each family at all sites (Zapata et al. 2013; Calleja et al. 2020). The molecular markers scores are then used to generate connectivity and predict the performances of all entries at all sites. In this scenario, field designs like augmented design or partially replicated (p-rep) are typically used to drive down costs.

9.1 Genomic Recurrent Scheme for Pyramiding Abiotic Stress Tolerance

The ICARDA durum wheat program has been called since its establishment in 1977 to deliver superior cultivars well adapted to tolerate the climatic stresses common in the drylands of the world. A recent assessment of 35 years of breeding has revealed that an overall rate of 0.9% annual genetic gain (Bassi and Nachit 2019) was achieved, mostly via the promotion of earlier flowering types, with increased spike density, while maintaining large grain size. This was done via a well-established breeding scheme that used shuttle breeding by selected bulks to reach Stage 1 yield trials in four years since the initial cross (Fig. 7). These Stage 1 trials were conducted at one main location to discard poor performing entries, to then advance to Stage 2 trials across locations with a manageable number of entries, and finally provide germplasm to partners around the world as Stage 3 trial (i.e., International Durum Observatory Nurseries, IDON). Parents were selected for recycling based on the data gathered as Stage 2.

It has been suggested that an effective breeding program would need to raise its rate of genetic gain to above 1% per year to face climate change. Since 2019, the ICARDA program has then shifted to a revised scheme that utilizes several of the considerations expressed in this chapter to deliver a hypothetical higher rate of annual genetic gain estimated at 1.5% (Fig. 5). The first aspect of this change is the incorporation of "speed breeding" by single seed descent to shrink the time required to advance from F_1 to F_4 from two years to just one (Alahmad et al. 2018). The second element is the use of 192 Kompetitive Allele-Specific PCR (KASP) markers to genotype all F_4 progenies. These breeding-informative markers represent regions of the genome that have proven high variability (Kabbaj et al. 2017; Mazzucotelli et al. 2020) and represent major loci known to be associated with specific traits of interest. The same markers are then used to also genotype the Stage 2 entries of the previous cycle, which become the training set to predict the performances.

All F_4 progenies are advanced to F_5 as plants-to-row and planted in the summer off-season for seed increase. Before they reach maturity, the genomic predictions models (GEBV) are run and only the entries most likely to succeed are advanced. The two best progenies of each family are included in the crossing block and recombined in a full-sib mating approach to generate the 1st recurrent cycle (see Bassi et al. 2016 for details). The top 12 progenies from each family are instead advanced to Stage 1 trials. Using the same genotypic information generated as F_4, these progenies are split across locations as explained previously in a sparse testing scheme. Physiological tools are deployed at the main breeding sites to predict performances for various stress-related responses, and across locations data are collected. Genomic predictions are then used to assign values for each trait for each location to all entries, and the best progenies are then selected based on a selection index that incorporates yield potential, yield stability, and stress tolerance. Stage 2 trials are then conducted across locations. By the time Stage 2 trials are completed, the progenies generated by the 1st recurrent cycle will reach the F_5 generation.

The Stage 2 results are then used to run predictions and select again 2 progenies for recy-

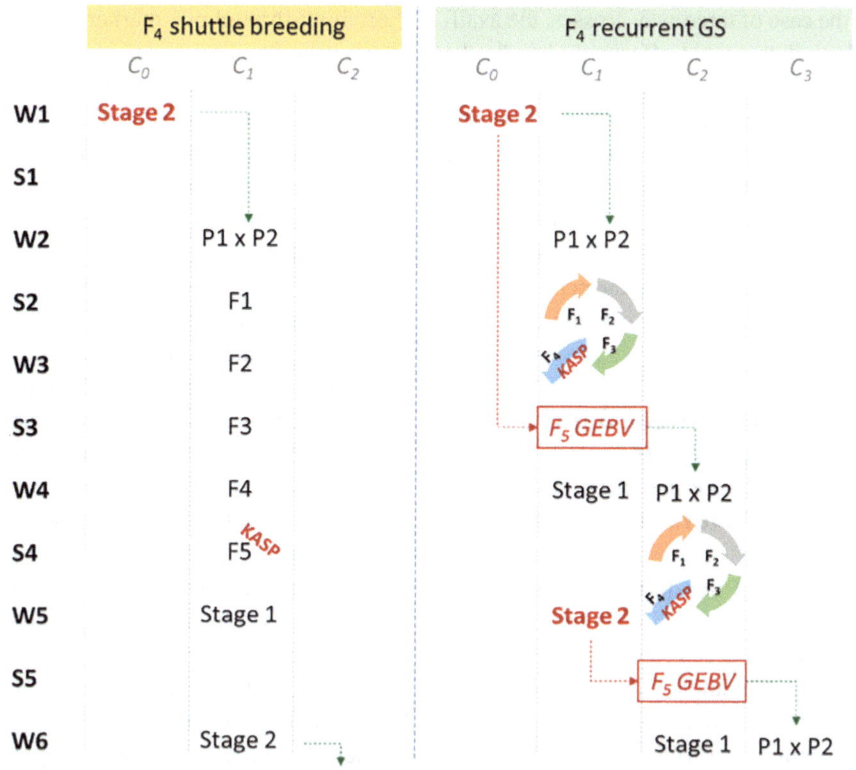

N Stage 1 done	1
Time	5
Accuracy	0.2
G. Gain annual	0.9%

N Stage 1 done	2
Time	3
Accuracy	0.4
G. Gain annual	1.5%

Fig. 7 Breeding scheme

cling and 6 for advancement as Stage 1. A total of three recurrent cycles are run for each family. There are two main scopes for using recurrent full sibs mating as explained by Bassi et al. (2016). First is to ensure that high relatedness remains between the training and breeding populations, which guarantees high rates of accuracy. Second is to increase the ability of maintaining an effective population size capable of identifying new progenies with a high number of segregating alleles fixed in the preferred states. In fact, the higher costs for genotyping and speed breeding allow us to advance only 48 progenies per cross per cycle, which per se results in an effective population size too small to handle the genetic diversity that exists among the parental material used by ICARDA. Hence, the recurrent approach vastly increases this population size.

In parallel to this recurrent full sib mating approach, new Stage 2 by Stage 2 crosses are also made each year as new cycle 0 progenies. The program also continues to use its traditional shuttle breeding approach via phenotypic selection for two type of crosses that cannot be effectively selected via genomic models: exotic introgressions and three-way crosses. In fact, in the case of introgressions, the exotic parent tends to have worst performances than the adapted one, meaning that the genomic models would assign negative effects to the exotic alleles, hence pushing for a selection against them, which is the opposite of the goal for this type of crosses (see Sharma et al. 2021 for details). In the case of three-way crosses, the available models to date cannot effectively handle the differences in genomic contributions of the three parents, nor the strong epistatic effect that typically results from combining three genomes into one progeny (Li et al. 2017; Sevilliano et al. 2017). As such, also this type of crosses is for now selected phenotypically until Stage 1, with the hope that these will be soon integrated into the genomic pipeline once better models become available.

10 The Durum Wheat Genome Assembly Unveils Accelerated Pyramiding of Stress Tolerance Alleles

The recent release of the durum wheat genome assembly (Maccaferri et al. 2019) unveiled several opportunities to accelerate the identification and transfer of useful alleles for stress tolerance. The main advantage is that the genome assembly offers one common scaffold onto which reports all QTL discoveries. In turn this allows to rapidly identify overlaps between discoveries, duplicates, and novel alleles, ensuring that breeders always have access to the most exploitable information (Maccaferri et al. 2022). In addition, this allows to compare objectively between different discovery initiatives to truly cross-validate QTL between programs and conditions.

The sequence used to generate array-based probes or the reads generated when doing genotyping by sequencing tend to be extremely short. As such, it is often complex to convert these into scalable PCR-based markers like Kompetitive Allele-Specific PCR (Kompetitive Allele-Specific PCR (KASP)). Thanks to the genome assembly, it is now fast to extend these short sequences with the surrounding genome, extending the space for designing effective primers and greatly increasing the chances of successful conversion.

This holds true also in the opposite direction, when attempting to impute the genome of progenies under selection starting from few markers. In fact, aligning PCR-based markers and the short high-throughput marker sequences to the Svevo genome allows us to order them correctly. Then the use of imputation models quickly converts the haplotype generated with few hundred markers to a complete full-scale haplotype with thousands of markers. This is of great interest when performing genomic selection as a strategic way to keep genotyping costs low.

Probably, the most immediate advantage of having a full genome sequence is the potential to greatly accelerate the identification of the candidate gene/s underlying a QTL of interest. Beyond the list of candidates, the genome assembly provides a map to guide forward- and reverse-genetic approaches for the fine mapping and cloning a gene/QTL of interest. In that sense, the resequencing initiatives have a major role to play to better define the available allelic diversity for several alleles of interest. Linking these to expression analysis of multiple tissues and genotype will also greatly accelerate the identification of loci of interest (Varshney et al. 2021).

Identifying, cloning, and deploying novel alleles for adaptation to climatic stresses will be the lifeline to ensure that durum wheat remains a crop of choice by future farmers despite the negative effect of climate change on yield. Achieving this result needs to be the priority of the next decade for the durum community. With the availability of the durum wheat genome, combined with the ongoing resequencing approaches linked to the Global Durum Panel set, and the ever-increasing set of genomic tools at our disposal, we can be confident that we will meet this challenge head on (Tuberosa et al. 2021).

Acknowledgments The authors wish to recognize the financial contribution to the discussion and write-up of this chapter of Accelerated Breeding Initiative by the OneConsultative Group on International Agricultural Research (CGIAR); by the Swedish Research Council VR Uforsk 2021-04956 "Delivering heat tolerant alleles to raise farm income along the Senegal River"; by "BreedMod" project funded by Arab Funds for Economic and Social Development (AFESD); and the contribution of the Indian Council of Agricultural research (ICAR) to ICARDA under the project "Deployment of high-yielding, stress tolerant and top industrial quality durum wheat elites to target growing regions of India."

References

Aberkane H, Amri A, Belkadi B et al (2021) Evaluation of durum wheat lines derived from interspecific crosses under drought and heat stress. Crop Sci 61:119–136. https://doi.org/10.1002/csc2.20319

Afzal F, Reddy B, Gul A, Khalid M, Subhani A, Shazadi K, Quraishi UM, Ibrahim AM, Rasheed A (2017) Physiological, biochemical and agronomic traits associated with drought tolerance in a synthetic-derived wheat diversity panel. Crop Pasture Sci 68:213–224

Al Khateeb W, Al Shalabi A, Schroeder D, Musallam I (2017) Phenotypic and molecular variation in drought tolerance of Jordanian durum wheat (*Triticum durum* Desf.) landraces. Physiol Mol Biol Plants 23(2):311–319. https://doi.org/10.1007/s12298-017-0434-y

Alahmad S, Dinglasan E, Leung KM et al (2018) Speed breeding for multiple quantitative traits in durum wheat. Plant Methods 14:36. https://doi.org/10.1186/s13007-018-0302-y

Alahmad S, El Hassouni K, Bassi FM, Dinglasan E, Youssef C, Quarry G, Aksoy A, Mazzucotelli E, Al J, Able JA, Christopher J, Voss-Fels KP, Hickey LT (2019) A major root architecture QTL responding to water limitation in durum wheat. Front Plant Sci 10:436. https://doi.org/10.3389/fpls.2019.00436

Alemu A, Feyissa T, Maccaferri M et al (2021) Genome-wide association analysis unveils novel QTLs for seminal root system architecture traits in Ethiopian durum wheat. BMC Genomics 22:1. https://bmcgenet.biomedcentral.com/articles/10.1186/s12863-020-0825-x

Alghabari F, Ihsan MZ (2018) Effects of drought stress on growth, grain filling duration, yield and quality attributes of Barley (*Hordeum vulgare* L.). Bangla J Bot 47:421–428. https://doi.org/10.3329/bjb.v47i3.38679

Ali Dib T, Monneveux P, Araus JL (1992) Adaptation à la sécheresse et notion d'ideotype chez le blé dur. II. Caractéres physiologiques d'adaptation. Agronomie 12:381–393

Al-Naggar AMM, Abd El-Maboud M, Abd El-Shafi A, El-Shal MH, Ali HA (2020) Evaluation of Egyptian wheat landraces (*Triticum aestivum* L) for drought tolerance, agronomic, grain yield and quality traits. Plant Arch 20(1):3487–3504

Amanda R, Haugrud P, Qijun Z, Andrew JG, Xu SS, Faris JD (2023) Identification of stable QTL controlling multiple yield components in a durum × cultivated emmer wheat population under field and greenhouse conditions. G3 Genes Genomes Genetics 13(2):1–19 jkac281. https://doi.org/10.1093/g3journal/jkac281

Amani I, Fischer RA, Reynolds MP (1996) Canopy temperature depression association with yield of irrigated spring wheat cultivars in hot climate. J Agron Crop Sci 176:119–129

Anuarbek S, Abugalieva S, Pecchioni N, Laidò G, Maccaferri M, Tuberosa R, Turuspekov Y (2020) Quantitative trait loci for agronomic traits in tetraploid wheat for enhancing grain yield in Kazakhstan environments. PLoS One 15(6):e0234863. https://doi.org/10.1371/J.pone.0234863

Araus JL, Cairns JE (2014) Field high-throughput phenotyping: the new crop breeding frontier. Trends Plant Sci 19:52–61. https://doi.org/10.1016/j.tplants.2013.09.008

Araus JL, Kefauver SC, Zaman-Allah M, Olsen MS, Cairns JE (2018) Translating High-Throughput Phenotyping into Genetic Gain. Trends Pl Sci 23(5): 451–466. DOI:https://doi.org/10.1016/j.tplants.2018.02.001

Arriagada O, Marcotuli I, Gadaleta A, Schwember AR (2020) Molecular mapping and genomics of grain yield in durum wheat: a review. Int J Mol Sci 21:7021. https://doi.org/10.3390/ijms21197021

Arriagada O, Gadaleta A, Marcotuli I, Maccaferri M, Campana M, Reveco S, Alfaro C, Matus I, Schwember AR (2022) A comprehensive meta-QTL analysis for yield-related traits of durum wheat (*Triticum turgidum* L. var. *durum*) grown under different water regimes. Front. Plant Sci 13:984269. https://doi.org/10.3389/fpls.2022.984269. PMID: 36147234; PMCID: PMC9486101

Aslam M, Maqbool MA, Cengiz R (2015) Mechanisms of drought resistance. In: Drought stress in maize (*Zea mays* L.). Springer, Cham, pp 19–36

Avni R, Oren L, Shabtay G, Assili S, Pozniak C, Hale I, Ben-Davis R, Peleg Z, Distelfeld A (2018) Genome based meta-QTL analysis of grain weight in tetraploid wheat identifies rare alleles of GRF4 associated with larger grains. Genes 9:636

Ayadi M, Brini F, Masmoudi K (2019) Overexpression of a wheat Aquaporin Gene, *TdPIP2;1*, enhances salt and drought tolerance in transgenic durum wheat cv. Maali. Int J Mol Sci 20:2389

Ayalew H, Liu H, Borner A, Kobiljski B, Liu C, Yan G (2018) Genome-wide association mapping of major root length QTLs under PEG induced water stress in wheat. Front Plant Sci 9:1–9

Bahar B, Yıldırım M, Celaleddin B, Ibrahim G (2008) Effect of Canopy temperature depression on grain yield and yield components in bread and Durum wheat. Not Bot Hort Agrobot 36(1):34–37. https://doi.org/10.15835/nbha36187

Baillot N, Girousse C, Allard V, Piquet-Pissaloux A, Le Gouis J (2018) Different grain-filling rates explain grain-weight differences along the wheat ear. PLoS One 13:e0209597

Balla K, Karsai I, Kiss T, Horváth A, Berki Z, Cseh A et al (2021) Single versus repeated heat stress in wheat: what are the consequences in different developmental phases? PLoS One 16(5):e0252070. https://doi.org/10.1371/journal.pone.0252070

Baloch MJ, Dunwell J, Dennet M, Rajpar I, Jatoi WA, Veesar NF (2012) Evaluating spring wheat cultivars for drought tolerance through yield and physiological parameters at booting and anthesis. Afr J Biotechnol 11:11559–11565

Baloch FS, Alsaleh A, Shahid MQ, Çiftçi VE, Saenz de Miera L, Aasim M et al (2017) A whole genome dartseq and Single Nucleotide Polymorphism (SNP) analysis for genetic diversity assessment in durum wheat from central fertile crescent. PLoS One 12(1):e0167821. https://doi.org/10.1371/journal.pone.0167821

Barber HM, Carney J, Alghabari F, Gooding MJ (2015) Decimal growth stages for precision wheat production in changing environments? Ann Appl Biol 166:355–371

Barlow KM, Christy BP, O'Leary GJ, Riffkin PA, Nuttall JG (2015) Simulating the impact of extreme heat and frost events on wheat crop production: a review. Field Crop Res 171:109–119

Bassi FM, Nachit MM (2019) Genetic gain for yield and allelic diversity over 35 years of durum wheat breeding at ICARDA. Crop Breed Genet Genom 1:1–19

Bassi FM, Sanchez-Garcia M (2017) Adaptation and stability analysis of ICARDA durum wheat elites across 18 countries. Crop Sci 57:2419–2430

Bassi FM, Ghavami F, Hayden MJ, Wang Y, Forrest KL, Kong S, Dizon R, Michalak de Jimenez MK, Meinhardt SW, Mergoum M et al (2016) Fast-forward genetics by radiation hybrids to saturate the locus regulating nuclear-cytoplasmic compatibility in *Triticum*. Plant Biotechnol J 14:1716–1726. https://doi.org/10.1111/pbi.12532

Bassi FM, Brahmi H, Sabraoui A, Amri A, Nsarellah N, Nachit MM, Al-Abdallat A, Chen MS, Lazraq A, El Bouhssini M (2019) Genetic identification of loci for Hessian fly resistance in durum wheat. Mol Breed 24:24–39

Ben Saad R, Harbaoui M, Ben Romdhane W, Zouari N, Giang Kh N, Ben Hsouna A, Brini F (2019) Overexpression of *Triticum durum TdAnn12* gene confers stress tolerance through scavenging reactive oxygen species in transgenic tobacco. Funct Plant Bio 46:1–11

Beres BL, Rahmani E, Clarke JM, Grassini P, Pozniak CJ, Geddes CM, Porker KD, May WE, Ransom JK (2020) A systematic review of durum wheat: enhancing production systems by exploring genotype, environment, and management (G × E × M) synergies. Front Plant Sci 11(1–18):568657. https://doi.org/10.3389/fpls.2020.568657

Berger B, Parent B, Tester M (2010) High-throughput shoot imaging to study drought responses. J Exp Bot 61:3519–3528

Bhavjot K, Mavi GS, Gill MS, Saini DK (2020) Utilization of Kompetitive Allele-Specific PCR (KASP) technology for wheat improvement. Cereal Res Commun 48:409–421. https://doi.org/10.1007/s42976-020-00057-6

Bheemanahalli R, John Sunoj VSJ, Saripalli G, Vara Prasad PV, Balyan HS, Gupta PK, Grant N, Gill KS, Jagadish SVK (2019) Quantifying the impact of heat stress on pollen germination, seed set, and grain filling in spring wheat. Crop Sci 59(2:684–696. https://doi.org/10.2135/cropsci2018.05.0292

Bilge B, Yildirim M, Barutcul C, Genc I (2008) Effect of canopy temperature depression on grain yield and yield components in bread and durum wheat. Not Bot Hort Agrobot Cluj 36(1):34–37

Blum A (2011) Plant breeding for water-limited environments, 254. Springer Science Business Media, LLC, New York

Blum A (2017) Osmotic adjustment is a prime drought stress adaptive engine in support of plant production. Plant Cell Environ 40:4–10. https://doi.org/10.1111/pce.12800

Blum A, Tuberosa R (2018) Dehydration survival of crop plants and its measurement. J Exp Bot 69(5):975–981. https://doi.org/10.1093/jxb/erx445

Bogale A, Tesfaye K (2016) Relationship between grain yield and yield components of the Ethiopian durum wheat genotypes at various growth stages. Trop Subtrop Agroecosyst 19:81–91

Borlaug NE (1968) Wheat breeding and its impact on world food supply. In: Finlay KW, Shephard KW (eds) Proceedings of the 3rd international wheat genetics symposium. Australian Academy of Sciences, Canberra, Australia, pp 1–36

Borrill P, Fahy B, Smith AM, Uauy C (2015) Wheat grain filling is limited by grain filling capacity rather than the duration of flag leaf photosynthesis: a case study using NAM RNAi plants. PLoS One 10:e0134947. https://doi.org/10.1371/journal.pone.0134947

Budak H, Hussain B, Khan Z, Ozturk NZ, Ullah N (2015) From genetics to functional genomics: improvement in drought signaling and tolerance in wheat. Front Plant Sci 6:1012. https://doi.org/10.3389/fpls.2015.01012

Cai H, Tian S, Liu C, Dong H (2011) Identification of a MYB3R gene involved in drought, salt and cold stress in wheat (*Triticum aestivum* L.). Gene 485:146–152

Calleja-Rodriguez A, Pan J, Funda T et al (2020) Evaluation of the efficiency of genomic versus pedigree predictions for growth and wood quality traits in Scots pine. BMC Genomics 21:796. https://doi.org/10.1186/s12864-020-07188-4

Camerlengo F, Frittelli A, Sparks C, Doherty A, Martignago D, Larré C, Lupi R, Sestili F, Masci F (2020) CRISPR-Cas9 multiplex editing of the α-amylase/trypsin inhibitor genes to reduce allergen proteins in durum wheat. Front Sustain Food Syst 4:104

Cattivelli L, Rizza F, Badeck FW, Mazzucotelli E, Mastrangelo AM, Francia E et al (2008) Drought tolerance improvement in crop plants: an integrated view from breeding to genomics. Field Crop Res 105:1–14. https://doi.org/10.1016/j.fcr.2007.07.004

Chaerle L, Leinonen I, Jones HG, Van Der Straeten D (2007) Monitoring and screening plant populations with combined thermal and chlorophyll fluorescence imaging. J Exp Bot 58:773–784. https://doi.org/10.1093/jxb/erl257

Chao S, Rouse MN, Acevedo M, Szabo-Hever A, Bockelman H, Bonman JM et al (2017) Evaluation of genetic diversity and host resistance to stem rust in USDA NSGC durum wheat accessions. Plant Genome 10. https://doi.org/10.3835/plantgenome2016.07.0071

Chenu K, Deihimfard R, Chapman SC (2013) Large-scale characterization of drought pattern: a continent-wide modelling approach applied to the Australian wheat belt-spatial and temporal trends. New Phatol 198:801–820. https://doi.org/10.1111/nph.12192

Christopher J, Manschadi A, Hammer G, Borrell A (2008) Developmental and physiological traits associated with high yield and stay-green phenotype in wheat. Aust J Agric Res 59:354–364. https://doi.org/10.1071/AR07193

Chung YS, Choi SC, Silva RR, Kang JW, Eom JH, Kim C (2017) Case study: estimation of sorghum biomass using digital image analysis with Canopeo. Biomass Bioenergy 105:207–210

Colasuonno P, Gadaleta A, Giancaspro A, Nigro D, Give S, Incerti O, Mangini G, Signorile A, Simeone R, Blanco A (2014) Development of a high-density Single Nucleotide Polymorphism (SNP)-based linkage map and detection of yellow pigment content QTLs in durum wheat. Mol Breed 34:1563–1578

Colasuonno P, Marcotuli I, Cutillo S, Simeone R, Blanco A, Gadaleta A (2020) Effect of barley chromosomes on the _-glucan content of wheat. Genet Resour Crop Evol 67:561–567

Collins NC, Tardieu F, Tuberosa R (2008) Quantitative trait loci and crop performance under abiotic stress: where do we stand? Plant Physiol 147(2):469–486. https://www.plantphysiol.org/cgi/doi/10.1104/pp.108.118117

Comastri A, Janni M, Simmonds J et al (2018) Heat in wheat: exploit reverse genetic techniques to discover new Alleles within the *Triticum durum* sHsp26 family. Front Plant Sci 9:1337. https://doi.org/10.3389/fpls.2018.01337

Condorelli GE, Maccaferri M, Newcomb M et al (2018) Comparative aerial and ground based high throughput phenotyping for the genetic dissection of NDVI as a proxy for drought adaptive traits in durum wheat. Front Plant Sci 9:893. https://doi.org/10.3389/fpls.2020.587093

Condorelli GE, Newcomb M, Groli EL, Maccaferri M, Forestan C, Babaeian E, Tuller M, White JW, Ward R, Mockler T et al (2022) Genome Wide Association study uncovers the QTLome for osmotic adjustment and related drought adaptive traits in Durum wheat. Genes 13:293. https://doi.org/10.3390/genes13020293

Cossani CM, Reynolds MP (2012) Physiological traits for improving heat tolerance in wheat. Plant Physiol 160(4):1710–1718. https://doi.org/10.1104/pp.112.207753

Crain J, Mondal S, Rutkoski J, Singh RP, Poland J (2018) Combining high-throughput phenotyping and genomic information to increase prediction and selection accuracy in wheat breeding. Plant Genom 11(1):1–14

Cruz ML, de los Campos G (2021) Optimal breeding-value prediction using a sparse selection index. Genetics 218(1):030. https://doi.org/10.1093/genetics/iyab030

Daryanto S, Wang L, Jacinthe PA (2016) Global synthesis of drought effects on maize and wheat production. PLoS One 11:1563–1562

Darzi-Ramandi H, Shariati JV, Tavakol E, Najafi-Zarini H, Bilgrami SS, Razavi K (2017) Detection of consensus genomic regions associated with root architecture of bread wheat on groups 2 and 3 chromosomes using QTL meta-analysis. Aust J Crop Sci 11:777

Davies JM (2014) Annexin-mediated calcium signalling in plants. Plants 3:128–140

Devate NB et al (2023) Strategies to develop heat and drought–tolerant wheat varieties following physiological breeding. In: Harohalli Masthigowda M, Gopalareddy K, Khobra R, Singh G, Pratap Singh G (eds) Translating physiological tools to augment crop breeding. Springer, Singapore. https://doi.org/10.1007/978-981-19-7498-4_3

Dhankher Om P, Foyer CH (2018) Climate resilient crops for improving global food security and safety. Plant Cell Environ 41(5):877–884. https://doi.org/10.1111/pce.13207

Di Fonzo N, Ravaglia S, De Ambrogio E, Blanco A, Troccoli A (2005) Durum wheat improvement in Italy. In: Royo C, Nachit M, Di Fonzo N, Araus JL, Pfeiffer WH, Slafer GA (eds) Durum wheat breeding: current approaches and future strategies, vol 1. Haworth Press, New York, pp 825–881

Ding J, Huang Z, Zhu M, Li C, Zhu X, Guo W (2018) Does cyclic water stress damage wheat yield more than a single stress? PLoS One 13:e0195535. https://doi.org/10.1371/journal.pone.0195535

Distelfeld A, Uauy C, Fahima T, Dubcovsky J (2006) Physical map of the wheat high-grain protein content gene *Gpc-B1* and development of a high-throughput molecular marker. New Phytol 169:753–763

Djemal R, Khoudi H (2016) TdSHN1, a WIN1/SHN1-type transcription factor, imparts multiple abiotic stress tolerance in transgenic tobacco. Environ Exp Bot 131:89–100

Dodig D, Zoric M, Knezevic D, King SR, Surlan-Momirovic G (2008) Genotype environment interaction for wheat yield in different drought stress conditions and agronomic traits suitable for selection. Aust J Agric Res 59:536–545. https://doi.org/10.1071/AR07281

Dolferus R (2014) To grow or not to grow: a stressful decision for plants. Plant Sci 229:247–261

Dolferus R, Powell N, Ji X, Ravash R, Edlington J, Oliver S, Van Dongen J, Shiran B (2013) The physiology of reproductive-stage abiotic stress tolerance in cereals. In: Rout GR, Das AB (eds) Mol stress physio plant, pp 193–216

Draeger T, Moore G (2017) Short periods of high temperature during meiosis prevent normal meiotic progression and reduce grain number in hexaploid wheat (*Triticum aestivum* L.). Theor Appl Genet 130:1785–1800. https://doi.org/10.1007/s00122-017-2925-1

Dreesen FE, De Boeck HJ, Janssens IA, Nijs I (2012) Summer heat and drought extremes trigger unexpected changes in productivity of a temperate annual/biannual plant community. Environ Exp Bot 79:21–30. https://doi.org/10.1016/j.envexpbot.2012.01.005

Dwivedi R, Prasad S, Jaiswal B, Kumar A, Tiwari A, Patel S et al (2017) Evaluation of wheat genotypes (*Triticum aestivum* L.) at grain filling stage for heat tolerance. Int J Pure App Biosci 5:971–975. https://doi.org/10.18782/2320-7051.2614

Ebel C, BenFeki A, Hanin M, Solano R, Chini A (2018) Characterization of wheat (*Triticum aestivum*) TIFY family and role of *Triticum durum TdTIFY11a* in salt stress tolerance. PLoS One 13:1–18

Edae EA, Byrne PF, Haley SD, Lopes MS, Reynolds MP (2014) Genome-wide association mapping of yield and yield components of spring wheat under contrasting moisture regimes. Theor Appl Genet 127: 791–807.https://doi.org/10.1007/s00122-013-2257-8

Ehdaie B, Alloush GA, Waines JG (2008) Genotypic variation in linear rate of grain growth and contribution of stem reserves to grain yield in wheat. Field Crop Res 106:34–43

El Haddad N, Sanchez-Garcia M, Visioni A, Jilal A, El Amil R, Sall AT, Lagesse W, Kumar S, Bassi FM (2021) Crop wild relatives crosses: multi-location assessment in durum wheat, barley, and lentil. Agronomy 11:2283. https://doi.org/10.3390/agronomy11112283

El Hassouni K, Alahmad S, Belkadi B, Filali-Maltouf A, Hickey LT, Bassi FM (2018) Root system architecture and its association with yield under different water regimes in Durum wheat. Crop Sci 58:2331–2346. https://doi.org/10.2135/cropsci2018.01.0076

El Hassouni K, Belkadi B, Filali-Maltouf A, Tidiane-Sall A, Al-Abdallat A, Nachit M, Bassi FM (2019) Loci controlling adaptation to heat stress occurring at the reproductive stage in durum wheat. Agronomy 9:414. https://doi.org/10.3390/agronomy9080414

Elazab A, Bort J, Zhou B, Serret MD, Nieto-Taladriz MT, Araus JL (2015) The combined use of vegetation indices and stable isotopes to predict durum wheat grain yield under contrasting water conditions. Agric Water Manag 158:196–208. https://doi.org/10.1016/j.agwat.2015.05.003

Elhadi GMI, Kamal NM, Gorafi YSA, Yamasaki Y, Takata K, Tahir ISA, Itam MO, Tanaka H, Tsujimoto H (2021) Exploitation of tolerance of wheat Kernel weight and shape-related traits from Aegilops tauschii under heat and combined heat-drought stresses. Int J Mol Sci 22:1830. https://doi.org/10.3390/ijms22041830

Elings A, Nachit MM (1991) Durum wheat landraces from Syria. I. Agro-ecological and morphological characterization. Euphytica 53:211–224. https://doi.org/10.1007/BF00023273

Elsayed S, Elhoweity M, Ibrahim HH, Dewir YH, Migdadic HM, Schmidhalter U (2017) Thermal imaging and passive reflectance sensing to estimate the water status and grain yield of wheat under different irrigation regimes. Agric Water Manag 189:98–110

Evers AD (2000) Grain size and morphology: implications for quality. Wheat structure, biochemistry and functionality, Schofield D. Royal Society of Chemistry, London, pp 19–24

Farooq M, Wahid A, Kobayashi N, Fujita D, Basra SMA (2009) Plant drought stress: effects, mechanisms and management. Agron Sustain Dev 29:185–212. https://doi.org/10.1051/agro:2008021

Farooq M, Bramley H, Palta JA, Siddique KHM (2011) Heat stress in wheat during reproductive and grain filling phases. Crit Rev Plant Sci 30:491–507. https://doi.org/10.1080/07352689.2011.615687

Fatiukha A, Filler N, Lupo I, Lidzbarsky G, Klymiuk V, Korol AB, Pozniak C, Tzion F, Krugmar T (2020) Grain protein content and thousand kernel weight QTLs identified in a durum X wild emmer wheat mapping population tested in five environments. Theor Appl Genet 133:119–131

Feldman M, Millet E (1993) Methodologies for identification, allocation and transfer of quantitative genes from wild emmer into cultivated wheat. In: Li ZS, Xin ZY (eds) Proceedings of the 8th international wheat genetics symposium. China Agricultural Scientech Press, Beijing, pp 19–27

Feng F, Han Y, Wang S, Yin S, Peng Z, Zhou M, Gao W, Wen X, Qin X, Siddique KHM (2018) The effect of grain position on genetic improvement of grain number and thousand grain weight in winter wheat in North China. Front Plant Sci 9:129. https://doi.org/10.3389/fpls.2018.00129

Fernandez-Gallego JA, Kefauver SC, Gutierrez NA et al (2018) Wheat ear counting in-field conditions: high throughput and low-cost approach using Red, Green, Blue (RGB) images. Plant Methods 14:22. https://doi.org/10.1186/s13007-018-0289-4

Fleury D, Jefferies S, Kuchel H, Langridge P (2010) Genetic and genomic tools to improve drought tolerance in wheat. J Exp Bot 61:3211–3222

Francki MG, Stainer GS, Walker E, Rebetzke GJ, Stefanova KT, French RJ (2021) Phenotypic evaluation and genetic analysis of seedling emergence in a global collection of wheat genotypes (*Triticum aestivum* L.) under limited water availability. Front. Plant Sci 12:796176 (1-17). https://doi.org/10.3389/fpls.2021.796176

Gale MD, Youssefian S (1985) In: Russell GE (ed) Dwarfing genes in wheat in progress in plant breeding. Butterworths, London, pp 1–35

Garcia del Moral LF, Rharrabti Y, Villegas D, Royo C (2003) Evaluation of grain yield and its components in durum wheat under Mediterranean conditions: An ontogenic approach. Agron J 95:266–274. https://doi.org/10.2134/agronj2003.2660

Gautam A, Agrawal D, Sai Prasad SV, Jajoo A (2014) A quick method to screen high and low yielding wheat cultivars exposed to high temperature. Phys Mol Biol Plant 20(4):533–537

Gautam A, Sai Prasad SV, Jajoo A, Ambati D (2015) Canopy temperature as selection parameter for grain yield and its components in durum wheat under terminal heat stress in late sown conditions. Agric Res 4:238–244

Gegas VC, Nazari A, Griffiths S, Simmonds J, Fish L, Orford S, Sayers L, Doonan JH, Snape JW (2010) A

genetic framework for grain size and shape variation in wheat. Plant Cell 22:1046–1056
Gevrek MN, Atasoy GD (2012) Effect of post anthesis drought on certain agronomical characteristics of wheat under two different nitrogen. Turkish J Field Crop 17:19–23
Ghimire B, Timsina D, Nepal J (2015) Analysis of chlorophyll content and its correlation with yield attributing traits on early varieties of maize (*Zea mays* L.). J Maize Res Dev 1(1):134–145
Giorno F, Wolters-Arts M, Mariani C, Rieu I (2013) Ensuring reproduction at high temperatures: the heat stress response during anther and pollen development. Plants 2:489–506. https://doi.org/10.3390/plants2030489
Giunta F, Motzo R, Pruneddu G (2007) Trends since 1900 in the yield potential of Italian-bred durum wheat cultivars. Eur J Agron 27:12–24
Goodwin AW, Lindsey LE, Harrison SK, Paul PA (2018) Estimating wheat yield with normalized difference vegetation index and fractional green canopy cover. Crop Forage Turfg Manag 4(1):1–6
Goudemand E, Laurent V, Duchalais L, Ghaffary SM, Kema GH, Lonnet P, Margalé E, Robert O (2013) Association mapping and meta-analysis: two complementary approaches for the detection of reliable Septoria tritici blotch quantitative resistance in bread wheat (*Triticum aestivum* L.). Mol Breed 32:563–584. https://doi.org/10.1007/s11032-013-9890-4
Gracia-Romero A, Kefauver SC, Fernandez-Gallego JA, Vergara-Diaz O, Nieto-Taladriz MT, Araus JL (2019) UAV and ground image-based phenotyping: a proof of concept with durum wheat. Remote Sens 11:1244. https://doi.org/10.3390/rs11101244
Graziani M, Maccaferri M, Royo C, Salvatorelli F, Tuberosa R (2014) QTL dissection of yield components and morpho-physiological traits in a durum wheat elite population tested in contrasting thermo-pluviometric conditions. Crop Pasture Sci 65(1):80–95
Griffiths S, Simmonds J, Leverington M, Wang Y, Fish LJ, Sayers L, Alibert L, Orford S, Wingen L, Snape J (2012) Meta-QTL analysis of the genetic control of crop height in elite European winter wheat germplasm. Mol Breed 29:159–171
Groli EL, Frascaroli E, Maccaferri M, Ammar K, Tuberosa R (2024) Dissecting the effect of heat stress on durum wheat under field conditions. Front Plant Sci 15. https://doi.org/10.3389/fpls.2024.1393349
Gupta NK, Gupta S, Kumar A (2001) Effect of water stress on physiological attributes and their relationship with growth and yield in wheat cultivars at different growth stages. J Agron 86:1437–1439
Gupta P, Kabbaj H, El Hassouni K, Maccaferri M, Sanchez-Garcia M, Tuberosa R, Bassi FM (2020) Genomic regions associated with the control of flowering time in Durum wheat. Plant 9:1628
Gutierrez-Rodriguez M, Reynolds MP, Escalante-Estrada JA, Rodriguez-Gonzalez MT (2004) Association between canopy reflectance indices and yield and physiological traits in bread wheat under drought and well-irrigated conditions. Aust J Agric Res 55(11):1139–1147. https://doi.org/10.1071/AR04214
Hanocq E, Laperche A, Jaminon O (2007) Most significant genome regions involved in the control of earliness traits in bread wheat as revealed by QTL meta-analysis. Theor Appl Genet 114:569–584
Harkel JT, Bartholomeus H, Kooistra L (2020) Biomass and crop height estimation of different crops using UAV-based Lidar. Remote Sens 12:17. https://doi.org/10.3390/rs12010017
Harrington SA, Cobo N, Karafiátová M, Doležel J, Borrill P, Uauy C (2019) Identification of a dominant chlorosis phenotype through a forward screen of the *Triticum turgidum* cv. Kronos TILLING population. Front Plant Sci 10:963. https://doi.org/10.3389/fpls.2019.00963
Hartung J, Wagener J, Ruser R et al (2019) Blocking and re-arrangement of pots in greenhouse experiments: which approach is more effective? Plant Meth 15:143. https://doi.org/10.1186/s13007-019-0527-4
Hayes SC (2019) Acceptance and commitment therapy: towards a unified model of behavior change. World Psychiatr 18(2):226–227. https://doi.org/10.1002/wps.20626
He D, Wang J, Pan Z, Dai T, Wang E, Zhang J (2017) Changes in wheat potential productivity and drought severity in Southwest China. Theor Appl Climatol 130:477–486
Hickey LT, Hafeez AN, Robinson H, Jackson SA et al (2019) Breeding crops to feed 10 billion. Nat Biotech 37(7):744–754. https://doi.org/10.1038/s41587-019-0152-9
Hlavacova M et al (2018) Interactive effects of high temperature and drought stress during stem elongation, anthesis and early grain filling on the yield formation and photosynthesis of winter wheat. Field Crop Res 221:182–195. https://doi.org/10.1016/j.fcr.2018.02.022
Hu H, Xiong L (2014) Genetic engineering and breeding of drought-resistant crops. Ann Rev Plant Bio 65:715–741
Hu D, Liang Li, Li Y, Li M, Zhang H, Zhao M (2014) Gas equilibrium regulation by closed-loop photo bioreactor built on system dynamics, fuzzy inference system and computer simulation. Com & Elect Agri. (103) 114–121. ISSN 0168–1699. https://doi.org/10.1016/j.compag.2014.02.002
Hu X, Rocheleau H, McCartney C, Biselli C, Bagnaresi P, Balcerzak M, Fedak G, Yan Z, Valè G, Khanizadeh S, Ouellet T (2019) Identification and mapping of expressed genes associated with the 2DL QTL for fusarium head blight resistance in the wheat line Wuhan 1. BMC Genet 20:47
Huang S, Spielmeyer W, Evans SL, James RA, Platten DJ, Elizabeth SD, Munns R (2006) A sodium transporter (HKT7) is a candidate for *Nax1*, a gene for salt tolerance in Durum wheat. Plant Physiol 142(4):1718–1727. https://doi.org/10.1104/pp.106.088864
Hyles J, Bloomfield MT, Hunt JR, et al (2020) Phenology and related traits for wheat adaptation.

Heredity 125:417–430. https://doi.org/10.1038/s41437-020-0320-1

Ibrahim SE, Schubert A, Pillen K, Leon J (2012) QTL analysis of drought tolerance for seedling root morphological traits in an advanced backcross of spring wheat. Int J Agric Sci 2:619–626

Ijaz R, Ejaz J, Gao S, Liu T, Imtiaz M, Ye Z, Wang T (2017) Overexpression of annexin gene AnnSp2, enhances drought and salt tolerance through modulation of ABA synthesis and scavenging ROS in tomato. Sci Rep 7:12087

Iqbal J (2019) Morphological, physiological and molecular markers for the adaptation of wheat in drought conditions. Asian J Biotechnol Genet Eng 2:1–13

Iqbal MA, Junaid R, Wajid N, Sabry H, Yassir K, Ayman S (2021) Rainfed winter wheat (Triticum aestivum L.) cultivars respond differently to integrated fertilization in Pakistan. Fresenius Environ Bull 30: 3115–3121

Iurlaro A, De Caroli M, Sabella E, De Pascali M, Rampino P, De Bellis L, Perrotta C, Dalessandro G, Piro G, Fry SC, Lenucci MS (2016) Drought and heat differentially affect xth expression and xet activity and action in 3-day-old seedlings of durum wheat cultivars with different stress susceptibility. Front Plant Sci 7:1686. https://doi.org/10.3389/fpls.2016.01686

Jauregui I, Aroca R, Garnica M, Zamarreno AM, Garcia-Mina JM, Serret D et al (2015) Nitrogen assimilation and transpiration: key processes conditioning responsiveness of wheat to elevated [CO2] and temperature. Physiol Plant 155:338–354. https://doi.org/10.1111/ppl.12345

Jimenez-Berni JA, Deery DM, Rozas-Larraondo P, Condon AG, Rebetzke GJ, James RA, Bovill WD, Furbank RT, Sirault XRR (2018) High throughput determination of plant height, ground cover, and above-ground biomass in wheat with LiDAR. Front Plant Sci 9:237. https://doi.org/10.3389/fpls.2018.00237

Jin X, Madec S, Dutartre D, de Solan B, Comar A, Baret F (2019) High-throughput measurements of stem characteristics to estimate ear density and above-ground biomass. Plant Pheno 10. https://doi.org/10.34133/2019/4820305

John JA, Williams ER (1995) Cyclic and computer-generated designs, vol 256. Chapman and Hall, London. https://doi.org/10.1201/b15075

John JA, Williams ER (1998) t–Latinized designs. Aust New Zeal J Stat 40:111–118

Jokar F, Karimizadeh R, Masoumiasl A, Amiri Fahliani R (2018) Canopy temperature and chlorophyll content are effective measures of drought stress tolerance in durum wheat. Notulae Scientia Biologicae 10(4):575–583. https://doi.org/10.15835/nsb10410288

Jovovic Z, Andjelkovic V, Przulj N, Mandic D (2020) Untapped genetic diversity of wild relatives for crop improvement. In: Rediscovery of genetic and genomic resources for future food security. Springer, Singapore, pp 25–65

Junhua P, Liu Z, Liu X, Yan J, Dongfa S, Nevo E (2021) Evolutionary agriculture domestication of wild emmer wheat. In: New horizons in evalution. Academic Press, pp 193–255

Kabbaj H, Sall AT, Al-Abdallat A, Geleta M, Amri A, Filali-Maltouf A et al (2017) Genetic diversity within a global panel of durum wheat (*Triticum durum* L) landraces and modern germplasm reveals the history of alleles exchange. Front Plant Sci 8:1277. https://doi.org/10.3389/fpls.2017.01277

Kadam S, Singh K, Shukla S, Goel S, Vikram P, Pawar V et al (2012) Genomic associations for drought tolerance on the short arm of wheat chromosome 4B. Funct Integr Genomics 12:447–464. https://doi.org/10.1007/s10142-012-0276-1

Kandel BP (2020) SPAD value varies with age and leaf of maize plant and its relationship with grain yield. BMC Res Notes 13:475. https://doi.org/10.1186/s13104-020-05324-7

Kandic V, Dodig D, Jovic M, Nikolic B, Prodanovic S (2009) The importance of physiological traits in wheat breeding under irrigation and drought stress. Genetic 41:11–20. https://doi.org/10.2298/GENSR0901011K

Kaya MD, Okcub G, Ataka M, Cikilic Y, Kolsaricia O (2006) Seed treatments to overcome salt and drought stress during germination in sunflower (*Helianthus annuus* L.). Eur J Agron 24:291–295. https://doi.org/10.1016/j.eja.2005.08.001

Kehel Z, Sanchez-Garcia M, El Baouchi A, Aberkane H, Tsivelikas A, Charles C, Amri A (2020) Predictive characterization for seed morphometric traits for genebank accessions using genomic selection. Front Ecol Evol 8:32. https://doi.org/10.3389/fevo.2020.00032

Keyvan S (2010) The effects of drought stress on yield, relative water content, proline, soluble carbohydrates and chlorophyll of bread wheat cultivars. J Ani Pl Sci 8(3):1051–1060

Khadka K, Earl HJ, Raizada MN, Navabi A (2020) A physio-morphological trait-based approach for breeding drought tolerant wheat. Front Plant Sci 11:715. https://doi.org/10.3389/fpls.2020.00715

Khalid M, Afzal F, Gul A, Amir R, Subhani A, Ahmed Z, He Z (2019) Molecular characterization of 87 functional genes in wheat diversity panel and their association with phenotypes under well-watered and water-limited conditions. Front Plant Sci 10:717. https://doi.org/10.3389/fpls.2019.00717

Khan MA, Iqbal M, Jameel M, Nazeer W, Shakir S, Aslam MT, Iqbal B (2011) Potentials of molecular based breeding to enhance drought tolerance in wheat (*Triticum aestivum* L.). Afr J Biotechnol 10:11340–11344

Khanghahi YM, Leoni B, Crecchio C (2021) Photosynthetic responses of durum wheat to chemical/microbiological fertilization management under salt and drought stresses. Acta Physiol Plant 43:123. https://doi.org/10.1007/s11738-021-03289-z

Kiani R, Arzani A, Habibi F (2015) Physiology of salinity tolerance in *Aegilops cylindrica*. Acta Physiol Plant 37:135–145. https://doi.org/10.1007/s11738-015-1881-0

Kilic H, Yagbasanlar T (2010) The effect of drought stress on grain yield and some quality traits of durum wheat (*Triticum turgidum* spp. *durum*) cultivars. Not Bot Hort Agrobot Cluj 38:164–170

Kirschner GK, Rosignoli S, Guo L, Vardanega I, Imani J, Altm€uller J, Milner SG, Balzano R, Nagel KA, Pflugfelder D et al (2021) *ENHANCED GRAVITROPISM 2* encodes a STERILE ALPHA MOTIF-containing protein that controls root growth angle in barley and wheat. Proc Natl Acad Sci USA 118:e2101526118

Kong B, Ma J, Zhang P, Chen T (2023) Deciphering key genomic regions controlling flag leaf size in wheat via integration of meta-QTL and in silico transcriptome assessment. BMC Genomics 24(1):1–17

Krasileva KV, Vasquez-Gross HA, Howell T, Bailey P, Paraiso F, Clissold L, Simmonds J, Ramirez-Gonzalez RH, Wang X, Borrill P, Fosker C, Ayling S, Phillips AL, Uauy C, Dubcovsky J (2017) Uncovering hidden variation in polyploid wheat. Proc Nat Acad Sci 114(6):913–921. https://doi.org/10.1073/pnas.1619268114

Kronstad WE, Rajaram S (1990) Winter X spring germplasm management and exploitation. In: Proceedings of the 6th assembly of wheat breeding. Society of Australia, Tamworth, Australia, pp 123–130

Kumar S, Choudhary AK, Rana KS, Sarker A, Singh M (2018) Bio-fortification potential of global wild annual lentil core collection. PLoS One 13:e0191122

Kumar S, Kumar H, Gupta V, Kumar A, Singh CM, Kumar M, Singh AK, Panwar GS, Kumar S, Singh AK, Kumar R (2023) Capturing agro-morphological variability for tolerance to terminal heat and combined heat-drought stress in landraces and elite cultivar collection of wheat. Front Plant Sci 14:1136455. https://doi.org/10.3389/fpls.2023.1136455. PMID: 37251757; PMCID: PMC10214469

Laido G, Marone D, Russo MA, Colecchia SA, Mastrangelo AM, De Vita P, Papa R (2014) Linkage disequilibrium and genome-wide association mapping in tetraploid wheat (*Triticum turgidum* L.). PLoS One 9:e95211

Langridge P, Alaux M, Almeida NF et al (2022) Meeting the challenges facing wheat production: the strategic research agenda of the global wheat initiative. Agronomy 12(11):2767

Lantican MA, Dubin HJ, Morris ML (2005) Impacts of International Wheat Breeding Research in the developing world, 1988–2002. CIMMYT, Mexico DF

Latican M, Braun HJ, Payne TS, Singh RP, Sonder K, Baum M, Van GM, Erenstein O (2016) Impacts of International Wheat improvement research: 1994–2014. CIMMYT, Mexico, p 72

Lepekhov SB (2022) Canopy temperature depression for drought- and heat stress tolerance in wheat breeding. Vavilovskii zhurnal genetiki i selektsii 26(2):196–201. https://doi.org/10.18699/VJGB-22-24

Lesk C, Rowhani P, Ramankutty N (2016) Influence of extreme weather disasters on global crop production. Nature 529:84–87

Li Q, Zhang Y, Liu T, Wang F, Liu K, Chen J et al (2015) Genetic analysis of kernel weight and kernel size in wheat (*Triticum aestivum* L.) using unconditional and conditional QTL mapping. Mol Breed 35:194. https://doi.org/10.1007/s11032-015-0384-4

Li Z, Philipp N, Spiller M, Stiewe G, Reif JC, Zhao Y (2017) Genome-wide prediction of the performance of three-way hybrids in barley. Plant Genom 10. https://doi.org/10.3835/plantgenome2016.05.0046

Li T, Deng G, Tang Y, Su Y, Wang J, Cheng J, Yang Z, Qiu X, Pu X, Zhang H, Liang J, Yu M, Wei Y, Long H (2021) Identification and validation of a novel locus controlling Spikelet number in bread wheat (*Triticum aestivum* L.). Front Plant Sci 12:611106. https://doi.org/10.3389/fpls.2021.611106

Lipiec J, Doussan C, Nosalewicz A, Kondracka K (2013) Effect of drought and heat stresses on plant growth and yield: a review. Int Agrophys 27:463–477. https://doi.org/10.2478/intag-2013-0017

Liu EK, Mei XR, Yan CR, Gong DZ, Zhang YQ (2016) Effects of water stress on photosynthetic characteristics, dry matter translocation and WUE in two winter wheat genotypes. Agric Water Manag 167:75–85

Liu L, Yuan CY, Wang MN, See DR, Zemetra RS, Chen XM (2019) QTL analysis of durable stripe rust resistance in the North American winter wheat cultivar Skiles. Theor Appl Genet 132(6):1677–1691

Liu Z, Wang Q, Wan H, Yang F, Li S, Yang M, Luo J, Zheng J, Deng Q, Long H et al (2020) Identification and validation of a novel major QTL controlling leaf pubescence in the Chinese wheat landrace 'Baimaomai'. Agronomy 11:2237

Loffler M, Schon C, Miedaner T (2009) Revealing the genetic architecture of FHB resistance in hexaploid wheat (*Triticum aestivum* L.) by QTL meta-analysis. Mol Breed 23:473–488

Lopes MS, Reynolds MP, Jalal-Kamali MR, Moussa M, Feltaous Y, Tahir ISA et al (2012) The yield correlations of selectable physiological traits in a population of advanced spring wheat lines grown in warm and drought environments. F Crop Res 128:129–136. https://doi.org/10.1016/j.fcr.2011.12.017

Lopes MS, El-Basyoni I, Baenziger PS, Singh S, Royo C, Ozbek K, Aktas H, Ozer E, Ozdemir F, Manickavelu A et al (2015) Exploiting genetic diversity from landraces in wheat breeding for adaptation to climate change. J Exp Bot 66:3477–3486

Ltaief S, Krouma A (2023) Functional dissection of the physiological traits promoting Durum wheat (*Triticum durum* Desf.). Tolerance to drought stress. Plants 12(7):1420. https://doi.org/10.3390/plants12071420

Ma J, Tu Y, Zhu J, Luo W, Liu H, Li C, Li S, Liu J, Ding P, Habib A, Mu Y, Tang H, Liu Y, Jiang Q, Chen G, Wang J, Li W, Pu Z, Zheng Y, Wei Y, Kang H, Chen G, Lan X (2020) Flag leaf size and posture of bread wheat: genetic dissection, QTL validation and their relationships with yield-related traits. Theor Appl Genet 133:297–315

Ma J, Liu Y, Zhang P et al (2022) Identification of quantitative trait loci (QTL) and meta-QTL analysis for kernel size-related traits in wheat (*Triticum aestivum* L.). BMC Plant Biol 22:607. https://doi.org/10.1186/s12870-022-03989-9

Macro ML, Heeney D, Binda S, Cifelli CJ, Cotter PD, Foligne B, Ganzle M, Kort R, Pasin G, Pihlanto A, Smid EJ, Hutkins R (2017) Health benefits of fermented foods: microbiota and beyond. Curr Opin Biotechnol 44: 94–102. doi:https://doi.org/10.1016/j.copbio.2016.11.010

Maccaferri M, Sanguineti MC, Corneti S et al (2008) Quantitative trait loci for grain yield and adaptation of durum wheat (*Triticum durum* Desf.) across a wide range of water availability. Genetics 178:489–511

Maccaferri M, Sanguineti MC, Demontis A et al (2011) Association mapping in durum wheat grown across a broad range of water regimes. J Exp Bot 62:409–438

Maccaferri M, Canè MA, Colalongo C, Massi A, Clarke F, Pozniak C et al (2014) A consensus framework map of durum wheat (*Triticum durum* Desf.) suitable for linkage disequilibrium and genome-wide association mapping. BMC. Genome 15:873. https://doi.org/10.1186/1471-2164-15-873

Maccaferri M, El-Feki W, Nazemi G et al (2016) Prioritizing quantitative trait loci for root system architecture in tetraploid wheat. J Exp Bot 67(4):1161–1178

Maccaferri M, Harris NS, Twardziok SO et al (2019) Durum wheat genome highlights past domestication signatures and future improvement targets. Nat Genet 51:885–895. https://doi.org/10.1038/s41588-019-0381-3

Maccaferri M, Bruschi M, Tuberosa R (2022) Sequence-based marker assisted selection in wheat. In: Wheat improvement: food security in a changing climate. Springer International Publishing, Cham, pp 513–538

Madec S, Jin X, Lu H et al (2019) Ear density estimation from high resolution Red, Green, Blue (RGB) imagery using deep learning technique. Agric Forest Meteorol 264:225–234

Mahato A (2014) Climate change and its impact on agriculture. Int J Sci Res 4:1–11

Mahdiyeh S, Roohollah A, Hassan P, Bahram H, Somayeh A (2021) Mapping quantitative trait loci for drought tolerance/susceptibility indices and estimation of breeding values of doubled haploid lines in wheat (*Triticum aestivum*). Crop Pastu Sci 72(7):500–513. https://doi.org/10.1071/CP20526

Manschadi AM, Christopher JT, deVoil P, Hammer GL (2006) The role of root architectural traits in adaptation of wheat to water-limited environments. Funct Plant Biol 33:823–837. https://doi.org/10.1071/FP06055

Manschadi AM, Hammer GL, Christopher JT, Devoil P (2008) Genotypic variation in seedling root architectural traits and implications for drought adaptation in wheat (*Triticum aestivum* L.). Plant Soil 303:115–129. https://doi.org/10.1007/s11104-007-9492-1

Malin GH, Cristiano El, Roberto D, Francesco V (2024) Durum wheat productivity today and tomorrow: A review of influencing factors and climate change effects. Res Env Sust. (17):100170. ISSN 2666–9161.https://doi.org/10.1016/j.resenv.2024.100170

Marcela Ayala-Benítez M, Carlos Guzmán C, Juan B, Alvarez Peña RJ (2013) Characterization of genetic diversity of puroindoline genes in Mexican wheat landraces. Euphytica 190(1):53–63. https://doi.org/10.1007/s10681-012-0773-2

Marco GVD, Carla C, Pierpaolo D (2017) Simulation of climate change impacts on production and phenology of Durum wheat in Mediterranean environments using CERES-wheat model. Field Crop Res 206:43–53

Marina M, Paola C, Costantino F, Francesco G, Antonio P, Georges H, Rosella M (2019) From ancient to old & modern durum wheat varieties: interaction among cultivar traits, management, and technological quality. J Sci Food Agric 99(5):2059–2067. https://doi.org/10.1002/jsfa.9388

Martínez BEDT, Schluepmann H, de Jong GJ, Somsen GW, Nunes C et al (2011) Wheat grain development is characterized by remarkable trehalose 6-phosphate accumulation pre-grain filling: tissue distribution and relationship to SNF1-related protein kinase1 activity. Plant Physiol 156:373–381. https://doi.org/10.1104/pp.111.174524

Maulana F, Huang W, Anderson JD, Ma XF (2020) Genome-Wide Association Mapping of seedling drought tolerance in winter wheat. Front Plant Sci 11:573786. https://doi.org/10.3389/fpls.2020.573786

Mazzucotelli E, Sciara G, Mastrangelo AM, Desiderio F, Xu SS, Faris JD, Hayden MJ, Tricker PJ, Ozkan H, Echenique V et al (2020) The Global Durum Wheat Panel: An international effort to identify and exchange beneficial alleles. Front Plant Sci 11:569905. https://doi.org/10.3389/fpls.2020.569905

Meister R, Rajani MS, Ruzicka D, Schachtman DP (2014). Challenges of modifying root traits in crops for agriculture. Tren Plan Sci 19:779–788

Mehmet Y, Hasan K, Enver K, Turan K (2011) Applicability of chlorophyll meter reading as yield predictor in durum wheat. J Plant Nutri 34(2):101–106

Merah O, Monneveux P (2015b) Contribution of different organs to grain filling in Durum wheat under Mediterranean conditions I. Contribution of post-Anthesis photosynthesis and remobilization. J Agron Crop Sci 201(5):344–352

MeRed, Green, Blue (RGB)y D, Hanin M, Saidi MN (2021) The durum wheat NAC transcription factor TtNAC2A enhances drought stress tolerance in *Arabidopsis*. Environ Exp Bot 186.104439

Miles C, Wayne M (2008) Quantitative trait locus (QTL) analysis. Nat Educ 1(1):208

Mkhabela SS, Shimelis H, Odindo AO, Mashilo J (2019) Response of selected drought tolerant wheat (*Triticum aestivum* L.) genotypes for agronomic traits and biochemical markers under drought-stressed and non-stressed conditions. Acta Agric Scand Sect B-Soil Plant Sci 69:674–689

Mkhabela SS, Shimelis H, Mashilo J (2020) Genetic differentiation of selected drought and heat tolerant wheat genotypes using simple sequence repeat markers and agronomic traits. S Afr J Plant Soil 37:211–219

Mohammadi R, Reza H, Sadeghzadeh B, Hasan A, Solaimani K, Amri A (2014) Adaptation patterns and yield stability of Durum wheat landraces to Highland Cold Rainfed areas of Iran. Crop Sci 54(3):944–954. https://doi.org/10.2135/cropsci2013.05.0343

Monneveux P, Zacharieva M, Rekika D (2000) Options Mediterraneennes. Serie A Seminares Mediterraneennes 40:71–81

Moragues M, Zarco-Hernandez J, Moralejo MA, Royo C (2006) Genetic diversity of glutenin protein subunits compositionin durum wheat landraces [Triticum turgidum ssp. turgidum convar. durum (Desf.) MacKey] from the Mediterranean basin. Genet Resour Crop Evol 53:993–1002

Moreira FF, Oliveira HR, Volenec JJ, Rainey KM, Brito LF (2020) Integrating High-Throughput Phenotyping and Statistical Genomic Methods to Genetically Improve Longitudinal Traits in Crops. Front Pl Sci. 26:11:681. doi: https://doi.org/10.3389/fpls.2020.00681

Mullan D, Garcia MB (2012) Chapter 10: crop ground cover. In: Pask A, Pietragalla J, Mullan D, Reynolds M (eds) Physiological breeding II: a field guide to wheat breeding. CIMMYT, Mexico, pp 46–50

Mwadzingeni L, Shimelis H, Tesfay S, Tsilo TJ (2016) Screening of bread wheat genotypes for drought tolerance using phenotypic and proline analyses. Front Plant Sci 7:1276. https://doi.org/10.3389/fpls.2016.01276

Mwadzingeni L, Shimelis H, Rees DJG, Tsilo TJ (2017) Genome-wide association analysis of agronomic traits in wheat under drought-stressed and non-stressed conditions. PLoS One 12:e0171692

N'Diaye A, Haile JK, Nilsen KT, Walkowiak S, Ruan Y, Singh AK et al (2018) Haplotype loci under selection in Canadian durum wheat germplasm over 60 years of breeding: association with grain yield, quality traits, protein loss, and plant height. Front Plant Sci 9:1589. https://doi.org/10.3389/fpls.2018.01589

Nachit MM (1992) Durum wheat breeding for Mediterranean drylands of North Africa and West Asia. In: Rajaram S, Saari EE, Hettel GP (eds) Durum wheats: challenges and opportunities, wheat special report No. 9. CIMMYT, Ciudad Obregon, Mexico, pp 23–25

Nachit M, Motawaj J, Kehel Z, Habash DZ, Elouafi I, Pagnotta M, Porceddu E, Bari A, Amri A, Mamluk OF, El-Bouhssini M (2016) Gene flow as a source of adaptation of durum wheat to changing conditions including climate change: double gradient selection technique. In: Raton B (ed) Applied mathematics and omics to assess crop genetic resources, pp 1–10

Naeem MK, Ahmad M, Kamran M, Shah MKN, Iqbal MS (2015) Physiological responses of wheat (*Triticum aestivum* L.) to drought stress. Int J Plant Soil Sci 6:1–9

Nagy-Reder D, Birinyi Z, Rakszegi M, Bekes F, Gell G (2021) The effect of abiotic stresses on the protein composition of four Hungarian wheat varieties. Plants (Basel) 11(1):1–19. https://doi.org/10.3390/plants11010001. PMID: 35009005; PMCID: PMC8747273

Nasehzadeh M, Ellis RH (2017) Wheat seed weight and quality differ temporally in sensitivity to warm or cool conditions during seed development and maturation. Ann Bot 120(3):479–493. https://doi.org/10.1093/aob/mcx074

Nass HG, Reiser B (1975) Grain filling period and grain yield relationships in spring wheat. Can J Plant Sci 55:673–678. https://doi.org/10.4141/cjps75-107

Nevo E, Beiles A, Gutterman Y, Storch N, Kaplan D (1984) Genetic resources of wild cereals in Israel and vicinity: I. Phenotypic variation within and between populations of wild wheat, *Triticum dicoccoides*. Euphytica 33:717–735

Nevo E, Korol AB, Beiles A, Fahima T (2002) Evolution of wild emmer and wheat improvement: population genetics, genetic resources, and genome organization of Wheat's Progenitor, *Triticum dicoccoides*. Springer-Verlag, Berlin

Ni Z, Li H, Zhao Y, Peng H, Hu Z, Xin M, Sun Q (2018) Genetic improvement of heat tolerance in wheat: recent progress in understanding the underlying molecular mechanisms. Crop J 6:32–41

Ober ES, Alahmad S, Cockram J, Forestan C, Hickey LT, Kant J, Maccaferri M, Marr E, Milner M, Pinto F et al (2021) Wheat root systems as a breeding target for climate resilience. Theor Appl Genet 134:1645–1662

Ogbonnaya FC, Rasheed A, Okechukwu EC, Jighly A, Makdis F, Wuletaw T, Hagras A, Uguru MI, Agbo CU (2017) Genome-wide association study for agronomic and physiological traits in spring wheat evaluated in a range of heat prone environments. Theor Appl Genet 130:1819–1835

Olsen OA (2020) The modular control of cereal endosperm development. Trends Plant Sci 25:279–290. https://doi.org/10.1016/j.tplants.2019.12.003

Onyemaobi I, Liu H, Siddique KHM, Yan GJ (2017) Both male and female malfunction contributes to yield reduction under water stress during meiosis in bread wheat. Front Plant Sci 7:2071

Pagnotta MA, Impiglia A, Tanzarella OA, Nachit MM, Porceddu E (2005) Genetic variation of the durum wheat landrace Haurani from different agro-ecological regions. Genet Resour Crop Evol 51:863–869. https://doi.org/10.1007/s10722-005-0775-1

Parmesan C, Hanley ME (2015) Plants and climate change: complexities and surprises. Ann Bot 116(6):849–864. https://doi.org/10.1093/aob/mcv169

Patrignani A, Ochsner TE (2015) Canopeo: a powerful new tool for measuring fractional green canopy cover. Agron J 107(6):2312–2320

Peleg Z, Fahima T, Abbo S, Krugman T, Nevo E, Yakir D, Saranga Y (2005) Genetic diversity for drought resistance in wild wheat and its ecogeographical association. Plant Cell Environ 28:176–191

Peters Haugrud AR, Achilli AL, Martínez-Peña R, Klymiuk V (2024) Future of durum wheat research and breeding: insights from early career researchers. Plant Genom 18:e20453

Piepho HP, Michel V, Williams ER (2018) Neighbor balance and evenness of distribution of treatment replications in row-column designs. Biom J 60:1172–1189

Pietragalla J, Vega AM (2012) Chapter 8. Normalized difference vegetation index. In: Pask A, Pietragalla J, Mullan D, Reynolds M (eds) Physiological breeding II: a field guide to wheat breeding. CIMMYT, Mexico, pp 37–40

Pineda M, Baron M, Perez-Bueno ML (2021) Thermal imaging for plant stress detection and phenotyping. Rem Sens 13:68. https://doi.org/10.3390/rs13010068

Pinter PJ Jr, Zipoli G, Jackson RJRD, Idso SB, Hohman JP (1990) Canopy temperature as an indicator of differential water use and yield performance among wheat cultivars. Agric Water Mang 18:35–48

Pinto RS, Reynolds MP, Mathews KL, McIntyre CL, Olivares-Villegas JJ, Chapman SC (2010) Heat and drought adaptive QTL in a wheat population designed to minimize confounding agronomic effects. Theor Appl Genet 121:1001–1021

Plamenov D (2003) Distribution and characterization of wild wheat species (Aegilops and Triticum) in Black Sea Coast—Autoreferate. TU, Varna (Bg)

Pour-Aboughadareh A, Omidi M, Naghavi MR, Etminan A, Mehrabi AA, Poczai P (2020) Wild relatives of wheat respond well to water deficit stress: a comparative study of antioxidant enzyme activities and their encoding gene expression. Agriculture 10:415. https://doi.org/10.3390/agriculture10090415

Qi Y, Yu H, Fu Q, Chen Q, Ran J, Yang Z (2022) Future changes in drought frequency due to changes in the mean and shape of the PDSI probability density function under RCP4.5 scenario. Front Earth Sci 10:857885

Quraishi UM, Murat F, Abrouk M, Pont C, Confolent C, Oury FX et al (2011) Combined meta-genomics analyses unravel candidate genes for the grain dietary fiber content in bread wheat (*Triticum aestivum* L.). Funct Integr Genomics 11:71–83. https://doi.org/10.1007/s10142-010-0183-2

Raj SRG, Nadarajah K (2023) QTL and candidate genes: techniques and advancement in abiotic stress resistance breeding of major cereals. Int J Mol Sci 24(1):6. https://doi.org/10.3390/ijms24010006

Rasheed A, Wen W, Gao F, Zhai S, Jin H, Liu J, Guo Q, Zhang Y, Dreisigacker S, Xia X, He Z (2016) Development and validation of Kompetitive Allele-Specific PCR (KASP) assays for genes underpinning key economic traits in bread wheat. Theor Appl Genet 10:1843–1860. https://doi.org/10.1007/s00122-016-2743-x

Rebetzke GJ, Richards RA, Fischer VM, Mickelson BJ (1999) Breeding long coleoptile, reduced height wheats. Euphytica 106:159–168. https://doi.org/10.1023/A:1003518920119

Rebetzke GJ, Rattey AR, Farquhar GD, Richards RA, Condon AG (2013) Genomic regions for canopy temperature and their genetic association with stomatal conductance and grain yield in wheat. Funct Plant Biol 40:14–33

Ren Y, He X, Liu D, Li J, Zhao X, Li B et al (2012) Major quantitative trait loci for seminal root morphology of wheat seedlings. Mol Breed 30:139–148. https://doi.org/10.1007/s11032-011-9605-7

Reynolds MP, Rebetzke G (2011) Application of plant physiology in wheat breeding. In: Bonjean AP, Angus WJ, Van Ginkel M (eds) The world wheat book: a history of wheat breeding, vol 2. Editions TEC, Paris, pp 877–906

Reynolds M, Tuberosa R (2008) Translational research impacting on crop productivity in drought-prone environments. Curr Opin Plant Biol 2(11):171–179

Reynolds MP, Rubeena S, Trethowan R (2004) Using "smart" physiological-trait based crossing strategies to accumulate drought-adaptive genes, in Resilient Crops for Water Limited Environments. In: Poland P, Sawkins M, Ribaut J, Hoisington D (eds) Proceedings of a workshop held at Cuernavaca. CIMMYT, Mexico D.F, pp 185–187

Reynolds MP, Pellegrineschi A, Skovmand B (2005) Sink-limitation to yield and biomass: a summary of some investigations in spring wheat. Ann Appl Biol 146:39–49. https://doi.org/10.1111/j.1744-7348.2005.03100.x

Richard CA, Hickey LT, Fletcher S, Jennings R, Chenu K et al (2015) High-throughput phenotyping of seminal root traits in wheat. Plant Methods 1:1–11. https://doi.org/10.1186/s13007-015-0055-9

Rharrabti Y, Garcia del Moral LF, Villegas D, Royo C (2003) Durum wheat quality in Mediterranean environments: III. Stability and comparative methods in analysing G×E interaction. Field Crop Res 80:141–146. https://doi.org/10.1016/S0378-4290(02)00178-8

Richards RA, Rebetzke GJ, Condon AG, Watt M (2011) Breeding to improve grain yield in water limited environments: the CSIRO experience with wheat. In: Araus JL, Slafer GA (eds) Crop stress management and global climate change. CABI, Wallington, UK, pp 105–121

Ristic Z, Urska Bukovnik U, Vara Prasad PV (2007) Correlation between heat stability of Thylakoid membranes and loss of chlorophyll in winter wheat under heat stress. Crop Sci 47:2067–2073. https://doi.org/10.2135/cropsci2006.10.0674

Robbana C, Kehel Z, Ben Naceur M, Sansaloni C, Bassi F, Amri A (2019) Genome-wide genetic diversity and population structure of Tunisian Durum wheat landraces based on DArTseq technology. Int J Mol Sci 20:1352. https://doi.org/10.3390/ijms20061352

Robbana C, Kehel Z, Ammar K, Guzmán C, Naceur MB, Amri A (2021) Unlocking the patterns of the Tunisian Durum wheat landraces genetic structure based on phenotypic characterization in relation to farmer's vernacular name. Agronomy 11:634. https://doi.org/10.3390/agronomy11040634

Roncallo PF, Akkiraju PC, Cervigni GL, Echenique VC (2017) QTL mapping and analysis of epistatic interactions for grain yield and yield-related traits in *Triticum turgidum* L. var. *durum*. Euphytica 213:277

Roncallo PF, Beaufort V, Larsen AO, Dreisigacker S, Echenique V (2019) Genetic diversity and linkage disequilibrium using Single Nucleotide Polymorphism (SNP) (Kompetitive Allele-Specific PCR (KASP)) and AFLP markers in a worldwide durum wheat (*Triticum turgidum* L. var. *durum*) collection. PLoS One 14:e0218562. https://doi.org/10.1371/journal.pone.0218562

Rosello M, Royo C, Sanchez-Garcia M, Soriano JM (2019) Genetic dissection of the seminal root system architecture in Mediterranean Durum wheat landraces by Genome-Wide Association Study. Agronomy 9(7):364. https://doi.org/10.3390/agronomy9070364

Royo C, Abaza M, Blanco R, Garcıa del Moral LF (2000) Triticale grain growth and morphometry as affected by drought stress, late sowing and simulated drought stress. Aust J Plant Physiol 27:1051–1059

Royo C, Villegas D, Del Moral LG, Elhani S, Aparicio N, Rharrabti Y (2002) Comparative performance of carbon isotope discrimination and canopy temperature depression as predictors of genotype differences in durum wheat yield in Spain. Aust J Agric Res 53:561–569. https://doi.org/10.1071/AR01016

Royo C, Villegas D, Rharrabti Y, Blanco R, Martos V, Garcia del Moral LF (2006) Grain growth and yield formation of durum wheat grown at contrasting latitudes and water regimes in a Mediterranean environment. Cereal Res Commun 34:1021–1028

Royo C, Alvaro F, Martos V, Ramdani A, Isidro J, Villegas D, Garcia del Moral LF (2007) Genetic changes in durum wheat yield components and associated traits in Italian and Spanish varieties during the 20th century. Euphytica 155:259–270

Royo C, Elias EM, Manthey FA (2009) Durum wheat breeding. In: Carena MJ (ed) Cereals—handbook of plant breeding, 3rd edn. Springer, Berlin, pp 199–226

Ruggiero A, Punzo P, Landi S, Costa A, Van Ooosten M, Grillo S (2017) Improving plant water use efficiency through molecular genetics. Horticulture 3:31. https://doi.org/10.3390/horticulturae3020031

Ruiz M, Giraldo P, González JM (2018) Phenotypic variation in root architecture traits and their relationship with eco-geographical and agronomic features in a core collection of tetraploid wheat landraces (*Triticum turgidum* L.). Euphytica 214:54

Rutkoski J, Poland J, Mondal S, Autrique E, Pérez LG, Crossa J et al (2016) Canopy temperature and vegetation indices from high-throughput phenotyping improve accuracy of pedigree and genomic selection for grain yield in wheat. G3 6:2799–2808. https://doi.org/10.1534/g3.116.032888

Ryser P, Eek L (2000) Consequences of phenotypic plasticity vs. interspecific differences in leaf and root traits for acquisition of aboveground and belowground resources. Am J Bot 87:402–411. https://doi.org/10.2307/2656636

Sabella E, Aprile A, Negro C, Nicolì F, Nutricati E, Vergine M, Luvisi A, De Bellis L (2020) Impact of climate change on durum wheat yield. Agronomy 10:793

Saeidi M, Ardalani S, Jalali-Honarmand S, Ghobadi ME, Abdoli M (2015) Evaluation of drought stress at vegetative growth stage on the grain yield formation and some physiological traits as well as fluorescence parameters of different bread wheat cultivars. Acta Biol Szeged 59:35–44

Saini DK, Srivastava P, Pal N (2022) Meta-QTLs, ortho-meta-QTLs and candidate genes for grain yield and associated traits in wheat (*Triticum aestivum* L.). Theor Appl Genet 135(3):1049–1081

Saini P, Kaur H, Tyagi V et al (2023) Genetic enhancement of nutritional and end-use quality in bread wheat through alien introgressions from wild relatives. Cereal Res Communi 51:295–314. https://doi.org/10.1007/s42976-022-00309-7

Sall AT, Bassi FM, Cisse M, Gueye H, Ndoye I, Filali-Maltouf A, Ortiz R (2018) Durum wheat breeding: in the heat of the Senegal River. Agriculture 8(7):99. https://doi.org/10.3390/agriculture8070099

Sall AT, Chiari T, Legesse W, Seid-Ahmed K, Ortiz R, van Ginkel M, Bassi FM (2019) Durum wheat (*Triticum durum* Desf.): origin, cultivation and potential expansion in Sub-Saharan Africa. Agronomy 9:263

Sallam A, Alqudah AM, Dawood M, Baenziger PS, Börner A (2019) Drought stress tolerance in wheat and barley: advances in physiology, breeding and genetics research. Int J Mole Sci 20(13):31–37. https://doi.org/10.3390/ijms20133137

Salvi S, Porfiri O, Ceccarelli S, Strampelli N (2013) The 'Prophet' of the green revolution J. Agric. Sci., 151. 1–5. https://doi.org/10.1017/S0021859612000214

Salvi S, Tuberosa R (2015) The crop QTLome comes of age. Curr Opin Biotechnol 32:179–185

Sarto MVM, Sarto JRW, Rampim L, Bassegio D, da Costa PF, Inagaki AM (2017) Wheat phenology and yield under drought: a review. Aust J Crop Sci 11:941–946. https://doi.org/10.21475/ajcs.17.11.08.pne351

Scarascia Mugnozza GT (2005) The contribution of Italian wheat geneticists: from Nazareno Strampelli to Francesco D'Amato. Accademia Nazionale delle Scienze, Rome

Schachtman DP, Goodger JQ (2008) Chemical root to shoot signaling under drought. Trends Plant Sci 13:281–287. https://doi.org/10.1016/j.tplants.2008.04.003

Schmidt J, Tricker PJ, Eckermann P, Kalambettu P, Garcia M, Fleury D (2020) Novel Alleles for Combined Drought and Heat Stress Tolerance in Wheat (2020) Fron Pl Sci 10. 1–10. ISSN=1664-462X. https://doi.org/10.3389/fpls.2019.01800

Schmidt J, Garcia M, et al (2020) Transcripts of wheat at a target locus on chromosome 6B associated with increased yield, leaf mass and chlorophyll index under combined drought and heat stress. PLoS ONE.15(11):e0241966, https://doi.org/10.1371/journal.pone.0241966

Semenov MA, Halford NG (2009) Identifying target traits and molecular mechanisms for wheat breeding under a changing climate. J Exp Bot 60:2791–2804

Sevillano CA, Vandenplas J, Bastiaansen JWM, Bergsma R, Calus MPL (2017) Genomic evaluation for a three-way crossbreeding system considering breed-of-origin of alleles. Genet Sel Evol 49:75. https://doi.org/10.1186/s12711-017-0350-1

Shackley BJ, Anderson WK (1995) Responses of wheat cultivars to time of sowing in the southern wheatbelt of Western Australia. Aus J Exp Agric 35:579–587. https://doi.org/10.1071/ea9950579

Shamuyaria KW, Shimelis H, Tapera T, Tsilo TJ (2019) Genetic advancement of newly developed wheat populations under drought-stressed and non-stressed conditions. J Crop Sci Biotechnol 22:169–176

Shanmugam S, Kjaer KH, Ottosen CO, Rosenqvist E, Kumari Sharma D, Wollenweber B (2013) The alleviating effect of elevated CO_2 on heat stress susceptibility of two wheat (*Triticum aestivum* L.) cultivars. J Agron Crop Sci 199:340–350

Sharma S, Schulthess AW, Bassi FM, Badaeva ED, Neumann K, Graner A, Ozkan H, Werner P, Knupffer H, Kilian B (2021) Introducing beneficial Alleles from plant genetic resources into the wheat germplasm. Biology 10:982. https://doi.org/10.3390/biology10100982

Sheoran S, Malik R, Narwal S, Tyagi BS, Mittal V, Kharub AS, Tiwari V, Sharma I (2015) Genetic and molecular dissection of drought tolerance in wheat and barley. J Wheat Res 7:1–13

Sheoran S, Jaiswal S, Kumar D, Raghav N, Sharma R, Pawar S, Paul S, Iquebal MA, Jaiswar A, Sharma P, Singh R, Singh CP, Gupta A, Kumar N, Angadi UB, Rai A, Singh GP, Kumar D and Tiwari R (2019) Uncovering Genomic Regions Associated With 36 Agro-Morphological Traits in Indian Spring Wheat Using GWAS. Front Plant Sci 10:527. https://doi.org/10.3389/fpls.2019.00527

Shirdelmoghanloo H, Taylor JD, Lohraseb I, Rabie H, Brien C, Timmins A, Martin P, Mather DE, Emebiri L, Collins NC (2016) A QTL on the short arm of wheat (*Triticum aestivum* L.) chromosome 3B affects the stability of grain weight in plants exposed to a brief heat shock early in grain filling. BMC Plant Biol 16:100

Slafer GA, Rawson HM (1994) Sensitivity of wheat phasic development to major environmental factors: a re-examination of some assumptions made by physiologists and modellers. Aust J Plant Physiol 21:393

Soriano JM, Villegas D, Aranzana MJ, García Del Moral LF, Royo C (2016) Genetic structure of modern durum wheat cultivars and Mediterranean landraces matches with their agronomic performance. PLoS One 11:e0160983. https://doi.org/10.1371/journal.pone.0160983

Soriano JM, Colasuonno P, Marcotuli I, Gadaleta A (2021) Meta-QTL analysis and identification of candidate genes for quality, abiotic and biotic stress in durum wheat. Sci Rep 11(1):11877. https://doi.org/10.1038/s41598-021-91446-2

Soriano JM, Malosetti M, Roselló M, Sorrells ME, Royo C (2017) Dissecting the old Mediterranean durum wheat genetic architecture for phenology, biomass and yield formation by association mapping and QTL meta-analysis. PLoS ONE 12(5): e0178290. https://doi.org/10.1371/journal.pone.0178290

Subira J, Alvaro F, Garcia del Moral LF, Royo C (2015) Breeding effects on the cultivar × environment interaction of durum wheat yield. Eur J Agron 68:78–88. https://doi.org/10.1016/j.eja.2015.04.009

Sukumaran S, Reynolds MP, Sansaloni C (2018) Genome-wide association analyses identify QTL hotspots for yield and component traits in durum wheat grown under yield potential, drought, and heat stress environments. Front Plant Sci 9:81

Swaminathan MS (2014) Norman Borlaug and a hunger-free world. Resonance 19:109–115. https://doi.org/10.1007/s12045-014-0014-1

Tabbita F, Ammar K, Ibba I et al (2024) Response to heat stress and glutenins allelic variation effects on quality traits in durum wheat. J Agron Crop Sci 210(1):e. https://doi.org/10.1111/jac.12686

Talebi, Reza, Fayyaz, Fakhar, Naji, Amir (2010) Genetic variation and interrelationships of agronomic characteristics in durum wheat under two constructing water regimes. Braz Arch Bio Tech. 53. 785–791.https://doi.org/10.1590/S1516-89132010000400006

Talebi R (2011) Evaluation of chlorophyll content and canopy temperature as indicators for drought tolerance in durum wheat (*Triticum durum* Desf.). Aust J Bas App Sci 5(11):1457–1462

Tedone L, Verdini L, De Mastro G (2023) Effects of different types of soil management on organic carbon and nitrogen contents and the stability index of a Durum wheat–Faba Bean rotation under a Mediterranean climate. Agronomy 13:1298. https://doi.org/10.3390/agronomy13051298

Tramblay Y, Llasat MC, Randin C, Coppola E (2020) Climate change impacts on water resources in the Mediterranean. Reg Environ Chang 20:83

Trethowan R (2014) Delivering drought tolerance to those who need it, from genetic resource to cultivar. Crop Pasture Sci 65:645–665

Tricker PJ, ElHabti A, Schmidt J, Fleury D (2018) The physiological and genetic basis of combined drought and heat tolerance in wheat. J Exp Bot 69(13):3195–3210

Tuberosa R (2012) Phenotyping for drought tolerance of crops in the genomics era. Front Physiol 3:347. https://doi.org/10.3389/fphys.2012.00347

Tuberosa R, Cattivelli L, Ceriotti A et al (2021) Proceedings of FSTP3 congress-a sustainable Durum wheat chain for food security and healthy lives. Front Plant Sci 12:675510–675510

Tura H, Edwards J, Gahlaut V, Garcia M, Sznajder B, Baumann U, Shahinnia F, Reynolds M, Langridge P, Balyan HS et al (2020) QTL analysis and fine mapping of a QTL for yield-related traits in wheat grown in dry and hot environments. Theor Appl Genet 133:239–257

Turner NC, Wright GC, Siddique KHM (2001) Adaptation of grain legumes (pulses) to water-limited environments. Adv Agron 71:193–231. https://doi.org/10.1016/S0065-2113(01)71015-2

Ur Rehman S, Wang J, Chang X, Zhang X, Mao X, Jing R (2019) A wheat protein kinase gene *TaSnRK2.9-5A* associated with yield contributing traits. Theor Appl Genet 132:907–919. https://doi.org/10.1007/s00122-018-3247-7

Ur Rehman S, Ali Sher M, Saddique MAB, Ali Z, Khan MA, Mao X, Irshad A, Sajjad M, Ikram RM, Naeem M, Jing R (2021) Development and exploitation of Kompetitive Allele-Specific PCR (KASP) assays for genes underpinning drought tolerance among wheat cultivars from Pakistan. Front Genet 12:684702. https://doi.org/10.3389/fgene.2021.684702

Varshney RK, Barmukh R, Roorkiwal M et al (2021) Breeding custom-designed crops for improved drought adaptation. Adv Genet 2(3):e202100017

Vetch JM, Stougaard RN, Martin JM, Giroux MJ (2019) Review: revealing the genetic mechanisms of pre-harvest sprouting in hexaploidy wheat (*Triticum aestivum* L.). Plant Sci 281:180–185. https://doi.org/10.1016/j.plantsci.2019.01.004

Vincent H, Amri A, Castañeda-Álvarez NP, Dempewolf H, Dulloo E, Guarino L, Hole D, Mba C, Toledo A, Maxted N (2019) Modeling of crop wild relative species identifies areas globally for in situ conservation. Commun Biol 2:136

Wahid A, Gelani S, Ashraf M, Foolad MR (2007) Heat tolerance in plants: an overview. Environ Exp Bot 61:199–223. https://doi.org/10.1016/j.envexpbot.2007.05.011

Wahn Y, Sanchis Gritsch C, Hawkesford MJ, Shewry PR (2014) Effects of nitrogen nutrition on the synthesis and deposition of the ω-gliadins of wheat. Ann Bot 113:607–615. doi: https://doi.org/10.1093/aob/mct291

Walter J, Edwards J, Cai J, McDonald G, Miklavcic SJ, Kuchel H (2019) High-throughput field imaging and basic image analysis in a wheat breeding programme. Front Plant Sci 10:449. https://doi.org/10.3389/fpls.2019.00449

Wang RX, Hai L, Zhang XY, You GX, Yan CS, Xiao SH (2009) QTL mapping for grain filling rate and yield-related traits in RILs of the Chinese winter wheat population Heshangmai× Yu8679. Theor Appl Genet 118:313–325. https://doi.org/10.1007/s00122-008-0901-5

Wang YQ, Xi WX, Wang ZM, Wang B, Xu XX, Han MK, Zhou SL, Zhang YH (2016) Contribution of ear photosynthesis to grain yield under rainfed and irrigation conditions for winter wheat cultivars released in the past 30 years in North China Plain. J Integr Agric 15:2247–2256

Wardlaw IF, Willenbrink J (2000) Mobilization of fructan reserves and changes in enzyme activities in wheat stems correlate with water stress during kernel filling. New Phytol 148:413–422

Wasaya A, Zhang X, Fang Q, Yan Z (2018) Root phenotyping for drought tolerance: a review. Agronomy 8:241. https://doi.org/10.3390/agronomy8110241

Wasson AP, Richards RA, Chatrath R, Misra SC, Prasad SVS, Rebetzke GJ et al (2012) Traits and selection strategies to improve root systems and water uptake in water-limited wheat crops. J Exp Bot 63:3485–3498. https://doi.org/10.1093/jxb/ers111

Watson A, Ghosh S, Williams MJ, Cuddy WS, Simmonds J, Rey MD, Adamski NM (2018) Speed breeding is a powerful tool to accelerate crop research and breeding. Nat Plants 4:23

Wei X, Jiao G, Lin H, Sheng Z, Shao G, Xie L, Tang S, Xu Q, Hu P (2017) Grain Incomplete Filling two regulates grain filling and starch synthesis during rice caryopsis development. J Integr Plant Biol 59(2):134–153. https://doi.org/10.1111/jipb.12510. PMID: 27957808

Weldearegay DF, Yan F, Jiang D, Liu F (2012) Independent and combined effects of soil warming and drought stress during anthesis on seed set and grain yield in two spring wheat varieties. J Agron Crop Sci. 198: 245–253. https://doi.org/10.1111/j.1439-037X.2012.00507.x

Whan AP, Smith AB, Cavanagh CR et al (2014) Grain-Scan: a low cost, fast method for grain size and colour measurements. Plant Meth 10:23. https://doi.org/10.1186/1746-4811-10-23

Williams ER, John JA, Whitaker D (2006) Construction of resolvable spatial row-column designs. Biometrics 62(1):103–108. http://www.jstor.org/stable/3695710

Wu Q, Chen Y, Zhou S, Fu L, Chen J, Xiao Y, Zhang D et al (2015) High-density genetic linkage map construction and QTL mapping of grain shape and size in the wheat population Yanda 1817 9 Beinong6. PLoS One 10:e0118144

Wurschum T, Leiser WL, Langer SM, Tucker MR, Longin CFH (2018) Phenotypic and genetic analysis of spike and kernel characteristics in wheat reveals long-term genetic trends of grain yield components. Theor Appl Genet 131:2071–2084

Xin F, Zhu T, Wei S, Han Y, Zhao Y, Zhang D, Ma L, Ding Q (2020) QTL mapping of Kernel traits and validation of a major QTL for Kernel length-width ratio using Single Nucleotide Polymorphism (SNP) and Bulked Segregant analysis in wheat. Sci Rep 10:1–12

Xu Y, Li S, Li L, Ma F, Fu X, Shi Z, Xu H, Ma P, An D (2017) QTL mapping for yield and photosynthetic related traits under different water regimes in wheat. Mol Breed 37:34

Xynias IN, Mylonas I, Korpetis EG, Ninou E, Tsaballa A, Avdikos ID, Mavromatis AG (2020) Durum wheat breeding in the Mediterranean region: current status and future prospects. Agronomy 10(3):432. https://doi.org/10.3390/agronomy10030432

Yang Y, Al-Baidhani HJ, Harris J, RiboniM YL, Mazonka I, Bazanova N, Chirkova L, Hussain S, Hrmova M, Haefele S, Lopato S, Kovalchuk N (2019) DREB/CBF expression in wheat and barley using the stress-inducible promoters of *HD-Zip I* genes: impact on plant development, stress tolerance and yield. Plant Biotech J 18:829–844

Yesayan AH, Grigorin KV, Danielian AM, Hovhannisyan NA (2009) Determination of salt tolerance in wild einkorn wheat (*Triticum boeoticum* Boiss.) under *in vitro* conditions. Crop Wild Relat 7:4–7

York LM, Slack S, Bennett MJ, Foulkes MJ (2018) Wheat shovelomics I: a field phenotyping approach for characterizing the structure and function of root systems in tillering species. bioRxiv [preprint]. https://doi.org/10.1101/280875

Zaharieva M, Gaulin E, Havaux M, Acevedo E, Monneveux P (2001) Drought and heat responses in the wild wheat relative *Aegilops geniculata* Roth: potential interest for wheat improvement. Crop Sci 41:1321–1329. https://doi.org/10.2135/cropsci2001.4141321X

Zaim M, El Hassouni K, Gamba FM, Filali-Maltouf A, Belkadi B, Sourour A, Amri A, Nachit M, Taghouti M, Bassi FM (2017) Wide crosses of durum wheat (*Triticum durum* Desf.) reveal good disease resistance, yield stability, and industrial quality across Mediterranean sites. Field Crop Res 214:219–227

Zaim M, Kabbaj H, Kehel Z, Gorjanc G, Filali-Maltouf A, Belkadi B, Nachit MM, Bassi FM (2020) Combining QTL analysis and genomic predictions for four durum wheat populations under drought conditions. Front Genet 11:316

Zaïm M, Kehel Z, Sanchez-Garcia M, Belkadi B, Filali-Maltouf A, Al-Abdallat A, Bassi FM (2023) Genomic regions of durum wheat involved in water productivity. bioRxiv 1-36:544022. https://doi.org/10.1101/2023.06.07.544022

Zang X, Geng X, He K, Wang F, Tian X, Xin M, Yao Y, Hu Z, Ni Z, Sun Q, Peng H (2018) Overexpression of the wheat (*Triticum aestivum* L.) *TaPEPKR2* gene enhances heat and dehydration tolerance in both wheat and *Arabidopsis*. Front Plant Sci 9:1710. https://doi.org/10.3389/fpls.2018.01710

Zapata-Valenzuela J, Whetten RW, Neale D, McKeand S, Isik F (2013) Genomic estimated breeding values using genomic relationship matrices in a cloned population of loblolly pine. G3 (Bethesda, Md.) 3(5):909–916. https://doi.org/10.1534/g3.113.005975

Zarei L, Cheghamirza K, Farshadfar E (2013) Evaluation of grain yield and some agronomic characters in durum wheat (*Triticum turgidum* L.) under rainfed conditions. Aus J Agri Res 7:609–661

Zhang L, Richards RA, Condon AG, Liu DC, Rebetzke GJ (2014) Recurrent selection for wider seedling leaves increases early biomass and leaf area in wheat (*Triticum aestivum* L.). J Exp Bot 66:1215–1226. https://doi.org/10.1093/jxb/eru468

Zhang H, Mittal N, Leamy LJ, Barazani O, Song BH (2016a) Back into the wild—apply untapped genetic diversity of wild relatives for crop improvement. Evol Appl 10:5–24

Zhang H, Li W, Mao X, Jing R, Jia H (2016b) Differential activation of the wheat SnRK2 family by abiotic stresses. Front Plant Sci 7:420

Zhang J, Zhang S, Cheng M, Jiang H, Zhang X, Peng C (2018a) Effect of drought on agronomic traits of rice and wheat: a meta-analysis. Int J Environ Res Public Health 15:839. https://doi.org/10.3390/ijerph15050839

Zhang C, Si Y, Lamkey J, Boydston RA, Garland-Campbell KA, Sankaran S (2018b) High-throughput phenotyping of seed/seedling evaluation using digital image analysis. Agronomy 8:63. https://doi.org/10.3390/agronomy8050063

Zhang WJ, Li T, Qin L, Zhao J, Zhao JJ, Liu HX (2018c) *TaDRO*, a gene associated with wheat root architectures, its global distribution and evolution in breeding. Sci Agric Sin 51:1813–1829

Zhikai Y, Xu G, Qi Z, Toshihiro O, Yang J (2022) Genome-wide mediation analysis: an empirical study to connect phenotype with genotype via intermediate transcriptomic data in maize. Genetics 221(2):1–9057. https://doi.org/10.1093/genetics/iyac057

Contribution of Genomics to Enhance Drought Stress Tolerance in Durum Wheat

Donatella Bianca Maria Ficco, Daniela Marone, Grazia Maria Borrelli, Antonia Mores, Giovanni Laidò, Maria Anna Russo, and Anna Maria Mastrangelo

Abstract

Drought stress is one of the major factors limiting yield and yield stability of durum wheat. Classical breeding schemes have used direct selection for yield over multiple locations or drought-related proxy traits, as physiological ones, but low heritability and high genotype x environment interaction have greatly limited the effectiveness of these approaches for durum wheat. During last years, genomics and other omic approaches have provided a considerable contribution to identify markers, transcripts, peptides, and metabolites associated with responses to drought. As whole genome sequences, physical maps, genetic and functional genomic tools are now available also for tetraploid wheats, integrated approaches leveraging molecular breeding, genetic engineering, and gene editing can be fully exploited to improve yield in drought-prone conditions. On one hand, the identification of quantitative trait loci (QTLs) with a major effect and the cloning of the underlying loci (coding and non-coding sequences) are expected to move the selection toward a "breeding by design" approach that will accumulate an increasing number of known useful alleles into elite genotypes. Additionally, genomic selection approaches are promising as they leverage genome-wide loci also with very small effects and, as such, with no previous knowledge on their map position and function. Efficient integration of all the genomic tools and approaches in breeding programs will ultimately improve durum wheat to better sustain yield and quality in drought-prone environments while reducing the gap between yield potential and actual yield.

Donatella Bianca Maria Ficco and Daniela Marone contributed equally to this work.

D. B. M. Ficco · D. Marone · G. M. Borrelli · A. Mores · G. Laidò · M. A. Russo · A. M. Mastrangelo (✉)
Consiglio per la ricerca in Agricoltura e l'Analisi dell'Economia Agraria - Centro di Ricerca Cerealicoltura e Colture Industriali, Foggia, Italy
e-mail: donatellabm.ficco@crea.gov.it; daniela.marone@crea.gov.it; graziamaria.borrelli@crea.gov.it; giovanni.laido@crea.gov.it; mariaanna.russo@crea.gov.it; annamaria.mastrangelo@crea.gov.it

1 Introduction

Environmental stresses are defined as soil or climatic conditions which limit the growth, development, and reproduction of plants. Drought stress is related to a limited availability of water for the plant. In the current scenario of climatic changes, the onset and magnitude of drought events are expected to be exacerbated by increas-

R. Tuberosa et al. (eds.), *The Durum Wheat Genome*, Compendium of Plant Genomes,
https://doi.org/10.1007/978-3-031-84689-2_11

ingly erratic rainfall and evapotranspiration rates due to growing temperatures (De Leonardis et al. 2012). A negative impact is, therefore, expected on agriculture, in terms of an increased gap between yield potential and actual production. This is particularly true for durum wheat [*Triticum turgidum* L. ssp. *durum* (Desf.)] which is grown mainly in rainfed semiarid environments (e.g., Mediterranean regions), where drought stress is a serious concern. In these environments, water stress occurs especially at flowering stage when spike fertility is established, grain filling, and in combination with other stresses such as high temperature (Groli et al. 2024); therefore, it causes severe damages to grain yield and quality, the main targets of durum wheat breeding. Indeed, durum wheat is used mainly for pasta, couscous, bulgur, and for bread production, and the quality of these products is strongly influenced by the genotype, the environment, management, and their interaction.

Plants usually react to a drought event by promoting numerous changes in their morphology, physiology, and metabolism, at organ and cellular levels (Levitt 1972; Blum 2005, 2010; Blum and Tuberosa 2018; Paudell et al. 2024). Three main drought response strategies may occur: (i) escape, mainly through the shortening of the growth cycle; (ii) avoidance, which helps the plant to maintain high water status during periods of stress, either by a more efficient water absorption from roots or by reducing evapotranspiration from aerial parts; and (iii) tolerance, in which active mechanisms such as osmotic adjustment and/or enhanced antioxidant capacity are activated to withstand water deficit with low tissue water potential (Reynolds and Tuberosa 2008; De Leonardis et al. 2012).

The plant response to drought depends on several factors, such as the duration and severity of water scarcity (Pinheiro and Chaves 2011), and the plant stage of development (De Leonardis et al. 2007). Moreover, the plant reaction to water stress is genotype dependent, and this is demonstrated at level of agronomic performance, stress-related physiological traits, and at molecular level, beginning from gene expression (De Leonardis et al. 2007). Studying the phenotypic and genetic diversity in durum wheat for drought stress tolerance is, therefore, the first step to improve its capacity to cope with drought hence reducing the stress damage, but it is very difficult because drought tolerance is a complex trait, controlled by many loci (Bernardo 2008; Collins et al. 2008; Tuberosa 2012; Tuberosa and Maccaferri 2015; Mwadzingeni et al. 2016; Roncallo 2017). Another problem is the difficulty to exactly evaluate the capacity of many genotypes to cope with drought in the same experiment. For this reason, great research has been carried out to identify proxy traits which describe the drought tolerance of a genotype as accurately as possible. Numerous studies pointed out that a good variation exists in durum wheat for traits related to drought tolerance (Graziani et al. 2014; Zemede et al. 2019). At phenotypic level, this is true when specific morphological and physiological traits are measured, and also when the drought tolerance is evaluated simply as grain yield and quality in control and drought conditions (Mohammadi and Abdulahi 2017).

Once the phenotypic features have been evaluated, they must be linked to specific genes/alleles to investigate the genetic bases of drought tolerance and to develop molecular tools to move the selection process from phenotype to DNA. The plant response to water stress can be analyzed in terms of gene sequence, transcriptome, proteome, and metabolome, and the integration of multiple omic approaches is important to maximize the understanding of the molecular mechanisms which regulate the physiological and agronomic performance in drought conditions. The recent availability of the high-quality genome sequences for bread wheat, wild emmer, and durum wheat (Avni et al. 2017; IWGSC 2018; Maccaferri et al. 2019; Walkowiak et al. 2020) has strongly improved the output of these studies also in durum wheat. Moreover, thanks to next-generation sequences approaches, high-throughput and cheap marker platforms are available for bread and durum wheat (Maccaferri et al. 2022; 2009), for the genotypic analysis of very large panels of lines, in order to map and project onto the genome the loci involved in the control of drought stress tolerance (Khadka et al. 2020; Colasuonno et al. 2021).

This genome-driven knowledge is then applied to breeding programs to develop new stress-resistant cultivars. Marker-based selection approaches can be used for the transfer of specific genes/quantitative trait loci (QTLs), once they have been identified through mapping, or on the prediction of a global breeding value estimated considering many markers with genome-wide coverage. In the last case, related to genomic selection (GS), no previous knowledge about the markers effectively involved in the control of the trait is needed (Mores et al. 2021).

The goal of genetic mapping studies and transcriptomic/proteomic approaches is to identify and clone the genes with an effective role in controlling drought-stress response. When the gene sequence and structure are known, they can be used to design perfect molecular markers which allow the transfer of the gene with no risk of segregation between the marker and the gene/QTL. Otherwise, by using the most recent biotechnological approaches, the genes can be introduced in an elite cultivar through genetic transformation (cisgenesis), or precisely modified with genome editing approaches (Colasuonno et al. 2021).

This chapter will consider the contribution provided by the progress made in durum wheat genomics to the understanding of response to drought stress in this species, from the setup and integration of different omic approaches to the study of phenotypic and genetic diversity for mapping loci/genes of interest for drought stress reaction, and the implementation of all these achievements in breeding programs to face the challenge represented by the ongoing climate changes.

2 Omic Approaches: Understanding Durum Wheat Responses to Drought Stress from Genome to Metabolome

In the last years, omic approaches provided a huge amount of data at level of DNA, QTLome, transcripts, proteins, and metabolites. The big challenge is to leverage this information to extract valuable clues on effective strategies for plants to cope with drought stress.

A first level of genomic investigation is the DNA. Many studies have been carried out in the past to identify molecular markers useful for genetic analyses (Colasuonno et al. 2021). The use of next-generation sequencing techniques has provided a strong impact on this kind of studies, allowing the identification of thousands of single-nucleotide polymorphism (SNP) markers for analysis in pre-built arrays, or through genotyping by sequencing (GBS) at different coverage levels. The availability of the high-quality sequence of the cv. Svevo genome has allowed to precisely map these markers to the 14 durum chromosomes (Maccaferri et al. 2019).

2.1 Study of Gene Families

The availability of DNA or transcript sequences is a prerequisite for the study of gene families with a role in response to drought stress in crops. Faraji et al. (2018) identified 122 *C2H2- zinc finger* (*C2H2-ZF*) genes, coding for transcription factors which play crucial roles in plant development and stress response in durum wheat. They were retrieved from transcriptome sequences obtained from multiple durum wheat tissues, and physically mapped onto the common wheat genome. Expression profiling of this gene family members showed a differential expression in response to dehydration and heat stresses.

A total of 113, 79, and 36 *Heat shock 70* (*Hsp70*) genes were identified in bread wheat, wild emmer wheat, and *Aegilops tauschii* by Lai et al. (2021). Their expression profile analysis showed that *TaHsp70* gene was highly expressed under water stress. The identification and classification of *Hsp70* genes in wheat and its relatives can provide a basis for understanding its evolution and its role in response to stress. Some gene families involved in drought stress response can also play a role in response to multiple abiotic stresses. As an example, the superoxide dismutase (*SOD*) family in durum wheat was recently found to be responsive under salt, drought, cold, and ABA treatments (Tounsi et al. 2023).

The description and characterization of gene families can be useful as a starting point to identify promising candidate genes to improve water stress tolerance in durum wheat. Starting from a previous study in which Saidi et al. (2017) identified 168 *NAM-ATAF1,2-CUC2* (*NAC*) genes from durum wheat transcriptome, Mergby et al. (2021) carried out a functional study on the *TtNAC2A* gene (see Sect. 4.2).

2.2 Analysis of Transcriptomic Changes in Response to Drought

Studies on durum wheat transcriptome following drought stress highlighted different expression strategies between bread and durum wheat and among durum wheat genotypes (Aprile et al. 2009). Interesting results have also been obtained when durum wheat segregating populations were investigated. Habash et al. (2014) studied the physiological and transcriptome responses of three durum breeding lines, parents of a mapping population (Lahn x Cham1), and a recombinant inbred line (RIL2219), varying for yield stability under drought. RIL2219 showed a defined physiological strategy for high yield stability under water stress which was different from that of both the parents. Indeed, the RIL lost less water and showed constitutively higher stomatal conductance, photosynthesis, transpiration, abscisic acid content, and enhanced osmotic adjustment at equivalent leaf water compared to parents. Aprile et al. (2013) studied the behavior of two durum wheat cultivars, parents of a mapping population, following water and heat stress. They found interesting differences in the water stress response between the two genotypes, as the cv. Ofanto, characterized by a lower water use efficiency, activated a large set of well-known drought-related genes after drought treatment, while cv. Cappelli, characterized by a higher water use efficiency, showed the constitutive expression of several genes that in cv. Ofanto were induced by drought and a modulation of a limited number of genes in response to stress. Moreover, the expression analysis in the whole RIL population of a stress-related gene differentially expressed in the parents allowed to map the expression QTL (eQTL) on chromosome 6B in a position coincident with the same stress-related gene (Aprile et al. 2013).

In the last years, the interest in transcriptomic studies shifted from mRNAs to small and microRNAs (miRNAs). The availability of the genome sequences for bread wheat, wild emmer wheat, and durum wheat (Avni et al. 2017; IWGSC 2018; Maccaferri et al. 2019; Walkowiak et al. 2020) has greatly improved the accuracy of transcriptomic studies thanks to the better quality of gene annotation. Recently, several studies led to the identification of miRNAs (and their candidate targets) involved in drought response and expression of drought-related physiological traits in durum wheat (Liu et al. 2015; Akpinar et al. 2015; Giusti et al. 2017; Liu et al. 2017). Giusti et al. (2017) identified nearly 80 miRNAs differentially expressed in response to heat and drought stress between Ofanto and Cappelli. The in silico analysis of predicted targets revealed that most of them were related to control of stomatal movements, a finding in agreement with previous results showing that Cappelli has a lower stomatal conductance than Ofanto (Rizza et al. 2012). Besides the immediate response following a stress event, miRNAs are also involved in stress transgenerational effect. Water-deficit and heat stress on parental durum wheat plants affected seed germination and seedling vigor of the subsequent generation in two Australian durum genotypes, respectively, stress-tolerant and stress-sensitive, with a weaker effect in the tolerant variety (Liu et al. 2020a). These findings could be related to miRNAs, which showed significant differences in the two genotypes (Liu et al. 2020a). This transgenerational effect of abiotic stress was confirmed for other traits of agronomic importance, as yield components, grain quality traits, and physiological traits (Liu et al. 2021a). Multi-omics analysis, in which mRNAs, miRNAs and degraded transcripts were investigated, allowed for a more complete picture of transcriptional response to drought in durum wheat. In particular, analysis of the degradome provides knowledge about transcripts which are effectively targeted by miRNA, offering a validation to putative targets previously identified by bioinformatic analysis (Liu et al. 2020a, 2021a).

2.3 Proteome Analysis in Drought Conditions

Very few studies focused on proteome changes following drought in durum wheat were carried out in the last years (Sehgal et al. 2023). Lists of protein sequences differentially regulated by water stress have been identified in the durum wheat cv. Ofanto (Caruso et al. 2009). This approach has also been used to reveal differences in proteome changes following water stress between wild emmer wheat accessions and durum cultivars: many of the proteins expressed in young plants subjected to water stress were common in both wild emmer and modern wheat proteomes, but differentially regulated proteins were also identified which were related to carbohydrate and amino acid metabolism (Budak et al. 2013). Differential accumulation of peptides in control versus drought conditions has been validated through Western blotting analysis (Kacem et al. 2016); moreover, their amount has been compared with that of the corresponding transcripts (Budak et al. 2013). Giuliani et al. (2015) investigated the effect of drought stress on the gluten protein composition in durum wheat grain, a feature of paramount importance for the quality of pasta, the main end-product. When two Italian durum wheat cultivars, Ciccio and Svevo, were assessed in this regard, differences in the proteins associated with immunological potential and involved in technological quality were observed between the two genotypes at all developmental stages.

2.4 Metabolome Analysis in Drought Conditions

Metabolomics is an important and complementary approach in systems biology to investigate the physiological responses of plants to biotic and abiotic stresses (Saito and Matsuda 2010; Chaouachi et al. 2024). Generally, the main plant responses to drought, some of which are common to other types of stress, affect both primary and secondary metabolism of plant tissues and organs (Nakabayashi and Saito 2015; Bowne et al. 2011). Drought tolerance is linked to the ability of maintaining crop performance in a condition of shortage of water, involving both carbon and nitrogen metabolism, leading to the accumulation of osmoregulators to maintain cell turgor, including amino acids, carbohydrates, and other osmolytes (Michaletti et al. 2018; Condorelli et al. 2022).

Similarly to what seen for transcriptomic analyses, first evidence coming from metabolomic studies is that the plant response to drought stress at metabolic level is dependent on genotype. A comparative targeted metabolic profiling was performed by Buffagni et al. (2020) on flag leaves both in control and in drought conditions of two durum wheat cultivars: Cappelli, more tolerant (Aprile et al. 2013) and Colosseo, more susceptible (Rampino et al. 2006) to drought. Some common mechanisms were observed in the two cultivars, such as a significantly increased amino acids like alanine, glycine, glutamate, proline (osmoprotectant), and gamma-aminobutyric acid (GABA) in water shortage, while a specific metabolic response was observed for GABA and its precursor glutamate, upregulated in Cappelli and downregulated in Colosseo in early phases of drought treatment (Buffagni et al. 2020).

Enhanced accumulation of amino acids was also observed following water stress in roots and leaves of wild emmer wheat (Ullah et al. 2017), and in advanced lines and commercial durum wheat varieties with some differences related to the tissue and growth stage (Vergara-Diaz et al. 2020). Kang et al. (2019) found higher levels of proline in both leaves and roots of the drought-tolerant genotype respect to the drought-sensitive one. Also, other amino acids, including proline, tryptophan, and branched-chain AAs (i.e., leucine, isoleucine, and valine) increased during drought stress (Michaletti et al. 2018). Besides amino acids with a protective role (i.e., hydroxyproline, 5-oxoproline, ornithine, threonine, tryptophan, GABA, and proline), aromatic amino acids were also upregulated due to an increasing demand for precursors for the synthesis of antioxidants, as flavonoids (Agati et al. 2020; Bowne et al. 2012) which, together with upregulation of antioxidant enzyme activities, can counteract the accumulation of reactive oxygen species (ROS) (Popova et al. 1998). Changes in accumulation of carbohydrates were also observed following exposure to drought (Fàbregas and Fernie 2019). Buffagni et al. (2020) showed sucrose accumulating differently in Cappelli and Colosseo following water stress, with a lower amount in the first genotype. A clear increase was described for other sugars with a specific role in preventing oxidative damage (Van der Ende 2013), as raffinose, maltose, isomaltose, and trehalose in durum wheat spikes subjected to stress (Vergara-Diaz et al. 2020). Similar results were found for these sugars in both leaf and root samples of *T. dicoccoides* (Hoekstra et al. 2001). A role in response to water stress has been proposed also for metabolites related to structural elements (cell wall related sugars, phenylpropanoids and hydroxyproline) that contribute to drought acclimatation modulating the cell wall and membranes (Zhao et al. 2015).

Considering other osmolytes, a drought-induced increase of some organic acids was observed in response to water stress. As an example, considering the leaves of Cappelli, an increase was reported in accumulation of malate, acting as a valve to balance cellular energy supply (Scheibe 2004), of acetate, involved in upregulation of the jasmonate signaling pathway (Kim et al. 2017), aand of choline, a precursor of glycine-betaine (Buffagni et al. 2020). In another study, levels of succinate, citrate, aspartate and gluconate were reported to increase in both leaf and root samples of *T. dicoccoides* (Ullah et al. 2017). In particular, the role of succinate in the efficient use of tricarboxylic acid cycle, for ATP production under water-limited conditions, has been shown (Cavalcanti et al. 2014; Ullah et al. 2017).

Analysis of metabolic changes following drought stress during flowering and grain development is crucial to limit the negative effects on durum wheat grain yield and quality (De Santis et al. 2021; Guzmán et al. 2016; Liu et al. 2020b). In a multi-omic approach, sequence-specific recognition of the stress-responsive miRNA-mRNA modules represents an efficient mechanism for the regulation of the accumulation of various grain nutrient components (starch, protein, carotenoids, and phenolics) in developing seeds. Starch accumulation decreased in drought conditions in two durum wheat genotypes, but in a lower extent in the tolerant genotype compared to the susceptible one (Liu et al. 2020b). This reduction was caused by a decrease in the expression and activity of starch synthesis-related proteins, as granule-bound starch synthase I (GBSSI), whose mRNA is targeted by miRNAs. Higher expression of *GBSSI* in stress-tolerant cultivar could alleviate the negative effects of stress on starch biosynthesis and contribute to superior grain quality (Liu et al. 2020a). Notably, miRNA can influence grain quality traits also regulating genes involved in the control of protein content, as *glutamate synthase* (*GOGAT*) and *glutamate receptor-like* (*GRL*), and accumulation of phenylpropanoids and carotenoids (Liu et al. 2020a).

Finally, a recent study revealed that the interaction with mycorrhizal fungi (*Funneliformis mosseae*) may enhance water and nutrients uptake in plants subjected to water stress, by reprogramming the root metabolism in terms of sugars, lipids, and plant phytohormones accumulation (Bernardo et al. 2019). Moreover, the analysis of the root secondary metabolome in a bread and a durum wheat genotype revealed that the bread wheat genotype benefited more than durum wheat from fungi colonization under limited water availability (Bernardo et al. 2019).

In conclusion, an integrated analysis of all the omics data publicly available would be of great interest and benefit to develop networks of response to drought at the different levels of expression, from RNAs to metabolites. Moreover, while many studies have been performed on effects of drought on primary metabolism in durum wheat, a significant gap is still present for secondary metabolites. A great effort is requested to fill it in the near future.

3 Dissection of Genetic Bases for Responses to Drought Stress in Durum Wheat

Genetic diversity refers to the variation occurring between different genotypes related to a single species or between species, and its study is of great importance as phenotypic differences for traits of agronomic importance are expected when diverse genetic materials are evaluated (Bhandari et al. 2017). Modern agriculture and conventional breeding have led to a reduction in genetic diversity of crop plants. In fact, many useful alleles for breeding could have been lost during domestication and/or selection process of crops, including durum wheat (Reif et al. 2005; Soriano et al. 2016). For this reason, it is important to evaluate and analyze very large germplasm collections with the aim to identify interesting genotypes for the expression of a trait, and the genetic loci/mechanisms controlling it. Wheat genetic resources are maintained and catalogued by several international research programs and gene banks (Marone et al. 2021), and large collections of tetraploid wheat genotypes, from wild emmer to durum wheat cultivars, are also publicly available together with the corresponding SNP data (Maccaferri et al. 2019; Mazzucotelli et al. 2020). Indeed, durum landraces, wild relatives, like *T. turgidum* ssp. *dicoccoides*, and the domesticated accessions, as *T. turgidum* spp. *dicoccum*, represent a valid gene pool to widen the genetic diversity available for durum wheat breeders (Dodig et al. 2012; Nevo and Chen 2010; Mengistu and Pe 2016; Mwadzingeni et al. 2016; Marone et al. 2021). Together with hexaploid wheats, all these genetic resources can be considered as a unique gene pool which provides a reservoir of desirable alleles for many traits, including those important for developing climate-resistant varieties (Mastrangelo and Cattivelli 2021).

Gene introgression from wild relatives has been shown to increase wheat resilience to climate change effects (Aberkane et al. 2020). Landraces derive predominantly from natural selection and are adapted to local environmental conditions according to their place of origin. Most of them are drought resilient and have higher yield stability in drought conditions than modern varieties (Reynolds et al. 2007). Several studies reported the adaptation capacity to drought of landraces (Abu-Zaitoun et al. 2018; Roselló et al. 2019) and the genetic variation existing in landrace groups with different geographic origin for this trait (Roselló et al. 2019).

Even if characterized by a low degree of genetic diversity, the germplasm composed of elite cultivars can also be a valuable source of useful alleles for drought tolerance. The advantage in this case is avoiding or strongly reducing the risk of linkage drag associated with the coinheritance of unwanted genes and rare alleles, often observed following introgressions from landraces or wild relatives. Salsman et al. (2021) evaluated the genetic diversity and agronomic traits of durum wheat germplasm (breeding lines from the North Dakota State University durum wheat breeding program and durum landraces) in western North Dakota and Montana, where drought is a major stress limiting grain yield. Despite a clear decrease in genetic diversity in breeding lines compared to landraces, in both groups interesting genotypes were identified which significantly outperformed local elite cultivars for grain yield and other agronomic traits in drought conditions. Elite cultivars can be characterized by different physiological mechanisms for responses to drought stress, as demonstrated for the durum cultivars Ofanto and Cappelli, and this opens the possibility of using this diversity to combine physiological adaptation capacity to drought and high productivity into the same genotype (Rizza et al. 2012).

In recent times, with the advent of genomic tools, and the availability of high-throughput and less expensive genotyping platforms, molecular markers have become the most widely used method for the assessment of genetic diversity (Poland et al. 2012; Wang et al. 2014; Avni et al. 2017; Maccaferri et al. 2019, 2022). These techniques have been successfully applied due to better coverage on the genome, high variability and reproducibility, possibility to automate the analysis, and absence of environmental fluctuations. The simple sequence repeat (SSR) markers are widely used to evaluate genetic diversity, thanks to their reproducibility and informativeness, with a high number of alleles detected per locus. As an example, from four to nine alleles per locus were found in a collection of 141 Ethiopian durum wheat landraces and 19 improved varieties analyzed with 12 SSRs (Asmamaw et al. 2019). A new generation of markers based on sequence or array platforms has been recently developed, including Diversity Array Technology (DArT) markers and SNP arrays, which have proven useful for genetic studies on panels of hundreds of genotypes (Kabbaj et al. 2017; Abu-Zaitoun et al. 2018; Mazzucotelli et al. 2020).

Genetic and genomic analyses are essential for the identification of loci/alleles with a key role in controlling drought tolerance in durum wheat, even if few studies are available for this species, despite durum wheat being mainly grown in semiarid environments (Table 1). Drought tolerance is a complex trait, with a strong effect of the environment on drought-related traits, and many genetic loci involved in its control. Moreover, high tolerance to drought should not only be combined with a good grain yield in stressed but also in non-stressed conditions (Sallam et al. 2019). Peleg et al. (2009) studied the genetic bases of responses to drought in a tetraploid RIL population under contrasting water availabilities. QTLs have been found for drought-adaptive traits associated with productivity in water-limited or well-watered conditions, suggesting the involvement of different strategies in wheat adaptation to drought stress. A mapping population derived from a cross between a durum wheat line and a wild emmer accession has been recently used for assessing QTLs for drought plasticity, known as the capacity to respond to environmental changes by altering the phenotype (Fatiukha et al. 2021). Seventeen traits related to yield, biomass, morphology, and phenology have been evaluated under water stress conditions, together with their plasticity in response to water deficit and 79 QTLs have been found, of which 33 had a plastic effect in response to drought. To note, the highest number of plastic QTLs were found for harvest index. A population of hybrids generated from crosses between Turkish durum wheat cultivars and wild tetraploid accessions has been characterized for osmotic stress, which is considered a component of abiotic stress conditions such as drought, and SNPs associated with stress tolerance have been identified (Lucas et al. 2017). In particular, these lines showed roots developing with the same or better efficiency during water deficit than in normal conditions, suggesting that they could carry genetic loci conferring drought tolerance. Very recently, a multi-reference parent nested-association mapping population obtained using speed breeding technology was used in durum wheat to dissect the quantitative nature of drought-adaptive traits (Alhamad et al. 2023). A comparative analysis of the genomic regions responsible for drought tolerance across many studies, different genetic backgrounds and environments, could help to better understand the genetics of this trait. A meta-QTL (MQTL) analysis for yield and agronomic traits in drought and heat-stressed environments has been performed in wheat, including data from five durum wheat populations, in order to identify consensus QTLs, refine the confidence interval, and find molecular markers strictly associated with the trait (Acuña-Galindo et al. 2015). Twenty MQTLs specifically related to drought were found starting from 502 single QTLs, whereas 43 MQTL regions were coincident for both drought and heat stress. More recently, genomic regions associated with grain yield-related traits in durum wheat under different water regimes have been obtained with meta-analysis, with 76 MQTLs associated with yield components under both irrigated and rainfed conditions (Arriagada et al. 2022). The recent availability of the sequence of complex genomes, as in the case of durum wheat, opened a new perspective in MQTL analysis through the anchoring of QTLs directly onto the genome, facilitating the search for relevant candidate genes (Salvi and Tuberosa 2015).

Linkage analysis in biparental mapping populations is traditionally used to map QTLs, even if it suffers from limited resolution, particularly in regions of the genome with low recombination frequencies. In contrast to classical linkage mapping approach, genome-wide association mapping studies (GWAS) involve panels of genetically diverse lines that are genotyped using many markers to identify alleles that are statistically associated with a trait (MTAs, marker trait associations). Few whole genome association studies are available in tetraploid wheats for the adaptive response to water stress (Merchuk-Ovnat et al. 2016; Maccaferri et al. 2011; Sukumaran et al. 2018a; Wang et al. 2019; Negisto et al. 2022; Condorelli et al. 2022). A panel of 208 durum wheat accessions representing a subset of CIMMYT gene bank accessions elite and exotic materials was phenotypically evaluated under various conditions, including drought stress, for two years and subjected to GWAS using DArTseq (Sukumaran et al. 2018a). More than 150 significant MTAs were identified for eight different traits (grain yield, thousand grain weight, grain number/m^2, days to anthesis, days to maturity, plant height, normalized difference vegetative index at vegetative, and grain filling stages), with QTL hotspots on chromosomes 2A and 2B. Similarly, co-locating MTAs associated with two or more traits evaluated under drought conditions were found in a panel of Ethiopian durum wheat landraces and on 493 USDA NSGC accessions by Negisto et al. (2022) and Wang et al. (2019), respectively. Moreover, these last authors carried out a haplotype analysis and found two minor haplotypes (hap1 and hap2) on chromosome 4B associated with both drought tolerance and short stature of the plant, due to the proximity to the *Rht-B1* semi-dwarfing gene. Haplotype maps may revolutionize GWAS for drought tolerance thanks to the rapid and accurate resequencing of many genomes which is useful for imputing genotypes and transferring sequence-level variation data across mapping projects. Alahmad et al. (2019) studied the root system as adaptive trait to drought through GWAS using DArT-seq markers on a subset of 393 durum RILs from a nested association mapping (NAM) population preselected for drought adaptation. Seven highly significant markers were identified on chromosome 6A associated with the major QTL for root architecture (*qSRA-6A*), the same region in which QTLs for yield components and quality have been mapped in durum wheat, suggesting a control on both root development and other agronomically important traits.

The occurrence of the combination of abiotic and biotic stress often affects the production and quality of durum wheat. An example is represented by crown rot disease (CR) caused by *Fusarium* species, especially when disease infection coincides with terminal drought. Alahmad et al. (2020) highlighted how the implementation of physiological traits for drought adaptation and optimal root system architecture to reduce water stress could minimize losses due to pathogen infection. Through weekly measurements of the vegetative index, the maintenance of the green (called stay-green) of each genotype was determined. The GWAS resulted in a QTL on chromosome 6B (qCR-6B) associated with CR tolerance and stay-green. Moreover, the authors found that in the presence of CR, the combination of favorable alleles for both *qCR-6B* and *qSRA-6A*, the major QTL for root architecture described above (Alahmad et al. 2019), resulted in a yield advantage.

Now it is possible to genotype hundreds of lines in a very short time and with reduced costs with respect to the past, but accurate phenotypic evaluation has become a bottleneck in genetic studies in durum wheat. New opportunities to effectively dissect the genetic basis of drought adaptive traits are provided by high-throughput phenotyping platforms (HTPPs). Condorelli et al. (2018) compared QTL results for normalized difference vegetation index (NDVI), an indicator of vegetation response to drought based on the relationships between NDVI and a meteorologically based drought index, measured with both aerial- and ground-based phenotyping platforms under both well-watered and water-deficit conditions of increasing severity. The GWAS was carried out on a panel of 248 elite durum wheat cultivars, and a larger number of individual QTLs (22) were identified with the aerial platforms with respect to the ground-based platforms (16 QTLs), in particular under terminal drought stress. A major NDVI QTL has been found on chromosome 1B, co-mapping with a QTL for total chlorophyll content, being a good start point for further characterization and positional cloning.

In conclusion, an efficient combination of genomic technologies and high-throughput phenotypic strategies could help to better understand the basis of drought tolerance and to improve durum wheat cultivars in an increasing climate changing scenario.

4 Genomic Applications to Durum Wheat Breeding

4.1 Drought Tolerance Improvement from Phenotypic Evaluation to Marker-Based Selection

Breeding programs must respond to growing challenges in the frame of the actual climate changes and increased food demands. Due to the difficulty of evaluating drought stress tolerance in a large set of genotypes, great research has been carried out to identify proxy traits related to tolerance. At seedling stage, leaf wilting, days to wilting, and stay green following water stress are indicators of tolerance to prolonged water deficit, while days to regrowth, regrowth biomass, and leaf recovery after drought address the ability to recover after water deficit and re-watering. Therefore, these traits might be used by breeders in the selection of drought tolerant varieties (Sallam et al. 2019). Nevertheless, reaction to drought stress at seedling stage can be independent from that occurring from flowering up to grain filling, which affects yield-related traits as the number of seeds per spike and kernel weight. Therefore, specific evaluation activities at plant adult stage are needed for traits connected to the ability to reduce evaporation loss and maintain photoassimilates production, as flag leaf persistence, leaf rolling, canopy temperature, and stomatal conductance, which show positive correlation with final grain yield.

High-throughput phenotyping with automated devices in greenhouses or in open field is now used for the rapid screening of thousands of genotypes for many traits (leaf temperature, plant water status, water use efficiency estimation, root traits, etc.), and this can accelerate plant breeding process because it generates a large amount of data simultaneously in a detailed and non-invasive manner. Plant breeding programs take many years for a new variety to be released. The process is even more difficult for drought tolerance being a complex trait, characterized by polygenic nature and low hereditability. For this reason, high-throughput phenotyping represents a huge opportunity, together with high-throughput genomics for mapping major and minor QTLs for drought tolerance that can be introgressed in durum wheat varieties. There are also additional methods to accelerate breeding process. In self-pollinating crop species as durum wheat, the introduction of advanced in-vitro tissue culture techniques to obtain doubled haploids can help breeders to accelerate trait fixation. Unfortunately, this approach has limited application in durum wheat which hardly responds to double haploid production in-vitro because the efficiency of the method is genotype dependent with low embryo production and high frequency of albino mutants (Cistué et al. 2009).

An alternative method to accelerate the development of breeding populations consists in increasing the number of generations per year. Speed breeding is a technique which significantly reduces the length of breeding cycles and accelerates genetic gain (Watson et al. 2018). The plants are grown in a controlled environment with extended photoperiod (22 hours light/2 hours dark), and this allows to reduce the vegetative cycle to achieve 6/8 generations/year for durum wheat (Alahmad et al. 2018). Speed breeding coupled with genomics-based technologies and other advances in phenomics could yield significant progress in the genetic gain in breeding schemes.

The first requirement for genomic assisted breeding is availability of molecular markers tightly linked with the loci controlling the trait of interest. The high-throughput genotyping of durum wheat varieties is now applied routinely, especially after the construction of specific genotyping arrays (from 9 to 820 K) (Cavanagh et al. 2013; Wang et al. 2014; Winfield et al. 2016;

Allen et al. 2017) and the sequencing of the durum wheat genome (Maccaferri et al. 2019). Molecular markers linked to the targeted trait/QTLs can be used for durum wheat improvement using genomic assisted breeding. There are several marker-assisted breeding strategies to transfer genes/alleles of interest, based on the targeted transfer of a particular QTL, such as marker-assisted selection (MAS) or backcrossing, or on the use of genome-wide markers platforms to predict the breeding value of a set of lines, genomic selection (GS). Drought stress is controlled by many QTLs/genes, usually with a small effect, which are difficult to use simultaneously in breeding. Recent activities have been carried out in bread wheat for the introgression of the genomic regions linked with physiological and yield-related traits for drought tolerance through marker-assisted backcross breeding (Gupta et al. 2017; Todkar et al. 2020), but very few studies are focused on durum wheat. Interestingly, some examples are available of genomic regions introgressed to elite durum wheat cultivars from wild emmer wheat through MAS. Merchuk-Ovnat et al. (2016) introgressed QTLs for plant productivity, particularly under drought, from wild emmer wheat to cultivars of durum and bread wheat. The near isogenic lines (NILs) developed from the introgression of two QTLs (on chromosomes 1BL and 2BS) into elite Israeli durum wheat cultivars (Inbar and Uzan) were evaluated in 2 years under well-watered and water-limited conditions and showed an increase in terms of grain yield and biomass under both treatments and a greater stability across treatments. Aberkane et al. (2020) assessed the contribution of wheat wild relatives to drought and heat tolerance through evaluation of 67 durum wheat lines derived from interspecific crosses between two durum cultivars (Cham 5 and Haurani) and *T. turgidum* ssp. *dicoccoides*, *T. monococcum*, *T. urartu* and *Aegilops speltoides*. Under drought two lines, derived from crosses Haurani*2/*T. urartu* and Cham5*2/*T. dicoccoides*, yielded 196 and 142% of their recurrent parents' yield, respectively. In the frame of a program aimed at developing a series of wild emmer (acc. Zavitan) introgressions into elite durum wheat (cv. Svevo), Bacher et al. (2021) identified an introgression line that exhibited a highly plastic response to water stress by shifting its root-to-shoot biomass ratio. Several candidate genes (including a root-specific kinase) were proposed, by combining genotypic information with root transcriptome analysis, that can contribute the shoot-to-root carbon resource allocation under water stress.

Genomic advances also can accelerate cloning of genes with a strong impact on agronomic traits. The *TaMATE1B* gene for aluminum tolerance, cloned in bread wheat, and responsible for constitutive citrate efflux from root tips (Delhaize et al. 2012; Tovkach et al. 2013) has been introgressed in durum wheat variety Jandaroi. Two durum near isogenic lines, one of which with the *TaMATE1B* gene (Jandaroi–*TaMATE1B*), have been subjected to terminal drought treatment following growth re-constructed field soil with Al^{3+}-rich acid subsoil. In these conditions, Jandaroi–*TaMATE1B* produced 25.3% higher grain yield than Jandaroi–null under well-watered conditions and 49.0% higher grain yield under terminal drought (Liu et al. 2021b).

Limited results in using MAS for durum wheat breeding for drought tolerance have been determined by the limited number of major QTLs available, the instability of QTLs across multiple environments and the QTL x environment interactions (Bernardo 2016; Sall et al. 2024). GS is a method that has promised to overcome the limitations of MAS for quantitative traits (Heffner et al. 2009), with the objective to determine the genetic potential of an individual instead of identifying the specific QTL. GS predicts the genetic value on lines under selection based on the estimated genomic breeding value (GEBV). High-density molecular markers are used, with high genome coverage, so that minor and medium effect QTLs are not left uncaptured unlike MAS, which focuses on the few markers linked to major genes (Newell and Jannink 2014). To implement GS, a training population and a validation population are first established with the availability of genotypic and phenotypic data for the target trait(s). Marker effects are first estimated from the genotypic and phenotypic data in the training population. Then marker effects are used to predict breeding values in the target population with only genotypic data, and selections are based on these GEBV (Meuwissen et al. 2001; Charmet et al. 2020). Some studies have been conducted on GS in durum wheat in drought conditions. Sukumaran et al. (2018b) tested the genomic-enabled prediction accuracy of genomic best linear unbiased predictor (GBLUP) models in a durum wheat panel grown under yield potential, drought stress, and heat stress conditions, obtaining better results with GS models incorporating G × E interaction.

Interestingly, GS can be coupled with classical QTL mapping or GWAS to improve the accuracy of prediction. A panel of 179 durum wheat lines was assessed under rainfed conditions in different environments in Southern Spain, and 37 marker-trait associations for three quality traits (specific weight, gluten, and sedimentation indexes) were identified (Mérida-García et al. 2019). The GS prediction ability values evaluated in the same panel showed promising results for the target, which can be used in plant breeding programs (Mérida-García et al. 2019). Zaïm et al. (2020) used four RILs populations for QTL discovery for grain yield and thousand kernel weight under drought conditions. The same populations were used to train genomic prediction models incorporating the relationship matrix, the genotype by environment interaction, and marker by environment interaction. The QTLs identified were also incorporated in the models as fixed effects, showing significant accuracy gain for all four populations. In general, the application of GS was successful for most of the traits in the breeding materials analyzed and showed promising results supporting its use in current plant breeding for key quality traits in durum wheat under Mediterranean rainfed conditions.

In order to improve water stress tolerance in durum wheat, an integrated breeding approach could be proposed in which MAS is used in early generations to fix QTLs with a stronger effect, followed by GS for drought tolerant traits sustained by many loci with a small effect (Heffner et al. 2010).

4.2 Biotech Approaches for Drought Resistance in Durum Wheat

Genetic modifications and eco-friendly new breeding technologies (NBT, namely cis/intragenesis, genome editing) have become very effective tools in the post-genomic era. Identification of genes that are induced or repressed under stress conditions is crucial for creating drought-tolerant plants through biotechnological approaches (Sahi et al. 2006). Structural and regulatory genes including dehydration-responsive element-binding (*DREB*) factors and NAC transcription factors were already successfully applied for this purpose in other crops (Gosala et al. 2009). Little efforts were performed to improve drought tolerance in durum wheat by using these strategies.

Until now, positive results have only been obtained in bread wheat (*Triticum aestivum* L.). Transgenic lines of wheat expressing a mutated version of the HaHB4 transcription factor from *Helianthus annuus* (Hahb-4) showed improved yield and increased drought tolerance when evaluated in multiple field trials with respect to the wild type (González et al. 2019, 2020). Recently, the HB4-genetically modified wheat developed by Bioceres, S.A. has been approved for food and feed direct use or processing in many countries, such as Brazil, Australia, Columbia, New Zealand, Nigeria, USA, and Argentina, also for cultivation (ISAAA 2023).

Furthermore, a recent study was conducted in wheat that investigates the possibility of using genome editing approaches to address drought tolerance (Abdallah et al. 2022).

Differently from bread wheat (Sivamani et al. 2000; Bahieldin et al. 2005; Saad et al. 2013; Mwadzingeni et al. 2016; Shavrukov et al. 2016; Gonzales et al. 2019, 2020; Joshi et al. 2019; Khan et al. 2019; Beznec et al. 2021), the main application of genetic transformation involving durum wheat genes was aimed at studying their functional role in response to drought, via their overexpression by *Agrobacterium*-mediated transformation. These studies are most often carried out in model species, *A. thaliana* or tobacco, with the advantages of rapid reproduction time and ease of genetic transformation. Here are some examples.

Transgenic *Arabidopsis* plants overexpressing the durum wheat gene *TtNAC2A,* a transcription activator of plant development, showed enhanced drought tolerance and hypersensitivity to ABA, together with greater water retention capacity, attenuated oxidative stress, lower electrolyte leakage, and higher proline contents, compared to wild-type plants. Moreover, many stress-responsive genes were induced or upregulated (Mergby et al. 2021).

Plant annexins are a family of Ca^{2+}-dependent phospholipid-binding proteins implicated in stress signaling pathways and in several cellular and plant developmental processes in response to numerous abiotic stresses and phytohormones (Konopka-Postupolska et al. 2009; Laohavisit et al. 2010; Harbaoui et al. 2018; Liu et al. 2021c). *Arabidopsis* plants expressing the durum wheat gene *TdAnn6* showed an improvement in salt/osmotic stress tolerance at germination and seedling stages, with an increase in antioxidant activities and in transcript levels of some stress-related genes (Harbaoui et al. 2021). Under greenhouse conditions, the transgenic plants exhibited increased tolerance to salt or drought stress. The positive role of annexin protein is also confirmed in a study on the functional role of the *TdAnn12* isolated from durum wheat and highly induced by different abiotic stresses (Ben Saad et al. 2019). Transgenic tobacco plants overexpressing *TdAnn12* showed lower amount of ROS together with increased antioxidant activities, higher chlorophyll content, lower lipid peroxidation levels, and upregulation of some stress-related genes, leading to a greater tolerance to drought and salt stresses.

Likewise, *TdDRF1*, a DREB-related gene involved in drought tolerance, was isolated from durum wheat (Latini et al. 2007) and introduced in bread wheat inducing improved survival but slower growth, delayed flowering, and lower grain yield in drought conditions (Morran et al. 2011). Durum wheat group 2 Late Embryogenesis Abundant (LEA) proteins, also known as dehydrins (DHNs), play a protective role under salt and drought stresses. Overexpression of *Dhn-5* in Arabidopsis contributed in improving tolerance to these stresses through osmotic adjustment due to proline accumulation (Brini et al. 2007).

Numerous studies have demonstrated the contribution of plant aquaporins (AQPs) to the water permeability of the root cortex cells and their involvement in osmotic transport of water in the root system and from the roots to the leaves, during transpiration (Maurel et al. 2008). The durum wheat genes *PIP1;1* and *PIP2;1*, encoding plasma membrane intrinsic AQPs, were overexpressed in *Nicotiana tabacum* confirming their function in regulating root and leaf growth under drought (Ayadi et al. 2011). In particular, the *TdPIP2;1* gene was successfully introduced in durum wheat under the control of native core *PrTdPIP2;1* or of constitutive PrCaMV35S promoters. Transgenic lines overexpressing *TdPIP2;1* exhibited improved germination rates and biomass production and retained low Na^{+} and high K^{+} concentrations in their shoots both in high salt and osmotic stress conditions (Ayadi et al. 2019). In a long-term study under greenhouse conditions, these lines showed an induction of the activities of antioxidant enzymes, as catalase and superoxide dismutase, and larger grains under drought stress with respect to wild-type plants. These results confirmed the high potential of the *TdPIP2;1* gene in reducing water evaporation from leaves in response to water deficit, and the role of the *PrTdPIP2;1* promoter as an abiotic and a tissue stress inducible promoter.

Besides *TdPIP2;1*, the *AlSAP* (stress-associated protein) gene from the halophyte grass *Aeluropus littoralis,* whose role in improving tolerance to continuous salt and drought stresses had already been demonstrated in transgenic tobacco (Ben-Saad et al. 2010), was successfully transferred in durum wheat. T3 homozygous lines of marker-free transgenic durum wheat plants expressing the *AlSAP* gene exhibited improved germination rates and biomass production under severe salinity and osmotic stress conditions, without apparent phenotype penalty

(Ben-Saad et al. 2012). In a long-term salt or drought stress greenhouse trial, AlSAP durum wheat lines produced normally filled grains and showed a lower water loss rate and a higher Na^+ accumulation in senescent-basal leaves, whereas wild-type plants either died at the vegetative stage under salt stress or showed markedly reduced grain filling under drought stress (Ben-Saad et al. 2012).

Transgenesis was applied also for optimizing protocols or detection of transgenes. Barley *HVA1* gene, a member of LEA3 closely related to water stress, was transferred in bread and durum wheat to optimize transformation protocols (Patnaik and Khurana 2003) and to develop an efficient method for simultaneous detection of *HVA1* transgene and of its transcript in transformed durum wheat, by conventional polymerase chain reaction (PCR) and quantitative real time PCR, respectively (Melloul et al. 2014). This is useful when a high similarity between the endogenous and transgene exists, as for barley *HVA1* and wheat gene.

All these studies provide useful information to identify genes able to provide attractive candidates for drought stress breeding of durum wheat. Given the scarce literature available on this subject, it becomes a priority to exploit all the data deriving from the sequencing of the durum wheat genome that can give a significant boost to the application of innovative biotechnological approaches.

5 Conclusion and Perspectives

In the present scenario of ongoing climate changes, more prolonged and severe drought events are expected. The damage on durum wheat grain yield and quality is further exacerbated by the fact that the geographical areas interested by this phenomenon are becoming larger, and that a concomitant increase in temperature, especially in the final growth stages of the plant, strongly enhances the effects of drought stress. Recent advances achieved in the knowledge and comprehension of plant genomics are very promising for the understanding of the genetic bases of drought tolerance and for using this knowledge in the practice of breeding programs to develop more resistant cultivars. Nevertheless, this research has been much more largely employed for bread wheat compared to durum wheat, despite this species being mainly grown in semi-arid environments. For many years, the durum scientific community has benefited from the genomic research carried out in bread wheat. As an example, the publication of the Chinese Spring genome (IWGSC 2018; Zhu et al. 2021) has been of great importance for genomic studies in durum wheat. At the same time, genomic platforms as the 90K wheat SNP array, carrying mainly bread wheat derived SNP markers, have been largely used in durum wheat mapping studies thanks to the great level of genetic similarity and marker collinearity between the A and B genomes of bread and durum wheat (Maccaferri et al. 2019). Nevertheless, in most recent times a new trend has been established towards the development of genomic knowledge and tools directly in durum wheat, as demonstrated by the availability of the genomes of wild emmer wheat (Avni et al. 2017) and of the durum wheat cv. Svevo (Maccaferri et al. 2019). This new path will certainly provide in the next years good results in genetic improvement in durum wheat resistance to drought and other environmental stresses and filling the scientific gap between this species and other crops.

Acknowledgments This work is supported by the CERERE project. CERERE (CEreals REsiliency REvolution for agile supply chain management in the Mediterranean) is funded by the PRIMA Programme 2023 – Section 1 – Food Value-chain 2023 - Topic 1.3.1 (RIA) - Increasing resilience of agri-food supply chain (cereal) in the MENA region. The views expressed belong to the authors alone and do not necessarily reflect the views of the European Union or the PRIMA Foundation.

References

Abdallah NA, Elsharawy H, Abulela HA et al (2022) Multiplex CRISPR/Cas9-mediatedgenome editing to address drought tolerance in wheat. GM Crops Food 6:1–17. https://doi.org/10.1080/21645698.2022.2120313

Aberkane H, Amri A, Belkadi B et al (2020) Evaluation of durum wheat lines derived from interspecific crosses under drought and heat stress. Crop Sci 61:119–136. https://doi.org/10.1002/csc2.20319

Abu-Zaitoun SY, Chandrasekhar K, Assili S et al (2018) Unlocking the genetic diversity within a Middle-East panel of durum wheat landraces for adaptation to semi-arid climate. Agronomy 8:233. https://doi.org/10.3390/agronomy8100233

Acuña-Galindo MA, Esten Mason R, Subramanian NK, Hays DB (2015) Meta-analysis of wheat QTL regions associated with adaptation to drought and heat stress. Crop Sci 55:477–492. https://doi.org/10.2135/cropsci2013.11.0793

Agati G, Brunetti C, Fini A, Gori A, Guidi L, Landi M, Sebastiani F, Tattini M (2020) Are flavonoids effective antioxidants in plants? Twenty years of our investigation. Antioxidants 9(11):1098. https://doi.org/10.3390/antiox9111098

Akpinar BA, Kantar M, Budak H (2015) Root precursors of microRNAs in wild emmer and modern wheats show major differences in response to drought stress. Funct Integr Genomics 15:587–598. https://doi.org/10.1007/s10142-015-0453-0

Alahmad S, Dinglasan E, Leung KM et al (2018) Speed breeding for multiple quantitative traits in durum wheat. Plant Methods 14:36. https://doi.org/10.1186/s13007-018-0302-y

Alahmad S, El Hassouni K, Bassi FM et al (2019) A major root architecture QTL responding to water limitation in durum wheat. Front Plant Sci 10:436. https://doi.org/10.3389/fpls.2019.00436

Alahmad S, Kang Y, Dinglasan E et al (2020) Adaptive traits to improve durum wheat yield in drought and crown rot environments. Int J Mol Sci 21:5260. https://doi.org/10.3390/ijms21155260

Alahmad S, Kang Y, Dinglasan E et al (2023) A multi-reference parent nested-association mapping population to dissect the genetics of quantitative traits in durum wheat. Genet Resour Crop Evol 70:1471–1485

Allen AM, Winfield MO, Burridge AJ et al (2017) Characterization of a wheat breeders' array suitable for high-throughput SNP genotyping of global accessions of hexaploid bread wheat (*Triticum aestivum*). Plant Biotechnol J 15:390–401. https://doi.org/10.1111/pbi.12635

Aprile A, Mastrangelo AM, De Leonardis AM et al (2009) Transcriptional profiling in response to terminal drought stress reveals differential responses along the wheat genome. BMC Genomics 10:279. https://doi.org/10.1186/1471-2164-10-279

Aprile A, Havlickova L, Panna R et al (2013) Different stress responsive strategies to drought and heat in two durum wheat cultivars with contrasting water use efficiency. BMC Genomics 14:821. https://doi.org/10.1186/1471-2164-14-821

Arriagada O, Gadaleta A, Marcotuli I, Maccaferri M, Campana M, Reveco S, Alfaro C, Matus I, Schwember AR (2022) A comprehensive meta-QTL analysis for yield-related traits of durum wheat (*Triticum turgidum* L. var. *durum*) grown under different water regimes. Front. Plant Sci 13:984269

Asmamaw M, Keneni G, Tesfaye K (2019) Genetic diversity of Ethiopian durum wheat (*Triticum durum* Desf.) landrace collections as reveled by SSR markers. Adv Crop Sci Tech 7:413. https://doi.org/10.4172/2329-8863.1000413

Avni R, Nave M, Barad O (2017) Wild emmer genome architecture and diversity elucidate wheat evolution and domestication. Science 357:93–97. https://doi.org/10.1126/science.aan0032

Ayadi M, Cavez D, Miled N et al (2011) Identification and characterization of two plasma membrane aquaporins in durum wheat (*Triticum turgidum* L. subsp. *durum*) and their role in abiotic stress tolerance. Plant Physiol Biochem 49:1029–1039. https://doi.org/10.1016/j.plaphy.2011.06.002

Ayadi M, Brini F, Masmoudi K (2019) Overexpression of a wheat aquaporin gene, *TdPIP2;1*, enhances salt and drought tolerance in transgenic durum wheat cv. Maali Int J Mol Sci 20:2389. https://doi.org/10.3390/ijms20102389

Bacher H, Zhu F, Gao T et al (2021) Wild emmer introgression alters root-to-shoot growth dynamics in durum wheat in response to water stress. Plant Physiol 187(3):1149–1162. https://doi.org/10.1093/plphys/kiab292

Bahieldin A, Hesham HH, Eissa HH et al (2005) Field evaluation of transgenic wheat plants stably expressing the *HVA1* gene for drought tolerance. Physiol Plant 123:421–427. https://doi.org/10.1111/j.1399-3054.2005.00470.x

Ben Saad R, Zouari N, Ben Ramdhan W, Azaza J, Meynard D, Guiderdoni E, Hassairi A (2010) Improved drought and salt stress tolerance in transgenic tobacco overexpressing a novel A20/AN1 zinc-finger "*AlSAP*" gene isolated from the halophyte grass *Aeluropus littoralis*. Plant Mol Biol 72:171–190. https://doi.org/10.1007/s11103-009-9560-4

Ben Saad R, Harbaoui M, Ben Romdhane W et al (2019) Overexpression of *Triticum durum TdAnn12* gene confers stress tolerance through scavenging reactive oxygen species in transgenic tobacco. Funct Plant Biol 46:885–895. https://doi.org/10.1071/FP18316

Ben-Saad R, Ben-Ramdhan W, Zouari N et al (2012) Marker-free transgenic durum wheat cv. Karim expressing the *AlSAP* gene exhibits a high level of tolerance to salinity and dehydration stresses. Mol Breed 30:521–533. https://doi.org/10.1007/s11032-011-9641-3

Bernardo R (2008) Molecular markers and selection for complex traits in plants: learning from the last 20 years. Crop Sci 48:1649–1664. https://doi.org/10.2135/cropsci2008.03.0131

Bernardo R (2016) Bandwagons I, too, have known. Theor Appl Genet 129:2323–2332. https://doi.org/10.1007/s00122-016-2772-5

Bernardo L, Carletti P, Badeck FW et al (2019) Metabolomic responses triggered by arbuscular mycorrhiza enhance tolerance to water stress in wheat cultivars. Plant Physiol Biochem 137:203–212. https://doi.org/10.1016/j.plaphy.2019.02.007

Beznec A, Faccio P, Miralles DJ et al (2021) Stress-induced expression of *IPT* gene in transgenic wheat reduces grain yield penalty under drought. J Genet Eng Biotechnol 19:67. https://doi.org/10.1186/s43141-021-00171-w

Bhandari HR, Nishant Bhanu A, Srivastava K et al (2017) Assessment of genetic diversity in crop plants - an overview. Adv Plants Agricult Res 7:279–286. https://doi.org/10.15406/apar.2017.07.00255

Blum A (2005) Drought resistance, water-use efficiency, and yield potential—are they compatible, dissonant, or mutually exclusive? Aust J Agric Res 56(11):159–1168

Blum A (2010) Plant breeding for water-limited environments. Springer Science & Business Media

Blum A, Tuberosa R (2018) Dehydration survival of crop plants and its measurement. J Exp Bot 69(5):975–981

Bowne J, Bacic A, Tester M, Roessner U (2011) Abiotic stress and metabolomics. Ann Plant Rev 43:61–85. https://doi.org/10.1002/9781444339956.ch3

Bowne JB, Erwin TA, Juttner J et al (2012) Drought responses of leaf tissues from wheat cultivars of differing drought tolerance at the metabolite level. Mol Plant 5:418–429. https://doi.org/10.1093/mp/ssr114

Brini F, Hanin M, Lumbreras V et al (2007) Overexpression of wheat dehydrin *DHN-5* enhances tolerance to salt and osmotic stress in *Arabidopsis thaliana*. Plant Cell Rep 26:2017–2026. https://doi.org/10.1007/s00299-007-0412-x

Budak H, Akpinar BA, Unver T, Turktas M (2013) Proteome changes in wild and modern wheat leaves upon drought stress by two-dimensional electrophoresis and nanoLC-ESI–MS/MS. Plant Mol Biol 83:89–103. https://doi.org/10.1007/s11103-013-0024-5

Buffagni V, Vurro F, Janni M et al (2020) Shaping durum wheat for the future: gene expression analyses and metabolites profiling support the contribution of *BCAT* genes to drought stress response. Front Plant Sci 11:891. https://doi.org/10.3389/fpls.2020.00891

Caruso G, Cavaliere C, Foglia P et al (2009) Analysis of drought responsive proteins in wheat (*Triticum durum*) by 2D-PAGE and MALDI-TOF mass spectrometry. Plant Sci 177:570–576. https://doi.org/10.1016/j.plantsci.2009.08.007

Cavalcanti JHF, Esteves-Ferreira AA, Quinhones CGS et al (2014) Evolution and functional implications of the tricarboxylic acid cycle as revealed by phylogenetic analysis. Genome Biol Evol 6:2830–2848. https://doi.org/10.1093/gbe/evu221

Cavanagh CR, Chao S, Wang S et al (2013) Genome-wide comparative diversity uncovers multiple targets of selection for improvement in hexaploid wheat landraces and cultivars. Proc Natl Acad Sci USA 110:8057–8062. https://doi.org/10.1073/pnas.1217133110

Chaouachi L, Marín-Sanz M, Barro F, Karmous C (2024) Genetic diversity of durum wheat (*Triticum turgidum* ssp. *durum*) to mitigate abiotic stress: drought, heat, and their combination. PLoS One 19(4):e0301018. https://doi.org/10.1371/journal.pone.0301018

Charmet G, Tran L-G, Auzanneau J et al (2020) BWGS: a R package for genomic selection and its application to a wheat breeding programme. PLoS One 15:e0222733. https://doi.org/10.1371/journal.pone.0222733

Cistué L, Romagosa I, Batlle F, Echávarri B (2009) Improvements in the production of doubled haploids in durum wheat (*Triticum turgidum* L.) through isolated microspore culture. Plant Cell Rep 28:727–735. https://doi.org/10.1007/s00299-009-0690-6

Colasuonno P, Marcotuli I, Gadaleta A, Soriano JM (2021) From genetic maps to QTL cloning: an overview for durum wheat. Plants 10:315. https://doi.org/10.3390/plants10020315

Collins NC, Tardieu F, Tuberosa R (2008) Quantitative trait loci and crop performance under abiotic stress: where do we stand? Plant Physiol 147(2):469–486. http://www.plantphysiol.org/cgi/doi/10.1104/pp.108.118117

Condorelli GE, Maccaferri M, Newcomb M et al (2018) Comparative aerial and ground based high throughput phenotyping for the genetic dissection of NDVI as a proxy for drought adaptive traits in durum wheat. Front Plant Sci 9:893. https://doi.org/10.3389/fpls.2018.00893

Condorelli GE, Newcomb M, Groli EL et al (2022) Genome Wide Association study uncovers the QTLome for osmotic adjustment and related drought adaptive traits in Durum wheat. Genes 13:293. https://doi.org/10.3390/genes13020293

De Leonardis AM, Marone D, Mazzucotelli E et al (2007) Durum wheat genes up-regulated in the early phases of cold stress are modulated by drought in a developmental and genotype dependent manner. Plant Sci 172:1005–1016. https://doi.org/10.1016/j.plantsci.2007.02.002

De Leonardis AM, Petrarulo M, De Vita P, Mastrangelo AM (2012) Genetic and molecular aspects of plant response to drought in annual crop species, advances in selected plant physiology aspects, Dr. Giuseppe Montanaro (Ed.), ISBN: 978-953-51-0557-2, InTech, Available from: http://www.intechopen.com/books/advances-in-selected-plant-physiology-aspects/genetic-and-molecular-aspects-of-plant-response-to-drought-stress

De Santis MA, Soccio M, Laus MN, Flagella Z (2021) Influence of drought and salt stress on Durum wheat grain quality and composition: a review. Plants 10:2599. https://doi.org/10.3390/plants10122599

Delhaize E, Ma JF, Ryan PR (2012) Transcriptional regulation of aluminium tolerance genes. Trends Plant Sci 17:341–348. https://doi.org/10.3389/fpls.2017.01358

Dodig D, Zorić M, Kandić V et al (2012) Comparison of responses to drought stress of 100 wheat accessions and landraces to identify opportunities for improving wheat drought resistance. Plant Breed 131:369–379. https://doi.org/10.1111/j.1439-0523.2011.01941.x

Fàbregas N, Fernie AR (2019) The metabolic response to drought. J Exp Bot 70:1077–1085. https://doi.org/10.1093/jxb/ery437

Faraji S, Rasouli SH, Kazemitabar SK (2018) Genome-wide exploration of C2H2 zinc finger family in durum wheat (*Triticum turgidum* ssp. *durum*): insights into the roles in biological processes especially stress response. Biometals 31:1019–1042. https://doi.org/10.1007/s10534-018-0146-y

Fatiukha A, Deblieck M, Klymiuk V et al (2021) Genomic architecture of phenotypic plasticity in response to water stress in tetraploid wheat. Int J Mol Sci 22:1723. https://doi.org/10.3390/ijms22041723

Giuliani MM, Palermo C, De Santis MA et al (2015) Differential expression of durum wheat gluten proteome under water stress during grain filling. J Agric Food Chem 63:29. https://doi.org/10.1021/acs.jafc.5b01635

Giusti L, Mica E, Bertolini E et al (2017) microRNAs differentially modulated in response to heat and drought stress in durum wheat cultivars with contrasting water use efficiency. Funct Integr Genomics 17:293–309. https://doi.org/10.1007/s10142-016-0527-7

González FG, Capella M, Ribichich KF et al (2019) Field-grown transgenic wheat expressing the sunflower gene HaHB4 significantly outyields the wild type. J Exp Bot 70:1669–1681. https://doi.org/10.1093/jxb/erz037

González FG, Rigalli N, Miranda PV et al (2020) An interdisciplinary approach to study the performance of second-generation genetically modified crops in field trials: a case study with soybean and wheat carrying the sunflower HaHB4 transcription factor. Front Plant Sci 11:178. https://doi.org/10.3389/fpls.2020.00178

Gosala SS, Wania HS, Kanga MS (2009) Biotechnology and drought tolerance. J Crop Improv 23:19–54. https://doi.org/10.1080/15427520802418251

Graziani M, Maccaferri M, Royo C, Salvatorelli F (2014) QTL dissection of yield components and morpho-physiological traits in a durum wheat elite population tested in contrasting thermo-pluviometric conditions. Crop Pasture Sci 65:80–95. https://doi.org/10.1071/CP13349

Groli EL, Frascaroli E, Maccaferri M, Ammar K, Tuberosa R (2024) Dissecting the effect of heat stress on durum wheat under field conditions. Front Plant Sci 15:1393349. https://doi.org/10.3389/fpls.2024.1393349

Gupta PK, Balyan HS, Gahlaut V (2017) QTL analysis for drought tolerance in wheat: present status and future possibilities. Agronomy 7:5. https://doi.org/10.3390/agronomy7010005

Guzmán C, Autrique JE, Mondal S et al (2016) Response to drought and heat stress on wheat quality, with special emphasis on bread-making quality, in durum wheat. Field Crop Res 186:157–165. https://doi.org/10.1016/j.fcr.2015.12.002

Habash DZ, Baudo M, Hindle M et al (2014) Systems responses to progressive water stress in durum wheat. PLoS One 9:e108431. https://doi.org/10.1371/journal.pone.0108431

Harbaoui M, Ben Saad R, Ben Halima N et al (2018) Structural and functional characterization of two novel durum wheat annexin genes in response to abiotic stress. Funct Plant Biol 45(5):542–552. https://doi.org/10.1071/FP17212

Harbaoui M, Ben Romdhane W, Ben Hsouna A et al (2021) The durum wheat annexin, TdAnn6, improves salt and osmotic stress tolerance in Arabidopsis via modulation of antioxidant machinery. Protoplasma 258(5):1047–1059. https://doi.org/10.1007/s00709-021-01622-4

Heffner EL, Sorrells ME, Jannink JL (2009) Genomic selection for crop improvement. Crop Sci 49:1–12. https://doi.org/10.2135/cropsci2008.08.0512

Heffner EL, Lorenz AJ, Jannink JL, Sorrells ME (2010) Plant breeding with genomic selection: gain per unit time and cost. Crop Sci 50:1681–1690. https://doi.org/10.2135/cropsci2009.11.0662

Hoekstra FA, Golovina EA, Buitink J et al (2001) Mechanisms of plant desiccation tolerance. Trends Plant Sci 6:431–438. https://doi.org/10.1016/S1360-1385(01)02052-0

ISAAA. GM Approval Database. Available online: http://www.isaaa.org/gmapprovaldatabase/default.asp. Accessed on 2 February 2023

IWGSC (2018) Shifting the limits in wheat research and breeding using a fully annotated reference genome. Science 361:7191. https://doi.org/10.1126/science.aar7191

Joshi S, Choukimath A, Isenegger D et al (2019) Improved wheat growth and yield by delayed leaf senescence using developmentally regulated expression of a cytokinin biosynthesis gene. Front Plant Sci 10:1285. https://doi.org/10.3389/fpls.2019.01285

Kabbaj H, Sall AT, Al-Abdallat A et al (2017) Genetic diversity within a global panel of durum wheat (*Triticum durum*) landraces and modern germplasm reveals the history of alleles exchange. Front Plant Sci 8:1277. https://doi.org/10.3389/fpls.2017.01277

Kacem NS, Mauro S, Muhovski Y et al (2016) Diagonal two-dimensional electrophoresis (D-2DE): a new approach to study the effect of osmotic stress induced by polyethylene glycol in durum wheat (*Triticum durum* Desf.). Mol Biol Rep 43:897–909. https://doi.org/10.1007/s11033-016-4028-5

Kang Z, Babar MA, Khan N, Guo J, Khan J, Islam S, Shrestha S, Shahi D (2019) Comparative metabolomic profiling in the roots and leaves in contrasting genotypes reveals complex mechanisms involved in post-anthesis drought tolerance in wheat. PLoS One 14:e0213502. https://doi.org/10.1371/journal.pone.0213502

Khadka K, Raizada MN, Navabi A (2020) Recent progress in germplasm evaluation and gene mapping to enable breeding of drought-tolerant wheat. Front Plant Sci 11:1149. https://doi.org/10.3389/fpls.2020.01149

Khan S, Anwar S, Yu S et al (2019) Development of drought-tolerant transgenic wheat: achievements and limitations. Int J Mol Sci 20:3350. https://doi.org/10.3390/ijms20133350

Kim J-M, To TK, Matsui A et al (2017) Acetate-mediated novel survival strategy against drought in plants. Nat Plants 3:17097. https://doi.org/10.1038/nplants.2017.97

Konopka-Postupolska D, Clark G, Goch G et al (2009) The role of annexin 1 in drought stress in *Arabidopsis*. Plant Physiol 150:1394–1410. https://doi.org/10.1104/pp.109.135228

Lai DL, Yan J, Fan Y et al (2021) Genome-wide identification and phylogenetic relationships of the *Hsp70* gene family of *Aegilops tauschii,* wild emmer wheat (*Triticum dicoccoides*) and bread wheat (*Triticum aestivum*). 3 Biotech 11:301. https://doi.org/10.1007/s13205-021-02639-5

Laohavisit A, Brown AT, Cicuta P, Davies JM (2010) Annexins: components of the calcium and reactive oxygen signaling network. Plant Physiol 152:1824–1829. https://doi.org/10.1104/pp.109.145458

Latini A, Rasi C, Sperandei M et al (2007) Identification of a DREB-related gene in *Triticum durum* and its expression under water stress conditions. Ann Appl Biol 150:187–195. https://doi.org/10.1111/j.1744-7348.2007.00128.x

Levitt (1972) Responses of plants to environmental stresses. Academic Press, New York

Liu H, Searle IR, Watson-Haigh NS et al (2015) Genome-wide identification of microRNAs in leaves and the developing head of four durum genotypes during water deficit stress. PLoS One 10:e0142799. https://doi.org/10.1371/journal.pone.0142799

Liu H, Able AJ, Able JA (2017) Water-deficit stress-responsive microRNAs and their targets in four durum wheat genotypes. Funct Integr Genomics 17:237–251. https://doi.org/10.1007/s10142-016-0515-y

Liu H, Able AJ, Able JA (2020a) Transgenerational effects of water-deficit and heat stress on germination and seedling vigour—new insights from durum wheat microRNAs. Plants 9:189. https://doi.org/10.3390/plants9020189

Liu H, Able AJ, Able JA (2020b) Multi-omics analysis of small RNA, transcriptome, and degradome in *T. turgidum*—regulatory networks of grain development and abiotic stress response. Int J Mol Sci 21:7772. https://doi.org/10.3390/ijms21207772

Liu H, Able AJ, Able JA (2021a) Small RNAs and their targets are associated with the transgenerational effects of water-deficit stress in durum wheat. Sci Rep 11:3613. https://doi.org/10.1038/s41598-021-83074-7

Liu L, Bai C, Chen Y et al (2021b) Durum wheat with the introgressed *TaMATE1B* gene shows resistance to terminal drought by ensuring deep root growth in acidic and Al^{3+}-toxic subsoils. Plant Soil 478:311–324. https://doi.org/10.1007/s11104-021-04961-6

Liu T, Du L, Li Q et al (2021c) AtCRY2 negatively regulates the functions of AtANN2 and AtANN3 in drought tolerance by affecting their subcellular localization and transmembrane Ca^{2+} flow. Front Plant Sci 12:754567. https://doi.org/10.3389/fpls.2021.754567

Lucas SJ, Salantur A, Yazar S et al (2017) High-throughput SNP genotyping of modern and wild emmer wheat for yield and root morphology using a combined association and linkage analysis. Funct Integr Genomics 17:667–685

Maccaferri M, Sanguineti MC, Giuliani S, Tuberosa R (2009) Genomics of tolerance to abiotic stress in the Triticeae. In: Muehlbauer G, Feuillet C (eds) Genetics and genomics of the Triticeae. Plant genetics and genomics: crops and models, vol 7. Springer, New York. https://doi.org/10.1007/978-0-387-77489-3_18

Maccaferri M, Sanguineti MC, Demontis A et al (2011) Association mapping in durum wheat grown across a broad range of water regimes. J Exp Bot 62:409–438. https://doi.org/10.1093/jxb/erq287

Maccaferri M, Harris NS, Twardziok SO et al (2019) Durum wheat genome highlights past domestication signatures and future improvement targets. Nat Genet 51:885–895. https://doi.org/10.1038/s41588-019-0381-3

Maccaferri M, Bruschi M, Tuberosa R (2022) Sequence-based marker assisted selection in wheat. In: Wheat improvement: food security in a changing climate. Springer International Publishing, Cham, pp 513–538. https://doi.org/10.1007/978-3-030-90673-3_28

Marone D, Russo MA, Mores A et al (2021) Importance of landraces in cereal breeding for stress tolerance. Plants 10:1267. https://doi.org/10.3390/plants10071267

Mastrangelo AM, Cattivelli L (2021) What makes bread and durum wheat different? Trends Plant Sci 26:677–684. https://doi.org/10.1016/j.tplants.2021.01.004

Maurel C, Verdoucq L, Luu DT, Santoni V (2008) Plant aquaporins: membrane channels with multiple integrated functions. Ann Rev Plant Biol 59:595–624. https://doi.org/10.1146/annurev.arplant.59.032607.092734

Mazzucotelli E, Sciara G, Mastrangelo AM et al (2020) The global durum wheat panel (GDP): an international platform to identify and exchange beneficial alleles. Front Plant Sci 11:569905. https://doi.org/10.3389/fpls.2020.569905

Melloul M, Iraqi D, Udupa SM et al (2014) Development of specific primers for the detection of *HVA1* from barley in transgenic durum wheat by polymerase chain reaction (PCR) technology. Afr J Biotechnol 13:581–592. https://doi.org/10.5897/AJB2013.13301

Mengistu DK, Pè ME (2016) Revisiting the ignored Ethiopian durum wheat (*Triticum turgidum* var. *durum*) landraces for genetic diversity exploitation in future wheat breeding programs. J Plant Breed Crop Sci 8:45–59. https://doi.org/10.5897/JPBCS2015.0542

Merchuk-Ovnat L, Barak V, Fahima T et al (2016) Ancestral QTL alleles from wild emmer wheat improve drought resistance and productivity in modern wheat cultivars. Front Plant Sci 7:452. https://doi.org/10.3389/fpls.2016.00452

Mergby D, Hanin M, Saidi MN (2021) The durum wheat NAC transcription factor TtNAC2A enhances drought stress tolerance in *Arabidopsis*. Environ Exp Bot 186:104439. https://doi.org/10.1016/j.envexpbot.2021.104439

Mérida-Garcìa R, Liu G, He S et al (2019) Genetic dissection of agronomic and quality traits based on association mapping and genomic selection approaches in durum wheat grown in Southern Spain. PLoS One 14:e0211718. https://doi.org/10.1371/journal.pone.0211718

Meuwissen TH, Hayes BJ, Goddard ME (2001) Prediction of total genetic value using genome-wide dense marker maps. Genetics 157:1819–1829. https://doi.org/10.1093/genetics/157.4.1819

Michaletti A, Naghavi MR, Toorchi M et al (2018) Metabolomics and proteomics reveal drought-stress responses of leaf tissues from spring-wheat. Sci Rep 8:5710. https://doi.org/10.1038/s41598-018-24012-y

Mohammadi R, Abdulahi A (2017) Evaluation of durum wheat genotypes based on drought tolerance indices under different levels of drought stress. J Agric Sci 62:1–14. https://doi.org/10.2298/JAS1701001M

Mores A, Borrelli GM, Laidò G et al (2021) Genomic approaches to identify molecular bases of crop resistance to diseases and to develop future breeding strategies. Int J Mol Sci 22:5423. https://doi.org/10.3390/ijms22115423

Morran S, Eini O, Pyvovarenko T et al (2011) Improvement of stress tolerance of wheat and barley by modulation of expression of DREB/CBF factors. Plant Biotechnol J 9:230–249. https://doi.org/10.1111/j.1467-7652.2010.00547.x

Mwadzingeni L, Shimelis H, Dube E et al (2016) Breeding wheat for drought tolerance: Progress and technologies. J Integr Agric 15:935–943. https://doi.org/10.1016/S2095-3119(15)61102-9

Nakabayashi R, Saito K (2015) Integrated metabolomics for abiotic stress responses in plants. Curr Opin Plant Biol 24:10–16. https://doi.org/10.1016/j.pbi.2015.01.003

Negisho K, Shibru S, Matros A, Pillen K, Ordon F, Wehner G (2022) Genomic dissection reveals QTLs for grain biomass and correlated traits under drought stress in Ethiopian durum wheat *(Triticum turgidum* ssp. *durum*). Plant Breed 141(3):338–354

Nevo E, Chen G (2010) Drought and salt tolerances in wild relatives for wheat and barley improvement. Plant Cell Environ 33:670–685. https://doi.org/10.1111/j.1365-3040.2009.02107.x

Newell MA, Jannink JL (2014) Genomic selection in plant breeding. In: Crop Breed. Humana Press, New York, pp 117–130

Patnaik D, Khurana P (2003) Genetic transformation of Indian bread (*T. aestivum*) and pasta (*T. durum*) wheat by particle bombardment of mature embryo-derived calli. BMC Plant Biol 3:5. https://doi.org/10.1186/1471-2229-3-5

Paudel P, Pandey MK, Subedi M et al (2024) Genomic approaches for improving drought tolerance in wheat (*Triticum aestivum* L.): a comprehensive review. Plant Arch 24(1):1289–1300

Peleg Z, Fahima T, Krugman T et al (2009) Genomic dissection of drought resistance in durum wheat x wild emmer wheat recombinant inbreed line population. Plant Cell Environ 32:758–779. https://doi.org/10.1111/j.1365-3040.2009.01956.x

Pinheiro C, Chaves MM (2011) Photosynthesis and drought: can we make metabolic connections from available data? J Exp Bot 62:869–882. https://doi.org/10.1093/jxb/erq340

Poland JA, Brown PJ, Sorrells ME, Jannink JL (2012) Development of high-density genetic maps for barley and wheat using a novel two-enzyme genotyping-by-sequencing approach. PLoS One 7. https://doi.org/10.1371/journal.pone.0032253

Popova T, De Carvalho MÂAP, De Carvalho M, Ângelo AP (1998) Citrate and isocitrate in plant metabolism. Biochim Biophys Acta—Bioenerg 1364:307–325. https://doi.org/10.1016/s0005-2728(98)00008-5

Rampino P, Pataleo S, Gerardi C et al (2006) Drought stress response in wheat: physiological and molecular analysis of resistant and sensitive genotypes. Plant Cell Environ 26:2143–2152. https://doi.org/10.1111/j.1365-3040.2006.01588.x

Reif JC, Zhang P, Dreisigacker S et al (2005) Wheat genetic diversity trends during domestication and breeding. Theor Appl Genet 110:859–864. https://doi.org/10.1007/s00122-004-1881-8

Reynolds M, Tuberosa R (2008) Translational research impacting on crop productivity in drought-prone environments. Curr Opin Plant Biol 2(11):171–179. https://doi.org/10.1016/j.pbi.2008.02.005

Reynolds M, Dreccer F, Trethowan R (2007) Drought-adaptive traits derived from wheat wild relatives and landraces. J Exp Bot 58:177–186. https://doi.org/10.1093/jxb/erl250

Rizza F, Ghashghaie J, Meyer S et al (2012) Constitutive differences in water use efficiency between two durum wheat cultivars. Field Crop Res 125:49–60. https://doi.org/10.1016/j.fcr.2011.09.001

Roncallo PF, Akkiraju PC, Cervigni GL et al (2017) QTL mapping and analysis of epistatic interactions for grain yield and yield-related traits in *Triticum turgidum* L. var. *durum*. Euphytica 213(12):277

Roselló M, Villegas D, Álvaro F et al (2019) Unravelling the relationship between adaptation pattern and yield formation strategies in Mediterranean durum wheat landraces. Eur J Agron 107:43–52. https://doi.org/10.1016/j.eja.2019.04.003

Saad ASI, Li X, Li HP et al (2013) A rice stress-responsive NAC gene enhances tolerance of transgenic wheat to drought and salt stresses. Plant Sci 203-204:33–40. https://doi.org/10.1016/j.plantsci.2012.12.016

Sahi C, Singh A, Blumwald E, Grover A (2006) Beyond osmolytes and transporters: novel plant salt-stress tolerance related genes from transcriptional profiling data. Physiol Plant 127:1–9. https://doi.org/10.1111/j.1399-3054.2005.00610.x

Saidi MN, Mergby D, Brini F (2017) Identification and expression analysis of the NAC transcription factor family in durum wheat (*Triticum turgidum* L. ssp. *durum*). Plant Physiol Biochem 112:117–128. https://doi.org/10.1016/j.plaphy.2016.12.028

Saito K, Matsuda F (2010) Metabolomics for functional genomics, systems biology, and biotechnology. Annu Rev Plant Biol 61:463–489. https://doi.org/10.1146/annurev.arplant.043008.092035

Sall AT, Kabbaj H, Menoum SOE et al (2024) Durum wheat heat tolerance loci defined via a north–south gradient. Plant Genome 17(1):e20414

Sallam A, Alqudah AM, Dawood MFA et al (2019) Drought stress tolerance in wheat and barley: advances in physiology, breeding and genetics research. Int J Mol Sci 20:3137. https://doi.org/10.3390/ijms20133137

Salsman E, Liu Y, Hosseinirad SA et al (2021) Assessment of genetic diversity and agronomic traits of durum wheat germplasm under drought environment of the northern Great Plains. Crop Sci 61:1194–1206. https://doi.org/10.1002/csc2.20449

Salvi S, Tuberosa R (2015) The crop QTLome comes of age. Curr Opin Biotechnol 100(32):179–185. https://doi.org/10.1016/j.copbio.2015.01.001

Scheibe R (2004) Malate valves to balance cellular energy supply. Physiol Plant 120:21–26. https://doi.org/10.1111/j.0031-9317.2004.0222.x

Sehgal D, Dhakate P, Ambreen H et al (2023) Wheat omics: advancements and opportunities. Plants 12:426. https://doi.org/10.3390/plants12030426

Shavrukov Y, Baho M, Lopato S, Langridge P (2016) The *TaDREB3* transgene transferred by conventional crossings to different genetic backgrounds of bread wheat improves drought tolerance. Plant Biotechnol J 14(1):313–322. https://doi.org/10.1111/pbi.12385

Sivamani E, Bahieldin A, Wraith JM et al (2000) Improved biomass productivity and water use efficiency under water deficit conditions in transgenic wheat constitutively expressing the barley *HVA1* gene. Plant Sci 155:1–9. https://doi.org/10.1016/s0168-9452(99)00247-2

Soriano JM, Villegas D, Aranzana MJ et al (2016) Genetic structure of modern durum wheat cultivars and Mediterranean landraces matches with their agronomic performance. PLoS One 11:e0160983. https://doi.org/10.1371/journal.pone.0160983

Sukumaran S, Reynolds MP, Sansaloni C (2018a) Genome-wide association analyses identify QTL hotspots for yield and component traits in durum wheat grown under yield potential, drought, and heat stress environments. Front Plant Sci 9:81. https://doi.org/10.3389/fpls.2018.00081

Sukumaran S, Jarquin D, Crossa J, Reynolds M (2018b) Genomic-enabled prediction accuracies increased by modeling genotype × environment interaction in durum wheat. Plant Genome 11. https://doi.org/10.3835/plantgenome2017.12.0112

Todkar L, Harikrishna SGP et al (2020) Introgression of drought tolerance QTLs through marker assisted backcross breeding in wheat (*Triticum aestivum* L.). Indian J Genet 80:209–212. https://doi.org/10.31742/IJGPB.80.2.12

Tounsi S, Jemli S, Feki K, Brini F, Najib Saïdi M (2023) Superoxide dismutase (SOD) family in durum wheat: promising candidates for improving crop resilience. Protoplasma 260:145–158. https://doi.org/10.1007/s00709-022-01767-w

Tovkach A, Ryan PR, Richardson AE et al (2013) Transposon-mediated alteration of *TaMATE1B* expression in wheat confers constitutive citrate efflux from root apices. Plant Physiol 161:880–892. https://doi.org/10.1104/pp.112.207142

Tuberosa R (2012) Phenotyping for drought tolerance of crops in the genomics era. Front Physiol 3:347. https://doi.org/10.3389/fphys.2012.00347

Tuberosa R, Maccaferri M (2015) Genomics approaches to dissect the genetic basis of drought resistance in durum wheat. In: Advances in wheat genetics: from genome to field: proceedings of the 12th international wheat genetics symposium. Springer Japan, pp 213–223

Ullah N, Yüce M, Neslihan Öztürk Gökçe Z, Budak H (2017) Comparative metabolite profiling of drought stress in roots and leaves of seven Triticeae species. BMC Genomics 18:969. https://doi.org/10.1186/s12864-017-4321-2

Van den Ende W (2013) Multifunctional fructans and raffinose family oligosaccharides. Front Plant Sci 4:247. https://doi.org/10.3389/fpls.2013.00247

Vergara-Diaz O, Vatter T, Vicente R et al (2020) Metabolome profiling supports the key role of the spike in wheat yield performance. Cells 9:1025. https://doi.org/10.3390/cells9041025

Walkowiak S, Gao L, Monat C et al (2020) Multiple wheat genomes reveal global variation in modern breeding. Nature 588:277–283. https://doi.org/10.1038/s41586-020-2961-x

Wang S, Wong D, Forrest K (2014) Characterization of polyploid wheat genomic diversity using a high-density 90,000 single nucleotide polymorphism array. Plant Biotechnol J 12:787–796. https://doi.org/10.1111/pbi.12183

Wang S, Xu S, Chao S et al (2019) A Genome-Wide association study of highly heritable agronomic traits in durum wheat. Front Plant Sci 10:919. https://doi.org/10.3389/fpls.2019.00919

Watson A, Ghosh S, Williams MJ et al (2018) Speed breeding is a powerful tool to accelerate crop research and breeding. Nat Plants 4:23–29. https://doi.org/10.1038/s41477-017-0083-8

Winfield MO, Allen AM, Burridge AJ et al (2016) High-density SNP genotyping array for hexaploid wheat and its secondary and tertiary gene pool. Plant Biotechnol J 14:1195–1206. https://doi.org/10.1111/pbi.12485

Zaïm M, Kabbaj H, Kehel Z et al (2020) Combining QTL analysis and genomic predictions for four durum wheat populations under drought conditions. Front Genet 11:316. https://doi.org/10.3389/fgene.2020.00316

Zemede A, Mekbib F, Assefa K, Bishaw Z (2019) Variability in Ethiopian durum wheat under rainfed environment subjected to drought at anthesis. Ethiop J Agric Sci 29:17–29

Zhao Y, Zhao J, Zhao C et al (2015) A metabolomics study delineating geographical location-associated primary metabolic changes in the leaves of growing tobacco plants by GC-MS and CE-MS. Sci Rep 5:16346. https://doi.org/10.1038/srep16346

Zhu T, Wang L, Rimbert H et al (2021) Optical maps refine the bread wheat *Triticum aestivum* cv. Chinese Spring genome assembly. Plant J 107:303–314. https://doi.org/10.1111/tpj.15289

Genomics of Nitrogen-Use Efficiency in Durum Wheat

Agata Gadaleta, Domenica Nigro, Maria Concetta de Pinto, Ilaria Marcotuli, Stefania Fortunato, Pierre Roumet, and Jacques Le Gouis

Abstract

Global agriculture is facing one of the biggest challenges of all time. As reported by the most recent forecasts, world population is expected to reach nearly 10 billion by 2050, thus causing higher demand for food, feed, and raw materials. Wheat is among the most widely cultivated crops, also considering the great variety of their derived end-use products. Great efforts have been made so far by breeders and research institutions to increase the productivity of these important crops, but better yield still need to be reached to guarantee global food and nutrition security in the near future. Nitrogen (N) is one of the most important elements for plant growth. It is a basic constituent of several molecules such as amino acids, proteins, nucleic acids, chlorophyll, and various primary and secondary metabolites, making it essential for the cellular growth and overall plant development. In the last decades, important amounts of fertilizer, especially nitrogenous ones, have been supplied to crops to increase final yield. A big issue is represented by the actual low percentage of N taken up and utilized by plants; as a result, high concentrations of those synthetic supplements are lost in the soil and groundwaters, thus resulting in negative effects on the environment and human health. Nevertheless, the actual climate changing scenario, inducing new stresses, is a main yield-constraining factor of most cereal crops. Drought and water stress have severe and different effects on plant growth; one is represented by the restriction of nutrient uptake, which may cause a deficiency of nutrient, especially nitrogen. One possible solution to prevent further environmental issues and still guarantee the necessary food supply may be the development of more efficient crops in terms of nitrogen-use efficiency (NUE). Thus, less nitrogen would be required as supplement and at the same time, more efficient plants could make the most out of lower inputs.

In this report, we will focus on the genomics of nitrogen-use efficiency in durum wheat, with a particular attention to the candidate genes which have been so far proved to be involved in NUE uptake, assimilation, and remobilization, as well as the relationship

A. Gadaleta (✉) · D. Nigro · I. Marcotuli
Department of Soil, Plant and Food Sciences, University of Bari "Aldo Moro", Bari, Italy
e-mail: agata.gadaleta@uniba.it

M. C. de Pinto · S. Fortunato
Department of Biology, University of Bari Aldo Moro, Bari, Italy

P. Roumet
UMR AGAP Institut, Univ Montpellier, CIRAD, INRAE, Institut Agro, Montpellier, France

J. Le Gouis
Jacques Le Gouis, INRAE, Université Clermont-Auvergne, Clermont-Ferrand, France

R. Tuberosa et al. (eds.), *The Durum Wheat Genome*, Compendium of Plant Genomes,
https://doi.org/10.1007/978-3-031-84689-2_12

with important agronomic traits such as grain protein content.

1 Introduction

The world's population is expected to reach nearly 10 billion by 2050, as the latest FAO forecast reports, with an estimated annual growth rate of 0.9% (OECD-FAO Agricultural Outlook 2020–2029, FAO). This rise will be resulting in a 70% boost in the global food demand, especially on cereals (Chen and Liao 2017). Cereals in their broad category are historically the major type of crops produced and traded across the world for many different purposes (FAO 2020), with a global cropping area of almost 700 million ha. They contribute to about 50% of world food (Singer et al. 2019).

Used both for food and feed, and given the wide range of their derived products, cereals are estimated to provide approximately 56 and 50% of caloric intake and protein needs, respectively, due to their high carbohydrate and protein contents (Muitire et al. 2021). Among them, wheat supplies globally about 20% of the human daily calories and protein intake and up to 45% in North Africa and Central Asia (Pena-Bautista et al. 2017). Most of the presently cultivated wheats belong either to the hexaploid *Triticum aestivum* L. or to the tetraploid *T. turgidum* L. var. *durum* species. Despite representing only 5% of the total wheat production, durum wheat represents an elite crop in dry lands, such as the Mediterranean basin, with a yearly production average of 40 million tons on a worldwide planting area of 17 million ha (International Grains Council 2020). Its productivity is harshly affected by drought, as durum wheat is mostly grown under rain-fed conditions, as well as other biotic (pests and diseases) and abiotic (cold, salinity) stresses.

Moreover, in the last 50 years, yield increase has been obtained by supplementing plants with high quantity of fertilizers, especially nitrogenous ones. Nitrogen (N) is indeed quantitatively the most important component of all fertilizers. N is together with carbon essential throughout the entire life cycle for plant growth. It is a crucial constituent of various amino acids, and then proteins, nucleotides, chlorophyll, cofactors, enzymes, vitamins, secondary metabolites, and growth hormones, all required elements for genome constituents, cellular structures, and overall growth (Krapp et al. 2005). Low or insufficient available N can cause damage to growth, photosynthetic and reproductive ability, limiting the yield (both total and grain biomasses) and quality, and can ultimately affect plant resistance against several biotic and abiotic stresses (Walters and Bingham 2007; Veresoglou et al. 2013; Ahanger et al. 2019).

As reviewed by Beatty and Good (2018), organic N fertilizers, such as livestock and green manure, have been used to increase crop production for centuries. In the early 1900s, the development and commercialization of the Haber–Bosch process allowed to synthesize ammonia from atmospheric N_2, resulting in the massive production of synthetic N fertilizers which have been so far extensively used (Smil 2001; Erisman et al. 2008; Follett et al. 2010; Cao et al. 2018). Agricultural fertilizer application has been one of the most important land management practices which has enhanced crop yield as well as soil fertility (Vitousek et al. 1997; Tilman et al. 2002; Lu and Tian 2017). From 1961 to 2019, the amount of N fertilizers supplied to crops mainly as ammonium nitrate or urea has risen from 13 to 123 million tons/year (FAOSTAT 2021). N fertilizers are mainly applied on the wheat crop with about 18% of global usage (Heffer et al. 2009).

Despite the dramatic increase in fertilizer production and application, crops yield, especially wheat, Soybean, and maize, have slowed to a growth rate of about 1% annually, and in some specific cases, as in developed countries, the growth rate is quite close to zero (Fischer et al. 2009; Arriagada et al. 2022).

Soil N cycle is complex and different processes such as volatilization, denitrification, leaching, and soil erosion, play important roles in reducing or depleting N levels in soil (Garnett et al. 2009), causing a number of environmental and ecological problems within and outside of farmlands, such as air pollution, soil acidification and degradation, water eutrophication, and crop yield reduction (Vitousek et al. 2009; Guo et al. 2010; Sutton et al. 2011; Lu and Tian 2013). Also, the increasing application of N fertilizer can rise N oxide emissions which can primarily cause ozone pollution and contribute to the release of nitrogen oxides NOX greenhouse gases (Zhang et al. 2015).

With increased concern for the environment and knowledge of excessive application of N fertilizers, increasing nitrogen-use efficiency (NUE) has become the center of ecological/agroecosystem research around the world. The biggest challenge to improve NUE is the ability to optimize genetic, agronomic, and environment factors in order to reduce N fertilizers inputs and still guarantee acceptable crop yield and quality levels (Hirel et al. 2007; Omara et al. 2019). Indeed, as highlighted by Bandyopadhyay and Prasad (2020), understanding how genetic differences affect N uptake, utilization, and remobilization into grains is fundamental to our understanding of NUE processes in crops, especially under contrasting N regimes. One of the main targets of plant breeders and researchers is to breed varieties with enhanced NUE (Hirel et al. 2007; Thompson 2012; Han et al. 2015; Cormier et al. 2016).

Several reviews on NUE have already been published especially in bread wheat (Hirel et al. 2007; Dawson et al. 2008; Le Gouis 2011; Cormier et al. 2016; Le Gouis and Hawkesford 2017; Swarbreck et al. 2019). In this chapter, we will mainly focus on durum wheat, which has been less studied than bread wheat, and on significant advances obtained in the last decade.

2 Nitrogen-Use Efficiency: Concept and Definitions

NUE has long been used as a concept to estimate crop production practices, although it is significantly influenced by fertilizer supply as well as soil, plant, and water interactions. Indeed, over the years, several scientific contributions have been focused on NUE, especially on how to improve crop NUE through innovative and precise agricultural management practices as well as through new breeding programs (Moll et al. 1982; Raun and Johnson 1999; Cormier et al. 2016; Martinez-Feria et al. 2018).

NUE is usually thought to be a straightforward expression. As a matter of fact, its definition has a notable complexity considering not only the different and potential plant N sources (soil organic matter, inorganic and organic fertilizer, manure, aerial deposition etc.) but also the multitude of factors influencing crop nutrient demand (crop management, genetics, environmental factors). Indeed, the bibliography dealing with NUE definitions involves a broad variety of NUE calculations and estimations and recognizes the establishment of different NUE indexes with distinctive purposes (agronomic, physiological, genetic, etc.) (Good et al. 2004; Ladha et al. 2005; Dobermann 2007; Fageria et al. 2008; Cormier et al. 2016; Ernst et al. 2020). Several authors attempted to extensively review and report more clearly the different NUE measures and their suggested applications (Dobermann 2007; Fixen et al. 2015), even though they principally focused on assessing crop responses to N or N balance and operate on a short-term frame. This further complicates the NUE concept, as this traditional interpretation of NUE does not consider how tightly N is cycled over time within cropping systems. It also focuses on yield, at the risk of misrepresenting the outcome of N, which is particularly important considering the actual climate change scenario and all the deleterious effects previously reported (Martinez-Feria et al. 2018). So far, it is extremely important to define the genetic bases, as well as the influencing factors, involved in NUE and N cycling to promote sustainable agricultural systems—a multifaceted objective that not only considers crop production but also the sustainability of soil fertility and the mitigation of environmental pollution.

Moll et al. (1982) defined NUE for cereal production and a breeding prospective as the grain production per unit of N available in the soil. They also proposed to separate it into two primary components, the plant uptake efficiency (NUpE) and the utilization efficiency (NUtE). The former, which reflects how effectively the plants capture N, is defined as the total amount of N acquired by plants divided by the entire available N sources within the soil. NUpE is influenced by both N availability (mass flow of soil water to the ground, timing and amount of N applications, and microbial competition), and plant potential N uptake ability (root morphology, transporter activity on the ground surface, regulation of the demand) (Garnett et al. 2009). Even if most of N is acquired through the root system, it should be mentioned that leaf N uptake has been demonstrated when liquid foliar N fertilizers are applied (Fernandez and Brown 2013). As noted by many authors, available N is difficult to assess (see Dawson et al. 2008 for a discussion) and is either not reported or differently estimated, which makes comparisons between studies also difficult.

NUtE, determining how well the plants use the N that is captured, is defined as the fraction of acquired N converted to grain or biomass (Moll et al. 1982; Xu et al. 2012). This grain N arises both from tissue N assimilated pre-anthesis and N absorbed post-anthesis (Moll et al. 1982; Hawkesford 2017). Several factors linked to nitrate assimilation, leaf and canopy photosynthesis per unit N, post-anthesis N remobilization, and senescence dynamics influence NUtE (Cormier et al. 2016).

As mentioned, these definitions of NUE mostly focus on grain yield. However, grain protein content (GPC) is particularly important for cereals, and specifically durum wheat, as it influences both processing efficiency and product end-use qualities. GPC is indeed a very important quality trait, likely the most, in both bread (*Triticum aestivum* L.) and durum wheat (*Triticum turgidum* L. ssp. *durum*), as it determines both the nutritional value and the baking quality and the pasta-making technology characteristics of flour and semolina, respectively. During the past decades, the increase in GPC has mainly been achieved by intensifying nitrogenous fertilization. It is known for a long time that there is a negative genetic correlation between grain yield and GPC (Simmonds 1995). It is then difficult for breeders to improve both these traits at the same time. Monaghan et al. (2001) proposed to use as a breeding criterion the residuals from the regression of grain yield on GPC which they termed grain protein deviation (GPD). Since then, this criterion has been the focus of several studies in wheat (e.g., Oury et al. 2003; Oury and Godin 2007; Nigro et al. 2019).

3 Genetic Variability in NUE Traits

As durum wheat is mainly cultivated under rainfed conditions, N nutrition interacts strongly with soil water availability, making more complex both experimental design and data analysis dedicated to cultivar responses to N applications (Abad et al. 2004; Basso et al. 2010; Mălinaş et al. 2022). Despite these difficulties, several field studies have been carried out in the last decade to quantify differences in NUE traits in durum wheat (Table 1). All the studies considered included contrasted N levels due to different N fertilizer applications. Large differences are observed between studies due to the genetic material assessed, various growing conditions and ways in which available N was estimated. Within studies, the genetic variability for NUE and its components is generally significant with ranges of variation in about 20% between extremes. This genetic variability is somewhat similar to that observed in bread wheat (Le Gouis and Hawkesford 2017). When factorial design allowed to quantify the interaction with water availability, a significant genotype × N treatment interaction was generally reported (Giuliani et al. 2011; Karrou and Nachit 2015; Ierna et al. 2016; Lupini et al. 2020), meaning that genotypes will behave differently depending on the N level with few exceptions, such as results reported by Ayadi et al. (2012) and Liang et al. (2014) where no significant interactions NUE x N availability have been observed, despite the large differences between treatments. The genotype × water regime is also generally significant whatever the trait concerned (Giuliani et al. 2011; Lupini et al. 2020). Lupini et al. (2020) also observed a significant genotype × N treatment × water regime interaction for NUE and its components, high-

Table 1 Summary of field studies conducted to analyze genetic variability of NUE traits in durum wheat

Paper	Country	Number of genotypes	Water regimes	N treatments kg N/ha	NUE kg grain/kg N	NUpE kg N/kg N	NUtE kg grain/kg N
Lupini et al. (2020)	Italy	8	90% intercepted/ Rainfed	0, 100	6.5–8.1	0.61–0.67	10.0–13.2
Ierna et al. (2016)	Italy	3	Rainfed	0, 80, 160	19.5–22.9		
Karrou et al. (2015)	Syria	10	Rainfed/Irrigated	40, 80,120)	33.2–60.2	1.15–1.53	20.7–35.6
Liang et al. (2014)	USA	6	Irrigated	0, 73, 123, 185, 269, 403	32.9–36.7	0.94[ns§]	39.7[ns]
Ayadi et al. (2012)	Tunisia	4	Rainfed	0, 75, 150	8.6–10.7	0.79[ns]	30.4–40.9
Giuliani et al. (2011)	Italy	4	Rainfed/Irrigated	0, 60, 120		2.01–2.15	

For NUE, NUpE and NUtE the range of variation is reported when the genotype effect was significant and the average on all genotypes when it is not significant
[a]ns = the genotype effect is not significant

lighting the complexity of N uptake and N utilization in relationship to water available to the plant.

Several authors have analyzed the relationship between NUE and its two main components, NUpE and NUtE. It is often reported that contributions of the two components to NUE depend on the N level. In bread wheat, N uptake generally explains variation in grain yield more than NUtE and at low N it accounted for even more of the genetic variation in NUE (Le Gouis 2011; Biradar et al. 2024). The same trend seems to be verified for durum wheat with NUpE being the more important contributor to NUE, especially when N availability is low (Giuliani et al. 2011; Liang et al. 2014), with some differences between seasons (Karrou and Nachit 2015).

4 Plant Traits Involved in NUE

To provide a better understanding of the different processes involved in an improved N recovering from soils and fertilizers (better NUpE) and its use to make more grain (better NUtE), traits related to the root system or to N resorption have been targeted.

4.1 Root Traits

Root traits have been classified in three classes (see Garnett et al. 2009 for a review): root growth (total biomass or root/shoot ratio), root architectural traits (length, number of seminal and nodal roots, insertion angle of seminal roots), and physiological traits involved in resource acquisition (tip number, root length density, percentage of fine roots). Since the first decade of the 2000s, studies exploring the relationships between these traits and NUE have been reported. A significant variability, like that observed in bread wheat (Xu et al. 2023), was described for most of these traits inside tetraploid *turgidum* species (Manschadi et al. 2008; Canè et al. 2014; Roucou et al. 2018; Puccio et al. 2021).

Based on studies conducted on different species, a root system ideotype was designed. High root surface area, via lateral branches and root hairs, allows plants to take up more N from the cultivated soil layer (Liao et al. 2006 on bread wheat; Gahoonia et al. 2006 on lentils). In addition, higher root length density and deeper root system have been described as positive traits to explore a greater soil volume and to take up residual N from the subsoil (Liu et al. 2009, rice; Barraclough et al. 2010; Liu et al. 2022 bread wheat). Under N limiting conditions, positive relationships between deep root system (i.e., longer roots, high root biomass) with N uptake suggested that breeding could lead to an improvement in N uptake efficiency (Brady et al. 1995; Garnett et al. 2009). As far as N availability and N spatial distribution change in the soil, relationships between these traits and NUpE are contrasting. Under non-limiting N conditions, plant investment in a costly root system (i.e., large root/shoot ratio, significant root length) is not required to maximize NUpE, whereas a good coupling between physiological root traits, such as high fine root density and N availability along soil depth, leads to increase N uptake (Nakhforoosh et al. 2021).

4.2 Plant Variability Related to N Uptake Time Course and N Remobilization

Post-flowering N uptake is generally low comparing to pre-flowering uptake or even null. It is both determined by environmental factors (such as N supply or water availability), plant N status at flowering (Taulemesse et al. 2015) and genetic control (Bogard et al. 2010). As this trait is negatively related to the efficiency of N remobilization during seed filling period, some studies report that it does not improve NUE (Suprayogi et al. 2011; Juraniec et al. 2017). However, it has been shown to be positively correlated to GPD in bread wheat (Monaghan et al. 2001; Bogard et al. 2010).

Post-flowering N remobilization to the grain is generally high, accounting to about 80% of total grain N. It is mainly computed as the percentage of total plant N at anthesis is not recovered from vegetative tissues at physiological maturity. This amount of N remobilized is affected both by environmental conditions (climate -temperature, hydric stress-, plant nutrition -N uptake after anthesis-; Gaju et al. 2014) and genetic background (Kade et al. 2005; Suprayogi et al. 2011).

As leaf N resorption represents the largest contribution to N remobilization process (Bertheloot et al. 2008), Vilmus et al. (2014) proposed to use near infrared spectroscopy, a non-destructive method, to monitor N resorption time course in flag leaf and to quantify durum wheat genetic variability available at each step of this kinetics. From flowering to physiological maturity, flag leaf N resorption has been modelled according to a logistic curve (plateau/drop/plateau) for each line.

The two non-correlated traits that expressed the largest genetic variability were the senescence onset, which marked the beginning of leaf N content decrease, and the N leaf content at maturity, whereas leaf N content at flowering remained quite similar between genotypes. Thus, decreasing the N leaf content at maturity appears as the main driver to improve NUtE. These two traits involved in N resorption impacted both the amount and the quality of grain harvested as reported previously on wheat. First, a delayed N resorption onset leads to increase in C flow and individual grain weight, as observed on bread wheat stay-green phenotypes (Thomas and Ougham 2014; Montazeaud et al. 2016), whereas an earlier N resorption increased GPC as the functional allele of the NAC gene (Uauy et al. 2006). Second, the positive correlation observed between grain number and N resorption efficiency is consistent with the results obtained on bread wheat by Gaju et al. (2014), which showed that the amount of N remobilized impacted positively the number of grains. In bread wheat, N remobilization and post-anthesis nitrogen uptake (NUP) contribute largely to grain yield and GPC. The relative importance of N remobilization and post-anthesis GPC that is environmental- and genotype-dependent is a function of the amount of available N for uptake, water availability, and plant health (Kichey et al. 2007; Arduini et al. 2006). So far, GPC has been indeed indicated as a NUE-related agronomic trait. Table 2 presents a summary of all GPC quantitative trait loci (QTL) reported in bread and durum wheat.

Table 2 Summary of QTL mapping studies for grain protein content (GPC) and NUE-associated QTL in wheat (ordered by year)

Cross	Population type and size	QTL chromosome localization	Reference
Durum and bread wheat lines	Hexaploid and tetraploid NILs	6B	Brevis and Dubkovski (2010)
Ning7840 x Clark	132 RIL	3A, 4B	Sun et al. (2010)
MN98550 x MN99394	139 RIL	2A, 2B, 5A,5B, 6B, 6D, 7A	Tsilo et al. (2010)
Huapei 3 xYumai57	168 DHL	3A, 3B, 5D, 6D	Zhao et al. (2010)
25R26 x Foster	171 RIL	2A, 2B, 7D	Smith et al. (2011)
Association mapping	207 accessions	1B, 2D, 3A, 5D	Reif et al. (2011)
UC1113 x Kofa	93 RIL	1B, 2A, 2B, 3B, 4A, 5A, 5B, 7A, 7B	Conti et al. (2011)
Association mapping	372 accessions	1A, 1B, 1D, 2A, 2B, 2D, 3D, 4B, 5B, 5D, 6A, 6B, 6D, 7B	Bordes et al. (2011)
Toisondor x CF9107	140 DHLs	2A, 2B, 2D, 3A, 3B, 5A, 7D	Bogard et al. (2011)
Svevo x Ciccio	120 RIL	1A, 2A, 2B, 3B, 4A, 4B, 5A, 6B	Blanco et al. (2012)
Am3 x Laizhou 953	82 RIL	1A, 2D, 4B, 5D, 6A, 6B, 6D, 7B	Li et al. (2012)
Xiaoyan 54/Jing 411	182 RIL	4B, 4D, 5A, 6A	Xu et al. (2012)
Weimai 8/Luohan 2	302 RIL	1D, 2A, 2B, 3B, 4A, 5B, 5D, 6B, 7A	Wang et al. (2012)
OS9 x Q36	164 RIL	6BS, 7DL, 2AS, 5DL, 1AL	Wang et al. (2012)
R34 x Grandin	118 RIL	5B	Simons et al. (2012)
Weimai 8 x imai 20, Weimai 8 'Yannong 19	485 RIL for P1, 229 RIL for P2	1A, 1B, 2A, 2B, 2D, 3A, 4A, 4B, 4D, 5A, 5B, 5D, 6B, 7A, 7B, 7D	Li et al. (2012)
Louise x Penawawa	188 RIL	3B	Carter et al. (2012)
Choteau/Yellowstone	97 RIL	3B, 5B	Heo and Sherman (2013)
Toisondor/Quebon, CF9107 x Quebon, Toisondor/ x CF9107	91 DH for P1, 90 DH for P2, 140 DH for P3	2A, 2D, 3A, 3B, 5D, 7B, 7D	Bogard et al. (2013)
Langdon x G18-16	152 RIL	2B, 4A, 7B	Yan et al. (2013)
CO940610 x Platte	185 DH	5B, 6A, 6B, 7B, 7D	El-Feki et al. (2013)
Association mapping	196 accessions	1A, 1B, 1D, 2B, 3B, 4A, 5B, 6A, 6B	Bordes et al. (2013)
Association mapping	376 cultivars	6A	Bentley et al. (2014)
Association mapping	214 European varieties	2A, 2B, 3A, 3B, 5A, 5B, 6A,6D,7A, 7B	Cormier et al. (2014)
Xiaoyan 54 x Jing 411	182 RIL	1B, 3A, 3D, 4B, 4D, 5A, 6A,	Xu et al. (2014)
Association mapping	118 accessions	3B, 5B	Tadesse et al. (2015)
Drysdale x Gladius	155 RIL	2B, 2D, 3B, 5A	Maphosa et al. (2015)
WCB414 xWCB617	163 RIL	1A, 1B, 2B, 2D, 3D, 4B, 5B, 6B, 7B	Echeverry-Solarte et al. (2015)
Huapei 3 x Yumai57, Nuomai 1 x Gaocheng 8901, Shannong 01-35 xGaocheng 9411	168 DH for P1, 256 RIL for P2, 182 RIL for P3	1B, 1D, 2A, 2B, 2D, 3B, 4B, 5B, 6D, 7A	Deng et al. (2015)

(continued)

Table 2 (continued)

Cross	Population type and size	QTL chromosome localization	Reference
Association mapping	407 varieties	5A, 5B, 6D	Würschum et al. (2016)
Berkut x Krichauff	138 DH	1A	Tiwari et al. (2016)
Yumechikara x Kitahonami	94 DH	2B	Terasawa et al. (2016)
Gaocheng8901 x Zhoumai16	176 RIL	3BL	Jin et al. (2016)
Norstar x winter Manitou, Cappelle Desprez x Norstar, Norstar x Manitou	161 DH for P1, 256 DH for P2, 152 DH for P3	3A, 4A, 5A, 6A, 6B, 7D	Fowler et al. (2016)
Ning7840/Clark	132 RIL	3A, 4BS, 5AL, 5BL	Li et al. (2016)
RAC875 and Kukri	156 DHLs	1B, 2D, 5A, 5B, 5D, 7A	Mahjourimajd et al. (2016)
WH542/synthetic wheat	286RIL	2A, 3A	Krishnappa et al. (2017)
Association mapping	299 accessions	1D, 2B, 2D, 4B	Guttieri et al. (2017)
Ning7840 x Clark	127 RIL	4B	Ding et al. (2018)
Association mapping	189 durum wheat (Central Europe)	5A, 5B, 6A, 6B, 7A, 7B	Rapp et al. (2018)
Association mapping	159 durum wheat (South Europe)	2A, 3B, 4A, 7A	Rapp et al. (2018)
Association mapping	236 accessions	2B, 3A, 3B, 4A, 4B, 5A, 5B 6B, 7A, 7B	Nigro et al. (2019)
NAM population	175 RIL	1A, 1B, 2A, 2B, 3A, 3B 4A, 4D, 5B, 7B, 7D	Sandh et al. (2021)
Cutler x AC Barrie Attila x CDC Go, Peace x Carberry, Peace x CDC Stanley	698 RIL	2D, 4B, 5A	Semagn et al. (2021)
Association mapping	150 accessions	1A, 2A, 2B, 3A, 3B, 6B	White et al. (2022)
HUW468 x C306	RIL	1B, 1D, 2A, 2B, 3A, 4B, 5A, 5B, 5D, 6A, 6D, 7A	Singh et al. (2023)
TN18 x LM6	184 RIL	1B, 3B, 6B, 5D, 7A, 7B	Han et al. (2023)
Meta-QTL (MQTL) Analysis		1A, 2B, 3A, 5A, 4A, 6A, 6B, TA, 7B	Cabas-Lühmann et al. (2024)
Association mapping	299 spring wheat cultivar	1D, 2B, 5A, 7B,	Aleliunas et al. (2024)

5 Genetic Determinism of NUE Uptake, Assimilation, and Remobilization: From QTLs to Candidate Genes

NUE is a complex quantitative trait affected by both multiple genetic and environmental factors and all these factors should be considered in wheat NUE QTL analysis. To date, while several papers have been published on bread wheat NUE QTL detection (see Cormier et al. 2016 for a review; Han et al. 2023), less information is available on durum wheat. Most studies have been focused on the effect of different nitrogen supply in Mediterranean rainfed condition (Lopez-Bellido et al. 2008; Giuliani et al. 2011; Ayadi et al. 2022; Ingraffia et al. 2023) and on the identification of candidate genes involved in nitrogen metabolism, protein content, and grain protein deviation (Nigro et al. 2019; Shi et al. 2022, 2024). Quantitative genetic studies pointed out many limitations and challenges related to NUE dissection, as there are multiple QTLs detected, each usually accounting for a small amount of the observed variation.

QTLs affecting NUE have been successfully mapped in bread wheat under different N levels (low, Optimum, and varying levels). Also, many reported QTL experiments have been carried out under different growing conditions, such as hydroponic culture (An et al. 2006; Laperche et al. 2006; Guo et al. 2012; Griffiths et al. 2022), pots (Habash et al. 2007) and field trials (e.g., Quarrie et al. 2005; An et al. 2006; Laperche et al. 2007; Le Gouis et al. 2008; Fontaine et al. 2009; Xu et al. 2014; Cui et al. 2016; Mahjourimajd et al. 2016).

QTL for activities of enzymes involved in N metabolism under varying levels of N application, such as Glutamine synthetase (GS) and Glutamate dehydrogenase (GDH), (Quarrie et al. 2005; Fontaine et al. 2009; Xu et al. 2014) were detected. The role of GS as a target gene to increase NUE was widely demonstrated by several authors using different genetic materials. QTLs for leaf GS activity co-localized with those for leaf protein and N contents and yield components, which were also found to exhibit genetic variations in wheat germplasm (Kichey et al. 2006; Habash et al. 2007; Bernard et al. 2008). Importantly, two QTL clusters for leaf GS activity were found to be associated with mapped *GS* genes (a *GS2* gene on chromosome 2A, and a *GS1* gene on 4A) and high GS activity was found to correlated with high GPC, with no negative effects on yield components (Habash et al. 2007). Other studies further highlighted the chromosomal regions containing GS2 in wheat that colocalized with a number of QTLs for agronomic traits related to N use and yield (Laperche et al. 2007; Bingham and Wu 2011; Fontaine et al. 2009; see also Bogard et al. 2013 for a meta-analysis), suggesting that the genomic region surrounding GS2 may be valuable for breeding wheat varieties with improved agronomic performance and NUE. However, to date, no one has been able to introduce a GS gene into a NUE-inefficient background and show either enhanced NUE or yield.

In durum wheat, fewer papers have been published on NUE, but a number of quantitative studies determining the involvement of *GS* and *GOGAT* candidate genes in the GPC control are available (Fortunato et al. 2023). Those studies were carried out either on bi-parental populations (Gadaleta et al. 2011, 2014) or broader genotype collections (Nigro et al. 2017, 2019). Interestingly, the plastidic GS2 and the cytosolic GS1.3 have been associated with QTLs for GPC (Gadaleta et al. 2011, 2014), specifically in homoeologous regions to the one previously reported in bread wheat on 2B and 4A chromosomes, respectively. Despite QTLs for GPC and NUE have been located on all chromosomes, several authors reported major ones on 2A and 2B chromosomes of bi-parental mapping populations, where *GS2* and *Fd-GOGAT* genes have been mapped (Boisson et al. 2005; Blanco et al. 2012; Rapp et al. 2018; Nigro et al. 2019). A recent paper from Nigro et al. (2020) reported the development of two distinct sets of heterogeneous inbred family-based NILs segregating for *GS2* and *Fd-GOGAT* genes which allow to validate the previously identified GPC QTL on 2A and 2B chromosomes, and the putative involvement of those key genes in GPC control.

Nigro et al. (2013) reported a QTL analysis and identification of candidate genes in a collection of almost 240 durum wheat genotypes; moreover, specific allelic variants for both *Fd-GOGAT* and *NADH-GOGAT* were found to be associated with GPC QTL (Nigro et al. 2017). A study from Nigro et al. (2014), reported a co-localization between GPC-QTL and Fd-GOGAT gene on 2A chromosome. Allelic variation in *Fd-GOGAT* in durum wheat has been identified associated with GPC that is potentially useful in breeding programs (Nigro et al. 2017).

Genome-Wide Association Studies (GWAS) on the same collection allowed to identify fourteen candidate genes for QTLs related to N metabolism, among which *NADH-GOGAT* (Nigro et al. 2019). *NADH-GOGAT* gene located in QTL on chromosome 3A and 3B was found to be a major candidate for cereal NUE (Cormier et al. 2016), as highlighted by the meta-QTL study carried out by Quraishi et al. (2011). Synteny analysis showed that a NUE QTL on 3B co-localized with GOGAT gene and was conserved across wheat, rice, sorghum, and maize (Quraishi et al. 2011). Interestingly, in wheat, the homeologs on 3A chromosome was found to co-localize with QTL of N at anthesis, N uptake at anthesis, and straw N concentration at maturity, while the homoeologous gene on chromosome 3B co-localized with QTL of N utilization efficiency to protein, GPC, and post-anthesis absorption (Cormier et al. 2016).

The other candidate genes involved alanine aminotransferase (*AlaAT*), asparagine synthetase (*ASN1*), Nitrate reductase (*NR*), Nitrite reductase (*NIR*), nitrate transporter 2, *NRT2*, and three GS (*GS1*, *GS2*, and *GSr*), all coding the major enzymes regulating N uptake and assimilation process. A schematic representation of the metabolism is reported in Fig. 1.

In wheat, QTLs contributed by these enzymes have not been widely used in breeding because of their minor effects.

5.1 Candidate Genes Involved in N Uptake

Since the first NUE step is N uptake from the soil, the transport proteins are key targets for improving N efficiency. N is present in the soil in organic, inorganic, or mineral form. The organic form is not directly utilizable by plants and must be chemically modified by the mineralization process, by which it is converted in ammoniacal N first and then nitric N (Robertson and Groffman 2015).

Higher plants are capable of absorbing inorganic nitrogen through their roots, primarily in the form of nitrate ion (NO_3^-), the most prevalent inorganic nitrogen form in the rhizosphere, as well as ammonium ion (NH_4^+), which is present at significantly lower concentrations compared to NO_3^-. Under certain conditions, plants can also absorb nitrite (NO_2^-) or simple organic molecules such as urea and certain amino acids (Crawford and Glass 1998; Schimel and Bennett 2004; Kant 2018). Furthermore, it has been widely demonstrated that leaves can absorb N from nitrogenous compounds present in the air, such as ureic N and N oxides, as well as from foliar-applied nutrients (Fageria et al. 2009; Fernandez and Brown 2013).

The type of inorganic N present and their concentrations depend on the type of soil; for these reasons, plants have developed different N absorption strategies, highly efficient in radical cells (Bernard and Habash 2009).

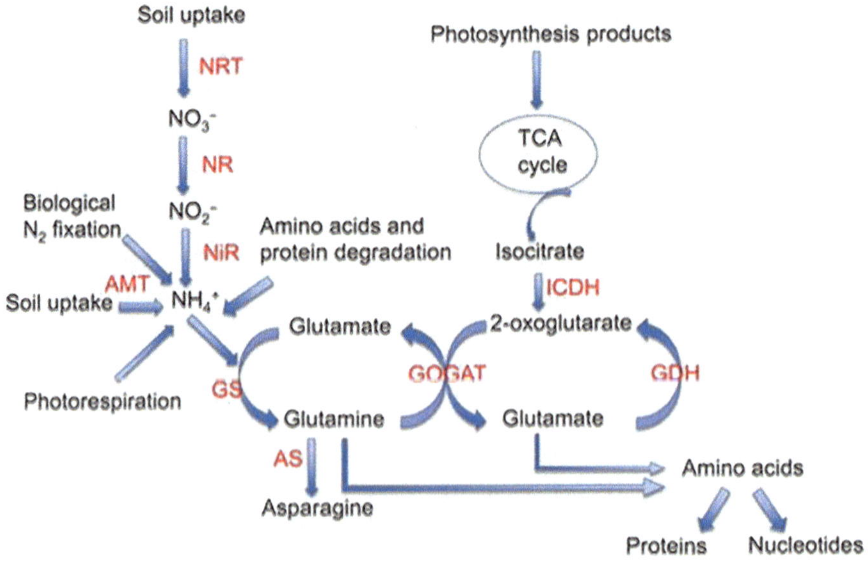

Fig. 1 Nitrogen-assimilation pathway in higher plants. Inorganic nitrogen in the form of nitrate or ammonia becomes incorporated into amino acids and other organic molecules as depicted. The specific steps shown include nitrate transporters (NRT), nitrate reductase (NR), nitrite reductase (NiR), ammonium transporters (AMT), glutamine synthetase (GS), glutamate synthase (GOGAT), asparagine synthetase (AS), glutamate dehydrogenase (GDH), and isocitrate dehydrogenase (ICDH). (From Lu et al. 2016)

When available, NH_4^+ is plant preferred nitrogen source, as it costs less energy to be assimilated compared to NO_3^-. Ammonium, however, being a positively charged ion, is rarely available for direct absorption, because, particularly in clay soils, it binds to negatively charged colloids. On the other hand, the NO_3^- ion, being almost completely dissolved in the circulating solution, can be easily adsorbed by the roots (Wang et al. 2018).

The primary uptake of both NO_3^- and NH_4^+ ions happens by diffusion and mass flow, respectively, thus ensuring that both inorganic forms penetrate root apoplasts (Mandal et al. 2018). Nevertheless, plants have developed many active transporters which further ensure that the absorbed nitrogenous compounds (both ammoniacal and nitric forms) reach the plant vascular system through the plasma membrane of cortical root cells and root epidermal cells.

Plants are able to accumulate nitrate to high concentrations, the majority of which is concentrated in the vacuole by an NO_3^-/H+ exchanger (De Angeli et al. 2006). Plants also have the capacity to grow successfully in a wide range of available nitrate concentrations. Because of these two properties, the uptake of nitrate by plant roots is not a simple process, utilizing at least four different transport systems: (a) constitutive high-affinity (cHATS), (b) nitrate-inducible high affinity (iHATS), (c) constitutive low-affinity (cLATS), and (d) nitrate-inducible low-affinity (iLATS) (Loqué and Wirén 2004; Okamoto et al. 2006; Glass 2009; Dechorgnat et al. 2011). HATS work at low N concentrations, while LATS work at high N concentrations. If the anion concentration is low, the plant absorbs it with a low flow, but high efficiency. HATS allow plants to maximize nitrate acquisition when limited or no N fertilizer is applied (Malagoli et al. 2004). On the contrary, when nitrate concentrations are high, the transport takes place with a higher flow, but with a lower efficiency (Noguero and Lacombe 2016).

Two major gene families encoding nitrate transporters (NRT) have been identified in higher plants, the low-affinity *NRT1* and the high-affinity *NRT2* (Tsay et al. 2007). Subsequent studies led these two gene families to be renamed: the *NRT1* gene group has been retitled as *NRT1 PTR* (Nitrate Transporter, Peptide Transporter) family (*NPF*) (Léran et al. 2014), while the *NRT2* family was referred as the Major Facilitator Superfamily (*MFS*) (Niño-González et al. 2019). The NRT1 proteins, including LATS and dual-affinity nitrate transporters, have a wide range of substrate affinities (Guo et al. 2002).

Studies focused on mutants and gene expression showed that, when the nitrate concentration in the soil is in the micromolar range, the NRT2 proteins contribute to the iHATS (Okamoto et al. 2003; Little et al. 2005). Overexpression of *NRT2.1* gene in *Nicotiana plumbaginifolia* (Fraisier et al. 2000) pointed out that, when transgenic plants were grown in high nitrate concentrations, the rate of nitrate uptake was three or four times higher than the wild type; however, the nitrate concentration in the wild-type and transgenic plants was found to be remarkably similar, irrespective of the level of nitrate supplied. Moreover, when *NRT2.1* overexpressing plants were supplied with ammonium nitrate, there was massive difference in *NRT2.1* mRNA expression between the wild-type and the transgenic plants; yet, the influx of ^{15}N-nitrate was very similar, thus suggesting that NRT2 proteins contribute to the iHATS.

Wang et al. (2011) isolated and characterized a *NRT2.1* transporter belonging to the HATS gene family in bread wheat and showed that its transcript abundances decreased in the roots in response to NO_3^- and NH_4^+.

Functional characterization of *NRT1* gene family has been so far conducted in few crop species, including bread wheat, where Buchner and Hawkesford (2014) identified 16 NPF genes, localized in all chromosomes except 4A. Furthermore, these genes were found having tissue specific expression and transcriptional responses to N supply (Puccio et al. 2023). For instance, some gene transcripts, such as *NPF6.1*, *NPF6.2,* and *NRT6.4* genes, were highly abundant in roots but rare in shoots, while *NPF6.3* had similar abundance in both roots and shoots. Such an intricate and variable expression pattern in different tissues, also depending on NO_3^- availabilities, suggested *NRT1* family's genes have different functional roles, which need to be further defined for wheat. Some authors (Cai et al. 2007; Akhtar et al. 2024) also reported that several genes that regulated the high affinity transport system in wheat (*TaNRT 2.1*, *TaNRT 2.2*, *TaNRT 2.3*, *TaNAR 2.1*, and *TaNAR2.2*) in the absence of NO_3^- are activated by the plant growth hormone abscisic acid.

In contrast to nitrate, plants in general tend to not accumulate high concentrations of ammonium ions. Toxicity symptoms frequently occur if crop plants are grown in ammonium in the absence of nitrate (Britto and Kronzucker 2002). NH_4^+ accumulation in cells occurs either by direct uptake from the rhizosphere via ammonium transporters (AMTs) or by reduction of NO_3^-. Both HATS and LATS for ammonium uptake are present in plant roots, which are constitutive and do not seem to be significantly induced by ammonium (Glass et al. 2002; Von Wirén and Merrick 2004). Like the NRT family, the AMT genes are divided into *AMT1* and *AMT2* based on affinity levels. Li et al. (2017) identified and characterized *TaAMTs* in wheat plants, reporting as many as 23 AMTs in bread wheat genome, specifically belonging to four AMT group (AMT1, AMT2, AMT3, and AMT4). AMT1 is a high affinity transporter, which exploits the membrane potential created by a H^+/ATP-ase pump (Tegeder 2014). A recent paper by Bajgain et al. (2018) focused on phylogenetic analyses and expression of ammonium and nitrate transporters in wheat, with the former being less extensively studied. Only two wheat AMT1 genes, namely *TaAMT1;1* and *TaAMT1;2,* have been reported. The production of AMT mutant lines and their association with some important phenotypic effect might be a valuable strategy to NUE improvement.

5.2 Candidate Genes Involved in N Assimilation: NR and Nir

In addition to genes involved in the primary N uptake, others involved in N assimilation have been identified as potential candidates to enhance NUE, such as nitrate reductase (NR) and nitrite reductase (NiR) (Fortunato et al. 2019; Gayatri et al. 2024). The nitrate reduction occurs both in the leaves and in the roots, particularly in the roots epidermis cells, in the cortical cells, and in the foliar mesophyll cells (Buchanan et al. 2015). All vegetative organs are able of assimilating N through the combined action of the two enzymes, but activity levels change in relation to the tissues considered; in cereals, such as wheat, the highest enzymatic activity is recorded at leaf level, even if it has been detected also in developing ears (Pathak et al. 2008).

NR catalyzes the transfer of two electrons from NAD(P)H to nitrate to produce nitrite, which is further reduced to NH_4^+ by NiR. Depending on the electron donor used in the reaction, three different NR genes have been reported: a NADH-specific, a NAD(P)H bispecific, and a NADPH specific form. In most plant species, as well as hexaploid wheat, two genes encoding NADH-NR have been identified, NADH and NAD(P)H bispecific NRs (Boisson et al. 2005).

Anjana et al. (2011) showed that wheat NR activity is genetically controlled and one genotype exhibiting a higher NR activity also had higher potential for N utilization in non-N limited conditions. The constitutive overexpression of the tobacco NR in two common wheat cultivars led to an increase in enzymatic activity, resulting in an increase in thousand seed weight and an increase in about 30% of seed protein content (Zhao et al. 2013), meaning a possible influence on NUE provided that the number of grains per plant, which was not measured on the plants grown in pots, was not affected. To date, no information is currently available for the allelic variation in the NR gene in wheat, but previous studies on both NR genes in rice (OsNR2) (Gao et al. 2019) and sorghum (Diatloff et al. 2017) did not detect any relation between allelic variation and difference to nitrate assimilation capacity and NUE.

At the second step of the nitrate assimilation process, ferredoxin-dependent nitrite reductase (Fd-NiR) catalyzes the reduction of nitrite to ammonium. NiR is a monomer of 60–70 kda consisting of two redox centers: a iron-siroeme and an iron–sulfur center (4Fe-4S). The ferredoxin's binding site, the enzyme's electron donor, is located in the N-terminal portion, while in the C-terminal there is the enzyme active site that binds the substrate (NO_2^-) and the two redox centers (Buchanan et al. 2015). In photosynthetic tissues, NiR is located on the chloroplasts thylakoid membranes, therefore, in order to be reduced to ammonium, NO_2^- ion has to be transferred from the cytosol to plastids (Masclaux-Daubrasse et al. 2010). There is less information about the gene encoding NiR compared to other genes involved in N metabolism. To date, no detailed characterization of this gene in wheat has been described, even though partial sequences have been reported (Boisson et al. 2005) and the gene has been cloned from other cereal species, including maize (Beatty and Good 2018).

Plants have a complex regulation system for NR and NiR, involving both transcriptional and post-transcriptional mechanisms; the former perform a long-term regulatory action (hours/days); post-transcriptional ones, on the other hand, are used by plants to manage sudden changes in enzyme activity (Buchanan et al. 2015). Nitrate is the main regulating signal for N assimilation, but several other exogenous and endogenous signals can be involved. For example, assimilation depends on CO_2 levels, carbon and nitrogen metabolism products (predominantly sucrose and glutamine), hormones availability, and circadian rhythms, that is, alternating light/dark cycles (Kaiser and Huber 1994; Crawford 1995; Pathak 2008).

Several studies have shown the presence of a regulatory region of 3 kb in the NR promoter region; its expression seems to be triggered by low concentrations of nitrate and sucrose and is otherwise inhibited by glutamine. A regulatory region has also been identified in the NiR promoter of spinach (Dorbe et al. 1992; Kaiser and Huber 1994; Crawford 1995; Pathak 2008) but not reported yet on wheat.

In many cases, however, the increase in NR and NiR mRNA is not associated with increased enzyme activity (Foyer and Noctor 2002), which might be due to downstream involvement of post-transcriptional regulation mechanisms.

5.3 Candidate Genes Involved in N Assimilation: The GS/GOGAT Cycle

NH_4^+ generated by nitrate reduction and/or absorbed by the root is incorporated into the amino acids by the action of two enzymes, GS and GOGAT (Glutamine-2-oxoglutarate starch-transferase), which act cooperatively in the GS/GOGAT cycle. GS, one of the most extensively studied enzymes, catalyzes the binding of an NH_4^+ to a molecule of glutamate, ending up with the formation of a glutamine molecule, which is then converted into two glutamate molecules by the GOGAT enzyme, as it transfers the amide group to a 2-oxoglutarate molecule. Of the two produced glutamate molecules, one is used by plants as a source of nitrogen for biosynthesis of other amino acids, nucleic acids and other organic molecules; the latter is used as a substrate by the GS for re-feeding the cycle (Lea and Azevedo 2007).

The two enzymes are also involved both in the secondary N assimilation. In plants, in fact, N is released during numerous metabolic processes, such as the amino acids catabolism (during germination or senescence) or photorespiration. Plants are not able to eliminate NH_4^+, which must therefore be reassimilated to prevent its accumulation at toxic levels. In some cases, the ammonium released during secondary metabolism may be ten times greater than ammonium produced with primary assimilation (Buchanan et al. 2015). Because of its capacity to efficiently incorporate ammonia into organic compounds, GS is considered a major NUE candidate gene, including in wheat, based on its key role for plant growth and production (Brestic et al. 2014; Thomsen et al. 2014).

The first studies on GS showed that it exists in two isoforms: the cytoplasmatic GS1 (38–40 kda), encoded by 3 to 5 nuclear genes (depending on the species), located predominantly in non-photosynthetic organs (Tabuchi et al. 2005;

Cánovas et al. 2007; Bernard et al. 2008), and a plastidic GS2 (44–45 kda), encoded by a single nuclear gene localized in chloroplasts/plastids (Taira et al. 2004). Cytosolic GS1 localized in roots and leaves is responsible for the organization of absorbed N from the external environment (Sukanya et al. 1994), while *GS2*, predominantly localized in chloroplasts, is responsible for the assimilation of NH_3 released during photorespiration (Keys 2006). So far, as differently compartmentalized, GS enzymes play different, not overlapping metabolic roles, such as assimilation and synthesis of ammonium derived from the reduction of nitrate, photorespiration, and N recycling, and are differentially expressed in relation to the different phenological phases and tissues considered.

In 2008, Bernard et al. identified more then 10 sequences coding GS enzymes in bread wheat. Of these 10 sequences, seven encode for the GS1 isoform; *TaGS1*(a, b, c) encode for GS1;1; *TaGSr1* and *TaGSr2* encode for GS1;2 (also called *GSr*); and *TaGSe1* and *TaGSe2* encode for GS1;3 (also called *GSe*). Three alleles encode the GS2, TaGS2 (a, b, c). This complexity probably reflects the hexaploid nature of the common wheat, as the three GS2 genes each come from different homoeologous chromosome sets. GS2 genes are localized on homoeologous chromosomes 2, GS1 genes on chromosomes 6 and *GSe*, and *GSr* genes on chromosome 4 (Habash et al. 2007; Li et al. 2024).

In 2011, Gadaleta et al. reported the *GS2* gene sequencing and genetic mapping in durum wheat. The homoeologous genes mapped on 2A and 2B chromosomes and are composed of 13 exons and 12 introns. The same RIL population was used to genetically map the *GSe* isoform on 4A chromosome, which was shown to have a similar genetic structure (Gadaleta et al. 2013). The same gene structure was reported for bread wheat *GS2* genes (Li et al. 2011); furthermore, Gadaleta et al. (2011) analyzed GS2 haplotype variations and investigated their associations with traits related to yield and N use, suggesting that certain haplotypes may be valuable in breeding wheat varieties with improved agronomic performance and NUE. Furthermore, a paper from Nigro et al. (2016) examined the role of GS2 in terms of NUE and GPC in durum wheat. Along with genotypic variations, in general it was observed that genotypes having higher GPC exhibited also higher values of GS activity.

GS1 appears to be involved in controlling some productivity parameters for cereals. In wheat plants, the overexpression of a *Phaseolus vulgaris* GS1 improved by fourfold the glutamine content in leaves, resulting in an increase in productivity in terms of dry weight of seeds and increase in root mass (Habash et al. 2001).

Depending on the electron donor used in the reaction, GOGAT exists in plants as two distinct forms: a NADH-dependent and a ferredoxin (Fd)-dependent GOGAT. Both GOGAT isoforms are located in plastids, as demonstrated by enzymatic activity (Lea and Miflin 2003) and immunolocalization studies (Cánovas et al. 2007; Tabuchi et al. 2007). The Fd-GOGAT is more active in photosynthetic tissue, where it can use the energy of light as a reducing source, but to a lesser extent was also detected at root level. The NADH-GOGAT is predominantly localized in plastids of non-photosynthetic tissue cells, where electrons necessary for reduction are provided by the pentose phosphate pathway (Bowsher et al. 2007). Genes for Fd-GOGAT have been mapped in durum wheat on 2A and 2B homoeologous chromosomes and their sequencing and characterization showed that they consist of 33 exons and 32 introns (Nigro et al. 2014). In durum wheat, the genes coding NADH-GOGAT have a gene structure consisting of 22 exons and 21 introns and are located on the homoeologous chromosomes 3A and 3B; qRT-PCR experiments found that the two genes have expression levels different in cultivars with different GPC (Nigro et al. 2013). Recently were identified the TaGOGATs gene family through bioinformatics analysis of which six genes located on chromosomes group 2 and 3. TaGOGAT2s was mainly expressed in leaves and flag leaves, while TaGOGAT3s was mainly expressed in roots and leaves (Li et al. 2024).

Literature studies have shown that Fd- and NADH-GOGAT have a localization and a not overlapping function. NADH-GOGAT is mainly involved in the primary assimilation of ammonium and in the re-synthesis of NH_4^+ issued by catabolism of amino acids. About 97% of all GOGAT activity measured in leaves is instead due to the Fd-dependent isoform (Buchanan et al. 2015). Its activity is induced by light, or by exogenously supplied sucrose, indicating that its function is closely associated with secondary N assimilation, dependent on photosynthetic processes and photoreflectors (Lam et al. 1996; Potel et al. 2009).

There also appears to be coordination in gene expression of NADH-GOGAT and GS1, and between Fd-GOGAT and GS2, suggesting a functional coordination between the isoforms of the two enzymes (Dragićević et al. 2016; Wang et al. 2024). Immuno-localization studies have allowed to detect the presence of NADH-GOGAT in vascular parenchyma cells, in cells of the young leaves in development, and in the dorsal cells of the caryopsis at the first developmental stages, assuming a role of the enzyme in the formation of reserve organs and in the reuse of glutamine (Hayakawa et al. 1994). Over-expression of NADH-GOGAT gene was found to be correlated to higher kernel weight in both wheat and rice, while its silencing determined the opposite effect (Yamaya et al. 2002; Lu and Hu 2011). These results highlighted the key role played by NADH-GOGAT in N utilization and grain filling, and ultimately in grain yield.

6 Conclusion and Future Prospects

Better insights into the genetic basis of N uptake and utilization in wheat will provide useful information for developing suitable breeding programs. Along the improved agronomical practices, for example, better timed applications of N fertilizer with optimized formulations, they will enable to increase NUE in farmers' fields.

To target molecular and physiological mechanisms in specific programs aiming to develop more efficient wheat varieties, the identification and characterization of NUE QTLs would be useful. However, there is still currently very limited information available on the genetic determinism of NUE, particularly in wheat, and specifically in durum wheat. Most studies about NUE have been focused on the two major cereal crops, maize and rice (Garnett et al. 2015). More than 40 papers have been published on wheat NUE (Hirel et al. 2007; Dawson et al. 2008; Le Gouis 2011; Cormier et al. 2016; Le Gouis and Hawkesford 2017; Swarbreck et al. 2019). As previously described, NUE-related traits are typical complex quantitative traits (Mandal et al. 2018). Indeed, not only they are determined by multiple genetic factors, involved in several biochemical pathways (genes involved in all nitrogen uptake, assimilation, and remobilization processes, as well as genes involved in C metabolism, redox metabolism, and root/shoot development), but they are also largely influenced by environmental factors, thus hindering NUE genetic components dissection. Thus, it is essential to document these NUE QTLs in multi-environments networks using wide genetic material to propose innovative strategies to improve NUE according to the targeted environments.

The discovery of molecular markers tightly linked to NUE loci would allow the design of marker-assisted selection (MAS) program and the selection for important NUE traits at early plant stages. Furthermore, knowledge of NUE candidate genes will be of primary importance in transgenic and genome editing studies (Kumar et al. 2024). For this, positional cloning is a valuable tool to narrow down a QTL of interest to a very small region on the wheat chromosome and clone the gene of interest (Govta et al. 2024). The high-density SNP arrays including the 90 K (Wang et al. 2014), 280 k (Rimbert et al. 2018), 660 K (Sun et al. 2020), and 820 K (Winfield et al. 2016) have provided a good foundation for wide applications of map-based cloning. The whole-genome sequence recently released for bread, durum and tetraploid wild emmer wheat "Zavitan" (Appels et al. 2018; Ramírez-González et al. 2018; Maccaferri et al. 2019; Avni et al. 2017), along with the recent ten reference-quality pseudomolecule assemblies and five scaffold-level assemblies (Walkowiak et al. 2020) are determinant to go further.

All these works are essential in the context of global changes. Future climatic scenarios predict an increase in average temperature (IPCC 2021) that will lead to shorter growth cycle with accelerated development rates and, even in the case of no change in the amount of rainfall, to less available water due to increased evapotranspiration. Furthermore, predicted regional changes include increases in pluvial flooding in northern Europe and agricultural droughts in the Mediterranean (IPCC 2021). Some studies also reported that the elevated atmospheric carbon dioxide concentration could interfere with the N metabolism resulting in less nitrates assimilation by wheat plants (Bloom et al. 2010). In addition, economic and ecological constraints may favor agricultural systems with less phytosanitary and fertilizer inputs. All these factors will probably increase the genotype by environmental interactions and the necessity to breed for durum wheat cultivars efficient and specifically adapted to growing regions and agricultural systems.

Acknowledgments This chapter represents an output from discussions between some of the authors as part of the expert working group on Nutrient Use Efficiency of the international Wheat Initiative. The International Wheat Initiative (www.wheatinitiative.org) provides a framework to establish strategic research and identify key challenges.

References

Abad A, Lloveras J, Michelena A (2004) Nitrogen fertilization and foliar urea effects on durum wheat yield and quality and on residual soil nitrate in irrigated Mediterranean conditions. Field Crop Res 87(2–3):257–269

Ahanger MA, Qin C, Begum N, Maodong Q, Dong XX, El-Esawi M, El-Sheikh MA, Alatar AA, Zhang L (2019) Nitrogen availability prevents oxidative effects of salinity on wheat growth and photosynthesis by up-regulating the antioxidants and osmolytes metabolism, and secondary metabolite accumulation. BMC Plant Biol 19(1):1–12

Akhtar K, Ain N, Prasad PVV, Naz M, Aslam MM, Djalovic I, Riaz M, Ahmad S, Varshney RK, He B, Wen R (2024) Physiological, molecular, and environmental insights into plant nitrogen uptake, and metabolism under abiotic stresses. Plant Genome 17:e20461. https://doi.org/10.1002/tpg2.20461

Aleliunas A, Gorash A, Armoniene R, Tamm I, Ingver A, Bleidere M, Fetere V, Kollist H, Mroz T, Lillemo M, Brazauskas G (2024) Genome-wide association study reveals 18 QTL for major agronomic traits in a Nordic–Baltic spring wheat germplasm. Front Plant Sci 15:1393170

An D, Su J, Liu Q, Zhu Y, Tong Y, Li J, Jing R, Li B, Li Z (2006) Mapping QTLs for nitrogen uptake in relation to the early growth of wheat (Triticum aestivum L.). Plant Soil 284(1):73–84

Anjana U, Abrol YP, Iqbal M (2011) Modulation of nitrogen-utilization efficiency in wheat genotypes differing in nitrate reductase activity. J Plant Nutr 34(6):920–933

Appels R, Eversole K, Stein N, Feuillet C, Keller B, Rogers J, Pozniak CJ, Choulet F, Distelfeld A, Poland J, Ronen G (2018) Shifting the limits in wheat research and breeding using a fully annotated reference genome. Science 361(6403):eaar7191

Arduini I, Masoni A, Ercoli L, Mariotti M (2006) Grain yield, and dry matter and nitrogen accumulation and remobilization in durum wheat as affected by variety and seeding rate. Eur J Agron 25(4):309–318

Arriagada O, Gadaleta A, Marcotuli I, Maccaferri M, Campana M, Reveco S, Alfaro C, Matus I, Schwember AR (2022) A comprehensive meta-QTL analysis for yield-related traits of durum wheat (*Triticum turgidum* L. var. *durum*) grown under different water regimes. Front Plant Sci 13:984269

Avni R, Nave M, Barad O, Baruch K, Twardziok SO, Gundlach H, Hale I, Mascher M, Spannagl M, Wiebe K, Jordan KW (2017) Wild emmer genome architecture and diversity elucidate wheat evolution and domestication. Science 357(6346):93–97

Ayadi S, Karmous C, Hammami Z, Trifa Y, Rezgui S (2014) Variation of durum wheat yield and nitrogen use efficiency under Mediterranean rainfed environment. Int J Agric Crop Sci 7(10):693

Ayadi S, Jallouli S, Chamekh Z, Zouari I, Landi S, Hammami Z, Azaiez FEB, Baraket M, Esposito S, Trifa Y (2022) Variation of grain yield, grain protein content and nitrogen use efficiency components under different nitrogen rates in Mediterranean Durum wheat genotypes. Agriculture 12(7):916. https://doi.org/10.3390/agriculture12070916

Bajgain P, Russell B, Mohammadi M (2018) Phylogenetic analyses and in-seedling expression of ammonium and nitrate transporters in wheat. Sci Rep 8(1):1–13

Bandyopadhyay T, Prasad M (2020) A precise method for analyzing nitrogen use in Foxtail Millet. In: Nitrogen metabolism in plants. Humana, New York, pp 113–118

Barraclough PB, Howarth JR, Jones J, Lopez-Bellido R, Parmar S, Shepherd CE, Hawkesford MJ (2010) Nitrogen efficiency of wheat: genotypic and environmental variation and prospects for improvement. Eur J Agron 33(1):1–11

Basso B, Cammarano D, Troccoli A, Chen D, Ritchie JT (2010) Long-term wheat response to nitrogen in a rainfed Mediterranean environment: field data and simulation analysis. Eur J Agron 33(2):132–138

Beatty PH, Good AG (2018) Improving nitrogen use efficient in crop plants using biotechnology approaches. In: Engineering nitrogen utilization in crop plants. Springer, Cham, pp 15–35

Bernard SM, Habash DZ (2009) The importance of cytosolic glutamine synthetase in nitrogen assimilation and recycling. New Phytol 182(3):608–620

Bernard SM, Møller ALB, Dionisio G, Kichey T, Jahn TP, Dubois F, Baudo M, Lopes MS, Tercé-Laforgue T, Foyer CH, Parry MA (2008) Gene expression, cellular localisation and function of glutamine synthetase isozymes in wheat (Triticum aestivum L.). Plant Mol Biol 67(1):89–105

Bingham IJ, Wu L (2011) Simulation of wheat growth using the 3D root architecture model SPACSYS: validation and sensitivity analysis. Eur J Agron 34(3):181–189

Biradar SS, Patil MK, Desai SA, Singh SK, Naik VR, Lamani K, Joshi AK (2024) Nitrogen use efficiency in bread wheat: genetic variation and prospects for improvement. PLoS One 19(4):e0294755. https://doi.org/10.1371/journal.pone.0294755

Blanco A, Mangini G, Giancaspro A, Giove S, Colasuonno P, Simeone R, Signorile A, De Vita P, Mastrangelo AM, Cattivelli L, Gadaleta A (2012) Relationships between grain protein content and grain yield components through quantitative trait locus analyses in a recombinant inbred line population derived from two elite durum wheat cultivars. Mol Breed 30(1):79–92

Bloom AJ, Burger M, Asensio JSR, Cousins AB (2010) Carbon dioxide enrichment inhibits nitrate assimilation in wheat and Arabidopsis. Science 328(5980):899–903

Bogard M, Allard V, Brancourt-Hulmel M, Heumez E, Machet JM, Jeuffroy MH, Gate P, Martre P, Le Gouis J (2010) Deviation from the grain protein concentration– grain yield negative relationship is highly correlated to post-anthesis N uptake in winter wheat. J Exp Bot 61(15):4303–4312

Bogard M, Allard V, Martre P, Heumez E, Snape JW, Orford S, Griffiths S, Gaju O, Foulkes J, Le Gouis J (2013) Identifying wheat genomic regions for improving grain protein concentration independently of grain yield using multiple inter-related populations. Mol Breed 31(3):587–599

Boisson M, Mondon K, Torney V, Nicot N, Laine AL, Bahrman N, Gouy A, Daniel-Vedele F, Hirel B, Sourdille P, Dardevet M (2005) Partial sequences of nitrogen metabolism genes in hexaploid wheat. Theor Appl Genet 110(5):932–940

Bowsher CG, Lacey AE, Hanke GT, Clarkson DT, Saker LR, Stulen I, Emes MJ (2007) The effect of Glc6P uptake and its subsequent oxidation within pea root plastids on nitrite reduction and glutamate synthesis. J Exp Bot 58(5):1109–1118

Brady DJ, Wenzel CL, Fillery IRP, Gregory PJ (1995) Root growth and nitrate uptake by wheat (Triticum aestivum L.) following wetting of dry surface soil. J Exp Bot 46(5):557–564

Brestic M, Zivcak M, Olsovska K, Shao HB, Kalaji HM, Allakhverdiev SI (2014) Reduced glutamine synthetase activity plays a role in control of photosynthetic responses to high light in barley leaves. Plant Physiol Biochem 81:74–83

Britto DT, Kronzucker HJ (2002) NH4+ toxicity in higher plants: a critical review. J Plant Physiol 159(6):567–584

Buchanan BB, Gruissem W, Jones RL (eds) (2015) Biochemistry and molecular biology of plants. Wiley

Buchner P, Hawkesford MJ (2014) Complex phylogeny and gene expression patterns of members of the NITRATE TRANSPORTER 1/PEPTIDE TRANSPORTER family (NPF) in wheat. J Exp Bot 65(19):5697–5710

Cabas-Lühmann P, Schwember AR, Arriagada O, Marcotuli I, Matus I, Alfaro C, Gadaleta A (2024) Meta-QTL analysis and candidate genes for quality traits, mineral content, and abiotic-related traits in wild emmer. Front Plant Sci 15:1305196

Cai C, Zhao XQ, Zhu YG, Li B, Tong YP, Li ZS (2007) Regulation of the high-affinity nitrate transport system in wheat roots by exogenous abscisic acid and glutamine. J Integr Plant Biol 49(12):1719–1725

Cane MA, Maccaferri M, Nazemi G, Salvi S, Francia R, Colalongo C, Tuberosa R (2014) Association mapping for root architectural traits in durum wheat seedlings as related to agronomic performance. Mol Breed 34(4):1629–1645

Cánovas FM, Avila C, Canton FR, Canas RA, de la Torre F (2007) Ammonium assimilation and amino acid metabolism in conifers. J Exp Bot 58(9):2307–2318

Cao P, Lu C, Yu Z (2018) Historical nitrogen fertilizer use in agricultural ecosystems of the contiguous United States during 1850–2015: application rate, timing, and fertilizer types. Earth Syst Sci Data 10(2):969–984

Chen L, Liao H (2017) Engineering crop nutrient efficiency for sustainable agriculture. J Integr Plant Biol 59(10):710–735

Cormier F, Foulkes J, Hirel B, Gouache D, Moënne-Loccoz Y, Le Gouis J (2016) Breeding for increased nitrogen-use efficiency: a review for wheat (T. aestivum L.). Plant Breed 135(3):255–278

Crawford NM (1995) Nitrate: nutrient and signal for plant growth. Plant Cell 7(7):859

Crawford NM, Glass AD (1998) Molecular and physiological aspects of nitrate uptake in plants. Trends Plant Sci 3(10):389–395

Cui F, Fan X, Chen M, Zhang N, Zhao C, Zhang W, Han J, Ji J, Zhao X, Yang L, Zhao Z (2016) QTL detection for wheat kernel size and quality and the responses of these traits to low nitrogen stress. Theor Appl Genet 129(3):469–484

Dawson JC, Huggins DR, Jones SS (2008) Characterizing nitrogen use efficiency in natural and agricultural ecosystems to improve the performance of cereal crops in low-input and organic agricultural systems. Field Crop Res 107(2):89–101

De Angeli A, Monachello D, Ephritikhine G, Frachisse JM, Thomine S, Gambale F, Barbier-Brygoo H

(2006) The nitrate/proton antiporter AtCLCa mediates nitrate accumulation in plant vacuoles. Nature 442(7105):939–942

Dechorgnat J, Nguyen CT, Armengaud P, Jossier M, Diatloff E, Filleur S, Daniel-Vedele F (2011) From the soil to the seeds: the long journey of nitrate in plants. J Exp Bot 62(4):1349–1359

Diatloff E, Mace ES, Jordan DR, Filleur S, Tai S, Schmidt S, Godwin ID (2017) The vegetative nitrogen response of sorghum lines containing different alleles for nitrate reductase and glutamate synthase. Mol Breed 37(11):1–14

Dobermann A (2007) Nutrient use efficiency-measurement and management. In: Proc. of International Fertilizer Industry Association (IFA) workshop on fertilizer best management practices, 7–9 March, Brussels, Belgium, p 22

Dorbe MF, Caboche M, Daniel-Vedele F (1992) The tomato nia gene complements a Nicotiana plumbaginifolia nitrate reductase-deficient mutant and is properly regulated. Plant Mol Biol 18(2):363–375

Dragićević M, Simonović A, Bogdanović M, Subotić A, Ghalawenji N, Dragićević I, Todorović S (2016) Differential regulation of GS-GOGAT gene expression by plant growth regulators in Arabidopsis seedlings. Arch Biol Sci 68(2):399–404

Erisman JW, Sutton MA, Galloway J, Klimont Z, Winiwarter W (2008) How a century of ammonia synthesis changed the world. Nat Geosci 1(10):636–639

Ernst OR, Kemanian AR, Mazzilli S, Siri-Prieto G, Dogliotti S (2020) The dos and don'ts of no-till continuous cropping: evidence from wheat yield and nitrogen use efficiency. Field Crop Res 257:107934

Fageria NK, Baligar VC, Li YC (2008) The role of nutrient efficient plants in improving crop yields in the twenty first century. J Plant Nutr 31(6):1121–1157

Fageria NK, Filho MB, Moreira A, Guimarães CM (2009) Foliar fertilization of crop plants. J Plant Nutr 32(6):1044–1064

FAO (2020) Statistical yearbook, available at: https://www.fao.org/3/cb1329en/online/cb1329en.html#chapter-2.

FAO (2021) Inorganic fertilizers 1961–2019. FAOSTAT Analytical Brief Series No 27. Rome.

Fernández V, Brown PH (2013) From plant surface to plant metabolism: the uncertain fate of foliar-applied nutrients. Front Plant Sci 4:289

Fischer RA, Byerlee DR, Edmeades GO (2009) Can technology deliver on the yield challenge to 2050? (No. 138-2016 2032). FAO expert meeting on how to feed the world in 2050. FAO, Rome, pp 24–26

Fixen P, Brentrup F, Bruulsema T, Garcia F, Norton R, Zingore S (2015) Nutrient/fertilizer use efficiency: measurement, current situation and trends. In: Managing water and fertilizer for sustainable agricultural intensification, 270

Follett JR, Follett RF, Herz WC (2010) Environmental and human impacts of reactive nitrogen. In: Advances in nitrogen management for water quality, pp 1–37

Fontaine JX, Ravel C, Pageau K, Heumez E, Dubois F, Hirel B, Le Gouis J (2009) A quantitative genetic study for elucidating the contribution of glutamine synthetase, glutamate dehydrogenase and other nitrogen-related physiological traits to the agronomic performance of common wheat. Theor Appl Genet 119(4):645–662

Fortunato S, Nigro D, Paradiso A, Cucci G, Lacolla G, Trani R, Agrimi G, Blanco A, Concetta M, de Pinto, and Agata Gadaleta. (2019) Nitrogen metabolism at Tillering stage differently affects the grain yield and grain protein content in two Durum wheat cultivars. Diversity 11(10):186. https://doi.org/10.3390/d11100186

Fortunato S, Nigro D, Lasorella C, Marcotuli I, Gadaleta A, Concetta M, de Pinto. (2023) The role of Glutamine Synthetase (GS) and Glutamate Synthase (GOGAT) in the improvement of nitrogen use efficiency in cereals. Biomolecules 13(12):1771. https://doi.org/10.3390/biom13121771

Foyer CH, Noctor G (2002) Photosynthetic nitrogen assimilation: inter-pathway control and signaling. In: Photosynthetic nitrogen assimilation and associated carbon and respiratory metabolism. Springer, Dordrecht, pp 1–22

Fraisier V, Gojon A, Tillard P, Daniel-Vedele F (2000) Constitutive expression of a putative high-affinity nitrate transporter in Nicotiana plumbaginifolia: evidence for post-transcriptional regulation by a reduced nitrogen source. Plant J 23(4):489–496

Gadaleta A, Nigro D, Giancaspro A, Blanco A (2011) The glutamine synthetase (*GS2*) genes in relation to grain protein content of durum wheat. Funct Integ Genom 11(4):665–670

Gadaleta A, Nigro D, Marcotuli I, Giancaspro A, Giove SL, Blanco A (2013) Isolation and characterisation of cytosolic glutamine synthetase (GSe) genes and association with grain protein content in durum wheat. Crop Pasture Sci 65(1):38–45

Gahoonia TS, Ali O, Sarker A, Nielsen NE, Rahman MM (2006) Genetic variation in root traits and nutrient acquisition of lentil genotypes. J Plant Nutr 29(4):643–655

Gaju O, Allard V, Martre P, Le Gouis J, Moreau D, Bogard M, Hubbart S, Foulkes MJ (2014) Nitrogen partitioning and remobilization in relation to leaf senescence, grain yield and grain nitrogen concentration in wheat cultivars. Field Crop Res 155:213–223

Gao Z, Wang Y, Chen G, Zhang A, Yang S, Shang L, Wang D, Ruan B, Liu C, Jiang H, Dong G (2019) The indica nitrate reductase gene OsNR2 allele enhances rice yield potential and nitrogen use efficiency. Nat Commun 10(1):1–10

Garnett T, Conn V, Kaiser BN (2009) Root based approaches to improving nitrogen use efficiency in plants. Plant Cell Environ 32(9):1272–1283

Garnett T, Plett D, Heuer S, Okamoto M (2015) Genetic approaches to enhancing nitrogen-use efficiency (NUE) in cereals: challenges and future directions. Funct Plant Biol 42(10):921–941

Gayatri, Ravi M, Chauhan H et al (2024) Characterization and expression analysis of nitrate reductase 6-1ABD

gene in hexaploid bread wheat under different nitrogen regimes. Plant Growth Regul 103:81–100. https://doi.org/10.1007/s10725-023-01082-1
Glass AD (2009) Nitrate uptake by plant roots. Botany 87(7):659–667
Glass AD, Britto DT, Kaiser BN, Kinghorn JR, Kronzucker HJ, Kumar A, Okamoto M, Rawat S, Siddiqi MY, Unkles SE, Vidmar JJ (2002) The regulation of nitrate and ammonium transport systems in plants. J Exp Bot 53(370):855–864
Good AG, Shrawat AK, Muench DG (2004) Can less yield more? Is reducing nutrient input into the environment compatible with maintaining crop production? Trends Plant Sci 9(12):597–605
Govta N et al (2024) Nitrogen deficiency tolerance conferred by introgression of a QTL derived from wild emmer into bread wheat. Theor Appl Genet 137(8):1–18
Griffiths M, Atkinson JA, Gardiner L-J, Swarup R, Pound MP, Wilson MH, Bennett MJ, Wells DM (2022) Identification of QTL and underlying genes for root system architecture associated with nitrate nutrition in hexaploid wheat. J Integr Agric 21(4):917–932. https://doi.org/10.1016/S2095-3119(21)63700-0
Guo FQ, Wang R, Crawford NM (2002) The Arabidopsis dual-affinity nitrate transporter gene AtNRT1. 1 (CHL1) is regulated by auxin in both shoots and roots. J Exp Bot 53(370):835–844
Guo JH, Liu XJ, Zhang Y, Shen JL, Han WX, Zhang WF, Christie P, Goulding KWT, Vitousek PM, Zhang FS (2010) Significant acidification in major Chinese croplands. Science 327(5968):1008–1010
Guo Y, Kong FM, Xu YF, Zhao Y, Liang X, Wang YY, An DG, Li SS (2012) QTL mapping for seedling traits in wheat grown under varying concentrations of N, P and K nutrients. Theor Appl Genet 124(5):851–865
Habash DZ, Massiah AJ, Rong HL, Wallsgrove RM, Leigh RA (2001) The role of cytosolic glutamine synthetase in wheat. Ann Appl Biol 138(1):83–89
Habash DZ, Bernard S, Schondelmaier J, Weyen J, Quarrie SA (2007) The genetics of nitrogen use in hexaploid wheat: N utilisation, development and yield. Theor Appl Genet 114(3):403–419
Han M, Okamoto M, Beatty PH, Rothstein SJ, Good AG (2015) The genetics of nitrogen use efficiency in crop plants. Annu Rev Genet 49:269–289
Han X, Zhang M, Gao M, Yuan Y, Yuan Y, Zhang G, An Y, Guo Y, Kong F, Li S (2023) QTL mapping and candidate gene identifying for N, P, and K use efficiency at the maturity stages in wheat. Genes 14(6):1168
Hawkesford MJ (2017) Genetic variation in traits for nitrogen use efficiency in wheat. J Exp Bot 68(10):2627–2632
Hayakawa T, Nakamura T, Hattori F, Mae T, Ojima K, Yamaya T (1994) Cellular localization of NADH-dependent glutamate-synthase protein in vascular bundles of unexpanded leaf blades and young grains of rice plants. Planta 193(3):455–460
Heffer P, Gruère A, Roberts T (2009) Assessment of fertilizer use by crop at the global level. International Fertilizer Industry Association, Paris. www.fertilizer.org/ifa/Home-Page/LIBRARY/Publication-database.html/Assessment-of-Fertilizer-Use-by-Crop-at-the-Global-Level-2006-07-2007-08.html2
Hirel B, Le Gouis J, Ney B, Gallais A (2007) The challenge of improving nitrogen use efficiency in crop plants: towards a more central role for genetic variability and quantitative genetics within integrated approaches. J Exp Bot 58(9):2369–2387
Ierna A, Lombardo GM, Mauromicale G (2016) Yield, nitrogen use efficiency and grain quality in durum wheat as affected by nitrogen fertilization under a Mediterranean environment. Exp Agric 52(2):314–329
Ingraffia R, Porto AL, Ruisi P, Amato G, Giambalvo D, Frenda AS (2023) Conventional tillage versus no-tillage: nitrogen use efficiency component analysis of contrasting durum wheat genotypes grown in a Mediterranean environment. Field Crop Res 296:108904
International Grains Council [IGC] (2020) World Grain Statistics 2016. Available: https://www.igc.int/en/subscriptions/subscription.aspx.
Juraniec M, Hermans C, Salis P, Geelen D, Verbruggen N (2017) Impact of post-flowering nitrate availability on nitrogen remobilization in hydroponically grown durum wheat. J Plant Nutr Soil Sci 180(3):273–278
Kade M, Barneix AJ, Olmos S, Dubcovsky J (2005) Nitrogen uptake and remobilization in tetraploid 'Langdon'durum wheat and a recombinant substitution line with the high grain protein gene Gpc-B1. Plant Breed 124(4):343–349
Kaiser WM, Huber SC (1994) Posttranslational regulation of nitrate reductase in higher plants. Plant Physiol 106(3):817
Kant S (2018) February. Understanding nitrate uptake, signaling and remobilisation for improving plant nitrogen use efficiency. In: Seminars in cell and developmental biology, vol 74. Academic Press, pp 89–96
Karrou M, Nachit M (2015) Durum wheat genotypic variation of yield and nitrogen use efficiency and its components under different water and nitrogen regimes in the Mediterranean region. J Plant Nutr 38(14):2259–2278
Keys AJ (2006) The re-assimilation of ammonia produced by photorespiration and the nitrogen economy of C 3 higher plants. Photosynth Res 87(2):165–175
Kichey T, Heumez E, Pocholle D, Pageau K, Vanacker H, Dubois F, Le Gouis J, Hirel B (2006) Combined agronomic and physiological aspects of nitrogen management in wheat highlight a central role for glutamine synthetase. New Phytol 169(2):265–278
Kichey T, Hirel B, Heumez E, Dubois F, Le Gouis J (2007) In winter wheat (Triticum aestivum L.), post-anthesis nitrogen uptake and remobilisation to the grain correlates with agronomic traits and nitrogen physiological markers. Field Crop Res 102(1):22–32
Krapp A, Saliba-Colombani V, Daniel-Vedele F (2005) Analysis of C and N metabolisms and of C/N interactions using quantitative genetics. Photosynth Res 83(2):251–263

Kumar A, Subbaiah M, Roy J et al (2024) Strategies to utilize genome editing for increasing nitrogen use efficiency in crops. Nucleus 67:205–225. https://doi.org/10.1007/s13237-024-00475-5

Ladha JK, Pathak H, Krupnik TJ, Six J, van Kessel C (2005) Efficiency of fertilizer nitrogen in cereal production: retrospects and prospects. Adv Agron 87:85–156

Lam HM, Coschigano KT, Oliveira IC, Melo-Oliveira R, Coruzzi GM (1996) The molecular-genetics of nitrogen assimilation into amino acids in higher plants. Annu Rev Plant Biol 47(1):569–593

Laperche A, Devienne-Barret F, Maury O, Le Gouis J, Ney B (2006) A simplified conceptual model of carbon/nitrogen functioning for QTL analysis of winter wheat adaptation to nitrogen deficiency. Theor Appl Genet 113(6):1131–1146

Laperche A, Brancourt-Hulmel M, Heumez E, Gardet O, Hanocq E, Devienne-Barret F, Le Gouis J (2007) Using genotype× nitrogen interaction variables to evaluate the QTL involved in wheat tolerance to nitrogen constraints. Theor Appl Genet 115(3):399–415

Le Gouis J (2011) Genetic improvement of nutrient use efficiency in wheat. In: Hawkesford MJ, Barraclough P (eds) Molecular and physiological basis of nutrient use efficiency in crops, pp 123–138

Le Gouis J, Hawkesford MJ (2017) Improving the uptake and assimilation of nitrogen in wheat plants. Peter Langridge. Achieving sustainable cultivation of wheat, vol 1: Breeding, quality traits, pests and diseases. Burleigh Dodds Science Publishing, p 686. 978-1-78676-019-7; 978-1-78676-016-6

Le Gouis J, Fontaine JX, Laperche A, Heumez E, Devienne-Barret F, Brancourt-Hulmel M, Dubois F, Hirel B (2008) Genetic analysis of wheat nitrogen use efficiency: coincidence between QTL for agronomical and physiological traits. Sydney University Press

Lea PJ, Azevedo RA (2007) Nitrogen use efficiency. 2. Amino acid metabolism. Ann Appl Biol 151(3):269–275

Lea PJ, Miflin BJ (2003) Glutamate synthase and the synthesis of glutamate in plants. Plant Physiol Biochem 41(6–7):555–564

Léran S, Varala K, Boyer JC, Chiurazzi M, Crawford N, Daniel-Vedele F, David L, Dickstein R, Fernandez E, Forde B, Gassmann W (2014) A unified nomenclature of NITRATE TRANSPORTER 1/PEPTIDE TRANSPORTER family members in plants. Trends Plant Sci 19(1):5–9

Li XP, Zhao XQ, He X, Zhao GY, Li B, Liu DC, Zhang AM, Zhang XY, Tong YP, Li ZS (2011) Haplotype analysis of the genes encoding glutamine synthetase plastic isoforms and their association with nitrogen-use-and yield-related traits in bread wheat. New Phytol 189(2):449–458

Li H, Hu B, Chu C (2017) Nitrogen use efficiency in crops: lessons from Arabidopsis and rice. J Exp Bot 68(10):2477–2488

Li S, Jiao B, Wang J, Zhao P, Dong F, Yang F, Zhou S (2024) Identification of wheat glutamate Synthetase gene family and expression analysis under nitrogen stress. Genes 15(7):827

Liang Z, Bronson KF, Thorp KR, Mon J, Badaruddin M, Wang G (2014) Cultivar and N fertilizer rate affect yield and N use efficiency in irrigated durum wheat. Crop Sci 54(3):1175–1183

Liao M, Palta JA, Fillery IR (2006) Root characteristics of vigorous wheat improve early nitrogen uptake. Aust J Agric Res 57(10):1097–1107

Little DY, Rao H, Oliva S, Daniel-Vedele F, Krapp A, Malamy JE (2005) The putative high-affinity nitrate transporter NRT2. 1 represses lateral root initiation in response to nutritional cues. Proc Natl Acad Sci 102(38):13693–13698

Liu S, Wang J, Wang L, Wang X, Xue Y, Wu P, Shou H (2009) Adventitious root formation in rice requires OsGNOM1 and is mediated by the OsPINs family. Cell Res 19(9):1110–1119

Liu H, Colombi T, Jäck O, Westerbergh A, Weih M (2022) Linking wheat nitrogen use to root traits: shallow and thin embryonic roots enhance uptake but reduce conversion efficiency of nitrogen. Field Crop Res 285:108603

López-Bellido RJ, Castillo JE, López-Bellido L (2008) Comparative response of bread and durum wheat cultivars to nitrogen fertilizer in a rainfed Mediterranean environment: soil nitrate and N uptake and efficiency. Nutr Cycl Agroecosyst 80(2):121–130

Loqué D, von Wirén N (2004) Regulatory levels for the transport of ammonium in plant roots. J Exp Bot 55(401):1293–1305

Lu MZ, Hu JJ (2011, December) A brief overview of field testing and commercial application of transgenic trees in China. In: BMC proceedings, vol 5, No. 7. BioMed Central, pp 1–1

Lu C, Tian H (2013) Net greenhouse gas balance in response to nitrogen enrichment: perspectives from a coupled biogeochemical model. Glob Chang Biol 19(2):571–588

Lu C, Tian H (2017) Global nitrogen and phosphorus fertilizer use for agriculture production in the past half century: shifted hot spots and nutrient imbalance. Earth Syst Sci Data 9(1):181–192

Lu J, Zhang L, Lewis RS, Bovet L, Goepfert S, Jack AM, Crutchfield JD, Ji H, Dewey RE (2016) Expression of a constitutively active nitrate reductase variant in tobacco reduces tobacco-specific nitrosamine accumulation in cured leaves and cigarette smoke. Plant Biotechnol J 14(7):1500–1510

Lupini A, Preiti G, Badagliacca G, Abenavoli MR, Sunseri F, Monti M, Bacchi M (2020) Nitrogen Use Efficiency (NUE) in Durum wheat under different nitrogen and water regimes in the Mediterranean Basin. Front Plant Sci 11:2304

Maccaferri M, Harris NS, Twardziok SO, Pasam RK, Gundlach H, Spannagl M, Ormanbekova D, Lux T, Prade VM, Milner SG, Himmelbach A (2019) Durum wheat genome highlights past domestication signatures and future improvement targets. Nat Genet 51(5):885–895

Mahjourimajd S, Taylor J, Rengel Z, Khabaz-Saberi H, Kuchel H, Okamoto M, Langridge P (2016) The genetic control of grain protein content under variable nitrogen supply in an Australian wheat mapping population. PLoS One 11(7):e0159371

Malagoli P, Lainé P, Le Deunff E, Rossato L, Ney B, Ourry A (2004) Modeling nitrogen uptake in oilseed rape cv Capitol during a growth cycle using influx kinetics of root nitrate transport systems and field experimental data. Plant Physiol 134(1):388–400

Mălinaş A, Vidican R, Rotar I, Mălinaş C, Moldovan CM, Proorocu M (2022) Current status and future prospective for nitrogen use efficiency in wheat (*Triticum aestivum* L.). Plants 11(2):217. https://doi.org/10.3390/plants11020217

Mandal VK, Sharma N, Raghuram N (2018) Molecular targets for improvement of crop nitrogen use efficiency: current and emerging options. In: Engineering nitrogen utilization in crop plants. Springer, Cham, pp 77–93

Manschadi AM, Hammer GL, Christopher JT, Devoil P (2008) Genotypic variation in seedling root architectural traits and implications for drought adaptation in wheat (*Triticum aestivum* L.). Plant Soil 303(1):115–129

Martinez-Feria RA, Castellano MJ, Dietzel RN, Helmers MJ, Liebman M, Huber I, Archontoulis SV (2018) Linking crop-and soil-based approaches to evaluate system nitrogen-use efficiency and tradeoffs. Agric Ecosyst Environ 256:131–143

Masclaux-Daubresse C, Daniel-Vedele F, Dechorgnat J, Chardon F, Gaufichon L, Suzuki A (2010) Nitrogen uptake, assimilation and remobilization in plants: challenges for sustainable and productive agriculture. Ann Bot 105(7):1141–1157

Moll RH, Kamprath EJ, Jackson WA (1982) Analysis and interpretation of factors which contribute to efficiency of nitrogen utilization 1. Agron J 74(3):562–564

Monaghan JM, Snape JW, Chojecki AJS, Kettlewell PS (2001) The use of grain protein deviation for identifying wheat cultivars with high grain protein concentration and yield. Euphytica 122(2):309–317

Montazeaud G, Karatoğma H, Özturk I, Roumet P, Ecarnot M, Crossa J, Özer E, Özdemir F, Lopes MS (2016) Predicting wheat maturity and stay–green parameters by modeling spectral reflectance measurements and their contribution to grain yield under rainfed conditions. Field Crop Res 196:191–198

Muitire C, Kamutando C, Moyo M (2021) Building stress resilience of cereals under future climatic scenarios: 'The Case of Maize, Wheat, Rice and Sorghum'. In: Cereal grains. IntechOpen

Nakhforoosh A, Nagel KA, Fiorani F, Bodner G (2021) Deep soil exploration vs. topsoil exploitation: distinctive rooting strategies between wheat landraces and wild relatives. Plant Soil 459(1):397–421

Nigro D, Gu YQ, Huo N, Marcotuli I, Blanco A, Gadaleta A, Anderson OD (2013) Structural analysis of the wheat genes encoding NADH-dependent glutamine-2-oxoglutarate amidotransferases and correlation with grain protein content. PLoS One 8(9):e73751

Nigro D, Blanco A, Anderson OD, Gadaleta A (2014) Characterization of ferredoxin-dependent glutamine-oxoglutarate amidotransferase (Fd-GOGAT) genes and their relationship with grain protein content QTL in wheat. PLoS One 9(8):e103869

Nigro D, Fortunato S, Giove SL, Paradiso A, Gu YQ, Blanco A, de Pinto MC, Gadaleta A (2016) Glutamine synthetase in durum wheat: genotypic variation and relationship with grain protein content. Front Plant Sci 7:971

Nigro D, Fortunato S, Giove SL, Mangini G, Yacoubi I, Simeone R, Blanco A, Gadaleta A (2017) Allelic variants of glutamine synthetase and glutamate synthase genes in a collection of durum wheat and association with grain protein content. Diversity 9(4):52

Nigro D, Gadaleta A, Mangini G, Colasuonno P, Marcotuli I, Giancaspro A, Giove SL, Simeone R, Blanco A (2019) Candidate genes and genome-wide association study of grain protein content and protein deviation in durum wheat. Planta 249(4):1157–1175

Nigro D, Fortunato S, Giove SL, Mazzucotelli E, Gadaleta A (2020) Functional validation of glutamine synthetase and glutamate synthase genes in durum wheat near isogenic lines with QTL for high GPC. Int J Mol Sci 21(23):9253

Niño-González M, Novo-Uzal E, Richardson DN, Barros PM, Duque P (2019) More transporters, more substrates: the Arabidopsis major facilitator superfamily revisited. Mol Plant 12(9):1182–1202

Noguero M, Lacombe B (2016) Transporters involved in root nitrate uptake and sensing by Arabidopsis. Front Plant Sci 7:1391

OECD-FAO Agricultural Outlook 2020–2029, FAO, available at https://www.fao.org/3/ca8861en/CA8861EN.pdf

Okamoto M, Vidmar JJ, Glass AD (2003) Regulation of NRT1 and NRT2 gene families of Arabidopsis thaliana: responses to nitrate provision. Plant Cell Physiol 44(3):304–317

Okamoto M, Kumar A, Li W, Wang Y, Siddiqi MY, Crawford NM, Glass AD (2006) High-affinity nitrate transport in roots of Arabidopsis depends on expression of the NAR2-like gene AtNRT3. 1. Plant Physiol 140(3):1036–1046

Omara P, Aula L, Oyebiyi F, Raun WR (2019) World cereal nitrogen use efficiency trends: review and current knowledge. Agrosyst Geosci Environ 2(1):1–8

Oury FX, Godin C (2007) Yield and grain protein concentration in bread wheat: how to use the negative relationship between the two characters to identify favourable genotypes? Euphytica 157(1):45–57

Oury FX, Berard P, Brancourt-Hulmel M, Heumez E, Pluchard P, Rousset M, Doussinault G, Rolland B, Trottet M, Giraud A, Lecomte C (2003) Yield and grain protein concentration in bread wheat: a review and a study of multi-annual data from a French breeding program [Triticum aestivum L.]. J Genet Breed (Italy) 57:59

Pathak RR, Ahmad A, Lochab S, Raghuram N (2008) Molecular physiology of plant nitrogen use efficiency and biotechnological options for its enhancement. Curr Sci 94:1394–1403

Pena-Bautista RJ, Hernandez-Espinosa N, Jones JM, Guzman C, Braun HJ (2017) CIMMYT series on carbohydrates, wheat, grains, and health wheat-based foods: their global and regional importance in the food supply, nutrition, and health. Cereal Foods World 62(5):231–249

Ping WANG, Wang ZL, Cai RG, Yong LI, Chen XG, Yin YP (2011) Physiological and molecular response of wheat roots to nitrate supply in seedling stage. Agric Sci China 10(5):695–704

Potel F, Valadier MH, Ferrario-Méry S, Grandjean O, Morin H, Gaufichon L, Boutet-Mercey S, Lothier J, Rothstein SJ, Hirose N, Suzuki A (2009) Assimilation of excess ammonium into amino acids and nitrogen translocation in Arabidopsis thaliana–roles of glutamate synthases and carbamoylphosphate synthetase in leaves. FEBS J 276(15):4061–4076

Puccio G, Ingraffia R, Giambalvo D, Amato G, Frenda AS (2021) Morphological and physiological root traits and their relationship with nitrogen uptake in wheat varieties released from 1915 to 2013. Agronomy 11(6):1149

Puccio G, Ingraffia R, Giambalvo D, Frenda AS, Harkess A, Sunseri F, Mercati F (2023) Exploring the genetic landscape of nitrogen uptake in durum wheat: genome-wide characterization and expression profiling of NPF and NRT2 gene families. Front Plant Sci 14:1302337. https://doi.org/10.3389/fpls.2023.1302337

Quarrie SA, Steed A, Calestani C, Semikhodskii A, Lebreton C, Chinoy C, Steele N, Pljevljakusić D, Waterman E, Weyen J, Schondelmaier J (2005) A high-density genetic map of hexaploid wheat (Triticum aestivum L.) from the cross Chinese Spring× SQ1 and its use to compare QTLs for grain yield across a range of environments. Theor Appl Genet 110(5):865–880

Quraishi UM, Abrouk M, Murat F, Pont C, Foucrier S, Desmaizieres G, Confolent C, Riviere N, Charmet G, Paux E, Murigneux A (2011) Cross-genome map based dissection of a nitrogen use efficiency ortho-metaQTL in bread wheat unravels concerted cereal genome evolution. Plant J 65(5):745–756

Ramírez-González RH, Borrill P, Lang D, Harrington SA, Brinton J, Venturini L, Davey M, Jacobs J, Van Ex F, Pasha A, Khedikar Y (2018) The transcriptional landscape of polyploid wheat. Science 361(6403):eaar6089

Rapp M, Lein V, Lacoudre F, Lafferty J, Müller E, Vida G, Bozhanova V, Ibraliu A, Thorwarth P, Piepho HP, Leiser WL (2018) Simultaneous improvement of grain yield and protein content in durum wheat by different phenotypic indices and genomic selection. Theor Appl Genet 131(6):1315–1329

Raun WR, Johnson GV (1999) Improving nitrogen use efficiency for cereal production. Agron J 91(3):357–363

Rimbert H, Darrier B, Navarro J, Kitt J, Choulet F, Leveugle M, Duarte J, Rivière N, Eversole K, International Wheat Genome Sequencing Consortium, Le Gouis J (2018) High throughput SNP discovery and genotyping in hexaploid wheat. PLoS One 13(1):e0186329

Robertson GP, Groffman PM (2015) Nitrogen transformations. In: Paul EA (ed) Soil microbiology, ecology and biochemistry, 4th edn. Academic Press, Burlington, pp 421–446

Roucou A, Violle C, Fort F, Roumet P, Ecarnot M, Vile D (2018) Shifts in plant functional strategies over the course of wheat domestication. J Appl Ecol 55(1):25–37

Sandhu KS, Mihalyov PD, Lewien MJ, Pumphrey MO, Carter AH (2021) Genomic selection and genome-wide association studies for grain protein content stability in a nested association mapping population of wheat. Agronomy 11:2528

Schimel JP, Bennett J (2004) Nitrogen mineralization: challenges of a changing paradigm. Ecology 85(3):591–602

Semagn K, Iqbal M, Chen H, Perez-Lara E, Bemister DH, Xiang R, Zou J, Asif M, Kamran A, N'Diaye A et al (2021) Physical mapping of QTL associated with agronomic and end-use quality traits in spring wheat under conventional and organic management systems. Theor Appl Genet 134:3699–3719

Shi H, Chen M, Gao L, Wang Y, Bai Y, Yan H, Xu C, Zhou Y, Xu Z, Chen J, Tang W, Wang S, Shi Y, Wu Y, Sun D, Jia J, Ma Y (2022) Genome-wide association study of agronomic traits related to nitrogen use efficiency in wheat. Theor Appl Genet 135(12):4289–4302. https://doi.org/10.1007/s00122-022-04218-5

Shi H, Wang W, Gao L, Wu J, Hu C, Yan H, Shi Y, Li N, Ma Y, Zhou Y, Xu Z, Chen J, Tang W, Chen K, Sun D, Wu Y, Chen M (2024) Genome-wide association study of seedling nitrogen-use efficiency-associated traits in common wheat (Triticum aestivum L.). Crop J 12(1):222–231. https://doi.org/10.1016/j.cj.2023.10.014

Simmonds NW (1995) The relation between yield and protein in cereal grain. J Sci Food Agric 67(3):309–315

Singer SD, Foroud NA, Laurie JD (2019) Molecular improvement of grain: target traits for a changing world. pp 545–555.

Singh R, Saripalli G, Kumar A, Gautam T, Singh SK, Gahlaut V et al (2023) QTL analysis for nitrogen use efficiency in wheat (*Triticum aestivum* L.). Euphytica 219(1):9

Smil V (2001) Enriching the earth: Fritz Haber, Carl Bosch, and the transformation of world food production, 1st edn. MIT press, Cambridge, MA

Sukanya R, Li MG, Snustad DP (1994) Root-and shoot-specific responses of individual glutamine synthetase genes of maize to nitrate and ammonium. Plant Mol Biol 26(6):1935–1946

Sun C, Dong Z, Zhao L, Ren Y, Zhang N, Chen F (2020) The wheat 660K SNP array demonstrates great potential for marker-assisted selection in polyploid wheat. Plant Biotechnol J 18(6):1354–1360

Suprayogi Y, Clarke JM, Bueckert R, Clarke FR, Pozniak CJ (2011) Nitrogen remobilization and post-anthesis nitrogen uptake in relation to elevated grain protein concentration in durum wheat. Can J Plant Sci 91(2):273–282

Sutton MA, Howard CM, Erisman JW (2011) The need to integrate nitrogen science and policies. In: Sutton MA, Howard CM, Erisman JW (eds) The European nitrogen assessment. Cambridge University Press, pp 82–86

Swarbreck SM, Wang M, Wang Y, Kindred D, Sylvester-Bradley R, Shi W, Bentley AR, Griffiths H (2019) A roadmap for lowering crop nitrogen requirement. Trends Plant Sci 24(10):892–904

Tabuchi M, Sugiyama K, Ishiyama K, Inoue E, Sato T, Takahashi H, Yamaya T (2005) Severe reduction in growth rate and grain filling of rice mutants lacking OsGS1; 1, a cytosolic glutamine synthetase1; 1. Plant J 42(5):641–651

Tabuchi M, Abiko T, Yamaya T (2007) Assimilation of ammonium ions and reutilization of nitrogen in rice (*Oryza sativa* L.). J Exp Bot 58(9):2319–2327

Taira M, Valtersson U, Burkhardt B, Ludwig RA (2004) Arabidopsis thaliana GLN2-encoded glutamine synthetase is dual targeted to leaf mitochondria and chloroplasts. Plant Cell 16(8):2048–2058

Taulemesse F, Le Gouis J, Gouache D, Gibon Y, Allard V (2015) Post-flowering nitrate uptake in wheat is controlled by N status at flowering, with a putative major role of root nitrate transporter NRT2. 1. PLoS One 10(3):e0120291

Tegeder M (2014) Transporters involved in source to sink partitioning of amino acids and ureides: opportunities for crop improvement. J Exp Bot 65(7):1865–1878

Thomas H, Ougham H (2014) The stay-green trait. J Exp Bot 65(14):3889–3900

Thompson H (2012) Food science deserves a place at the table–US agricultural research chief aims to raise the profile of farming and nutrition science. Nature 2012:1476

Thomsen HC, Eriksson D, Møller IS, Schjoerring JK (2014) Cytosolic glutamine synthetase: a target for improvement of crop nitrogen use efficiency? Trends Plant Sci 19(10):656–663

Tilman D, Cassman KG, Matson PA, Naylor R, Polasky S (2002) Agricultural sustainability and intensive production practices. Nature 418(6898):671–677

Tsay YF, Chiu CC, Tsai CB, Ho CH, Hsu PK (2007) Nitrate transporters and peptide transporters. FEBS Lett 581(12):2290–2300

Uauy C, Brevis JC, Dubcovsky J (2006) The high grain protein content gene Gpc-B1 accelerates senescence and has pleiotropic effects on protein content in wheat. J Exp Bot 57(11):2785–2794

Veresoglou SD, Barto EK, Menexes G, Rillig MC (2013) Fertilization affects severity of disease caused by fungal plant pathogens. Plant Pathol 62(5):961–969

Vilmus I, Ecarnot M, Verzelen N, Roumet P (2014) Monitoring nitrogen leaf resorption kinetics by near-infrared spectroscopy during grain filling in Durum wheat in different nitrogen availability conditions. Crop Sci 54(1):284–296

Vitousek PM, Mooney HA, Lubchenco J, Melillo JM (1997) Human domination of earth's ecosystems. Science 277(5325):494–499

Vitousek PM, Naylor R, Crews T, David MB, Drinkwater LE, Holland E, Johnes PJ, Katzenberger J, Martinelli LA, Matson PA, Nziguheba G (2009) Nutrient imbalances in agricultural development. Science 324(5934):1519–1520

Von Wirén N, Merrick M (2004) Regulation and function of ammonium carriers in bacteria, fungi, and plants. In: Molecular mechanisms controlling transmembrane transport. Springer, Berlin, Heidelberg, pp 95–120

Walkowiak S, Gao L, Monat C, Haberer G, Kassa MT, Brinton J, Ramirez-Gonzalez RH, Kolodziej MC, Delorean E, Thambugala D, Klymiuk V (2020) Multiple wheat genomes reveal global variation in modern breeding. Nature 588(7837):277–283

Walters DR, Bingham IJ (2007) Influence of nutrition on disease development caused by fungal pathogens: implications for plant disease control. Ann Appl Biol 151(3):307–324

Wang S, Wong D, Forrest K, Allen A, Chao S, Huang BE, Maccaferri M, Salvi S, Milner SG, Cattivelli L, Mastrangelo AM (2014) Characterization of polyploid wheat genomic diversity using a high-density 90 000 single nucleotide polymorphism array. Plant Biotechnol J 12(6):787–796

Wang YY, Cheng YH, Chen KE, Tsay YF (2018) Nitrate transport, signaling, and use efficiency. Annu Rev Plant Biol 69:85–122

Wang M, Wang Y, Wang X, Wei G, Yang H, Yang X et al (2024) Integrated physiological, biochemical, and transcriptomics analyses reveal the underlying mechanisms of high nitrogen use efficiency of black sesame. Plant Physiol Biochem 206:108205

White J, Sharma R, Balding D, Cockram J, Mackay IJ (2022) Genome-wide association mapping of Hagberg falling number, protein content, test weight, and grain yield in U.K. wheat. Crop Sci 62:965–981

Winfield MO, Allen AM, Burridge AJ, Barker GL, Benbow HR, Wilkinson PA, Coghill J, Waterfall C, Davassi A, Scopes G, Pirani A (2016) High-density SNP genotyping array for hexaploid wheat and its secondary and tertiary gene pool. Plant Biotechnol J 14(5):1195–1206

Xu G, Fan X, Miller AJ (2012) Plant nitrogen assimilation and use efficiency. Annu Rev Plant Biol 63:153–182

Xu Y, Wang R, Tong Y, Zhao H, Xie Q, Liu D, Zhang A, Li B, Xu H, An D (2014) Mapping QTLs for yield and nitrogen-related traits in wheat: influence of nitrogen and phosphorus fertilization on QTL expression. Theor Appl Genet 127(1):59–72

Xu Y, Yang Y, Si W, Liu D, Ren Y (2023) QTL mapping for root traits and their effects on nutrient uptake and yield performance in common wheat (*Triticum aestivum* L.). Agriculture 13(1):210. https://doi.org/10.3390/agriculture13010210

Yamaya T, Obara M, Nakajima H, Sasaki S, Hayakawa T, Sato T (2002) Genetic manipulation and quantitative-trait loci mapping for nitrogen recycling in rice. J Exp Bot 53(370):917–925

Zhang FF, Si GAO, Zhao YY, Zhao XL, Liu XM, Kai XIAO (2015) Growth traits and nitrogen assimilation-associated physiological parameters of wheat (*Triticum aestivum* L.) under low and high N conditions. J Integr Agric 14(7):1295–1308

Zhao XQ, Nie XL, Xiao XG (2013) Over-expression of a tobacco nitrate reductase gene in wheat (*Triticum aestivum* L.) increases seed protein content and weight without augmenting nitrogen supplying. PLoS One 8(9):e74678

Genomics and Improvement in Fusarium Head Blight Resistance in Durum Wheat

Dhondup Lhamo, Yong-Qiang Gu, Xuehui Li, Shengming Yang, Shaobin Zhong, Elias M. Elias, and Steven S. Xu

Abstract

Fusarium head blight (FHB), caused by several *Fusarium* species, is the most devastating disease in durum wheat. Modern durum varieties are highly susceptible to FHB, and thus a massive effort has been dedicated to understanding and enhancing FHB resistance in durum wheat in the past two decades. In this chapter, we reviewed the important sources of FHB resistance for durum breeding, including the major quantitative trait loci (QTL) from exotic sources and the minor QTL from native sources. We elaborated on recent advances in uncovering the genetic and molecular basis of FHB resistance, with an emphasis on functional genes for *Fhb1* and *Fhb7*, and genes that inhibit virulence factors such as mycotoxins and cell wall degrading enzymes, reinforce cell wall structures, transduce signaling cascades to promote defense responses, and release antifungal metabolites. We reviewed the strategies for improving FHB resistance in modern durum germplasm, such as the genetic manipulation of known susceptible and resistance genes, genomic selection using genomic estimated breeding values, and mutagenesis and epigenetics to generate novel FHB-resistant alleles. We also discussed the recent advances in developing durum varieties with increased FHB resistance in the United States. The future of FHB resistance breeding in durum wheat holds a great promise because many elite durum germplasms carrying the major QTL including *Fhb1* and *Fhb7* from bread wheat and wild relatives now become available and are being used in breeding. In addition, expanding genomic and genetic resources provides useful tools to dissect the factors affecting FHB resistance expression and susceptibility in durum wheat.

Keywords

Durum wheat · *Triticum turgidum* ssp. *Durum* · Tetraploid wheat · Fusarium head blight · Resistance · *Fusarium graminearum* · Genomics

D. Lhamo · Y.-Q. Gu · S. S. Xu (✉)
USDA-ARS, Crop Improvement and Genetics Research Unit, Western Regional Research Center, Albany, CA, USA
e-mail: dhondup.lhamo@usda.gov; yong.gu@usda.gov; steven.xu@ars.usda.gov

X. Li · E. M. Elias
Department of Plant Sciences, North Dakota State University, Fargo, ND, USA
e-mail: xuehui.li@ndsu.edu; elias.elias@ndsu.edu

S. Yang
USDA-ARS, Cereal Crops Improvement Research Unit, Edward T. Schafer Agricultural Research Center, Fargo, ND, USA
e-mail: shengming.yang@usda.gov

S. Zhong
Department of Plant Pathology, North Dakota State University, Fargo, ND, USA
e-mail: shaobin.zhong@ndsu.edu

R. Tuberosa et al. (eds.), *The Durum Wheat Genome*, Compendium of Plant Genomes,
https://doi.org/10.1007/978-3-031-84689-2_13

1 Introduction

Wheat is one of the most important cereal crops in the world. The large portion of the global wheat production is occupied by the hexaploid common wheat (*Triticum aestivum* L., $2n = 6x = 42$, AABBDD) used mainly for making bread, whereas a small portion (~8%) by the tetraploid durum wheat [*T. turgidum* ssp. *durum* (Desf.) Husn., $2n = 4x = 28$, AABB] for pasta products (Dubcovsky and Dvorak 2007; Kuzmanović et al. 2019). However, durum wheat has a great potential for future expansion due to the escalating demands worldwide. At present, the wheat production in many wheat-growing areas around the world is seriously threatened by *Fusarium* species, the fungal pathogens that cause Fusarium head blight (FHB), Fusarium crown rot (FCR), and Fusarium root rot (FRR). The incidence of these diseases depends on the prevailing *Fusarium* species and their favorable environments. For instance, FHB is caused predominantly by *Fusarium graminearum* Schwabe [teleomorph *Gibberella zeae* (Schw.) Petch.] during warm and humid climates, whereas *F. culmorum* (W.G. Smith) Sacc. and *F. avenaceum* (Fr.) Sacc. during cooler climates (Bushnell et al. 2003).

The FHB-associated pathogens infect spikelets at anthesis and thereafter colonize the entire head systematically, thus producing extensive blight symptoms and shriveled seeds (Sella et al. 2014), as shown for the susceptible durum variety "Lebsock" in Fig. 1. On the other hand, FCR is prominently caused by *F. pseudograminearum* (teleomorph *Gibberella coronicola* T. Aoki & O'Donnell) that triggers necrosis and dry rot of the crown during drier climates (Aoki and O'Donnell 1999a, b; Kazan and Gardiner 2018). Serious epidemics of these *Fusarium* diseases often lead to a drastic yield loss and a poor grain quality with a high level of trichothecene mycotoxins such as

Fig. 1 *Fusarium graminearum*-infected spikes of hexaploid wheat line PI 277012, durum wheat variety Lebsock, and their derived durum wheat line 10Ae564 (Lebsock/PI 277012//Lebsock) in a mist-irrigated nursery (Fargo, North Dakota) in 2015

deoxynivalenol (DON), nivalenol (NIV), and zearalenone (ZEA) in grains that constitute a serious threat to food safety (Chilaka et al. 2017). Therefore, many countries have placed strict regulations on the maximum level of mycotoxins in consumable food and feed products.

Agronomic practices and fungicide applications are moderately effective in minimizing the growth and spread of *Fusarium* diseases. A more sustainable approach is the genetic resistance for controlling disease severity and associated mycotoxin contamination consistently and durably. As such, the major global efforts have been dedicated toward finding new sources of resistance to develop robust wheat varieties. Many breeding programs have made a significant progress in breeding bread wheat varieties for FHB resistance, resulting in numerous, newly developed varieties with moderate FHB resistance that are currently grown in the regions with serious epidemics such as North America. Contrarily, breeding durum varieties for FHB resistance has been gradual in progress because sources with a high level of FHB resistance are almost absent in native durum germplasm. The current endeavor of many durum breeding and pre-breeding programs is to transfer FHB resistance from bread wheat and related species into modern durum varieties. So far, a durum variety with a high level of FHB or FCR resistance has not been available worldwide. Therefore, many durum breeding programs are making substantial efforts to find diverse resistance sources, which is essential to dissect the genetics and genomics of *Fusarium* resistance.

In this chapter, we focus on the recent advancement in the resistance breeding, especially for FHB in durum wheat. We explore the available resistance sources present in the close and wild relatives of wheat and the adapted durum germplasm, and success in developing genomic resources for incorporating the identified resistance genes and alleles into modern durum wheat varieties. We also discuss the susceptibility factors or genes that could hinder the improvement of FHB resistance in durum wheat. In addition, we elaborate on known resistance genes with the potential to enhance FHB resistance and mycotoxin detoxification in the susceptible durum wheat. Moreover, we review the strategies to improve FHB resistance in durum wheat, including chromosome engineering, genomic selection, mutagenesis, and epigenetics. While the progress of resistance breeding in durum wheat has been slow, the future holds a great promise with the release of genomic footprints in the form of high-density SNP-based consensus maps (Maccaferri et al. 2015), 10.45 Gigabase Svevo reference genome (Maccaferri et al. 2019), and the enlarged collection of a global durum panel consisting of 1100 tetraploid wheat accessions (Mazzucotelli et al. 2020).

2 Sources of FHB Resistance

Resistance to FHB in wheat is a complex trait and it has been classified into five types of resistance based on the evaluation methodology and resistance mode. Type I, II, III, and V are defined as resistance to the initial infection, fungal spread within the spike, kernel infection, and mycotoxin accumulation, respectively, and Type IV for tolerance (Schroeder and Christensen 1963; Mesterhazy 1995). As FHB has reemerged as one of the most devastating diseases in many wheat-producing regions since the 1990s, effective resistance sources have been desperately needed to develop highly resistant varieties in both durum and bread wheat. Therefore, enormous efforts have been taken to search for useful sources of FHB resistance in the wheat breeding community by extensively evaluating the world core germplasm collections of wheat and its relatives (see review by McMullen et al. 2012; Zhang et al. 2008). According to a recent review by Ma et al. (2020), about 70,000 accessions of durum wheat, common wheat, and their relatives have been evaluated for FHB resistance, with about 7000 being identified to have various levels of resistance. Approximately 600 quantitative trait loci (QTL) for FHB resistance have been identified through genetic linkage mapping and genome-wide association studies (GWAS) (see reviewed by Buerstmayr et al. 2009, 2020; Venske et al. 2019). Nine of the QTL have been formally designated as *Fhb1* through *Fhb9* (McIntosh et al. 2013, 2015, 2024).

The FHB-resistant QTL reported in the literatures were mostly identified in common wheat. However, many of these QTL have only minor effects and lack consistency across different environments and genetic backgrounds. Therefore, most of the QTL have not been used in wheat breeding. So far, *Fhb1* on chromosome arm 3BS derived from Chinese wheat variety, "Sumai 3" (PI 481542), is one QTL that has been effectively used in the wheat breeding programs around the world. It has been successfully deployed in many common wheat varieties in the hotspots with severe FHB epidemics such as the northern United States (U.S.) and Canada (Hao et al. 2020). In addition, several other QTLs, including *Fhb2*, *Fhb4*, *Fhb5*, *Fhb7*, *Fhb8*, *Fhb9*, and hexaploid wheat PI 277012-derived 5A QTL (Chu et al. 2011; Ma et al. 2020; Wang et al. 2024; Zhang et al. 2024), are currently being used in many common wheat breeding programs.

Compared with common wheat, durum wheat is more susceptible to FHB. In the past two decades, numerous durum wheat germplasm collections have rigorously been evaluated for FHB resistance in several durum breeding programs. In the U.S., approximately 12,000 durum accessions from worldwide collections, including 6000 accessions from the International Center for Agricultural Research in the Dry Areas (ICARDA), have been screened for FHB reactions (Elias et al. 2018). However, none of the durum varieties or lines with a high level of resistance have been identified yet, and only a small number of accessions, including four Syrian and five Tunisian durum landraces were found to be moderately resistant (Talas et al. 2011; Huhn et al. 2012; Zhang et al. 2014; Elias et al. 2018). Several QTL with minor effects on chromosomes 2A, 2B, 3BL, and 5BL have been identified in a susceptible durum variety (Zhang et al. 2014, 2017) and Tunisian durum lines (Ghavami et al. 2011; Fakhfakh et al. 2011; Huhn et al. 2012; Zhang et al. 2014; Pirseyedi et al. 2019) (Table 1).

Because durum wheat lacks sources of acceptable levels of FHB resistance in its base germplasm (Zhang et al. 2014), exotic sources such as the close relatives in the genus *Triticum* including bread wheat, various tetraploid subspecies (*T. turgidum* ssp.), and distant species in other genera under tribe Triticeae have been exploited as sources of high levels of FHB resistance for durum wheat. Several closely related tetraploid subspecies such as cultivated emmer (*T. turgidum* subsp. *dicoccum*), wild emmer (*T. turgidum* subsp. *dicoccoides*), and Persian wheat (*T. turgidum* subsp. *carthlicum*) have been identified to have moderate-to-high FHB resistance (Buerstmayr et al. 2003; Oliver et al. 2007, 2008). Several FHB resistance QTL on chromosomes 3A, 5A, and 7A were identified from wild and cultivated emmer wheat (Table 1). These tetraploids may share a common genetic basis of FHB resistance with hexaploid bread wheat since most of their QTL showed strong overlaps (Prat et al. 2014).

In addition to wheat (*Triticum*) species, wild relatives of wheat represent valuable genetic resources for the improvement of FHB resistance (Cai et al. 2005; Ceoloni et al. 2014; Buerstmayr et al. 2020). These include species from genera *Thinopyrum (*syn. *Lophopyrum or Agropyron), Roegneria, Elymus,* and *Leymus.* Several major FHB resistance QTL such as *Fhb3* (Qi et al. 2008), *Fhb6* (Cainong et al. 2015), and *Fhb7* (Wang et al. 2020; Zhang et al. 2022; Cai et al. 2024) were successfully transferred from the respective wild species, *Leymus racemosus*, *Elymus tsukushiensis*, *Th. elongatum,* and *Th. ponticum* into the genome of bread wheat. *Fhb7* was also transferred from tall wheatgrass *Th. ponticum* into durum wheat (Forte et al. 2014; Kuzmanović et al. 2019) (Table 1).

3 Susceptibility Genes to *Fusarium* Species

The efforts in improving FHB resistance have been primarily focused on the identification and introgression of resistance QTL or genes. Recently, identification and elimination of susceptibility genes have been recognized as an alternative strategy for enhancing FHB resistance in wheat (Chhabra et al. 2021; Gorash et al.

Table 1 Native and exotic sources of FHB resistance for durum wheat varieties

Donor species	Resistance lines	Remarks	Pedigrees of mapping populations or alien gene introgression lines	Chromosome/QTL from resistance lines	Closest markers	References
Bread wheat (*T. aestivum*)	10Ae564	A durum line with PI 277012-derived resistance	"Joppa"/10Ae564	5A (*Qfhb.ndwp-5A*); 7A (*Qfhb.ndwp-7A*)	*IWB26525; IWB74024*	Chu et al. (2011), Zhao et al. (2018a)
	02-5B-318	Sumai 3-derived resistance	"Saragolla"/02-5B-318	2AS; 2BS; 4BL	*IWB63138; IWB55365; IWB48353*	Giancaspro et al. (2016)
	DBC-480	Semperdur durum experimental lines with Sumai 3-derived resistance	"Karur"/DBC-480; "Durobonus"/ DBC-480; SZD1029K//DBC-480	3BS (*Fhb1*)	*Xbarc147*	Prat et al. (2017)
Durum wheat (*T. durum*)	Tun7, 18, 34, 36, 108	Moderate resistant Tunisian landraces	"Lebsock"/Tun34	5BL (*Qfhs.ndsu-5BL*)	*wPt-0054*	Ghavami et al. (2011), Fakhfakh et al. (2011), Huhn et al. (2012)
			"Ben"/Tun108	2B (*Qfhb.ndsu-2B*); 3BL (*Qfhb.ndsu-3BL*)	*gwm71; Xbarc229*	Pirseyedi et al. (2019)
	DT696	A Canadian durum breeding line with moderate resistance	"DT707"/"DT696"	5A	*gwm156 / wmc110*	Singh et al. (2008)
Wild emmer wheat (*T. dicoccoides*)	LDN-DIC(ISA-3A)	Langdon durum disomic substitution line with chromosome 3A from wild emmer line Israel A	Langdon/LDN-DIC(ISA-3A); LDN-DIC(ISA-3A)/ "Buck Esmeralda"; LDN-DIC(ISA-3A)/"Buck Candisur"	3AS (*Qfhs.ndsu-3AS*)	*Xgwm2*	Joppa and Cantrell (1990), Otto et al. (2002), Stack et al. (2002), Xu et al. (2004), Soresi et al. (2017)
	LDN-DIC(742-7A)	Langdon durum disomic substitution line with chromosome 7A from wild emmer accession PI478742	Langdon/LDN-DIC(742-7A)	7AL (*Qfhs.fcu-7AL*)	*Xbarc121*	Xu et al. (2004), Kumar et al. (2007)
	"Mt. Hermon#22"	Moderate-resistant variety from Israel	"Helidur"/Mt. Hermon#22	3A; 4A	*Xgwm2; Xgwm610*	Buerstmayr et al. (2003), Gladysz et al. (2007)
	"Mt. Gerizim#36"	Moderate-resistant variety from Israel	Helidur durum/Mt. Gerizim#36	3A; 6B	*Xgwm2; Xgwm626*	Buerstmayr et al. (2003, 2013)

(continued)

Table 1 (continued)

Donor species	Resistance lines	Remarks	Pedigrees of mapping populations or alien gene introgression lines	Chromosome/QTL from resistance lines	Closest markers	References
Cultivated emmer wheat (*T. dicoccum*)	PI 41025	Moderate-resistant accession	"Ben"/ PI 41025	3A (*QFhb.rwg-3A*); 5AL (*QFhb.rwg-5A.3*)	*IWA8624; Xfcp650*	Zhang et al. (2014)
	Td161	An Austrian most resistant tetraploid wheat line	DS-131621/Td161; "Floradur"/ Td161; Helidur/ Td161	4B; 6A; 6B; 7B	*Rht-B1; Xgwm356 Xgwm816; Xs24m12_f6h5*	Buerstmayr et al. (2012)
	BGRC3487	A line from Germany, with higher resistance than DT735	"DT735"/BGRC3487/2	3B (*QFhb.usw-3B*); 7AL (*QFhb.usw-7A2*)	*wPt-5105; barc165.2*	Ruan et al. (2012)
Persian wheat (*T. carthlicum*)	Blackbird	Moderate-resistant Persian wheat line from France	"Strongfield"/Blackbird	1AS; 6BS	*cfa2153; WMC397*	Somers et al. (2006), Singh et al. (2008)
Tall wheatgrass (*Th. ponticum*)	R193; R129; R216	Durum introgression lines carrying *Fhb7* derived from *Th. ponticum*	KS24-1/durum 7AS·7AL-7el$_1$L lines (R5, R112, R23)//2*durum "Simeto"	7AS·7AL7el$_1$L/7el$_2$L (*Fhb7*)	*Xcfa2240*; *XBE445653*	Forte et al. (2014)
Diploid wheatgrass (*Th. elongatum*)	R74-10/R112; R69-9/R112; R74-10/R5	Durum introgression lines carrying *Fhb-7EL* derived from *Th. elongatum*	Hexaploid wheat lines (R69-9, R74-10)/durum lines (R5, R112)//2*Simeto	7AS·7AL7el$_1$L/7EL (*Fhb-7EL*)	*BE405003; BE445653; GWM344*	Kuzmanović et al. (2019)
	DGE-1 (PI 645483)	Durum wheat-*Th. elongatum* chromosome 1E disomic addition line	Langdon/*Th. elongatum* PI 531719//2*Langdon	1E		Jauhar et al. (2008), Jauhar and Peterson (2008)
	DGE-2 (PI 663216)	Durum wheat- *Th. elongatum* chromosome 1E(1A) disomic substitution line	DGE-1/Langdon 1D(1A)	1E		Jauhar and Peterson (2012)
	DGE-3 (PI 665473)	Durum wheat- *Th. elongatum* chromosome 1E(1B) disomic substitution line	DGE-1/Langdon 1D(1B)	1E		Jauhar and Peterson (2012)

Adapted from Haile et al. (2019) with permission

2021). The susceptibility genes in plants are characterized by promoting pathogen infection and disease development, which have been extensively studied in model plant species, *Arabidopsis thaliana* (see reviewed by Gorash et al. 2021). Many susceptibility genes to various pathogens have been identified in *Arabidopsis*, which might be conserved across other plant species (Sun et al. 2016; Huibers et al. 2013; Gorash et al. 2021). For example, *Arabidopsis AtDMR1* (Downy Mildew Resistance 1) encoding homoserine kinase is a susceptible gene to the biotrophic fungus, *Hyaloperonospora parasitica*, that causes downy mildew (van Damme et al. 2009). The loss-of-function mutation of *AtDMR1* in *Arabidopsis* was found to provide resistance against multiple fungal pathogens including *H. parasitica*, *F. graminearum*, and *F. culmorum* (Brewer et al. 2014). Similarly, RNAi (RNA interference) silencing of orthologs of six different *Arabidopsis* susceptible genes, including *AtDMR1* and *AtDMR6,* in potato largely resulted in complete resistance to the late blight caused by *Phytophthora infestans* (Sun et al. 2016).

Susceptibility genes to *Fusarium* species have not been extensively investigated in wheat. Based on the evaluation of 30 Chinese Spring (CS) ditelosomic lines, Ma et al. (2006) found that chromosome arms 7AS, 3BL, 6BL, 7BS, and 4DS possibly carry FHB susceptibility factors. Hales et al. (2020) and Chhabra et al. (2021) confirmed the presence of susceptibility factors on 4DS and 7AS, respectively, by analyzing the CS chromosome deletion lines. Chhabra et al. (2021) identified a susceptibility factor *Sf-Fhb-7AS*, and mapped it to 48.5–50.5 Mb region on chromosome arm 7AS. Deletion of *Sf-Fhb-7AS* resulted in increased type II FHB resistance, up to 50–60%. Substitution of CS chromosome 7A with 7A from four common wheat varieties and wild emmer did not change its susceptibility levels, indicating that the susceptibility factor might be commonly present in both tetraploid and hexaploid wheat (Chhabra et al. 2021). Hales et al. (2020) mapped the susceptibility factor on 4DS at 31.7 Mb interval. Garvin et al. (2015) reported that deletion of a region on chromosome arm 3DL in common wheat variety "Apogee" increased FHB resistance, suggesting that the deleted 3DL region may harbor FHB susceptibility genes that promote disease spread in infected spikes.

It has been hypothesized that either presence of suppressers or absence of the D genome may contribute to high FHB susceptibility in durum wheat (Stack et al. 2002; Garvin et al. 2009; Fakhfakh et al. 2011; Buerstmayr et al. 2012; Zhang et al. 2014; Szabo-Hever et al. 2018). Based on the literature discussed above, we speculate that susceptibility genes may be additional factors causing high FHB susceptibility of durum wheat as their expression in the favorable environments may mask the effects of the resistance QTL. This might be the reason for disappearance or decrease in FHB resistance of some QTL such as *Fhb1* when transferred into certain durum backgrounds. Although FHB susceptibility genes and/or resistance suppresser genes in durum wheat have not been reported in the literature, identification of the FHB resistant QTL on chromosomes 2A, 2B, 3B, and 5B from the susceptible durum varieties "Ben" (Zhang et al. 2014) and "Joppa" (Zhao et al. 2018a), "Strongfield" (Somers et al. 2006), "Floradur" (Buerstmayr et al. 2012), and Lebsock (Ghavami et al. 2011), respectively, might indicate the presence of FHB susceptibility genes and/or resistance suppressers in these varieties.

4 Genetics and Genomics of Fusarium Head Blight Resistance

Identification and isolation of the functional genes for FHB resistance are essential to understand the genetic basis and mechanism underlying *Fusarium* resistance. However, such studies are hindered by the complexity of FHB resistance and polyploid nature of the wheat genomes. So far, only the gene candidates for two QTL, *Fhb1* and *Fhb7*, have been isolated using map-based cloning. In addition, numerous candidate genes involved in FHB and DON resistance have been identified using other approaches such as profiling of transcripts, proteins, metabolites, and cell wall compositions in resistant and susceptible wheat spikes infected with *F. graminearum* or treated with DON (El-Gendy et al. 2001; Walter et al. 2008; Walter and Doohan 2011; Ding et al. 2011; Lionetti et al. 2015). Silencing of some of these genes in resistant lines and/or overexpression or complementation in susceptible wheat have validated their roles in disease resistance. These open a new avenue to improve FHB resistance in durum, common wheat, and other crops. Below we review and discuss the genetic and molecular basis of FHB resistance, with a focus on genes underlying *Fhb1* and *Fhb7* resistance, and genes involved in DON detoxification, inhibition of cell wall degrading enzymes (CWDEs), cell wall reinforcement, signal transduction, synthesis of antifungal metabolites and activation of defense responses.

4.1 Genes Underlying *Fhb1* Resistance

Fhb1 is the first prominent QTL (*Qfhs.ndsu-3BS*) known to provide Type II resistance in bread wheat (Bai et al. 1999; Waldron et al. 1999). This QTL was mapped to the distal segment of chromosome 3BS in Sumai 3 and Wangshuibai (Waldron et al. 1999; Anderson et al. 2001; Lin et al. 2004). Map-based cloning was initiated early to identify the candidate gene conferring *Fhb1* resistance, with limited success (Liu and Anderson 2003; Liu et al. 2006, 2008). It is only recently that a high-resolution, map-based cloning of *Fhb1* was achieved by three different studies. Briefly, Rawat et al. (2016) reported *Fhb1* to be a pore-forming toxin-like gene (*TaPFT*), whose expression was exclusively detected in the FHB-resistant populations. However, this claim was later disputed since *TaPFT* was also found to be expressed in many FHB-susceptible varieties (Zhu et al. 2018; He et al. 2018b). Subsequently, two independent studies simultaneously found a nuclear-localized, histidine-rich calcium-binding protein (*TaHRC/TaHis*) gene to confer *Fhb1* resistance (Li et al. 2019; Su et al. 2019). Both studies admitted that a large deletion (752 bp) in *TaHRC/TaHis* was responsible for FHB resistance, however, their conclusions on the causative allele differed substantially. Su et al. (2019) referred to the loss-of-function of the susceptible *TaHRC-S/TaHis* ^{S} allele as the contributing factor for *Fhb1* resistance, whereas Li et al. (2019) supported the gain-of-function of resistant *TaHRC-R/TaHis*^{R} allele. Paudel et al. (2020) further identified and cloned a wheat gene *WFhb1-1*, encoding a putative membrane protein of 127 amino acids to be involved in FHB resistance with a potential antifungal function, suggesting that it may be a key functional component of *Fhb1*. In addition, TaPFT-interacting TaRBL (ricin B-like lectin) protein was identified using a yeast two-hybrid system, and knockdown of *TaRBL* expression in Sumai 3 (FHB resistant) and "Jimai 22" (FHB susceptible) resulted in reduced FHB resistance (Song et al. 2021). Overall, these studies allowed the identification of at least three resistance genes for *Fhb1* and provided the possibility to uncover more genes and mechanisms in host *Fhb1* resistance against *Fusarium* pathogens.

4.2 Trichothecene Detoxifying Genes

Fusarium species produce various trichothecene mycotoxins, such as DON, 3-acetyl-deoxynivalenol (3-ADON), 15-acetyl-deoxynivalenol (15-ADON), T-2 toxin (T-2), HT-2 toxin (HT-2), fusarenon-X (Fus-x), NIV, and diacetoxyscirpenol (DAS), that act as inhibitors of host protein biosynthesis and as virulence factors of FHB disease development (McCormick et al. 2011; Li et al. 2017; Wang et al. 2020). DON is predominantly found in North America and Europe, and NIV in Asia and South America (van der Lee et al. 2015). The accumulation of these toxins in wheat worsens the disease severity and threatens food safety. Therefore, detoxification of these toxins in wheat varieties is a promising step for FHB resistance breeding to minimize yield and quality losses. Below are some of the host genes identified to play a role in detoxifying the trichothecene mycotoxins.

Glutathione S-Transferase Gene Conferring Fhb7 Resistance

Previous attempts were made to fine-map the *Fhb7* locus derived from *Th. ponticum* to determine the candidate gene conferring FHB resistance (Shen and Ohm 2007; Zhang et al. 2011; Guo et al. 2015a, b). However, the map-based cloning was impeded by the lack of a reference genome for the *Thinopyrum* E genome. Recently, Wang et al. (2020) assembled a diploid *Th. elongatum* E reference genome, which enabled fine-mapping of *Fhb7* to a short region (~1.2 Mb) between *XSdauK79* and *XSdauK80* markers. Transcriptomic analysis and further genetic mapping using newly developed markers from BAC (bacterial artificial chromosome) clones identified eight candidate genes within the 245 kb region of *Fhb7* between *XsdauK86* and *Xsdua88,* among which, only a gene coding for glutathione S-transferase (GST) enzyme was expressed in both the resistant parent and the *Thinopyrum* E genome. Wang et al. (2020) demonstrated that the enzyme conferred resistance to both FHB and FCR by conjugating glutathione (GSH) to the epoxy group of DON, resulting in non-toxic, de-epoxidated DON-GSH adducts. *Fhb7* could also detoxify the other seven trichothecenes (3-ADON, 15-ADON, T-2, HT-2, Fus-X, NIV, and DAS) through de-epoxidation. A striking discovery is that wheatgrass genome probably hijacked *Fhb7* from an endophytic fungal *Epichloë* species (Wang et al. 2020). In a recent study, Yang et al. (2024) acquired the first crystal structure of GST and revealed the enzyme's catalytic mechanism in DON detoxification, showing that GST transforms DON into a nontoxic product by opening the C12/C13 epoxide (Yang et al. 2024). The *GST* gene was recently further confirmed to be *Fhb7* through the evaluation of near-isogenic, transgenic, mutant, and CRISPR-Cas9 knockout lines (Zhao et al. 2024).

Uridine Diphosphate-Dependent Glucosyltransferase Gene

Like GST, other enzymes can detoxify trichothecenes via toxin conjugation. Uridine diphosphate-dependent glycosyltransferase (UGT) can convert DON and NIV into the respective non-toxic glucoside forms, DON-3-*O*-glucoside (D3G) and NIV-3-*O*-glucoside (NIV3G) via glycosylation (Nakagawa et al. 2011; Shin et al. 2012; Yoshinari et al. 2014), which was primarily demonstrated in *Arabidopsis* (Poppenberger et al. 2003). Its importance in FHB resistance was illustrated in several genetic studies. For example, mutation in *Brachypodium* UGT, *Bradi5g03300,* resulted in reduced resistance to *Fusarium* infection (Pasquet et al. 2016). On the other hand, ectopic expression of a barley UGT gene, *HvUGT13248,* in the transgenic bread wheat varieties was effective in detoxifying DON and NIV, with enhanced resistance to FHB (Li et al. 2015, 2017; Mandalà et al. 2019). A similar result was observed in the *HvUGT13248*-transgenic durum wheat with high DON to D3G conversion and significant improvement in FHB and FCR resistance (Mandalà et al. 2019). Recently, *UGT* from *Aegilops tauschii* (AET5Gv20385300) was identified on chromosome 5D using *k*-mer-based GWAS to confer DON resistance, functionally validated by complementation in toxin-sensitive yeast strain (Kirana et al. 2023). Several wheat UGTs such as *TaUGT3, TaUGT5,* and *TaUGT6* have also been proven to enhance DON resistance in susceptible wheat varieties (Gatti et al. 2018; Xing et al. 2018; Zhao et al. 2018b; He et al. 2020). Nevertheless, many *TaUGTs* (179) were identified in wheat (He et al. 2018a), whose functions in DON resistance remain unelucidated. Several of those *TaUGT* genes exhibited high expression in spikes upon DON treatment (He et al. 2018a), which are strong candidates for further functional studies. Co-expression of these *TaUGTs* in durum wheat may maximize DON detoxification and minimize FHB severity via additive effect.

Cytochrome P450 Gene

Cytochrome P450s (CYPs) are heme-containing membrane-bound enzymes that catalyze oxidation-reduction reactions and participate in plant defense by synthesizing secondary metabolites including defense compounds and engaging in detoxification processes (Schuler and Werck-Reichhart 2003; Morant et al. 2003; Schuler et al. 2006). Wheat *TaCYPs* might be involved in DON toxification as their expressions (e.g.,

TaCYP709C1 and *TaCYP72A*) were induced early in response to *Fusarium* infection and DON treatment (Walter et al. 2008; Li et al. 2010; Walter and Doohan 2011; Gunupuru et al. 2018). Silencing of *TaCYP72A* in wheat heads further confirmed its contribution to DON resistance (Gunupuru et al. 2018). Although the mechanism of *TaCYPs* in DON resistance remains unclear, these enzymes are predicted to be involved in conjugating different moieties (e.g., glutathione and hydroxyl) to DON, converting it into less-toxic forms (Schuler and Werck-Reichhart 2003; Ito et al. 2013; Kluger et al. 2013).

ATP-Binding Cassette Transporter Genes
The ABCC subfamily of adenosine triphosphate (ATP)-binding cassette (ABC) transporters, also known as multidrug resistance-associated proteins (MRPs), is potential vacuolar pumps that might sequester glutathione conjugates (generated by GSTs and CYPs) into vacuoles in the detoxification process (Li et al. 1997; Lu et al. 1998; Rea 1999; Wanke and Kolukisaoglu 2010). Hexaploid wheat genome contains 18 TaABCC transporters, among which TaABCC13 was shown to transport glutathione-conjugates in yeast (Bhati et al. 2015). In addition, a *TaABCC/MRP* homolog was suggested to be a potential candidate for QTL controlling FHB and DON resistance on chromosome 2DS since it was found near the associated markers in the linkage analysis (Handa et al. 2008). Expression of such an ABCC family gene was also induced in wheat spikes treated with DON (Walter et al. 2008). The silencing of *TaABCC3.1* in wheat increased disease severity upon DON treatment, which validated its role in FHB resistance (Walter et al. 2015). Similar was observed for three wheat homoeologous of pleiotropic drug resistance (PDR)-type ABC transporter (*TaPDR7*) located on group-6 chromosomes (Wang et al. 2016). *TaABC4* was identified as one of the putative resistance genes located within the *QTL-Fhb2* region (Dhokane et al. 2016). These studies imply the significance of different types of ABC transporters in disease resistance and DON detoxification although their modes of action remain unresolved.

Signaling Genes—Ta-FROG-TaSNRK1α and Ta-FROG-TaNACL-D
Triticum aestivum Fusarium resistance orphan gene (*TaFROG*) was identified as a nuclear-localized orphan gene present only in grass subfamily *Pooideae*, which was primarily found as a gene responsive to DON (Perochon et al. 2015). Silencing and overexpression of *TaFROG* in wheat supported its role in host resistance to DON and FHB. Because TaFROG is a new intrinsically disordered protein (IDP), it is suggested to be a component of the signaling response to DON (Wright and Dyson 2015; Perochon et al. 2015). Its interaction with signaling machinery such as α-subunit of the sucrose non-fermenting 1-related kinase 1 (TaSNRK1α) implicates their shared roles in signaling events to promote plant resistance to the fungus and the toxin (Perochon et al. 2015). The function of TaSNRK1α in FHB resistance was later confirmed by gene silencing, and its kinase activity was proven to be responsive to DON (Perochon et al. 2019b). In addition to TaSNRK1α, TaFROG formed distinct protein complexes with a NAC-like transcriptional factor, TaNACL-D1, whose ectopic expression in wheat conferred better resistance to FHB disease than wild type (Perochon et al. 2019a). Despite their known contributions to FHB resistance, the downstream targets of TaFROG-TaSNRK1α and TaFROG-TaNACL-D1 complexes remain unelucidated. Finding their targets will allow a better understanding of their underlying mechanism in DON and FHB resistance.

4.3 Cell Wall-Related Genes in *Fusarium* Resistance

The plant cell wall (CW), composed of polysaccharides such as cellulose, hemicellulose, and pectin, represents a main barrier that restricts the entry of pathogenic fungi into the cells. To penetrate through the barrier, fungal pathogens produce cell wall degrading enzymes (CWDEs) as virulence factors to degrade cell wall polysaccharides for colonization (Wanyoike et al. 2002). The extent of such cell wall degradation has often been associated with the severity of disease (King et al. 2011). In response, plants inhibit the activities of fungal and plant CWDEs to prevent degradation and strengthen the wall by changing its composition and releasing pathogenesis-related (PR) proteins and defense compounds to further resist pathogen invasion. Identifying genes involved in these processes are important strategies to enhance FHB resistance in durum and bread wheat varieties.

Resistance Genes Encoding Fungal Cell Wall Degrading Enzyme Inhibitors

Lionetti et al. (2015) performed a comparative analysis of cell wall components between the resistant hexaploid wheat (Sumai 3-derived breeding line 02-5B-318) and the highly susceptible durum variety "Saragolla." Higher levels of syringyl and p-hydroxyphenyl monolignols, xylan arabinosylation, and pectin methyl esterification were discovered in the resistant bread wheat than the susceptible durum, which possibly limits the enzymatic digestibility of cell walls to improve FHB resistance. Expression of such CW-related genes especially CWDE inhibitors in modern wheat varieties might improve FHB resistance. This is evident in susceptible durum and bread wheat overexpressing the respective kiwi *pectin methylesterase inhibitor* (*AcPMEI*) (Volpi et al. 2011) and bean *polygalacturonase-inhibiting proteins* (*PvPGIPs*) (Ferrari et al. 2012), which retained highly methyl-esterified pectins that are less prone to degradation (Volpi et al. 2013). Overexpression of other CWDE inhibitors such as *T. aestivum* xylanase inhibitors (TAXIs) in durum (cv. Svevo) inhibited the activity of pathogenic xylanases and delayed FHB symptoms (Moscetti et al. 2013). Other xylanase inhibitor proteins (XIPs) and thaumatin-like XI (TLXI) in bread wheat might also be associated with FHB resistance (McLauchlan et al. 1999; Igawa et al. 2005; Fierens et al. 2007; Takahashi-Ando et al. 2007; Pollet et al. 2009). In addition, finding inhibitors of pathogenic fungal cutinases, lipases, pectinases, xylanases, hemicellulases, cellulases, oxidases, and esterases might prevent the degradation of plant physical barriers such as cuticle, middle lamella, and cell walls (Walter et al. 2010; Bellincampi et al. 2014; Blümke et al. 2014).

Susceptible Genes Encoding Plant Cell Wall Degrading Enzymes

The repression of susceptible genes is also a strategy to enhance FHB resistance. Plant CW enzymes involved in wall loosening and degradation may serve as susceptible genes upon pathogen attack. For example, *pectin methylesterase* (*WheatPME-1*) is known to be involved in reducing pectin methyl-esterification, making the wall more susceptible to degradation by fungal pectinases (Raiola et al. 2011). *WheatPME-1* expression was highly induced in the susceptible durum wheat and ^{S}RILs, and repressed in the resistant bread wheat and ^{R}RILs upon *Fusarium* infection (Giancaspro et al. 2018). Its mutation in durum wheat (Gadàleta et al. 2019) conferred higher FHB resistance than its control, confirming the FHB susceptibility role of *WheatPME-1*. Other CW enzymes such as polygalacturonases (PGs), expansins, and endoglucanases may also increase wall susceptibility when in contact with necrotrophic pathogens (Flors et al. 2007; Cantu et al. 2009).

Plant Cell Wall Reinforcing Genes

During *Fusarium* infection, counteraction to plant and fungal CWDEs require biosynthesis and cross-linking of CW structural components to strengthen the barrier, therefore, proteins involved in these processes are also potential candidates for disease improvement. Structural components such as glycine-rich, serine-rich, and hydroxyproline-rich proteins, arabinogalactans, lignin, and phosphatidic acids were accumulated in FHB-resistant wheat upon *Fusarium* infection, which is needed for cell wall thickening to restrict the pathogen invasion (Carpita 1996; El-Gendy et al. 2001; Kang and Buchenauer 2003; Deepak et al. 2010; Ding et al. 2011; Kage et al. 2017). Nonstructural proteins such as hydrolases, transferases, esterases, and peroxidases may also play a role in FHB resistance since these proteins participate in the modification of CW polymers (Carpita 1996). For example, peroxidases involved in cross-linking phenolic compounds to cell wall proteins and polysaccharides were expressed and their activities enhanced in wheat infected with *Fusarium* species (Caruso et al. 1999; Ansari et al. 2007). Phenolics such as ferulic acid, *p*-coumaric acid, and sinapic acid can act as CW reinforcers and antifungals since they are precursors for lignin and defense compounds such as flavonoids and stilbenes (Whetten and Sederoff 1995), which can also contribute to FHB resistance (Friebe et al. 1998; McKeehen et al. 1999; Siranidou et al. 2002; Søltoft et al. 2008; Kage et al. 2017). Hence, genes involved in the synthesis of these phenolics are potential CW reinforcers to resist *Fusarium* invasion. Other genes participating in appositions, formation of papillae, and lignification may engage in FHB resistance (Ding et al. 2011; Dhokane et al. 2016).

Wall-Associated Receptor-Like Kinase Gene
Another potential CW-related resistance gene involved in signaling response to fungal pathogen is *wall-associated receptor-like kinase 2* (*WAK2*), which was located near the marker (IWB63138) associated with the major QTL, *QFhb.mgb-2AS*, for incidence and severity (Giancaspro et al. 2016; Gadaleta et al. 2019). Its loss-of-function mutation in the durum wheat variety "Kronos" resulted in higher susceptibility to fungal attack (Gadaleta et al. 2019). The gene structure analysis of *WAK2* in the resistant bread wheat and the susceptible durum wheat detected a polymorphic fragment at the last exon, which could be used as a functional marker in marker-assisted selection for FHB resistance. Such signaling genes associated with the cell wall might trigger the modification of CW components needed to resist a pathogen attack.

4.4 Defense Response Genes and Regulatory Genes

Pathogenesis-related (*PR*) genes participate in pathogen recognition, degrading the fungal structural barrier, and exhibiting antifungal activity (Walter et al. 2010). These include *PR-1*, *PR-2* (*β-1,3-glucanase*), *PR-3* (*chitinase*), *PR-4*, and *PR-5* (*thaumatin-like protein*), whose transcripts accumulated in spikes of resistant wheat varieties infected with *F. graminearum* (Pritsch et al. 2000; Kruger et al. 2002; Anand et al. 2003; Golkari et al. 2009; Ding et al. 2011). In fact, *PR-2*, *PR-5*, *α-1-purothionin*, and the regulator of defense responses, *NPR1* (*Non-expressor of Pathogenesis-Related Gene1*) were effective in heightening FHB resistance in susceptible wheat varieties (Makandar et al. 2006; Mackintosh et al. 2007). Other cysteine-rich defence-related genes encoding defensins, thionins, nonspecific lipid transfer proteins (nsLTPs), puroindolines, and hevein- and knottin-type proteins target the fungal membrane through interactions with phospholipids and sphingolipids (Lay and Anderson 2005; van Loon et al. 2006; Walter et al. 2010). Among these, the role of a *nsLTP* gene in trichothecene resistance was demonstrated in *Arabidopsis* (McLaughlin et al. 2015). Wheat pore-forming antimicrobial peptides such as puroindonolines PINA and PINB exert antifungal activities (Krishnamurthy et al. 2001). Wheat transcriptional factor, *TaWRKY70*, was identified and mapped within FHB resistance *QTL-2DL*, which imparted resistance against FHB by inducing expression of resistance genes such as *TaDGK*, *TaGLI1*, and *TaACT* and accumulating metabolites such as hydroxycinnamic acid amides and phosphatic acids (Kage et al. 2017). Genes involved in jasmonate (JA) and ethylene (ET) signaling were identified to be responsive to necrotrophic *Fusarium* pathogen, which might be involved in FHB resistance (Golkari et al. 2009; Jia et al. 2009; Ding et al. 2011; Gottwald et al. 2012; Xiao et al. 2013).

5 Improvements in Durum Wheat for FHB Resistance

Development of FHB-resistant varieties has been the top priority in most of durum wheat breeding programs globally in the past two decades. A large effort has been dedicated towards the utilization of a limited amount of native resistance and the introgression of the highly effective QTL such as *Fhb1* and *Qfhb.rwg-5A* from common wheat, and *Fhb7* from *Thinopyrum* wheatgrass into durum wheat varieties. Meanwhile, various new breeding strategies and genomic tools have been explored or integrated into the breeding and pre-breeding processes for FHB resistance. Below, we review the progress and achievements in improving the FHB resistance of modern durum germplasm and varieties.

5.1 Utilization of Native Sources of Resistance

A grand challenge for breeding durum wheat for FHB resistance is to combine required adaptability, productivity, and end-use quality with the resistance. When domesticated and wild relatives are used as the FHB resistant donors, it generally takes a long process to restore the required adaptability and end-use quality in durum wheat, often leading to loss of resistance due to linkage drag or environmental constraints. Therefore, the limited availability of native resistance in the adapted germplasm has become an important resource to meet a need in the short-term durum breeding.

In the U.S., the North Dakota (ND) durum breeding program has developed and registered 17 durum varieties using adapted durum germplasm including breeding lines and previously released varieties since 1995 (Table 2). These varieties have been grown on more than 90% of the U.S. durum wheat acreage, mainly in North Dakota and Montana. Except for the "Belzer" (Elias et al. 1999) with moderate resistance or intermediate reaction to FHB, most of the durum varieties released prior to 2003 such as Ben (Elias and Miller 1998), "Maier" (Elias and Miller 2000a), "Mountrail" (Elias and Miller 2000b), "Plaza" (Elias et al. 2001b), "Dilse" (Elias et al. 2004a) and "Pierce" (Elias et al. 2004b) are either moderately susceptible or susceptible (Table 2) (Ransom et al. 2015, 2016, 2017, 2018, 2019, 2020). On the contrary, most of the varieties released since 2005, including "Divide" (Elias and Manthey 2007a), "Carpio" (Elias et al. 2015), "Joppa" (Elias and Manthey 2016), and "ND Riveland" (Elias and Manthey 2019) had moderate resistance or intermediate reaction to FHB, clearly enhancing resistance over their parental varieties and lines (Table 2) (Ransom et al. 2017, 2018, 2019, 2020). Similarly, several Canadian durum varieties with improved FHB resistance, such as "Brigade" (DT513/DT696) (Clarke et al. 2009), "Transcend" (DT707/DT696) (Singh et al. 2012), "CDC Credence" (Strongfield/DT763//Brigade) (Pozniak et al. 2020a), and "CDC Defy" ("CDC Verona"/DT563) (Pozniak et al. 2020b), were developed by using the adapted germplasm. These varieties have performed well under the low and moderate disease pressures, but they are vulnerable to severe FHB epidemics. Therefore, it is necessary to further improve FHB resistance in these varieties using other resources and tools.

5.2 Transfer of FHB Resistance from Hexaploid Wheat into Durum Wheat

Transfer of major FHB resistance QTL from common wheat into durum wheat has been one of the most attractive strategies. However, the previous attempts to transfer FHB resistance QTL such as *Fhb1* from Sumai 3 and Wangshuibai into adapted durum germplasm had been hampered due to the complexity of resistance expression in different ploidy levels and genetic backgrounds and the unavailability of diagnostic markers in the early days (Elias et al. 2005; Prat et al. 2017; Ruan et al. 2020). Giancaspro et al. (2016) generated a durum population of 137 RILs by crossing a resistant bread wheat line, 02-5B-318, derived from Sumai 3 to the susceptible durum variety, "Saragolla." Several major QTLs were detected on chromosome arm 2AS, 3AL, 2BS, 4BL, but *Fhb1* was not identified in the population (Table 1). Until recently, Prat et al. (2017) reported the successful introgression of *Fhb1* and other QTL from Sumai 3 into the durum wheat. During the introgression process, an FHB-resistant durum line DBC-480 was first developed by marker-assisted backcrossing of Sumai 3 with durum variety "Semperdur." Subsequent QTL mapping using three RIL populations, which were developed by crossing DBC-480 to a

Table 2 Durum wheat varieties developed using adapted durum germplasm in North Dakota durum breeding program since 1995

Variety	Year released	Pedigree	Reaction to FHB[a]	References
Munich	1995	D8030/D8016	NA	Elias et al. (1997)
Ben	1996	D8024/Monroe	8	Elias and Miller (1998)
Belzer	1997	D7798/DT367	NA	Elias et al. (1999)
Maier	1998	D8193/D8335	8	Elias and Miller (2000a)
Mountrail	1998	D8479/Renville	8	Elias and Miller (2000b)
Lebsock	1999	Munich/D8469	6	Elias et al. (2001a)
Plaza	1999	Plenty/D8291	NA	Elias et al. (2001b)
Dilse	2002	Maier/D88273	NA	Elias et al. (2004a)
Pierce	2001	D86117/D88289	8	Elias et al. (2004b)
Divide	2005	Ben/D901282//Belzer	5	Elias and Manthey (2007a)
Grenora	2005	D901260/D901419	6	Elias and Manthey (2007b)
Alkabo	2005	D901247/D89263	6	Elias and Manthey (2007c)
Tioga	2010	Plaza/Maier	6	Elias and Manthey (2013)
Carpio	2012	D95580/D95595	5	Elias et al. (2015)
Joppa	2012	Maier/D97643	5	Elias and Manthey (2016)
ND Riveland	2017	DT764/DH01146	5	Elias and Manthey (2019)
ND Grano	2017	D00752/DH01146	6	Elias et al. (2021)

[a]Reaction to FHB were extracted from North Dakota Durum Variety Trial Results (Ransom et al. 2017, 2018, 2019, 2020). The FHB disease reaction scores from 1 to 9, with 1 = resistant and 9 = very susceptible. *NA* Not tested. Four varieties Munich, Belzer, Plaza, and Dilse were not included in the variety trials after 2017. In the variety trials conducted in 2015 and 2016, Belzer had moderately resistant (MR) or intermediate reactions, whereas both Plaza and Dilse showed moderately susceptible (MS) reaction (Ransom et al. 2015, 2016)

durum line (SZD1029K) and varieties "Karur" and "Durobonus," confirmed the expression of *Fhb1* and other QTL from Sumai 3 in the three durum backgrounds (Table 1).

PI 277012, previously misclassified as cultivated emmer (*T. dicoccum*), is a hexaploid wheat line that shows a high level of FHB resistance comparable to Sumai 3 (Fig. 1), thus, representing another effective source for disease improvement in durum and bread wheat (Xu et al. 2010; Chu et al. 2010). Chu et al. (2011) found its resistance to be contributed by the two major QTL, *Qfhb.rwg-5A.1*, and *Qfhb.rwg-5A.2*, on chromosome arms 5AS and 5AL, respectively. Both QTLs were successfully introgressed into several durum wheat varieties in North Dakota, exhibiting better FHB resistance (Chu et al. 2010). One of the durum lines, 10Ae564 (Fig. 1, Lebsock/PI 277012//Lebsock) carrying the two QTL, was used to develop a RIL population from its hybrid with Joppa (Table 1). The strong expression of the PI 277012-derived 5AL QTL was confirmed in durum background through subsequent linkage mapping (Zhao et al. 2018a). In addition, a new minor QTL (*Qfhb.ndwp-7A*) was identified on chromosome arm 7AL of 10Ae564 that was not reported in earlier studies. The success in developing novel durum germplasm carrying two PI 277012-derived 5A QTL, *Fhb1*, and other major QTL represents a key milestone toward using hexaploid wheat sources for durum FHB resistance breeding. In addition, the recent identification of *TaHRC/TaHis* as the candidate gene for *Fhb1* and our ongoing fine mapping of PI 277012-derived 5AL QTL (Leng et al. 2022) would speed up the breeding process to deploy and stack these major QTL into various durum varieties.

5.3 Transfer of FHB Resistance from Wild Relatives into Durum Wheat

Various wild grass species in the Triticeae have been exploited for decades to introduce different beneficial alleles into durum and bread wheat (reviewed in Cai et al. 2005; Ceoloni et al. 2014). However, alien gene introgression for FHB resistance in durum wheat has been focused mostly on the *Thinopyrum* species. Jauhar et al. (2009) made an early attempt to transfer the FHB resistance to durum wheat directly from diploid wheatgrass *Th. elongatum* [*Lophopyrum elongatum* (Host) Á. Löve, $2n = 2x = 14$, EE genome). They made an intergeneric cross between *Th. elongatum* accession PI 531719 and durum variety "Langdon" (LDN) and then extensively backcrossed the resultant F_1 hybrid with Langdon. From the backcross progeny, they isolated an FHB-resistant durum line DGE-1 (Table 1), which was identified as a durum-*Th. elongatum* chromosome 1E disomic addition line (Jauhar and Peterson 2008; Jauhar et al. 2009). Two durum-*Th. elongatum* chromosome 1E(1A) and 1E(1B) disomic substitution lines, designated as DGE-2 and DGE-3 (Table 1), respectively, were further developed using DGE-1 as the donor and Langdon chromosome 1D(1A) and 1D(1B) disomic substitution lines as the recipients (Jauhar and Peterson 2011, 2012, 2013; Jauhar 2014). These lines (DGE-2 and DGE-3) will facilitate introgression of the novel FHB resistance gene(s) from 1E into wheat genome using chromosome engineering.

Like *Fhb1* and PI 277012 5AL QTL, the *Fhb7* from decaploid (10*x*) tall wheatgrass *Th. ponticum* also has a great potential for improving FHB resistance in durum wheat. *Fhb7* was initially identified in the bread wheat-*Th. ponticum* $7D/7el_2$ translocation lines (e.g., KS10-2 and KS24-1) developed previously (Kibirige-Sebunya and Knott 1983; Kim et al. 1993). Successful introgression of *Fhb7* into durum wheat has been reported in several studies. Forte et al. (2014) backcrossed $7DS{\cdot}7el_2L$ translocation line KS24-1 with three durum lines containing a pair of $7AS{\cdot}7AL\text{-}7el_1L$ translocation chromosomes carrying leaf rust resistance gene *Lr19*, stem rust resistance gene *Sr25*, and yellow pigment (*Yp*) genes. The recombination between those closely related *Th. ponticum* $7el_1L$ and $7el_2L$ chromatins in the backcross progenies resulted in pyramiding *Fhb7*, *Sr25*, *Lr19*, and *Yp* genes in the new durum $7el_1L+7el_2L$ recombinant lines (R129, R193, and R216), which showed 70–85% increase in FHB resistance (Forte et al. 2014). By using a similar procedure, Kuzmanović et al. (2019) transferred a *Th. elongatum* derived QTL *Fhb-7E* from a bread wheat stock into durum wheat using recombination between 7EL and $7el_1L$ chromatins. The durum lines carrying *Fhb-7E* were found to have much lower disease and DON content than the susceptible controls (Kuzmanović et al. 2019). Zhang et al. (2022) identified that the *Fhb7* from *Th. elongatum* differed for 10 nucleotides from the *Fhb7* cloned by Wang et al. (2020), indicating that the *Fhb7* from *Th. elongatum* is a new haplotype or allele. They transferred *Fhb7* from *Th. elongatum* chromosome 7E to the common wheat 7B using *ph1b*-induced homoeologous recombination. We are currently transferring the *Fhb7* on chromosome 7B in common wheat into ND durum varieties using marker-assisted backcross.

5.4 Genomic Selection

Genomic selection is a selection approach that relies on genomic estimated breeding values (GEBVs) predicted from genome-wide markers, which can capture most genes involved in a complex trait (Meuwissen et al. 2001). Genomic selection can enhance genetic improvement for complex traits by shortening breeding cycles as well as increasing selection intensity and accuracy compared to phenotypic selection (Heffner et al. 2010). Therefore, genomic selection may provide a solution for gaining high and durable resistance to FHB in durum and common wheat by integrating genes with medium and minor effects combined with major resistant genes like *Fhb1* and *Fhb7*. In genomic selection, a prediction model is developed using phenotypic values and marker data collected from a training population; then the resulted model and marker data from the population under selection are used to predict GEBVs. As advances in high-throughput and low-cost genotyping technologies, genomic selection becomes more and more feasible. Guzman et al. (2016) reported that implementing genomic selection on early generation lines as a pre-selection could effectively eliminate poor quality lines prior to the expensive field trials.

In common wheat, several studies have evaluated genomic prediction for FHB resistance using data collected from active breeding programs (Hoffstetter et al. 2016; Miradita et al. 2015; Liu et al. 2019), from which prediction accuracies of 0.22–0.60 were reported. Adeyemo et al. (2020) reported that genomic selection on F_5 lines could improve the breeding efficiency of FHB resistance in hard red spring wheat, where a subset of F_5 lines were phenotypically evaluated and served as the training population to predict a large number of unphenotyped F_5 lines from the same breeding cycle. They also found that including the parents in the training population could increase prediction accuracy (Adeyemo et al. 2020). In durum wheat, Steiner et al. (2019) reported a prediction accuracy of 0.39 for FHB severity using a collection of 288 international elite varieties. From the same study, the authors found that genomic selection to eliminate the most susceptible lines followed by phenotypic selection could facilitate the improvement of FHB resistance, compared to phenotypic selection only. For the development of durum wheat varieties with FHB resistance, what's even more promising is to employ genomic selection on breeding populations from active breeding programs.

Genomic selection can also play an important role in pre-breeding efforts to develop FHB-resistant germplasms. In addition to known major genes like *Fhb1*, unknown resistance genes from hexaploid common wheat and other tetraploid wheat relatives are great resources. Phenotypic backcross selection is a common approach to introgress desirable genes from exotic lines or relative species, which, however, is challenged by the complex genetic architecture and high-cost phenotyping of FHB resistance. A simulation study showed that genomics-assisted backcross, using GEBVs to select progenies for backcrossing to recurrent parents, is more efficient than phenotypic backcross selection for a complex trait (Bernardo 2016). Therefore, genomics-assisted backcross introgression is worth experimenting for the development of FHB-resistant durum wheat germplasms.

Recurrent selection can be another approach for the development of durum FHB-resistant germplasms, given its effectiveness in increasing frequencies of diverse desirable alleles of a population. Wang et al. (2021) observed that FHB resistance in a durum population was improved after three cycles of phenotypic selection, where ~200 S_1 families were evaluated in a field nursery during summer and the top 20 families were selected and recombined in greenhouse during winter in each cycle. In the same study using the recurrent selection population, they developed a genomic prediction model with a prediction accuracy of 0.36 for FHB severity (Wang et al. 2021). While only one cycle of recurrent phenotypic selection for FHB resistance can be completed per year, genomic selection can be carried out prior to flowering so that three cycles of selection can be completed in greenhouse within a year. Therefore, genomics-assisted recurrent selection holds a promise to speed up the development of FHB-resistant germplasms in durum wheat.

5.5 Mutagenesis and Epigenetics

The scarcity of FHB resistance and the presence of resistance inhibitor and susceptibility factors within the existing germplasm may render mutagenesis as a promising strategy to generate new stable genetic resources in the native gene pools of durum wheat by random mutagenesis involving either chemical (ethyl methanesulfonate, EMS), physical (X-ray, ion beam, or gamma rays), and biotechnological approaches (Haile et al. 2019). This may generate novel alleles for FHB resistance and may silence resistance suppressors or susceptibility factors present in durum wheat. In bread wheat, a new resistance gene *MNR220* derived from EMS mutagenesis was identified to enhance resistance against four biotrophic fungal pathogens that cause leaf rust, stem rust, stripe rust, and powdery mildew (Campbell et al. 2012). Generation of such EMS mutagenized or targeting induced local lesions in genome (TILLING) resources can also be used to screen new resistance genes against diverse biotrophic pathogens. The TILLING resources are available for durum (cv. Kronos) and hexaploid wheat (cv. Cadenza) (Krasileva et al. 2017). Such materials have been used in discovery and functional characterization of the resistance genes, *TaPFT* and *TaWAK2*, and the susceptible gene, *WheatPME-1* (Rawat et al. 2016; Gadaleta et al. 2019).

Other mutagenesis approaches to improve disease resistance include the RNA interference (RNAi) and CRISPR-Cas9 system, which can be used to inactivate specific or known susceptible genes in durum wheat. CRISPR-mediated mutation of *TaNFXL1* (*nuclear transcription factor, X-box binding like 1*) in susceptible bread wheat resulted in better resistance to *Fusarium* infection (Brauer et al. 2020), probably by relieving the suppression of DON-induced defense responses (Asano et al. 2007). RNAi-based silencing of susceptible genes has been demonstrated in other plant species (Rayapuram et al. 2012; Huibers et al. 2013; Pessina et al. 2016a, b; Sun et al. 2016). In addition, CRISPR-mediated and RNAi silencing of *TaHRC/TaHis* validated its positive role in providing *Fhb1* resistance (Li et al. 2019; Su et al. 2019). However, the use of these gene-editing tools for variety development remains controversial and subject to certain limitations and regulations for commercialization in most countries (Brinegar et al. 2017; Zhang et al. 2019).

Besides mutagenesis, epigenetic regulation is another strategy to introduce metastable changes in gene activity and fine-tune gene expression patterns in response to different environmental stresses (Pikaard and Mittelsten Scheid 2014). Cytosine methylation is one such epigenetic mechanism that exists in both CG and non-CG contexts (Wang et al. 2014). Plants can adjust methylation patterns based on the changing environments (Chinnusamy and Zhu 2009). In general, demethylation at promoter regions and methylation in gene-body DNA have been associated with the active expression of stress-responsive genes (Simmen et al. 1999; Zhang et al. 2006; Zilberman et al. 2007; Choi and Sano 2007; Li et al. 2012; Yu et al. 2013). Cytosine demethylation can be induced by 5-azadeoxycytidine, an effective inhibitor of methyltransferase (Tatra et al. 2000; Christman 2002). Such chemical treatment in susceptible rice plants resulted in high levels of resistance to a bacterial pathogen (Akimoto et al. 2007). Similarly, cytosine demethylation has been attempted to improve FHB resistance in durum wheat. Kumar et al. (2020) treated eight durum advanced-breeding lines using 5-azadeoxycytidine and obtained five mutant (M_4) lines with increased FHB resistance over their parental lines. However, these mutants were not observed to be significantly different from their parental lines in methylation levels. The improvement of FHB resistance in these lines, as revealed by transcriptome analysis of one M_4 line, may be attributed to the changes in differentially expressed genes involved in defense, phosphorylation cascade, hormone signaling, transcriptional regulation, proteolysis, secondary metabolism, cell wall, and oxidative stress (Kumar et al. 2020). Introducing such a stable and novel source of FHB resistance in adapted durum varieties via epigenetic manipulations may be beneficial for future breeding efforts.

6 Conclusion and Perspectives

A significant progress has been made in the past two decades to understand the genetics and genomics of FHB resistance despite the complexity and limited sources for FHB resistance in durum wheat. Figure 2 summarizes the resistance sources and strategies implemented to identify and characterize putative QTL/genes that confer FHB resistance, and breeding strategies used to introduce the resistant QTL/genes into modern durum wheat germplasm. A large number of FHB-resistant QTLs have been found using GWAS and linkage analysis in closely related and wild species that can be used as resistance sources in durum breeding. However, only two major QTL, *Fhb1* and *Fhb7*, have been successfully cloned, which can be used to develop more resilient durum wheat varieties able to withstand *Fusarium* diseases. In bread wheat, a number of elite varieties with different combinations of

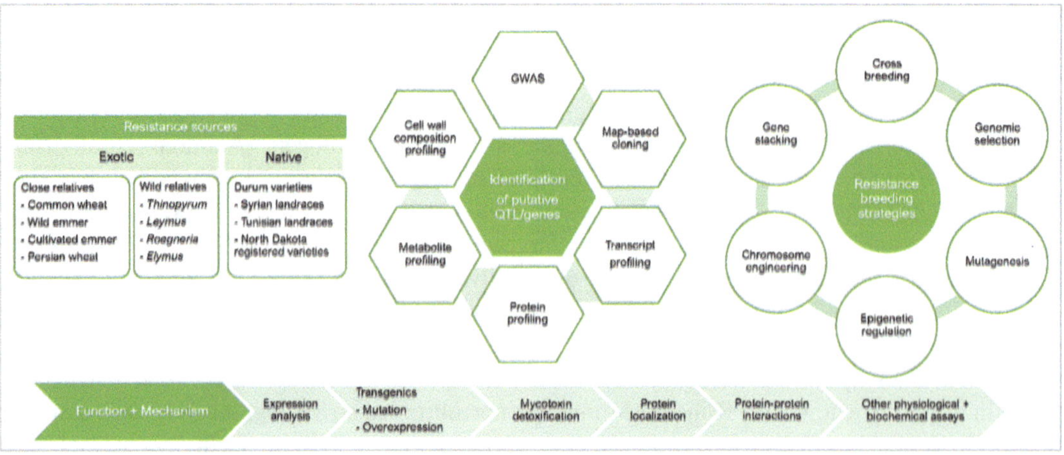

Fig. 2 Sources and strategies to improve and dissect genetics and genomics of *Fusarium* head blight resistance in durum wheat

Fhb1/2/4/5 QTL from Wangshubai were generated via marker-assisted backcross, proving more robust and durable resistance (Ma et al. 2020). Fine-mapping these major QTL (*Fhb2/3/4/5/8/9*) is also essential to identify the causal genes behind FHB resistance, which will facilitate a proper cloning, functional characterization, and expression of these genes in elite durum varieties.

Although durum germplasms lack major QTL present in the hexaploid wheat, several minor QTL are found, which can be introgressed into the new varieties in combination with major QTL using marker-assisted selection or genomic selection to achieve higher levels of FHB resistance. Several Syrian and Tunisian durum landraces and North Dakota registered durum varieties are important native resources to initiate further improvement in disease resistance. The high susceptibility of durum wheat is speculated to be contributed by the presence of susceptible factors in durum genome that suppress the effect of FHB-resistant QTL. Identification and deactivation of these susceptible factors using genomic tools such as CRISPR-cas9 and RNAi are alternative strategies to improve FHB resistance in durum wheat. Another more stable approach to identify novel resistance and susceptible genes is using EMS mutagenesis. In addition, epigenetic regulation such as cytosine demethylation in susceptible durum varieties may allow the proper expression of resistance genes and suppression of susceptible genes to better resist *Fusarium* infection.

A vast number of known resistance genes have been identified by a comparative profiling of transcripts, proteins, metabolites, and cell wall components in the resistant and susceptible wheat varieties infected with *Fusarium* pathogens or treated with DON. These are important targets for functional genomics to dissect the mechanism underlying FHB resistance and mycotoxin detoxification. Transgenic expression of these resistance genes in durum wheat might enhance FHB resistance. Additionally, a wealth of resources are currently available to better understand the genetics and genomics of FHB resistance and susceptibility factors in durum wheat, including high-density consensus maps of durum and tetraploid wheat (Maccaferri et al. 2014, 2015), durum (cv. Svevo) reference genome (Maccaferri et al. 2019) and 90K SNP genotyping data for global durum wheat panel (GDP) (Mazzucotelli et al. 2020). Ultimately, a better understanding of the resistance mechanisms is needed to develop tools and strategies to prevent mycotoxin accumulation and yield loss caused by FHB. The prospects of resistance breeding are tremendous with the advancement in sequencing technologies for transcriptomics, proteomics, metabolomics, and genome-wide association studies in addition to forward and reverse genetic tools.

Disclaimer All opinions expressed in this paper are the author's and do not necessarily reflect the policies and views of USDA, ARS, DOE, or ORAU/ORISE. USDA is an equal opportunity provider and employer.

Conflict of Interest Statement All authors have no conflict of interest.

Funding This material is based upon work supported by the US Department of Agriculture, under Agreement No. FY22-DU-011 and the CRIS Project No. 2030-21430-015-00D. This is a cooperative project with the US Wheat & Barley Scab Initiative. This research was supported in part by an appointment to the Agricultural Research Service (ARS) Research Participation Program administered by the Oak Ridge Institute for Science and Education (ORISE) through an interagency agreement between the U.S. Department of Energy (DOE) and the U.S. Department of Agriculture (USDA). ORISE is managed by ORAU under DOE contract number DE-SC0014664.

References

Adeyemo E, Bajgain P, Conley E, Sallam AH, Anderson JA (2020) Optimizing training population size and content to improve prediction accuracy of FHB-related traits in wheat. Agronomy 10(4):543. https://doi.org/10.3390/agronomy10040543

Akimoto K, Katakami H, Kim H-J et al (2007) Epigenetic inheritance in rice plants. Ann Bot 100:205–217. https://doi.org/10.1093/aob/mcm110

Anand A, Schmelz EA, Muthukrishnan S (2003) Development of a lesion-mimic phenotype in a transgenic wheat line overexpressing genes for pathogenesis-related (PR) proteins is dependent on salicylic acid concentration. Mol Plant-Microbe Interact 16:916–925. https://doi.org/10.1094/MPMI.2003.16.10.916

Anderson JA, Stack RW, Liu S et al (2001) DNA markers for *Fusarium* head blight resistance QTL in two wheat populations. Theor Appl Genet 102:1164–1168. https://doi.org/10.1007/s001220000509

Ansari KI, Walter S, Brennan JM et al (2007) Retrotransposon and gene activation in wheat in response to mycotoxigenic and non-mycotoxigenic-associated *Fusarium* stress. Theor Appl Genet 114:927–937. https://doi.org/10.1007/s00122-006-0490-0

Aoki T, O'Donnell K (1999a) Morphological characterization of *Gibberella coronicola* sp. nov., obtained through mating experiments of *Fusarium pseudograminearum*. Mycoscience 40:443–453. https://doi.org/10.1007/BF02461021

Aoki T, O'Donnell K (1999b) Morphological and molecular characterization *of Fusarium pseudograminearum* sp. nov., formerly recognized as the Group 1 population of *F. graminearum*. Mycologia 91:597–609. https://doi.org/10.1080/00275514.1999.12061058

Asano T, Masuda D, Yasuda M et al (2007) *AtNFXL1*, an Arabidopsis homologue of the human transcription factor NF-X1, functions as a negative regulator of the trichothecene phytotoxin-induced defense response. Plant J 53:450–464. https://doi.org/10.1111/j.1365-313X.2007.03353.x

Bai G, Kolb FL, Shaner G, Domier LL (1999) Amplified fragment length polymorphism markers linked to a major quantitative trait locus controlling scab resistance in wheat. Phytopathology 89:343–348. https://doi.org/10.1094/PHYTO.1999.89.4.343

Bellincampi D, Cervone F, Lionetti V (2014) Plant cell wall dynamics and wall-related susceptibility in plant–pathogen interactions. Front Plant Sci 5:228. https://doi.org/10.3389/fpls.2014.00228

Bernardo R (2016) Genomewide predictions for backcrossing a quantitative trait from an exotic to an adapted line. Crop Sci 56:1067–1075. https://doi.org/10.2135/cropsci2015.09.0586

Bhati KK, Sharma S, Aggarwal S et al (2015) Genome-wide identification and expression characterization of ABCC-MRP transporters in hexaploid wheat. Front Plant Sci 6:488. https://doi.org/10.3389/fpls.2015.00488

Blümke A, Falter C, Herrfurth C et al (2014) Secreted fungal effector lipase releases free fatty acids to inhibit innate immunity-related callose formation during wheat head infection. Plant Physiol 165:346–358. https://doi.org/10.1104/pp.114.236737

Brauer EK, Balcerzak M, Rocheleau H et al (2020) Genome editing of a Deoxynivalenol-induced transcription factor confers resistance to *Fusarium graminearum* in wheat. Mol Plant-Microbe Interact 33:553–560. https://doi.org/10.1094/MPMI-11-19-0332-R

Brewer HC, Hawkins ND, Hammond-Kosack KE (2014) Mutations in the *Arabidopsis* homoserine kinase gene *DMR1* confer enhanced resistance to *Fusarium culmorum* and *F. graminearum*. BMC Plant Biol 14:317. https://doi.org/10.1186/s12870-014-0317-0

Brinegar K, Yetisen KA, Choi S et al (2017) The commercialization of genome-editing technologies. Crit Rev Biotechnol 37:924–932. https://doi.org/10.1080/07388551.2016.1271768

Buerstmayr H, Stierschneider M, Steiner B et al (2003) Variation for resistance to head blight caused by *Fusarium graminearum* in wild emmer (*Triticum dicoccoides*) originating from Israel. Euphytica 130:17–23. https://doi.org/10.1023/A:1022324727780

Buerstmayr H, Ban T, Anderson JA (2009) QTL mapping and marker-assisted selection for Fusarium head blight resistance in wheat: a review. Plant Breed 128:1–26. https://doi.org/10.1111/j.1439-0523.2008.01550.x

Buerstmayr M, Huber K, Heckmann J et al (2012) Mapping of QTL for Fusarium head blight resistance and morphological and developmental traits in three backcross populations derived from *Triticum dicoccum* × *Triticum durum*. Theor Appl Genet 125:1751–1765

Buerstmayr M, Alimari A, Steiner B, Buerstmayr H (2013) Genetic mapping of QTL for resistance to *Fusarium* head blight spread (type 2 resistance) in a *Triticum dicoccoides* × *Triticum durum* backcross-derived population. Theor Appl Genet 126:2825–2834

Buerstmayr M, Steiner B, Buerstmayr H (2020) Breeding for *Fusarium* head blight resistance in wheat—Progress and challenges. Plant Breed 139:429–454. https://doi.org/10.1111/pbr.12797

Bushnell WR, Hazen BE, Pritsch C (2003) Histology and physiology of Fusarium head blight. In: Leonard KJ, Bushnell WR (eds) Fusarium head blight of wheat and barley. American Phytopathological Society, St. Paul, pp 44–83

Cai X, Chen PD, Xu SS et al (2005) Utilization of alien genes to enhance Fusarium head blight resistance in wheat—a review. Euphytica 142:309–318. https://doi.org/10.1007/s10681-005-2437-y

Cai X, Danilova T, Charif A et al (2024) Registration of WGC002 spring wheat containing wild grass-derived Fusarium head blight resistance gene $Fhb7^{The2}$. J Plant Regist 18:179–186. https://doi.org/10.1002/plr2.20342

Cainong JC, Bockus WW, Feng Y et al (2015) Chromosome engineering, mapping, and transferring of resistance to Fusarium head blight disease from *Elymus tsukushiensis* into wheat. Theor Appl Genet 128:1019–1027. https://doi.org/10.1007/s00122-015-2485-1

Campbell J, Zhang H, Giroux MJ et al (2012) A mutagenesis-derived broad-spectrum disease resistance locus in wheat. Theor Appl Genet 125:391–404. https://doi.org/10.1007/s00122-012-1841-7

Cantu D, Blanco-Ulate B, Yang L et al (2009) Ripening-regulated susceptibility of tomato fruit to *Botrytis cinerea* requires NOR but not RIN or ethylene. Plant Physiol 150:1434–1449. https://doi.org/10.1104/pp.109.138701

Carpita NC (1996) Structure and biogensis of the cell walls of grasses. Annu Rev Plant Physiol Plant Mol Biol 47:445–476. https://doi.org/10.1146/annurev.arplant.47.1.445

Caruso C, Chilosi G, Caporale C et al (1999) Induction of pathogenesis-related proteins in germinating wheat seeds infected with Fusarium culmorum. Plant Sci 140:87–97. https://doi.org/10.1016/S0168-9452(98)00199-X

Ceoloni C, Kuzmanović L, Gennaro A et al (2014) Genomes, chromosomes and genes of the wheatgrass genus *Thinopyrum*: the value of their transfer into wheat for gains in Cytogenomic knowledge and sustainable breeding. In: Tuberosa R, Graner A, Frison E (eds) Genomics of plant genetic resources: volume 2. Crop productivity, food security and nutritional quality. Springer Netherlands, Dordrecht, pp 333–358. https://doi.org/10.1007/978-94-007-7575-6_14

Chhabra B, Tiwari V, Gill BS, Dong Y, Rawat N (2021) Discovery of a susceptibility factor for Fusarium head blight on chromosome 7A of wheat. Theor Appl Genet 134:2273–2289. https://doi.org/10.1007/s00122-021-03825-y

Chilaka CA, De Boevre M, Atanda OO, De Saeger S (2017) The status of Fusarium mycotoxins in sub-Saharan Africa: a review of emerging trends and post-harvest mitigation strategies towards food control. Toxins 9. https://doi.org/10.3390/toxins9010019

Chinnusamy V, Zhu J-K (2009) Epigenetic regulation of stress responses in plants. Curr Opin Plant Biol 12:133–139. https://doi.org/10.1016/j.pbi.2008.12.006

Choi C-S, Sano H (2007) Abiotic-stress induces demethylation and transcriptional activation of a gene encoding a glycerophosphodiesterase-like protein in tobacco plants. Mol Gen Genomics 277:589–600. https://doi.org/10.1007/s00438-007-0209-1

Christman JK (2002) 5-Azacytidine and 5-aza-2′-deoxycytidine as inhibitors of DNA methylation: mechanistic studies and their implications for cancer therapy. Oncogene 21:5483–5495. https://doi.org/10.1038/sj.onc.1205699

Chu CG, Zhong S, Chao S, Friesen TL, Halley S, Elias EM, Faris JD, Xu SS (2010) Fighting against FHB - an excellent novel resistance source for future wheat breeding. In: Canty S, Clark A, Anderson-Scully A, Ellis D, Van Sanford D (eds) Proceedings of the National Fusarium Head Blight Forum, Milwaukee, WI, 7–9 December, 2010. University of Kentucky, Lexington

Chu C, Niu Z, Zhong S et al (2011) Identification and molecular mapping of two QTL with major effects for resistance to Fusarium head blight in wheat. Theor Appl Genet 123:1107–1119. https://doi.org/10.1007/s00122-011-1652-2

Clarke JM, Knox RE, DePauw RM, Clarke FR, Fernandez MR, McCaig TN, Singh AK (2009) Brigade durum wheat. Can J Plant Sci 89:505–509. https://doi.org/10.4141/CJPS08168

Deepak S, Shailasree S, Kini RK et al (2010) Hydroxyproline-rich glycoproteins and plant Defence. J Phytopathol 158(9):585–593. https://doi.org/10.1111/j.1439-0434.2010.01669.x

Dhokane D, Karre S, Kushalappa AC, McCartney C (2016) Integrated Metabolo-Transcriptomics reveals Fusarium head blight candidate resistance genes in wheat QTL-Fhb2. PLoS One 11:e0155851. https://doi.org/10.1371/journal.pone.0155851

Ding L, Xu H, Yi H et al (2011) Resistance to hemi-biotrophic *F. graminearum* infection is associated with coordinated and ordered expression of diverse defense signaling pathways. PLoS One 6:e19008. https://doi.org/10.1371/journal.pone.0019008

Dubcovsky J, Dvorak J (2007) Genome plasticity a key factor in the success of Polyploid wheat under domestication. Science 316:1862–1866. https://doi.org/10.1126/science.1143986

El-Gendy W, Brownleader MD, Ismail H (2001) Rapid deposition of wheat cell wall structural proteins in response to Fusarium-derived elicitors. J Exp Bot 52(354):85–90. https://doi.org/10.1093/jexbot/52.354.85

Elias EM, Manthey FA (2007a) Registration of 'Divide' durum wheat. J Plant Regist 1:7–8. https://doi.org/10.3198/jpr2006.05.0281crc

Elias EM, Manthey FA (2007b) Registration of 'Grenora' durum wheat. J Plant Regist 1:8–9. https://doi.org/10.3198/jpr2006.05.0282crc

Elias EM, Manthey FA (2007c) Registration of 'Alkabo' durum wheat. J Plant Regist 1:10–11. https://doi.org/10.3198/jpr2006.05.0279crc

Elias EM, Manthey FA (2013) Registration of 'Tioga' durum wheat. J Plant Regist 7:69–74. https://doi.org/10.3198/jpr2012.04.0244crc

Elias EM, Manthey FA (2016) Registration of 'Joppa' durum wheat. J Plant Regist 10:139–144. https://doi.org/10.3198/jpr2015.11.0071crc

Elias EM, Manthey FA (2019) Registration of 'ND Riveland' durum wheat. J Plant Regist 13:387–391. https://doi.org/10.3198/jpr2019.01.0001crc

Elias EM, Miller JD (1998) Registration of 'Ben' durum wheat. Crop Sci 38:895

Elias EM, Miller JD (2000a) Registration of 'Maier' durum wheat. Crop Sci 40:1498–1499

Elias EM, Miller JD (2000b) Registration of 'Mountrail' durum wheat. Crop Sci 40:1499–1500

Elias EM, Miller JD, Francl LJ (1997) Registration of 'Munich' durum wheat. Crop Sci 17:290

Elias EM, Stack RW, Miller JD (1999) Registration of 'Belzer' durum wheat. Crop Sci 39:881–882

Elias EM, Miller JD, Manthey FA (2001a) Registration of 'Lebsock' durum wheat. Crop Sci 41:2007–2008

Elias EM, Miller JD, Manthey FA (2001b) Registration of 'Plaza' durum wheat. Crop Sci 41:2008

Elias EM, Manthey FA, Miller JD (2004a) Registration of 'Dilse' durum wheat. Crop Sci 44:1024

Elias EM, Manthey FA, Miller JD (2004b) Registration of 'Pierce' durum wheat. Crop Sci 44:1025

Elias EM, Manthey FA, Stack RW, Kianian SF (2005) Breeding efforts to develop Fusarium head blight resistant durum wheat in North Dakota. In: Canty SM, Boring T, Wardwell J, Siler L, Ward RW (eds) Proc 2005 National Fusarium Head Blight Forum. Milwaukee, WI, 11–13 Dec 2005, Michigan State University, East Lansing, pp 25–26

Elias EM, Manthey FA, Abuhammad WA (2015) Registration of 'Carpio' durum wheat. J Plant Regist 9:78–82. https://doi.org/10.3198/jpr2014.05.0030crc

Elias ME, Xu SS, Manthey FA (2018) Decades of breeding scab resistant durum wheat in North Dakota: challenges and successes. In: Meeting abstract international conference from seed to Pasta III: a sustainable Durum wheat chain for food security and healthy lives, September 19–21, 2018, Bologna, Italy

Elias EM, Manthey FA, Kumar A, Salsman E (2021) Registration of 'ND Grano' durum wheat. J Plant Regist 15:535–541

Fakhfakh MM, Yahyaoui A, Rezgui S et al (2011) Inheritances of Fusarium head blight resistance in a cross involving local and exotic durum wheat cultivars. Crop Sci 51:2517–2524. https://doi.org/10.2135/cropsci2011.03.0140

Ferrari S, Sella L, Janni M et al (2012) Transgenic expression of polygalacturonase-inhibiting proteins in Arabidopsis and wheat increases resistance to the flower pathogen *Fusarium graminearum*. Plant Biol 14(Suppl 1):31–38. https://doi.org/10.1111/j.1438-8677.2011.00449.x

Fierens E, Rombouts S, Gebruers K et al (2007) TLXI, a novel type of xylanase inhibitor from wheat (*Triticum aestivum*) belonging to the thaumatin family. Biochem J 403:583–591. https://doi.org/10.1042/BJ20061291

Flors V, de la Leyva OM, Vicedo B et al (2007) Absence of the endo-β-1,4-glucanases Cel1 and Cel2 reduces susceptibility to *Botrytis cinerea* in tomato. Plant J 52:1027–1040. https://doi.org/10.1111/j.1365-313X.2007.03299.x

Forte P, Virili ME, Kuzmanović L et al (2014) A novel assembly of *Thinopyrum ponticum* genes into the durum wheat genome: pyramiding Fusarium head blight resistance onto recombinant lines previously engineered for other beneficial traits from the same alien species. Mol Breed 34:1701–1716. https://doi.org/10.1007/s11032-014-0175-3

Friebe A, Vilich VV, Hennig L et al (1998) Detoxification of Benzoxazolinone Allelochemicals from wheat by *Gaeumannomyces graminis* var. *tritici*, *G. graminis* var. *graminis*, *G. graminis* var. *avenae*, and *Fusarium culmorum*. Appl Environ Microbiol 64:2386–2391. https://doi.org/10.1128/AEM.64.7.2386-2391.1998

Gadaleta A, Colasuonno P, Giove SL et al (2019) Map-based cloning of *QFhb.mgb-2A* identifies a *WAK2* gene responsible for Fusarium Head Blight resistance in wheat. Sci Rep 9:6929. https://doi.org/10.1038/s41598-019-43334-z

Garvin DF, Stack RW, Hansen JM (2009) Quantitative trait locus mapping of increased Fusarium head blight susceptibility associated with a wild emmer wheat chromosome. Phytopathology 99:447–452

Garvin DF, Porter H, Blankenheim ZJ, Chao S, Dill-Macky R (2015) A spontaneous segmental deletion from chromosome arm 3DL enhances Fusarium head blight resistance in wheat. Genome 58:479–488. https://doi.org/10.1139/gen-2015-0088

Gatti M, Choulet F, Macadré C et al (2018) Identification, molecular cloning, and functional characterization of a wheat UDP-glucosyltransferase involved in resistance to Fusarium Head Blight and to mycotoxin accumulation. Front Plant Sci 9:1853. https://doi.org/10.3389/fpls.2018.01853

Ghavami F, Elias EM, Mamidi S et al (2011) Mixed model association mapping for fusarium head blight resistance in tunisian-derived durum wheat populations. G3 1:209–218. https://doi.org/10.1534/g3.111.000489

Giancaspro A, Giove SL, Zito D et al (2016) Mapping QTL for Fusarium Head Blight resistance in an interspecific wheat population. Front Plant Sci 7:1381. https://doi.org/10.3389/fpls.2016.01381

Giancaspro A, Lionetti V, Giove SL et al (2018) Cell wall features transferred from common into durum wheat to improve Fusarium Head Blight resistance. Plant Sci 274:121–128. https://doi.org/10.1016/j.plantsci.2018.05.016

Gladysz C, Lemmens M, Steiner B, Buerstmayr H (2007) Evaluation and genetic mapping of resistance to Fusarium head blight in *Triticum dicoccoides*. Isr J Plant Sci 55:263–266. https://doi.org/10.1560/IJPS.55.3-4.263

Golkari S, Gilbert J, Ban T, Procunier JD (2009) QTL-specific microarray gene expression analysis of wheat resistance to Fusarium head blight in Sumai-3 and two susceptible NILs. Genome 52:409–418. https://doi.org/10.1139/g09-018

Gorash A, Armonienė R, Kazan K (2021) Can effectoromics and loss-of-susceptibility be exploited for improving Fusarium head blight resistance in wheat? Crop J 9:1–16. https://doi.org/10.1016/j.cj.2020.06.012

Gottwald S, Samans B, Lück S, Friedt W (2012) Jasmonate and ethylene dependent defence gene expression and suppression of fungal virulence factors: two essential mechanisms of Fusarium head blight resistance in wheat? BMC Genomics 13:369. https://doi.org/10.1186/1471-2164-13-369

Gunupuru LR, Arunachalam C, Malla KB et al (2018) A wheat cytochrome P450 enhances both resistance to deoxynivalenol and grain yield. PLoS One 13:e0204992. https://doi.org/10.1371/journal.pone.0204992

Guo J, He F, Cai J-J et al (2015a) Molecular and cytological comparisons of chromosomes $7el_1$, $7el_2$, $7E^e$, and $7E^i$ derived from *Thinopyrum*. Cytogenet Genome Res 145:68–74. https://doi.org/10.1159/000381838

Guo J, Zhang X, Hou Y et al (2015b) High-density mapping of the major FHB resistance gene Fhb7 derived from *Thinopyrum ponticum* and its pyramiding with *Fhb1* by marker-assisted selection. Theor Appl Genet 128:2301–2316. https://doi.org/10.1007/s00122-015-2586-x

Guzman C, Peña RJ, Singh R, Autrique E et al (2016) Wheat quality improvement at CIMMYT and the use of genomic selection on it. Appl Transl Genom 11:3–8. https://doi.org/10.1016/j.atg.2016.10.004

Haile JK, N'Diaye A, Walkowiak S et al (2019) Fusarium Head Blight in Durum wheat: recent status, breeding directions, and future research prospects. Phytopathology 109:1664–1675. https://doi.org/10.1094/PHYTO-03-19-0095-RVW

Hales B, Steed A, Giovannelli V, Burt C, Lemmens M, Molnár-Láng M, Nicholson P (2020) Type II Fusarium head blight susceptibility conferred by a region on wheat chromosome 4D. J Exp Bot 71:4703–4714. https://doi.org/10.1093/jxb/eraa226

Handa H, Namiki N, Xu D, Ban T (2008) Dissecting of the FHB resistance QTL on the short arm of wheat chromosome 2D using a comparative genomic approach: from QTL to candidate gene. Mol Breed 22:71–84. https://doi.org/10.1007/s11032-008-9157-7

Hao Y, Rasheed A, Zhu Z, Wulff BBH, He Z (2020) Harnessing wheat *Fhb1* for fusarium resistance. Trends Plant Sci 25:1–3. https://doi.org/10.1016/j.tplants.2019.10.006

He Y, Ahmad D, Zhang X et al (2018a) Genome-wide analysis of family-1 UDP glycosyltransferases (UGT) and identification of UGT genes for FHB resistance in wheat (*Triticum aestivum* L.). BMC Plant Biol 18:67. https://doi.org/10.1186/s12870-018-1286-5

He Y, Zhang X, Zhang Y et al (2018b) Molecular characterization and expression of PFT, an FHB resistance gene at the *Fhb1* QTL in wheat. Phytopathology 108:730–736. https://doi.org/10.1094/PHYTO-11-17-0383-R

He Y, Wu L, Liu X et al (2020) *TaUGT6*, a novel UDP-glycosyltransferase gene enhances the resistance to FHB and DON accumulation in wheat. Front Plant Sci 11:574775. https://doi.org/10.3389/fpls.2020.574775

Heffner EL, Lorenz AJ, Jannink J, Sorrells ME (2010) Plant breeding with genomic selection: gain per unit time and cost. Crop Sci 50:1681–1690. https://doi.org/10.2135/cropsci2009.11.0662

Hoffstetter A, Cabrera A, Huang M, Sneller C (2016) Optimizing training population data and validation of genomic selection for economic traits in soft winter wheat. G3 6:2919–2928. https://doi.org/10.1534/g3.116.032532

Huhn MR, Elias EM, Ghavami F et al (2012) Tetraploid Tunisian wheat germplasm as a new source of Fusarium Head Blight resistance. Crop Sci 52:136–145. https://doi.org/10.2135/cropsci2011.05.0263

Huibers RP, Loonen AEHM, Gao D et al (2013) Powdery mildew resistance in tomato by impairment of *SlPMR4* and *SlDMR1*. PLoS One 8:e67467. https://doi.org/10.1371/journal.pone.0067467

Igawa T, Tokai T, Kudo T et al (2005) A wheat xylanase inhibitor gene, *Xip-I*, but not *Taxi-I*, is significantly induced by biotic and abiotic signals that trigger plant defense. Biosci Biotechnol Biochem 69:1058–1063. https://doi.org/10.1271/bbb.69.1058

Ito M, Sato I, Ishizaka M et al (2013) Bacterial Cytochrome P450 system catabolizing the Fusarium toxin Deoxynivalenol. Appl Environ Microbiol 79:1619–1628. https://doi.org/10.1128/AEM.03227-12

Jauhar PP (2014) Durum wheat genetic stocks involving chromosome 1E of diploid wheatgrass: resistance to Fusarium head blight. Nucleus 57:19–23. https://doi.org/10.1007/s13237-014-0102-3

Jauhar PP, Peterson TS (2008) Registration of DGE-1, durum wheat alien disomic addition line with resistance to Fusarium head blight. J Plant Regist 2:167–168. https://doi.org/10.3198/jpr2007.01.0048crgs

Jauhar PP, Peterson TS (2011) Cytological and molecular characterization of homoeologous group-1 chromosomes in hybrid derivatives of a durum disomic alien addition line. Plant Genome 4:102–109. https://doi.org/10.3835/plantgenome2011.01.0002

Jauhar PP, Peterson TS (2012) Registration of DGE-2, a Durum wheat Disomic Alien substitution line 1E(1A) involving a diploid wheatgrass chromosome. J Plant Regist 6:221–223

Jauhar PP, Peterson TS (2013) Registration of DGE-3, a durum wheat disomic substitution line

1E(1B) involving a wheatgrass chromosome. J Plant Regist 7:257–259. https://doi.org/10.3198/jpr2012.09.0030crgs
Jauhar PP, Peterson TS, Xu SS (2009) Cytogenetic and molecular characterization of a durum alien disomic addition line with enhanced tolerance to Fusarium head blight. Genome 52:467–483. https://doi.org/10.1139/g09-014
Jia H, Cho S, Muehlbauer GJ (2009) Transcriptome analysis of a wheat near-isogenic line pair carrying Fusarium head blight-resistant and -susceptible alleles. Mol Plant-Microbe Interact 22:1366–1378. https://doi.org/10.1094/MPMI-22-11-1366
Joppa LR, Cantrell RG (1990) Chromosomal location of genes for grain protein content of wild tetraploid wheat. Crop Sci 30:1059. https://doi.org/10.2135/cropsci1990.0011183x003000050021x
Kage U, Yogendra KN, Kushalappa AC (2017) TaWRKY70 transcription factor in wheat QTL-2DL regulates downstream metabolite biosynthetic genes to resist *Fusarium graminearum* infection spread within spike. Sci Rep 7:42596. https://doi.org/10.1038/srep42596
Kang Z, Buchenauer H (2003) Immunocytochemical localizationofCellWall-BoundThioninsandHydroxyproline-rich glycoproteins in *Fusarium culmorum*-infected wheat spikes. J Phytopathol 151:120–129. https://doi.org/10.1046/j.1439-0434.2003.00693.x
Kazan K, Gardiner DM (2018) Fusarium crown rot caused by *Fusarium pseudograminearum* in cereal crops: recent progress and future prospects. Mol Plant Pathol 19:1547–1562. https://doi.org/10.1111/mpp.12639
Kibiridge-Sebunya I, Knott DR (1983) Transfer of stem rust resistance to wheat from an *Agropyron* chromosome having a gametocidal effect. Can J Genet Cytol 25:215–221. https://doi.org/10.1139/g83-035
Kim NS, Armstrong K, Knott DR (1993) Molecular detection of *Lophopyrum* chromatin in wheat-Lophopyrum recombinants and their use in the physical mapping of chromosome 7D. Theor Appl Genet 85:561–567. https://doi.org/10.1007/BF00220914
King BC, Waxman KD, Nenni NV et al (2011) Arsenal of plant cell wall degrading enzymes reflects host preference among plant pathogenic fungi. Biotechnol Biofuels 4:4. https://doi.org/10.1186/1754-6834-4-4
Kirana RP, Gaurav K, Arora S et al (2023) Identification of a UDP-glucosyltransferase conferring deoxynivalenol resistance in *Aegilops tauschii* and wheat. Plant Biotechnol J 21:109–121. https://doi.org/10.1111/pbi.13928
Kluger B, Bueschl C, Lemmens M et al (2013) Stable isotopic labelling-assisted untargeted metabolic profiling reveals novel conjugates of the mycotoxin deoxynivalenol in wheat. Anal Bioanal Chem 405:5031–5036. https://doi.org/10.1007/s00216-012-6483-8
Krasileva KV, Vasquez-Gross HA, Howell T et al (2017) Uncovering hidden variation in polyploid wheat. Proc Natl Acad Sci USA 114:E913–E921. https://doi.org/10.1073/pnas.1619268114
Krishnamurthy K, Balconi C, Sherwood JE, Giroux MJ (2001) Wheat puroindolines enhance fungal disease resistance in transgenic rice. Mol Plant-Microbe Interact 14:1255–1260. https://doi.org/10.1094/MPMI.2001.14.10.1255
Kruger WM, Pritsch C, Chao S, Muehlbauer GJ (2002) Functional and comparative bioinformatic analysis of expressed genes from wheat spikes infected with Fusarium graminearum. Mol Plant-Microbe Interact 15:445–455. https://doi.org/10.1094/MPMI.2002.15.5.445
Kumar S, Stack RW, Friesen TL, Faris JD (2007) Identification of a novel Fusarium head blight resistance quantitative trait locus on chromosome 7A in tetraploid wheat. Phytopathology 97:592–597. https://doi.org/10.1094/PHYTO-97-5-0592
Kumar J, Rai KM, Pirseyedi S et al (2020) Epigenetic regulation of gene expression improves Fusarium head blight resistance in durum wheat. Sci Rep 10:17610. https://doi.org/10.1038/s41598-020-73521-2
Kuzmanović L, Mandalà G, Tundo S et al (2019) Equipping Durum wheat—*Thinopyrum ponticum* recombinant lines with a *Thinopyrum elongatum* Major QTL for resistance to Fusarium diseases through a cytogenetic strategy. Front Plant Sci 10:1324. https://doi.org/10.3389/fpls.2019.01324
Lay FT, Anderson MA (2005) Defensins-components of the innate immune system in plants. Curr Protein Pept Sci 6:85–101. https://doi.org/10.2174/1389203053027575
Leng ZM, Liu Y, Xi P, et al (2022) Development of high-density single nucleotide polymorphisms (SNPs) markers for a major QTL for Fusarium head blight resistance in wheat line PI277012. International Plant and Animal Genome Conference XXIX, San Diego, CA. January 8–12, 2022
Li ZS, Alfenito M, Rea PA et al (1997) Vacuolar uptake of the phytoalexin medicarpin by the glutathione conjugate pump. Phytochemistry 45:689–693. https://doi.org/10.1016/s0031-9422(97)00031-9
Li X, Zhang JB, Song B et al (2010) Resistance to Fusarium head blight and seedling blight in wheat is associated with activation of a cytochrome p450 gene. Phytopathology 100:183–191. https://doi.org/10.1094/PHYTO-100-2-0183
Li X, Zhu J, Hu F et al (2012) Single-base resolution maps of cultivated and wild rice methylomes and regulatory roles of DNA methylation in plant gene expression. BMC Genomics 13:300. https://doi.org/10.1186/1471-2164-13-300
Li X, Shin S, Heinen S et al (2015) Transgenic wheat expressing a barley UDP-glucosyltransferase detoxifies Deoxynivalenol and provides high levels of resistance to *Fusarium graminearum*. Mol Plant-Microbe Interact 28:1237–1246. https://doi.org/10.1094/MPMI-03-15-0062-R
Li X, Michlmayr H, Schweiger W et al (2017) A barley UDP-glucosyltransferase inactivates nivalenol and provides Fusarium Head Blight resistance in trans-

genic wheat. J Exp Bot 68:2187–2197. https://doi.org/10.1093/jxb/erx109
Li G, Zhou J, Jia H et al (2019) Mutation of a histidine-rich calcium-binding-protein gene in wheat confers resistance to Fusarium head blight. Nat Genet 51:1106–1112. https://doi.org/10.1038/s41588-019-0426-7
Lin F, Kong ZX, Zhu HL et al (2004) Mapping QTL associated with resistance to Fusarium head blight in the Nanda2419 x Wangshuibai population. I. Type II resistance. Züchter Genet Breed Res 109:1504–1511. https://doi.org/10.1007/s00122-004-1772-z
Lionetti V, Giancaspro A, Fabri E et al (2015) Cell wall traits as potential resources to improve resistance of durum wheat against *Fusarium graminearum*. BMC Plant Biol 15:6. https://doi.org/10.1186/s12870-014-0369-1
Liu S, Anderson JA (2003) Targeted molecular mapping of a major wheat QTL for Fusarium head blight resistance using wheat ESTs and synteny with rice. Genome 46:817–823. https://doi.org/10.1139/g03-066
Liu S, Zhang X, Pumphrey MO et al (2006) Complex microcolinearity among wheat, rice, and barley revealed by fine mapping of the genomic region harboring a major QTL for resistance to Fusarium head blight in wheat. Funct Integr Genomics 6:83–89. https://doi.org/10.1007/s10142-005-0007-y
Liu S, Pumphrey M, Gill B et al (2008) Toward positional cloning of *Fhb1*, a major QTL for Fusarium head blight resistance in wheat. Cereal Res Commun 36:195–201. https://doi.org/10.1556/crc.36.2008.suppl.b.15
Liu Y, Salsman E, Fiedler JD et al (2019) Genetic mapping and prediction analysis of FHB resistance in a Hard Red Spring wheat breeding population. Front Plant Sci 10:1007. https://doi.org/10.3389/fpls.2019.01007
Lu Y-P, Li Z-S, Drozdowicz YM et al (1998) AtMRP2, an Arabidopsis ATP binding cassette transporter able to transport glutathione S-conjugates and chlorophyll catabolites: functional comparisons with AtMRP1. Plant Cell 10:267–282. https://doi.org/10.1105/tpc.10.2.267
Ma HX, Bai GH, Gill BS, Hart LP (2006) Deletion of a chromosome arm altered wheat resistance to Fusarium head blight and deoxynivalenol accumulation in Chinese Spring. Plant Dis 90:1545–1549. https://doi.org/10.1094/PD-90-1545
Ma Z, Xie Q, Li G et al (2020) Germplasms, genetics and genomics for better control of disastrous wheat Fusarium head blight. Theor Appl Genet 133:1541–1568. https://doi.org/10.1007/s00122-019-03525-8
Maccaferri M, Cane' MA, Sanguineti MC et al (2014) A consensus framework map of durum wheat (*Triticum durum* Desf.) suitable for linkage disequilibrium analysis and genome-wide association mapping. BMC Genomics 15:873. https://doi.org/10.1186/1471-2164-15-873
Maccaferri M, Ricci A, Salvi S et al (2015) A high-density, SNP-based consensus map of tetraploid wheat as a bridge to integrate durum and bread wheat genomics and breeding. Plant Biotechnol J 13:648–663. https://doi.org/10.1111/pbi.12288
Maccaferri M, Harris NS, Twardziok SO et al (2019) Durum wheat genome highlights past domestication signatures and future improvement targets. Nat Genet 51:885–895. https://doi.org/10.1038/s41588-019-0381-3
Mackintosh CA, Lewis J, Radmer LE et al (2007) Overexpression of defense response genes in transgenic wheat enhances resistance to Fusarium head blight. Plant Cell Rep 26:479–488. https://doi.org/10.1007/s00299-006-0265-8
Makandar R, Essig JS, Schapaugh MA et al (2006) Genetically engineered resistance to Fusarium head blight in wheat by expression of *Arabidopsis NPR1*. Mol Plant-Microbe Interact 19:123–129. https://doi.org/10.1094/MPMI-19-0123
Mandalà G, Tundo S, Francesconi S et al (2019) Deoxynivalenol detoxification in transgenic wheat confers resistance to Fusarium Head Blight and Crown Rot diseases. Mol Plant-Microbe Interact 32:583–592. https://doi.org/10.1094/MPMI-06-18-0155-R
Mazzucotelli E, Sciara G, Mastrangelo AM et al (2020) The Global Durum wheat Panel (GDP): an international platform to identify and exchange beneficial alleles. Front Plant Sci 11:569905. https://doi.org/10.3389/fpls.2020.569905
McCormick SP, Stanley AM, Stover NA, Alexander NJ (2011) Trichothecenes: from simple to complex mycotoxins. Toxins 3:802–814. https://doi.org/10.3390/toxins3070802
McIntosh RA, Yamazaki Y, Dubcovsky J, et al (2013) Catalogue of gene symbols for wheat. http://www.shigen.nig.ac.jp/wheat/komugi/genes/download.jsp. Accessed on 28 December 2021. Committee for the National BioResource Project (NBRP)/KOMUGI, Japan
McIntosh RA, Dubcovsky J, Rogers WJ, et al (2015) Catalogue of gene symbols for wheat: 2015–2016 supplement. http://www.shigen.nig.ac.jp/wheat/komugi/genes/download.jsp. Accessed on 28 December 2021. Committee for the National BioResource Project (NBRP)/KOMUGI, Japan
McIntosh RA, Dubcovsky J, Rogers WJ, et al (2024) Catalogue of gene symbols for wheat: 2024 edition. https://graingenes.org//GG3/Wheat_Gene_Catalog_Documents. Accessed on 02 September 2025. Committee for the National BioResource Project (NBRP)/KOMUGI, Japan
McKeehen JD, Busch RH, Fulcher RG (1999) Evaluation of wheat (*Triticum aestivum* L.) phenolic acids during grain development and their contribution to Fusarium resistance. J Agric Food Chem 47:1476–1482. https://doi.org/10.1021/jf980896f
McLauchlan WR, Garcia-Conesa MT, Williamson G et al (1999) A novel class of protein from wheat which

inhibits xylanases. Biochem J 338:441–446. https://doi.org/10.1042/bj3380441

McLaughlin JE, Bin-Umer MA, Widiez T et al (2015) A lipid transfer protein increases the glutathione content and enhances Arabidopsis resistance to a Trichothecene mycotoxin. PLoS One 10:e0130204. https://doi.org/10.1371/journal.pone.0130204

McMullen M, Bergstrom G, De Wolf E et al (2012) A unified effort to fight an enemy of wheat and barley: Fusarium Head Blight. Plant Dis 96:1712–1728. https://doi.org/10.1094/PDIS-03-12-0291-FE

Mesterhazy A (1995) Types and components of resistance to Fusarium head blight of wheat. Plant Breed 114:377–386. https://doi.org/10.1111/j.1439-0523.1995.tb00816.x

Meuwissen TH, Hayes BJ, Goddard ME (2001) Prediction of total genetic value using genome-wide dense marker maps. Genetics 157:1819–1829. https://doi.org/10.1093/genetics/157.4.1819

Mirdita V, He S, Zhao Y et al (2015) Potential and limits of whole genome prediction of resistance to Fusarium head blight and *Septoria tritici* blotch in a vast Central European elite winter wheat population. Theor Appl Genet 128(12):2471–2481

Morant M, Bak S, Møller BL, Werck-Reichhart D (2003) Plant cytochromes P450: tools for pharmacology, plant protection and phytoremediation. Curr Opin Biotechnol 14:151–162. https://doi.org/10.1016/s0958-1669(03)00024-7

Moscetti I, Tundo S, Janni M et al (2013) Constitutive expression of the xylanase inhibitor TAXI-III delays Fusarium head blight symptoms in durum wheat transgenic plants. Mol Plant-Microbe Interact 26:1464–1472. https://doi.org/10.1094/MPMI-04-13-0121-R

Nakagawa H, Ohmichi K, Sakamoto S et al (2011) Detection of a new Fusarium masked mycotoxin in wheat grain by high-resolution LC–Orbitrap™ MS. Food Addit Contam Part A Chem Anal Control Expo Risk Assess 28:1447–1456. https://doi.org/10.1080/19440049.2011.597434

Oliver RE, Stack RW, Miller JD, Cai X (2007) Reaction of wild Emmer wheat accessions to Fusarium head blight. Crop Sci 47:893–897. https://doi.org/10.2135/cropsci2006.08.0531

Oliver RE, Cai X, Friesen TL et al (2008) Evaluation of Fusarium head blight resistance in tetraploid wheat (*Triticum turgidum* L.). Crop Sci 48:213–222. https://doi.org/10.2135/cropsci2007.03.0129

Otto CD, Kianian SF, Elias EM et al (2002) Genetic dissection of a major Fusarium head blight QTL in tetraploid wheat. Plant Mol Biol 48:625–632. https://doi.org/10.1023/a:1014821929830

Pasquet J-C, Changenet V, Macadré C et al (2016) A *Brachypodium* UDP-glycosyltransferase confers root tolerance to Deoxynivalenol and resistance to Fusarium infection. Plant Physiol 172:559–574. https://doi.org/10.1104/pp.16.00371

Paudel B, Zhuang Y, Galla A et al (2020) *WFhb1-1* plays an important role in resistance against Fusarium head blight in wheat. Sci Rep 10:7794. https://doi.org/10.1038/s41598-020-64777-9

Perochon A, Jianguang J, Kahla A et al (2015) *TaFROG* encodes a Pooideae orphan protein that interacts with *SnRK1* and enhances resistance to the Mycotoxigenic fungus Fusarium graminearum. Plant Physiol 169:2895–2906. https://doi.org/10.1104/pp.15.01056

Perochon A, Kahla A, Vranić M et al (2019a) A wheat NAC interacts with an orphan protein and enhances resistance to Fusarium head blight disease. Plant Biotechnol J 17:1892–1904. https://doi.org/10.1111/pbi.13105

Perochon A, Váry Z, Malla KB et al (2019b) The wheat *SnRK1α* family and its contribution to Fusarium toxin tolerance. Plant Sci 288:110217. https://doi.org/10.1016/j.plantsci.2019.110217

Pessina S, Angeli D, Martens S et al (2016a) The knockdown of the expression of *MdMLO19* reduces susceptibility to powdery mildew (*Podosphaera leucotricha*) in apple (*Malus domestica*). Plant Biotechnol J 14:2033–2044. https://doi.org/10.1111/pbi.12562

Pessina S, Lenzi L, Perazzolli M et al (2016b) Knockdown of *MLO* genes reduces susceptibility to powdery mildew in grapevine. Hortic Res 3:16016. https://doi.org/10.1038/hortres.2016.16

Pikaard CS, Mittelsten Scheid O (2014) Epigenetic regulation in plants. Cold Spring Harb Perspect Biol 6:a019315. https://doi.org/10.1101/cshperspect.a019315

Pirseyedi SM, Kumar A, Ghavami F et al (2019) Mapping QTL for Fusarium Head Blight resistance in a Tunisian-derived Durum wheat population. Cereal Res Commun 47:78–87. https://doi.org/10.1556/0806.46.2018.053

Pollet A, Beliën T, Fierens K et al (2009) Fusarium graminearum xylanases show different functional stabilities, substrate specificities and inhibition sensitivities. Enzym Microb Technol 44:189–195. https://doi.org/10.1016/j.enzmictec.2008.12.005

Poppenberger B, Berthiller F, Lucyshyn D et al (2003) Detoxification of the Fusarium mycotoxin deoxynivalenol by a UDP-glucosyltransferase from *Arabidopsis thaliana*. J Biol Chem 278:47905–47914. https://doi.org/10.1074/jbc.M307552200

Pozniak CJ, Clarke JM, Haile JK, Haile TA (2020a) CDC Credence durum wheat. Can J Plant Sci 100:720–724. https://doi.org/10.1139/cjps-2020-0092

Pozniak CJ, Clarke JM, Haile JK, Haile TA (2020b) CDC Defy durum wheat. Can J Plant Sci 100:725–730. https://doi.org/10.1139/cjps-2020-0091

Prat N, Buerstmayr M, Steiner B et al (2014) Current knowledge on resistance to Fusarium head blight in tetraploid wheat. Mol Breed 34:1689–1699. https://doi.org/10.1007/s11032-014-0184-2

Prat N, Guilbert C, Prah U et al (2017) QTL mapping of Fusarium head blight resistance in three related durum wheat populations. Theor Appl Genet 130:13–27. https://doi.org/10.1007/s00122-016-2785-0

Pritsch C, Muehlbauer GJ, Bushnell WR et al (2000) Fungal development and induction of defense response genes during early infection of wheat spikes by *Fusarium graminearum*. Mol Plant-Microbe Interact 13:159–169. https://doi.org/10.1094/MPMI.2000.13.2.159

Qi LL, Pumphrey MO, Friebe B et al (2008) Molecular cytogenetic characterization of alien introgressions with gene *Fhb3* for resistance to Fusarium head blight disease of wheat. Theor Appl Genet 117:1155–1166. https://doi.org/10.1007/s00122-008-0853-9

Raiola A, Lionetti V, Elmaghraby I et al (2011) Pectin methylesterase is induced in Arabidopsis upon infection and is necessary for a successful colonization by necrotrophic pathogens. Mol Plant-Microbe Interact 24:432–440. https://doi.org/10.1094/MPMI-07-10-0157

Ransom J, Elias E, Acevedo M, et al (2015) North Dakota durum wheat variety trial results for 2015 and selection guide. https://www.ag.ndsu.edu/publicationS/crops/north-dakota-durum-wheat-variety-trial-results-for-2015-and-selection-guide. Accessed 28 December 2021

Ransom J, Elias E, Friskop A, et al (2016) North Dakota durum wheat variety trial results for 2016 and selection guide. https://www.ag.ndsu.edu/publications/crops/north-dakota-durum-wheat-variety-trial-results-for-2016-and-selection-guide. Accessed 28 December 2021

Ransom J, Elias E, Friskop A, et al (2017) North Dakota durum wheat variety trial results for 2017 and selection guide. https://www.ag.ndsu.edu/publications/crops/north-dakota-durum-wheat-variety-trial-results-for-2017-and-selection-guide. Accessed 28 December 2021

Ransom J, Elias E, Friskop A, et al (2018) North Dakota durum wheat variety trial results for 2018 and selection guide. https://www.ag.ndsu.edu/publications/crops/north-dakota-durum-vareity-trial-result-for-2018-and-selection-guide/a1067-durum-18.pdf. Accessed December 2021

Ransom J, Elias E, Friskop A, et al (2019) North Dakota durum wheat variety trial results for 2019 and selection guide. https://www.ag.ndsu.edu/publications/crops/north-dakota-durum-variety-trial-results-for-2019-and-selection-guide. Accessed 28 December 2021

Ransom J, Elias E, Friskop A, et al (2020) North Dakota durum wheat variety trial results for 2020 and selection guide. https://www.ag.ndsu.edu/publications/crops/north-dakota-durum-wheat-variety-trial-results-for-2020-and-selection-guide. Accessed 28 December 2021

Rawat N, Pumphrey MO, Liu S et al (2016) Wheat *Fhb1* encodes a chimeric lectin with agglutinin domains and a pore-forming toxin-like domain conferring resistance to Fusarium head blight. Nat Genet 48:1576–1580. https://doi.org/10.1038/ng.3706

Rayapuram C, Jensen MK, Maiser F et al (2012) Regulation of basal resistance by a powdery mildew-induced cysteine-rich receptor-like protein kinase in barley. Mol Plant Pathol 13:135–147. https://doi.org/10.1111/j.1364-3703.2011.00736.x

Rea PA (1999) MRP subfamily ABC transporters from plants and yeast. J Exp Bot 50:895–913

Ruan Y, Comeau A, Langevin F et al (2012) Identification of novel QTL for resistance to Fusarium head blight in a tetraploid wheat population. Genome 55:853–864. https://doi.org/10.1139/gen-2012-0110

Ruan Y, Zhang W, Knox RE et al (2020) Characterization of the genetic architecture for Fusarium head blight resistance in durum wheat: the Complex Association of Resistance, Flowering Time, and Height Genes. Front Plant Sci 11:592064. https://doi.org/10.3389/fpls.2020.592064

Schroeder HW, Christensen JJ et al (1963) Factors affecting resistance of wheat to scab caused by *Gibberella zeae*. Phytopathology 53:831–838

Schuler MA, Werck-Reichhart D (2003) Functional genomics of P450s. Annu Rev Plant Biol 54:629–667. https://doi.org/10.1146/annurev.arplant.54.031902.134840

Schuler MA, Duan H, Bilgin M, Ali S (2006) Arabidopsis cytochrome P450s through the looking glass: a window on plant biochemistry. Phytochem Rev 5:205–237. https://doi.org/10.1007/s11101-006-9035-z

Sella L, Gazzetti K, Castiglioni C et al (2014) Fusarium graminearum possesses virulence factors common to Fusarium Head Blight of wheat and seedling rot of soybean but differing in their impact on disease severity. Phytopathology 104:1201–1207. https://doi.org/10.1094/PHYTO-12-13-0355-R

Shen X, Ohm H (2007) Molecular mapping of *Thinopyrum*-derived Fusarium head blight resistance in common wheat. Mol Breed 20:131–140. https://doi.org/10.1007/s11032-007-9079-9

Shin S, Torres-Acosta JA, Heinen SJ et al (2012) Transgenic *Arabidopsis thaliana* expressing a barley UDP-glucosyltransferase exhibit resistance to the mycotoxin deoxynivalenol. J Exp Bot 63:4731–4740. https://doi.org/10.1093/jxb/ers141

Simmen MW, Leitgeb S, Charlton J et al (1999) Nonmethylated transposable elements and methylated genes in a chordate genome. Science 283:1164–1167. https://doi.org/10.1126/science.283.5405.1164

Singh AK, Knox RE, Clarke FR et al (2008) Fusarium head blight QTL mapping in durum wheat and *Triticum carthlicum* sources of resistance. In: Appels R, Eastwood R, Lagudah E, Langridge P, Mackay M, McIntyre L, Sharp P (eds) The 11th international wheat genetics symposium proceedings. Sydney University Press, Sydney

Singh AK, Clarke JM, Knox RE, DePauw RM, McCaig TN, Fernandez MR, Clarke FR (2012) Transcend durum wheat. Can J Plant Sci 92:809–813. https://doi.org/10.4141/cjps2011-255

Siranidou E, Kang Z, Buchenauer H (2002) Studies on symptom development, phenolic compounds and morphological defence responses in wheat cultivars differing in resistance to Fusarium head blight. J Phytopathol 150:200–208. https://doi.org/10.1046/j.1439-0434.2002.00738.x

Søltoft M, Jørgensen LN, Svensmark B, Fomsgaard IS (2008) Benzoxazinoid concentrations show correlation with Fusarium Head Blight resistance in Danish wheat varieties. Biochem Syst Ecol 36:245–259. https://doi.org/10.1016/j.bse.2007.10.008

Somers DJ, Fedak G, Clarke J, Cao W (2006) Mapping of FHB resistance QTL in tetraploid wheat. Genome 49:1586–1593. https://doi.org/10.1139/g06-127

Song P, Zhang L, Wu L et al (2021) A Ricin B-Like Lectin protein physically interacts with TaPFT and is involved in resistance to Fusarium Head Blight in wheat. Phytopathology 111:2309–2316. https://doi.org/10.1094/PHYTO-11-20-0506-R

Soresi D, Zappacosta D, Garayalde A et al (2017) A valuable QTL for Fusarium head blight resistance from *Triticum turgidum* L. ssp. *dicoccoides* has a stable expression in durum wheat cultivars. Cereal Res Commun 45:234–247. https://doi.org/10.1556/0806.45.2017.007

Stack RW, Elias EM, Fetch JM et al (2002) Fusarium head blight reaction of Langdon durum- *Triticum dicoccoides* chromosome substitution lines. Crop Sci 42:637–642. https://doi.org/10.2135/cropsci2002.6370

Steiner B, Michel S, Maccaferri M et al (2019) Exploring and exploiting the genetic variation of Fusarium head blight resistance for genomic-assisted breeding in the elite durum wheat gene pool. Theor Appl Genet 132(4):969–988. https://doi.org/10.1007/s00122-018-3253-9

Su Z, Bernardo A, Tian B et al (2019) A deletion mutation in *TaHRC* confers *Fhb1* resistance to Fusarium head blight in wheat. Nat Genet 51:1099–1105. https://doi.org/10.1038/s41588-019-0425-8

Sun K, Wolters AMA, Vossen JH et al (2016) Silencing of six susceptibility genes results in potato late blight resistance. Transgenic Res 25:731–742. https://doi.org/10.1007/s11248-016-9964-2

Szabo-Hever A, Zhang Q, Friesen T et al (2018) Genetic diversity and resistance to Fusarium head blight in synthetic hexaploid wheat derived from *Aegilops tauschii* and diverse *Triticum turgidum* subspecies. Front Plant Sci 9:1829. https://doi.org/10.3389/fpls.2018.01829

Takahashi-Ando N, Inaba M, Ohsato S et al (2007) Identification of multiple highly similar XIP-type xylanase inhibitor genes in hexaploid wheat. Biochem Biophys Res Commun 360:880–884. https://doi.org/10.1016/j.bbrc.2007.06.151

Talas F, Longin F, Miedaner T (2011) Sources of resistance to Fusarium head blight within Syrian durum wheat landraces. Plant Breed 130:398–400. https://doi.org/10.1111/j.1439-0523.2011.01867.x

Tatra GS, Miranda J, Chinnappa CC, Reid DM (2000) Effect of light quality and 5-azacytidine on genomic methylation and stem elongation in two ecotypes of *Stellaria longipes*. Physiol Plant 109:313–321. https://doi.org/10.1034/j.1399-3054.2000.100313.x

van Damme M, Zeilmaker T, Elberse J et al (2009) Downy mildew resistance in Arabidopsis by mutation of homoserine kinase. Plant Cell 21:2179–2189. https://doi.org/10.1105/tpc.109.066811

van der Lee T, Zhang H, van Diepeningen A, Waalwijk C (2015) Biogeography of Fusarium graminearum species complex and chemotypes: a review. Food Addit Contam Part A Chem Anal Control Expo Risk Assess 32:453–460. https://doi.org/10.1080/19440049.2014.984244

van Loon LC, Rep M, Pieterse CMJ (2006) Significance of inducible defense-related proteins in infected plants. Annu Rev Phytopathol 44:135–162. https://doi.org/10.1146/annurev.phyto.44.070505.143425

Venske E, dos Santos RS, da Farias DR et al (2019) Meta-analysis of the QTLome of Fusarium head blight resistance in bread wheat: refining the current puzzle. Front Plant Sci 10:727. https://doi.org/10.3389/fpls.2019.00727

Volpi C, Janni M, Lionetti V et al (2011) The ectopic expression of a pectin methyl esterase inhibitor increases pectin methyl esterification and limits fungal diseases in wheat. Mol Plant-Microbe Interact 24:1012–1019. https://doi.org/10.1094/MPMI-01-11-0021

Volpi C, Raiola A, Janni M et al (2013) Claviceps purpurea expressing polygalacturonases escaping PGIP inhibition fully infects PvPGIP2 wheat transgenic plants but its infection is delayed in wheat transgenic plants with increased level of pectin methyl esterification. Plant Physiol Biochem 73:294–301. https://doi.org/10.1016/j.plaphy.2013.10.011

Waldron BL, Moreno-Sevilla B, Anderson JA et al (1999) RFLP mapping of QTL for Fusarium head blight resistance in wheat. Crop Sci 39:805–811. https://doi.org/10.2135/cropsci1999.0011183x003900030032x

Walter S, Doohan F (2011) Transcript profiling of the phytotoxic response of wheat to the Fusarium mycotoxin deoxynivalenol. Mycotox Res 27:221–230. https://doi.org/10.1007/s12550-011-0099-2

Walter S, Brennan JM, Arunachalam C et al (2008) Components of the gene network associated with genotype-dependent response of wheat to the Fusarium mycotoxin deoxynivalenol. Funct Integr Genomics 8:421–427. https://doi.org/10.1007/s10142-008-0089-4

Walter S, Nicholson P, Doohan FM (2010) Action and reaction of host and pathogen during Fusarium head blight disease. New Phytol 185:54–66. https://doi.org/10.1111/j.1469-8137.2009.03041.x

Walter S, Kahla A, Arunachalam C et al (2015) A wheat ABC transporter contributes to both grain formation and mycotoxin tolerance. J Exp Bot 66:2583–2593. https://doi.org/10.1093/jxb/erv048

Wang M, Qin L, Xie C et al (2014) Induced and constitutive DNA methylation in a salinity-tolerant wheat introgression line. Plant Cell Physiol 55:1354–1365. https://doi.org/10.1093/pcp/pcu059

Wang G-P, Hou W-Q, Zhang L et al (2016) Functional analysis of a wheat pleiotropic drug resistance gene involved in Fusarium head blight resistance. J Integr Agric 15:2215–2227. https://doi.org/10.1016/S2095-3119(16)61362-X

Wang H, Sun S, Ge W et al (2020) Horizontal gene transfer of *Fhb7* from fungus underlies Fusarium head

blight resistance in wheat. Science 368. https://doi.org/10.1126/science.aba5435
Wang R, Axtman J, Salsman E et al (2021) Recurrent selection to develop Fusarium head blight resistance germplasm for durum wheat. In: Canty S, Hoffstetter A, Dill-Macky R (eds) Proceedings of the National Fusarium Head Blight Forum, Virtual meeting, 6–7 December, 2021
Wang X, Li G, Jia H, Cheng R et al (2024) Breeding evaluation and precise mapping of *Fhb8* for Fusarium head blight resistance in wheat (*Triticum aestivum*). Plant Breed 143:26–33. https://doi.org/10.1111/pbr.13113
Wanke D, Kolukisaoglu HU (2010) An update on the ABCC transporter family in plants: many genes, many proteins, but how many functions? Plant Biol 12(Suppl 1):15–25. https://doi.org/10.1111/j.1438-8677.2010.00380.x
Wanyoike MW, Kang Z, Buchenauer H (2002) Importance of cell wall degrading enzymes produced by *Fusarium graminearum* during infection of wheat heads. Eur J Plant Pathol; Dordrecht 108:803–810. https://doi.org/10.1023/A:1020847216155
Whetten R, Sederoff R (1995) Lignin biosynthesis. Plant Cell 7:1001–1013. https://doi.org/10.1105/tpc.7.7.1001
Wright PE, Dyson HJ (2015) Intrinsically disordered proteins in cellular signalling and regulation. Nat Rev Mol Cell Biol 16:18–29. https://doi.org/10.1038/nrm3920
Xiao J, Jin X, Jia X et al (2013) Transcriptome-based discovery of pathways and genes related to resistance against Fusarium head blight in wheat landrace Wangshuibai. BMC Genomics 14:197. https://doi.org/10.1186/1471-2164-14-197
Xing L, Gao L, Chen Q et al (2018) Over-expressing a UDP-glucosyltransferase gene (Ta-UGT 3) enhances Fusarium Head Blight resistance of wheat. Plant Growth Regul 84:561–571. https://doi.org/10.1007/s10725-017-0361-5
Xu SS, Khan K, Klindworth DL et al (2004) Chromosomal location of genes for novel glutenin subunits and gliadins in wild emmer wheat (*Triticum turgidum* L. var. *dicoccoides*). Theor Appl Genet 108:1221–1228. https://doi.org/10.1007/s00122-003-1555-y
Xu SS, Chu CG, Friesen TL et al (2010) Introgression of two major FHB-resistance QTL into durum and hard red spring wheat. In: Canty S, Clark A, Anderson-Scully A, Ellis D, Van Sanford D (eds) Proceedings of the National Fusarium Head Blight Forum, Milwaukee, WI, 7–9 December, 2010. University of Kentucky, Lexington
Yang J, Liang K, Ke H, Zhang Y et al (2024) Enzymatic degradation of deoxynivalenol with the engineered detoxification enzyme *Fhb7*. JACS Au 4:619–634. https://doi.org/10.1021/jacsau.3c00696
Yoshinari T, Sakuda S, Furihata K et al (2014) Structural determination of a nivalenol glucoside and development of an analytical method for the simultaneous determination of nivalenol and deoxynivalenol, and their glucosides, in wheat. J Agric Food Chem 62:1174–1180. https://doi.org/10.1021/jf4048644
Yu A, Lepère G, Jay F et al (2013) Dynamics and biological relevance of DNA demethylation in *Arabidopsis* antibacterial defense. Proc Natl Acad Sci USA 110:2389–2394. https://doi.org/10.1073/pnas.1211757110
Zhang X, Yazaki J, Sundaresan A et al (2006) Genome-wide high-resolution mapping and functional analysis of DNA methylation in *Arabidopsis*. Cell 126:1189–1201. https://doi.org/10.1016/j.cell.2006.08.003
Zhang JX, Jin Y, Rudd JC, Bockelman HE (2008) New Fusarium head blight resistant spring wheat germplasm identified in the USDA national small grains collection. Crop Sci 48:223–235. https://doi.org/10.2135/cropsci2007.02.0116
Zhang X, Shen X, Hao Y et al (2011) A genetic map of *Lophopyrum ponticum* chromosome 7E, harboring resistance genes to Fusarium head blight and leaf rust. Theor Appl Genet 122:263–270. https://doi.org/10.1007/s00122-010-1441-3
Zhang Q, Axtman JE, Faris JD et al (2014) Identification and molecular mapping of quantitative trait loci for Fusarium head blight resistance in emmer and durum wheat using a single nucleotide polymorphism-based linkage map. Mol Breed 34:1677–1687. https://doi.org/10.1007/s11032-014-0180-6
Zhang Q, Faris JD, Chao S et al (2017) Identification and molecular mapping of quantitative trait loci for resistance to Fusarium head blight in cultivated emmer PI 272527 [abstract]. In: Proceedings of the 2017 National Fusarium Head Blight Forum. December 3–5, 2017, Milwaukee, Wisconsin, vol 97
Zhang X, Lai T, Zhang P et al (2019) Mini review: revisiting mobile RNA silencing in plants. Plant Sci 278:113–117. https://doi.org/10.1016/j.plantsci.2018.10.025
Zhang W, Danilova T, Zhang M et al (2022) Cytogenetic and genomic characterization of a novel tall wheatgrass-derived *Fhb7* allele integrated into wheat B genome. Theor Appl Genet 135:4409–4419. https://doi.org/10.1007/s00122-022-04228-3
Zhang F, Zhang H, Liu J et al (2024) *Fhb9*, a major QTL for Fusarium head blight resistance improvement in wheat. J Integ Agric J Pre-proof. https://doi.org/10.1016/j.jia.2024.03.045
Zhao M, Leng Y, Chao S et al (2018a) Molecular mapping of QTL for Fusarium head blight resistance introgressed into durum wheat. Theor Appl Genet 131:1939–1951. https://doi.org/10.1007/s00122-018-3124-4
Zhao L, Ma X, Su P et al (2018b) Cloning and characterization of a specific UDP-glycosyltransferase gene induced by DON and Fusarium graminearum. Plant Cell Rep 37:641–652. https://doi.org/10.1007/s00299-018-2257-x
Zhao L, Bernardo A, Kong F et al (2024) A glutathione S-transferase from *Thinopyrum ponticum* con-

fers *Fhb7* resistance to Fusarium head blight in wheat. Phytopathology 114. https://doi.org/10.1094/PHYTO-03-24-0106-SC

Zhu ZW, Xu DA, Cheng SH et al (2018) Characterization of Fusarium head blight resistance gene *Fhb1* and its putative ancestor in Chinese wheat germplasm. Acta Agron Sin 44(04):473–482. https://doi.org/10.3724/sp.j.1006.2018.00473

Zilberman D, Gehring M, Tran RK et al (2007) Genome-wide analysis of Arabidopsis thaliana DNA methylation uncovers an interdependence between methylation and transcription. Nat Genet 39:61–69. https://doi.org/10.1038/ng1929

The Genomic Landscape of Wheat Prolamin Genes

Claudia Maria Liberatore, Aldo Ceriotti, and Massimiliano Lauria

Abstract

Wheat is one of the world's major crops and is at the basis of human nutrition in several countries. The main protein components of the wheat seed are gliadins and glutenins, which belong to the prolamin protein family and accumulate in endosperm cells during seed development. Although these proteins have been widely studied for decades, some aspects of their genomic organisation have been revealed only recently by genomic studies that used both classical and next-generation experimental approaches. Since prolamins influence technological characteristics of wheat flours which are important for the production of a wide range of foods, this information is functional to the development of breeding strategies focused on grain quality. The aim of this chapter is to discuss studies on the structural organization of wheat prolamin loci, on the transcriptional differences between different family members and on their evolution. Available information relating to the durum wheat genome is discussed in the general framework provided by studies on diploid, tetraploid and hexaploid wheats and on donor species.

C. M. Liberatore · A. Ceriotti · M. Lauria (✉)
Istituto di Biologia e Biotecnologia Agraria, CNR, Via Corti 12, Milan, Italy
e-mail: claudia.liberatore@ibba.cnr.it; ceriotti@ibba.cnr.it; lauria@ibba.cnr.it

1 Introduction

Gluten is defined as the rubbery mass that remains after washing the dough to remove starch and water-soluble components. The composition of the remaining dry part depends on wash intensity but is mainly constituted by proteins (75–85%) and lipids (5–10%; Wieser 2007). The main components of protein fraction are two classes of seed storage proteins called gliadins and glutenins, which belong to the prolamin protein family. Based on their electrophoretic mobility in sodium dodecyl sulphate gel electrophoresis, glutenins are classified as high and low molecular weight glutenin subunits (HMW-GS and LMW-GS), whereas gliadins are classified as α-, γ-, ω- and δ-types (Gianibelli et al. 2001; Anderson et al. 2012). At the same time, a classification that takes into account their structural and evolutionary relationships divides them into sulphur-rich (S-rich), sulphur-poor (S-poor) and HMW prolamins (Shewry and Tathamt 1990). An additional class of wheat seed proteins that has an impact on dough quality is represented by atypical gluten proteins defined as avenin-like proteins (ALPs). These proteins are characterized by a high number of cysteine residues, but unlike glutenins and gliadins they lack repetitive domains and possess

R. Tuberosa et al. (eds.), *The Durum Wheat Genome*, Compendium of Plant Genomes,
https://doi.org/10.1007/978-3-031-84689-2_14

short proline and glutamine sequences (Chen et al. 2016; Clarke et al. 2003; Ma et al. 2013; Zhang et al. 2018).

Wheat prolamins have been intensively studied not only because they have a fundamental biological role in supporting germination and seedling development but also because of their importance for human nutrition. Crucially, they determine the biochemical properties of dough and, consequently, the processing capacity of wheat flour for the production of food products such as bread and pasta. In this respect, gliadins are responsible for dough viscosity and extensibility, whereas glutenins are responsible for dough elasticity (Shewry and Halford 2002). Glutenins are polymeric proteins that form intermolecular disulphide bonds, whereas gliadins mainly exist as monomeric proteins that form intramolecular disulphide bonds, although examples of gliadins that act as glutenins have been reported (Masci et al. 2002; D'Ovidio and Masci 2004). Interaction between polymeric and monomeric proteins occurs by noncovalent hydrogen bonds. The glutenin/gliadin ratio and the type of glutenin and gliadin polypeptides have a profound effect on the main characteristic of gluten: that is, strength and extensibility, and therefore on the final end use of dough (Rasheed et al. 2014; Veraverbeke and Delcour 2002; Wieser 2007; Wieser et al. 2006). The biochemical properties of prolamins and their role in wheat dough characteristics are out the scope of this chapter, and readers can refer to several reviews for a more detailed description of the topic (Shewry and Tathamt 1990; D'Ovidio and Masci 2004; Wieser 2007; Li et al. 2020; Wang et al. 2020; Pourmohammadi et al. 2023).

Although important for human nutrition, gluten proteins are also known to induce a T cell-mediated autoimmune disease (Sapone et al. 2012), that is, the celiac disease (CD) which can affect about 1% of the human population, with prevalence depending also on the geographical area. The prevalence of CD is continually rising possibly due to the increased consumption of wheat-based foods (Lionetti et al. 2015). The T-cell epitopes that trigger CD are localized in the repetitive domains of both glutenins and gliadins, in which the high proline levels confer resistance to proteolytic degradation in the gastrointestinal lumen. Among CD epitopes, those with high immune response are located in α- and ω-gliadins (Scherf et al. 2016; Juhász et al. 2018). However, not all CD epitopes show the same immunogenicity and not all gliadins contain the same number of epitopes (Sollid et al. 2012; Juhász et al. 2018). For example, the highly immunogenic 33-mer peptide, which arise from the overlapping of six different epitopes (DQ2.5-glia-a1a, DQ2.5-glia-a2 and DQ2.5-glia-a1b), was found in only two of the α-gliadin sequences analysed in the Chinese Spring reference genome (Juhász et al. 2018). In general, gliadin sequences containing peptides with strong immunoreactivity were found to be abundant in the D genome and less frequent in B genome (Wang et al. 2017; Juhász et al. 2018). However, copy number variation within different wheat varieties has been observed for the highly immunogenic α-gliadin genes on chromosome 6D (White et al. 2024). Interestingly, the 33-mer peptide, which was thought to have originated during hexaploid wheat formation, was recently identified in 13 accessions of *Aegilops tauschii (Ae. tauschii)* following amplicon sequencing screening of 51 different accessions, dating its origin before wheat allopolyploidization (Schaart et al. 2021).

Because both gliadins and glutenins, with the exception of the HMW-GS, are encoded by families of tightly linked genes, the selection of favourable alleles, in term of absence of coeliac-active T-cell epitopes/low immunoreactivity, can be complicated as recombination events are rare. Therefore, breeding efforts in this direction have been focused to produce wheat lines that completely lack specific prolamin members, for example, wheat mutant lines that lack a complete α-gliadin set encoded by one *Gli-1* locus (Shewry and Tatham 2016). However, lines with large chromosome deletions result in the loss of many genes, which make them not always suitable for commercial breeding programmes (Smulders et al. 2015). An alternative approach is based on the genetic characterization of the existing wheat germplasm to identify varieties characterized by the presence of low-immunogenic CD epitopes

(Smulders et al. 2015; Camerlengo et al. 2017). Moreover, given that the D genome of bread wheat possesses the highest CD load, new varieties can be obtained following hybridization between durum wheat (*Triticum turgidum* spp. *durum*) and *Ae. tauschii* accessions characterized by low immunogenic CD epitopes (Smulders et al. 2015; Schaart et al. 2021). Apart gene silencing mediated by RNAi, which has been successfully used to down-regulate prolamin gene expression, and consequently the immunoreactivity of gliadins and glutenins (Gil-Humanes et al. 2010; Guzmán-López et al. 2024), a more recent and effective approach to achieve a similar result is based on CRISPR/Cas9 gene editing system, which can precisely modify toxic epitopes (Altenbach et al. 2020; Jouanin et al. 2019a, b; Sánchez-León et al. 2018) Clearly, all these approaches can benefit from a complete knowledge of the genomic organization of prolamin genes, information about their expression patterns, and the correct match between genes, transcripts and proteins (Altenbach et al. 2020). This latter aspect has been quite challenging due to high level of sequence identity that characterizes these genes. In this chapter, data arising from studies focused on the genomic characterization of these loci, including aspect related to their expression and evolution will be discussed.

2 Genomic Organization of Prolamin Genes

Genes encoding glutenins and gliadins map on different chromosomal locations known as *Glu-1*, *Glu- 3*, *Gli-1/3* and *Gli-2* loci (Fig. 1). The *Glu-1* loci are located on the long arm of the group 1 homoeologous chromosomes (A, B and D in the hexaploid wheat) and encode the HMW-GS (Lawrence and Shepherd 1981; Payne and Lawrence 1983). The short arm of the group 1 homoeologous chromosomes harbours the *Glu-3* and *Gli-1/3* loci, which are tightly linked and encode the LMW-GS and the γ-, ω- and δ-gliadins respectively (Dong et al. 2016). Remarkably, a recent manual annotation of γ-gliadins in the genome of durum wheat Svevo revealed the presence of avenin-like proteins (ALPs) in this region (Paris et al. 2021), which correlate with those identified on chromosome arms 4AL, 7AS and 7DS (Chen et al. 2016; Zhang et al. 2018; Fig. 1). Finally, the *Gli-2* loci are located on the group 6 chromosomes and encode α-gliadins (Huo et al. 2018a). Given the importance of prolamins for wheat flour characteristics, several efforts in the past years have been made to elucidate the genomic organization of these loci in order to facilitate the improvement in grain quality (Blechl and Anderson 1996; Rooke et al. 1999).

2.1 The *Glu-1* Locus

Among all prolamin loci, the HMW-GS are the best studied in term of the effect on bread-making qualities (Wrigley 1996). The genomic organization of the *Glu-1* locus has been extensively studied through BAC clone sequencing of orthologous HMW-GS regions from the *Ae. tauschii* diploid D genome (Anderson et al. 2003), the *T. turgidum* tetraploid A and B genomes (Kong et al. 2004; Gu et al. 2004a) and the *T. aestivum* hexaploid A, B and D genomes (Gu et al. 2006). At genomic level, the *Glu-1* locus is relatively simple, when compared to other prolamin loci, since HMW-GS genes are present in a low copy number, generally two copies per subgenome, with some exceptions related to a duplication event involving the *Bx7* allele that has been identified in several wheat varieties (D'Ovidio et al. 1997; Cloutier et al. 2005; Ragupathy et al. 2008).

Sequence analysis of a BAC clone covering 102 kb of the *Glu-D1* locus (*Ae. tauschii*) confirmed that the two paralogues HMW-GS sequences, x-type and y-type subunits, are linked together and in the same orientation (Anderson et al. 2003). This probably explains the difficulty of retrieving recombination events at this locus (Shewry et al. 2003). The analysis also revealed the presence of addition genes encoding two globulin storage proteins, a Leu-rich repeat receptor kinase and a Ser/Thr protein kinase (Fig. 2). Comparison of this region with orthologous regions of the tetraploid and hexaploid wheat subgenomes revealed a general conserva-

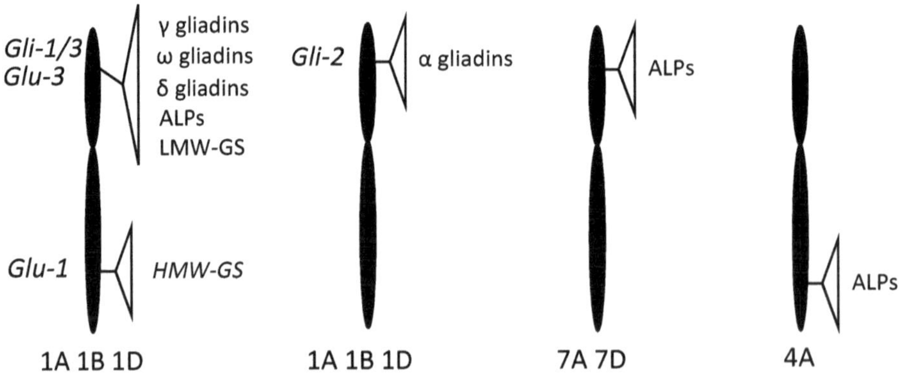

Fig. 1 Chromosomal localization of prolamin loci described in this chapter

tion of gene content and order (Kong et al. 2004; Gu et al. 2004a, 2006): that is, a Leu-rich receptor kinase precedes a globulin and a y-type HMW-GS, while a duplicated globulin precedes an x-type HMW-GS and a Ser/Thr protein kinase. Differences were, however, observed at intragenic level. For instance, while the x-type and y-type subunits are about 51 kb apart on the D genome (*Ae. tauschii*), this distance is expanded to 168 kb and 200 kb on the B and A genome respectively of the tetraploid wheat (*T. turgidum*) mainly due to insertion of clusters of nested retrotransposons within the intergenic regions (Kong et al. 2004; Gu et al. 2004a). Interestingly, when the *Glu-1* region between the B and D genomes was compared, the type of retroelements identified in these regions was quite different. The WIS long-terminal-repeat (LTR) retrotransposons were found to be the most abundant in the B genome, whereas the ANGELA LTR type was the most representative in the D genome. Despite these differences, a comparison of each pair of two homologous genomes, for example, A, B and D genome from hexaploid, tetraploid and diploid wheat, evidenced that they share a considerable portion of conserved sequences. The possible link between homologous genomes is further evidenced by the analysis of gene inactivation that occurred at the *Glu-1* loci. Gu et al. (2006) found that inactivation of one of the two globulin genes in both the hexaploid and tetraploid B genomes was due to the same type of retrotransposon insertion event. In the diploid D and tetraploid A genome, the same globulin gene was inactivated by an identical deletion event (Anderson et al. 2003; Gu et al. 2004a), which was conserved in both the A and D genomes of the hexaploid wheat (Gu et al. 2006). Likewise, the receptor kinase gene in the hexaploid A genome contains the same retroelements and miniature inverted repeat sequences as those present in the tetraploid A genome (Gu et al. 2004a, 2006). This suggests that in the hexaploid wheat genome the majority of retroelements and mutational events were likely inherited from their diploid and tetraploid ancestral genomes (Gu et al. 2006). The above studies also depict some interesting features related to the evolution of *Glu-1* locus. The two globulin sequences present in the *Glu-1* region are adjacent to the x- and y-type HMW-GS genes. While the globulin linked to y-type HMW-GS is active in all subgenomes investigated, the one linked to the x-type is interrupted by several transposition/mutational events that are particularly pronounced in the A and D genome (Anderson et al. 2003; Gu et al. 2004a; Kong et al. 2004). Based on sequence similarity of the two globulin sequences in both the promoter and part of the coding regions, and considering their position in respect to the two HMW-GS genes, it was hypothesized that the globulin and HMW-GS genes duplicated giving origin to the paralogous x- and y-type HMW-GS genes and to paralogous globulin genes (Anderson et al. 2003). This duplication is not present in the orthologous region of the barley genome, but apparently present in rye, which possesses two copies of the HMW-GS genes. This suggested

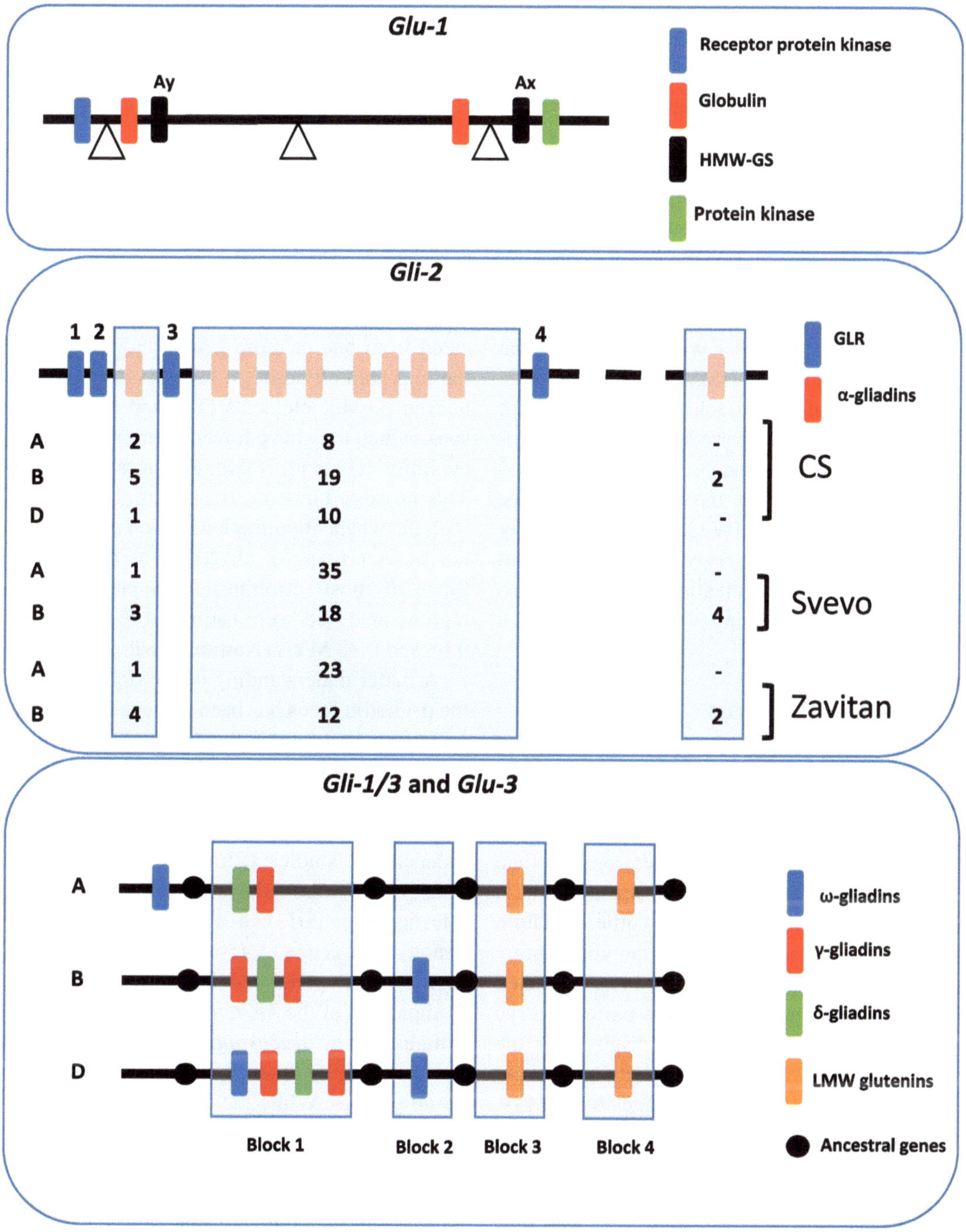

Fig. 2 Graphical representation of the *Glu-1*, *Gli-2*, *Gli-1/3* and *Glu-3* prolamin loci. Empty triangles indicate that these intergenic regions are variable in size among different subgenomes. Numbers below the *Gli-2* graph indicate the copy number of α-gliadins observed in different wheat genomes (Huo et al. 2018a; Huo et al. 2019; Lauria and Ceriotti unpublished data). All graphs are representative of the prolamins organization and do not reflect the real nucleotide distance among different gene types. The *Gli-1/3* and *Glu-3* graph reflects only the genomic organization of different prolamin families in the three subgenomes of Chinese Spring (CS) as deduced from Huo et al. 2018b

that such duplication occurred after the divergence of wheat and barley, but before the separation of wheat and rye from their common ancestor, which is estimated have occurred about 7 million years (De Bustos et al. 2001; Huang et al. 2002; Kong et al. 2004). Further comparison of different grass genomes of the *Poaceae* family evidenced that the globulin gene is highly conserved, whereas the HMW-GS genes are only present in the *Triticeae* and *Brachypodium* genomes (Xu and Messing 2009;Gu et al. 2010). This observation, together with the fact that the N-terminal regions of HMW-GS and globulin share substantial amino acid identity, that both genes encode for endosperm specific storage proteins, and the presence of conserved cis-regulatory elements in the promoter regions, suggested that the HMW-GS gene was likely duplicated from the conserved globulin gene, and underwent neofunctionalization, thus becoming a new gene (Xu and Messing 2009; Gu et al. 2010).

2.2 The *Gli-2* Locus

The loci encoding the α-gliadins are located on the short arm of homoeologous group 6 chromosomes (Payne 1987; Gu et al. 2004b), and they can account for up to 30% of total seed storage protein in the wheat seed (Altenbach et al. 2011; Shewry et al. 1992). Compared to other prolamin genes, α-gliadins are considered the youngest by evolution, as they are absent in closely related species in *Triticeae* tribe, such as barley and rye (Huo et al. 2017) dating their origin after the divergence of wheat and rye and before the separation of the wheat genome lineages 6.5-7 MYA (Levy and Feldman 2022). It has been proposed that α gliadins are derived from a duplication event of a γ-gliadin gene in the *Gli- 1/3* region of chromosome 1. The duplicated copy was then translocated to chromosome 6, where it duplicated and gave rise to the α-gliadins family (Huo et al. 2017). The number of α-gliadin genes varies in the different wheat accessions so far analysed, ranging from 25 to 150, likely as the result of multiple rounds of tandem gene duplications or deletion events, which may also involve an entire cluster of α-gliadin genes (Lafiandra et al. 1984; Harberd et al. 1985; Reeves and Okita 1987; D'Ovidio et al. 1991; Anderson et al. 1997; Huo et al. 2017, 2018a, 2019; Halstead-Nussloch et al. 2021).

Given the complexity of the *Gli-2* locus, BAC sequencing approach has only partially depicted its full organization (Gu et al. 2004b; Kawaura et al. 2012). Studies have shown that within the *Gli-2* locus, α-gliadin genes are clustered or present as singletons, that multiple genes were duplicated from one or more genes, and that most of the loci are composed of repetitive regions, that is, transposable elements (TE) and retrotransposons, which may have forced their expansion and evolution (Gu et al. 2004b; Kawaura et al. 2012). This process, however, is not closely associated with the wheat allotetraploidization that occurred 0.5 MYA (Huang et al. 2002), as the insertion times of most retrotransposons present in the regions analysed were estimated to be between 9.65 and 0.30 MYA (Kawaura et al. 2012).

A better understanding of the organization of the α-gliadin locus has been achieved for the diploid grass *Ae. tauschii*, the donor of the D subgenome during the allopolyploidization event that produced the hexaploid wheat (Feldman 2001). By combing genome sequencing data of two independent studies (Zimin et al. 2017b; Luo et al. 2017) and BioNano genome map data (Hastie et al. 2013), a high-quality contiguous genomic sequence (1.75-Mb) of the *Gli-2* locus region was generated (Huo et al. 2017). Comparison of the *Ae. tauschii Gli-2* region with orthologous of *Brachypodium*, rice and sorghum species, identified seven ancestral genes shared by all grasses. While the orthologous regions of the three grasses show similar genome size, in *Ae. tauschii* the *Gli-2* region has greatly expanded and a total of 39 genes were annotated, most of which were non-syntenic genes and only present in *Ae. tauschii*. Interestingly, one of the ancient conserved genes, a glutamate-like receptor (GLR), has duplicated to have four copies in *Ae. tauschii* and define the boundaries of the *Gli-2* locus within a region of ~550 kb. Twelve α-gliadins between the *GLR2* and *GLR4* genes were identified, five of which were pseudogenes due to the presence of deletions and premature stop codons. In agreement with previous studies, the α-gliadins were either clustered or present as singletons: the distances between two adjacent genes were quite variable and ranged from 6 to 163 kb (Huo et al. 2017). An *Hawi* LTR element was found in the majority of the intergenic regions that separate two adjacent α- gliadins. On the basis of the integrity of the *Hawi* element, and the types and conservation of additional repeat insertions that occurred within the intergenic regions, the authors were able to trace the original sequence that gave rise to the duplication events and their subsequent evolution on the D genome of *Ae. tauschii.* (Huo et al. 2017). A high-quality sequence and a detailed characterization of the homologous *Gli-2* loci regions have been also reported for the hexaploid Chinese Spring (CS) and the tetraploid wild emmer wheat (WEW) Zavitan (Huo et al. 2018a, 2019). Moreover, an annotation draft of *Gli-2* loci has also been reported for the tetraploid durum wheat Svevo (Maccaferri et al. 2019), and for 11 accessions, 10 bread wheat and one spelt, released by the '10+ Wheat Genomes Project' (Walkowiak et al. 2020; Halstead-Nussloch et al. 2021).

A first emerging aspect from the studies of Huo et al. (2018a, 2019) is that all subgenomes in CS and Zavitan contain the four duplicated *GLR* genes, and that the majority of α-gliadin genes lie between the *GRL2* and *GRL4* genes (Fig. 2). This conservation pattern suggests that duplication of α- gliadins occurred after that of the *GRLs* and before the separation of the three subgenomes. However, an exception was observed for the B chromosome of CS where two copies of α-gliadins map outside the GRL region (Huo et al. 2018a). Halstead-Nussloch et al. (2021) recently identified a clear bipartite structure of the *Gli-2B locus* in all hexaploid wheat varieties analysed, in which two α-gliadin genes were separated 12–21 Mb from the GRL region (Fig. 2). The presence of the *Gli-2B* bipartite structure has been also observed for the tetraploid Svevo, and it is further confirmed by a recent reannotation of α-gliadins on an improved genome version (Lauria and Ceriotti unpublished data). Moreover, when a BLAST analysis was performed against the Zavitan genome with representative α-gliadin sequences, we observed the presence of α-gliadins outside the GRL region of the B genome. The lack of this bipartite structure in the A and D genomes, if confirmed by further studies, would date its origin, following a transposition event, after the separation of the three subgenomes and before the allopolyploidization events that generated tetraploid and hexaploid wheat varieties.

The above studies revealed additional aspects suggesting that α-gliadins have been subjected to independent ways of evolution on homoeologous chromosomes, as evidenced by differences in copy number, level of pseudogenization, and characteristics of the intergenic regions. In WEW genome (variety Zavitan), the progenitor of both the cultivated tetraploid and hexaploid wheat (Levy and Feldman 2022), 24 and 16 α-gliadins were identified in the region defined by *GLR* genes in the A and B genomes respectively (Huo et al. 2019, Fig. 2). In the modern durum wheat Svevo, the α-gliadins copy number observed was still higher in the A genome (36 vs. 21; Maccaferri et al. 2019; Lauria and Ceriotti unpublished data; Fig. 2). However, marked differences in the distribution of α-gliadins genes were observed in 11 bread wheat genomes (Halstead-Nussloch et al. 2021). Although on average α-gliadins were more abundant on the A genome (17.6 vs. 9.7), with at least two genotypes harbouring 5 times the number of α-gliadins present on the B genome, examples of genotypes harbouring slightly more copies on the B genome were also identified. In this regard, the best evidence comes from CS where the copy number on the B genome is more than twice those on the A genome, that is, 24 versus 10 (Huo et al. 2018a, Fig. 2). Less differences in copy number were observed for the α- gliadins present on the D genome of different accessions, which also showed a relatively small genetic distances when a phylogenetic analysis was performed (Halstead-Nussloch et al. 2021). In line with this observation, a comparison between CS and *Ae. tauschii* D genomes revealed that the α-gliadins copy number differed by only one sequence. Indeed, comparison of the α-gliadin regions of *Ae. tauschii* and CS showed a level of sequence identity greater than 98%, the main differences being due to differential TE insertions (Huo et al. 2018a).

Despite the high copy number, only a portion of α-gliadin genes contains a full-length (ORF) Open Reading Frame. Analysis of 230 unique α-gliadin sequences identified in different diploid wheat accessions evidenced a high level (87%) of pseudogenization (van Herpen et al. 2006). The majority of pseudogenes identified (77%) were the results of in frame stop codon generated by C-T transition of the glutamine codons (CAG or CAA) into stop codons (TAG or TAA). Because glutamine codons are abundant in these genes, the chance that single point mutations can generate in frame stop codons is therefore relatively high. Ozuna et al. (2015) analysed the N-terminal repetitive region of α -gliadins in 96 accessions of *Triticum* and *Aegilops* species, including hexaploid, tetraploid and diploid accessions, and observed that the level of pseudogenes tends to increase with the level of polyploidy. However, considering that tetraploid lines can possess more pseudogenes than hexaploid ones (Ozuna et al. 2015; Huo et al. 2019), and that the level of pseudogenization can be quite high also in diploid accessions (van Herpen et al. 2006), including members of the *Triticeae* family such as *Thinopyrum elongatum* (84.2%; Ge et al. 2021), the polyploidy level alone cannot fully explain the percentage of pseudogenes observed within different wheat genomes. Huo et al. (2019) pointed out that a lower level of pseudogenes presents in the hexaploid CS genome compared to that of tetraploid Zavitan may be explained by human selection of favourable traits related to end-use qualities of gluten.

Different level of pseduogenization has also been observed among homoeologous genomes. In CS, 45% of all α gliadin genes identified were classified as pseudogenes, and the level of pseudogenization was particularly pronounced in the B genome: that is, 20% from the A genome, 57% from the B genome, ad 36% from the D genome (Huo et al. 2018b). However, although this trend is confirmed by the study of Halstead-Nussloch et al. (2021), with 34, 51 and 44% of pseudogenes found on average in the A, B and D genomes respectively, examples of high levels of pseduogenization were also observed for the A genome in some of the varieties investigated. Indeed, in the WEW Zavitan, the level of pseudogenes reached 83 and 69% in the A and B genomes respectively. Therefore, as with the copy number, it seems difficult to identify a specific evolutionary path of pseduogenization during the evolution of wheat genomes.

Repeat sequences such as TE and LTR are widespread in cereal genomes, and their genomic signature has helped to depict some evolutionary aspect of wheat prolamin loci. In the CS A genome, the intergenic region that separates two adjacent α-gliadins, in several cases (7/9), forms a base structure composed by three conserved repetitive elements: a DNA transposable element (Manor) and two nested LTRs (Ada and Sabrina; Huo et al. 2018a). This base structure likely originated before α- gliadins expansion, and successively became the target of differential insertion of additional repetitive elements that have expanded the region to a final size of e 696 kb, with a distance between adjacent α-gliadins ranging from 37 to 147 kb. The *Gli-2B* locus spans a region of 765 kb and contains, as mentioned above, the highest number of α-gliadins in CS. However, differently from the A genome, most of the genes are close together, and are characterized by a base structure of around 10 kb that is devoid of TE insertions with a few exceptions: seven out of 23 intergenic regions contain repetitive elements that in only two cases were shared, supporting the idea that majority of repeat insertions occurred after α-gliadins duplication in this region. In one case, the region between two α-gliadins expanded to 245 kb due to the insertion of multiple TEs, and three α-gliadins showed an interrupted coding sequence as result of TEs insertion (Huo et al. 2018a). The *Gli-2D* region is 387 kb in size and contains a conserved base structure characterized by the presence of a *Hawi* LTR element that was expanded by an insertion of a 27 kb sequence, and is shared with the one observed for the *Ae. tauschii* D genome (see above). Thus, as for the A genome, the bulk of α-gliadins duplication in the D genome occurred after the original insertion of specific repeat elements. Collectively these observations reveal that the genomic organizations of α-gliadin loci, and in particular their intergenic regions, are quite different in size and TE insertions in the A, B and D genomes, and that the duplication and expansion of α-gliadin genes in the three subgenomes occurred independently in the last ~2.4 million years. Indeed, a phylogenetic analysis of CS α-gliadins evidenced that the majority of paralogous genes from the same genome are more closely related than orthologues from different genomes, and that only a clade contained genes from all three homeologous genomes, which could represent ancestral α-gliadin genes (Huo et al. 2018a).

Comparative analysis of the *Gli-2A* and *-2B* loci in WEW basically confirmed that these regions underwent an independent evolution: as observed for the hexaploid CS, the intergenic regions between adjacent α-gliadins in the two genomes were structurally different as was their gene content (Huo et al. 2019).

Interestingly, consistent sequence differences were also observed among homologous regions of the A genome. When compared to CS, the *Gli-2A* locus in Zavitan has greatly expanded to a final size of 2.5 Mb, as consequence of both α-gliadin genes duplication and TE insertion. Nine out of 23 intergenic regions share the same three repetitive elements (Ada, Manor and Sabrina), already identified in CS A genome. Although these elements are slightly differently organized in Zavitan, at least two of them (Ada and Manor LTRs) represent repeats originally present in the progenitor of the CS and Zavitan A genome. Therefore, the insertion of additional repeats occurred after the separation of the two A genomes and contributed to the observed sequence diversity. Differently, comparison of the homologous *Gli-2B* regions evidenced less differences, likely the results of a lower repeat content (Huo et al. 2019).

2.3 The *Glu-3* and *Gli-1/3* Loci

The *Glu-3* and *Gli-1/3* loci are genetically tightly linked on the short arm of chromosome 1, and harbour different prolamin gene families, including the LMW-GS-glutenins and the γ-, ω- and δ-gliadins (Anderson et al. 2012; Fig. 2). Within these genomic regions, these gene families are present in multiple copies. In hexaploid wheat, Southern blot analysis revealed the existence of up to 40 copies of LMW-GS and up to 39 and 19 copies of γ- and ω-gliadins respectively (Lew et al. 1992; Sabelli and Shewry 1991; Cassidy et al. 1998). However, as already observed for α-gliadins the copy number of different families can vary depending on the genetic background analysed (Ikeda et al. 2006; Huang and Cloutier 2008).

Different attempts have been made to elucidate the genomic organization of these loci through BAC clone sequencing (Wicker et al. 2003; Johal et al. 2004; Gao et al. 2007; Huang and Cloutier 2008; Dong et al. 2010). A first comparative analysis of homeologues regions containing both gliadins and LMW-GS genes was provided for the A and B genomes of tetraploid wheat (Gao et al. 2007). A first emerging aspect of the study was the different organization of these gene families. Both the γ- and ω- gliadins were present as cluster of multiple copies as consequence of several rounds of segmental duplications which, in the case of ω-gliadins, resulted in eight copies included in a region of ~100 kb (Gao et al. 2007). Differently, the two LMW-GS genes identified in two overlapping BAC clones of the A genome were separated by more than 100 kb. This observation was in line with previous studies conducted in *T. monococcum* and *Ae. tauschii*, the former related to the A genome and the latter the progenitor of the D genomes respectively. In the former study, two LMW-GS genes were separated by more than 150 kb, whereas in the latter all LMW-GS genes identified were found on different BAC clones. (Wicker et al. 2003; Johal et al. 2004). Similarly, analysis of BAC clones of the hexaploid wheat cultivars Glenlea and Xiaoyan 54 showed that most of BAC clones analysed contained a single copy of LMW-GS genes, reinforcing the idea that large intergenic distances may be quite common between two adjacent LMW-GS genes (Huang and Cloutier 2008; Dong et al. 2010). Beside gene duplication, all studies also evidenced a key role of repeat elements in driving both the expansion and intergenic diversification of prolamin loci among homologues genomes. For instance, in the BAC contig region (285 kb) overlapping the *Glu-3A* locus in *T. monococcum* 83% of the sequence is repetitive DNA. Comparison of this region with its orthologous counterpart in *T. durum* revealed that only gene regions were highly conserved, while the rest of the total sequences analysed (90%) were substantially different (Wicker et al. 2003). A second interesting aspect of the study by Gao et al. (2007) is that γ- and ω-gliadins showed a different micro-collinearity in the two orthologous genomes. Although the whole *Glu-3* and *Gli-1/3* regions were only partially resolved, it was clear that the ω- and γ-gliadins had an inverted position in the A and B genomes. For instance, in the A genome the γ-gliadin cluster is linked to an LMW-GS gene that has an orthologous in the B genome, whereas in the B genome this position is occupied by ω-gliadins.

A more comprehensive view of the genomic organization of the *Glu-3* and *Gli-1/3* loci has been recently provided for the diploid and hexaploid wheat genomes (Dong et al. 2016; Huo et al. 2018a). Using a combinational approach, which includes BAC sequencing, physical mapping and BioNano genome mapping (Hastie et al. 2013; Luo et al. 2013), a region of 2.8-Mb of the short arm of 1D chromosome was sequenced in *Ae. tauschii* (Dong et al. 2016). The whole set of prolamin genes, within a region of about 2.3 Mb, was characterized and a total of 8 ω-, 4 γ-, 2 δ-gliadins and 5 LMW-GS genes were identified. Consistent with previous studies, all gliadin gene families were found mainly clustered together with a few examples of interruption by non-prolamin genes, whereas the intergenic distance between adjacent LMW-GS genes was more variable, reaching up to 1Mb, and characterized by the insertion of additional genes (Dong et al. 2016).

Syntenic analysis of orthologous regions of *Brachypodium*, rice and sorghum genomes, which represents an evolutionary period of about 52 MY (Massa et al. 2011), showed that, as observed for the *Gli-2* locus, in *Ae. tauschii* this genomic region underwent a rapid evolution that greatly modified and expanded the region. A total of 103 genes were identified in *Ae. tauschii* and more than 80 were non-syntenic with the other three grass genomes. Thirteen genes were found to be conserved among the four species, likely representing ancient genes that were present in an ancestral species. Although these genes were conserved, their micro-collinearity was modified in *Ae. tauschii* as a result of four inversion events that occurred after the separation of *Triticeae* and *Brachypodieae* lineages. Gene expansion in the *Glu-3* and *Gli-1/3* and loci involved different gene types, including prolamins and two families of resistance genes (Dong et al. 2016). Interestingly, three prolamin-related genes, which were absent in the syntenic region of sorghum and rice, were identified in the *Brachypodium* orthologous region and found to be phylogenetically closely related to *Ae. tauschii* LMW-GSs. This finding suggests that ancestral prolamin genes were already present in a common progenitor of *Brachypodieae* and *Triticeae* lineages before their separation, and that in the *Triticeae* lineage the original genes have been subjected to extensive amplifications and structural rearrangements that generated different prolamin families (Dong et al. 2016).

The *Glu-3* and *Gli-1/3* loci have been characterized in detail in CS by combining genomic sequence data generated by PacBio long reads (Zimin et al. 2017a) and BioNano genome mapping (Huo et al. 2018b). Three regions with lengths of 5.3, 6.5 and 5.6 Mbps were sequenced for the A, B and D genomes respectively. A total of 96, 164 and 129 genes were annotated in the A, B and D genome regions respectively including multiple copies of prolamin genes: that is, 14 γ-, 19 ω-, 5 δ-gliadins and 17 LMW-GS genes. A first comparison between the orthologous regions in the D genomes from *Ae. tauschii* and CS revealed the presence of substantial structural variation, which was mainly localized in five regions of variable size, ranging from 100 kb to 2 Mb (Huo et al. 2018b). This structural variation largely impacted on the shared regions of the two genomes, as only 56 and 29% of the *Ae. tauschii* and CS sequences were found to be conserved respectively including at least 79 genes (Huo et al. 2018b). Despite this variation, the copy number of prolamins between the two D genomes was similar, and only three genes in CS (one ω-gliadin and two LMW glutenins) lacked an orthologue in *Ae. tauschii*. Likewise, their general organization was well maintained, with ω-, γ- and δ-gliadins mainly clustered together and the LMW glutenins dispersed on a larger region.

Using ancient genes (Dong et al. 2016) shared among the A, B and D genomes in CS, Huo and collaborators (Huo et al. 2018b) identified four regions, or blocks (Fig. 2), that were used to simplify collinearity analysis. Among these orthologous regions, considerable variation was observed in terms of gene content. For example, in the first block, the D genome contains more than twice as many genes as the A genome (51 vs. 18). When focusing solely on prolamin genes, one of the main differences was the position of ω-gliadin genes. In both the B and D genomes, the ω-gliadins are present in the second block, whereas they are completely missing in the corresponding region of the A genome (Fig. 2). This observation is in line with a previous study (Gao et al. 2007), reporting an inverted position of these gene families in the durum wheat Langdon, and by a comparative analysis of ω- and γ-gliadins location in the A and B genomes of Svevo (Maccaferri et al. 2019; Lauria and Ceriotti unpublished data).

Phylogenetic analysis of each group of prolamin genes provided also examples of genes that were either conserved or differentially represented among genomes. For instance, γ-gliadins from the CS D genome were all clustered with the homologous copies from *Ae. tauschii*, evidencing that these genes are basically conserved between the hexaploid wheat and its progenitor. Differently, while the D and B genomes contain γ-gliadins that clustered in two different clades, γ-gliadins from the A genome were only present in one clade, suggesting that at least a set of genes might have been lost. Moreover, although ω-gliadin genes could be divided into two main clades, only genes from the D genome were present in both of them, whereas those from the A and B genomes were clade specific, suggesting either a differentially loss/duplication of some genes or a hybrid origin of the D genome from a cross between A and B (Marcussen et al. 2014; Huo et al. 2018b). Collectively, these observations evidenced that likewise other prolamin gene families, the *Glu-3* and *Gli-1/3* regions evolved independently after the divergence of the A, B and D genomes.

3 Expression of Prolamin Genes

The characterization of prolamin transcripts has been extensively analysed over the years, and an impressive amount of data is available to date. This information is based on different experimental strategies, such as the use of expressed sequence tags (Ogihara et al. 2004; Kawaura et al. 2005; Salentijn et al. 2009; Altenbach et al. 2010), cDNA and PCR sequencing (Pistón et al. 2006; van Herpen et al. 2006; Noma et al. 2016) and, more recently, the use of next-generation sequencing approaches (Salentijn et al. 2013). When considering second sequencing technology, such as Illumina HiSeq's short reads, the main problem in analysing multigene families is related to the high degree of sequence identity that characterizes the different members of a gene family and the presence, in the case of prolamins, of highly repetitive nucleotide sequence elements. This can lead to an incorrect assembly between orthologues or paralogous members and as well as incorrect mapping, if a high-quality genome reference is not available. In this regard, the development of third-generation sequencing, such as the PacBio long-read, has greatly improved this aspect and enabled a more correct assembly of the different members of this gene family (Anderson et al. 2013; Dong et al. 2015; Wang et al. 2017; White et al. 2024). The complete portrait of the genomic organization of prolamin genes in different wheat genomes has made possible an accurate transcriptome profiling of these genes, providing insight about their genomic landscape and the evolution of the prolamin family.

Seed transcriptome analysis of *Ae. tauschii* revealed that all 12 α-gliadin genes present in this genome, including pseudogenes, were transcribed, albeit to varying degrees. When transcription of intact genes was examined a 10-fold difference was observed between paralogues. Overall, the expression level was lower for pseudogenes than for intact genes, but much higher in comparison to non-prolamin genes present in the region (Huo et al. 2017). Sequence analysis of the promoter region of pseudogenes revealed that in most cases these sequences possess the classical TATA box and cis- regulatory motifs responsible for endosperm-specific expression such as the prolamin-box, the GCN4-like motif and RY repeat motif (Bäumlein et al. 1992; Albani et al. 1997; Mena et al. 1998). Thus, the low expression of pseudogenes may be related to the nonsense-mediated mRNA decay pathway that selectively degrades mRNAs containing premature termination codons (Hug et al. 2016; Huo et al. 2017).

A detailed expression analysis of α-gliadin genes was also performed for the hexaploid *Triticum aestivum* CS and the tetraploid wild emmer Zavitan (Huo et al. 2018a, 2019). Consistent with what was observed for *Ae. tauschii*, RNA-seq transcriptome data confirmed a certain degree of variation in the expression of intact α-gliadin genes in the endosperm, while pseudogenes, with a few exceptions, were generally expressed at a lower level. Comparison of α-gliadins expression from the three homoeologous loci of the hexaploid wheat is consistent with the notion that the B genome gives a large contribution to α-gliadin deposition in CS. Differently, in WEW Zavitan, α-gliadin transcripts from the A genome are almost double the transcripts from the B genome (Huo et al. 2019). Individual α- gliadin genes were found to have different expression profiles during seed development, with genes that maintained a similar expression level at different developmental stages and others that were subjected to upregulation from 10 to 30 days post-anthesis. Interestingly, expression differences were also observed among homologous α-gliadin pairs from CS and *Ae. tauschii* D genomes, despite the high sequence identity of two D genomic regions (see above).

Expression analysis of prolamins at the *Gli-1/3* locus provides information on additional prolamin members such as the δ-, ω-, γ-gliadins and LMW glutenin present in the CS genome (Huo et al. 2018b). Of the five δ-gliadins present in CS only the intact ones were found to be expressed, and in particular the one present on the D genome. This trend was also observed for ω-gliadins, whose contribution at the transcriptional level was mainly due to intact genes from the D genome. The expression of γ-gliadins was approximately equivalent among the different copies of the three subgenomes, and the transcripts of intact LMW genes came mainly from the B and D genomes (Huo et al. 2018b). At 30 days post-anthesis, within the CS transcriptome, there is a large contribution of prolamin transcripts, and in particular of γ-gliadin, α-gliadin and members of the LMW-GS class. The ω-gliadin and HMW-GS transcripts each accounted for over 7% of total, while δ-gliadin contributed in minimal part (0.3%.). Overall, the contribution of the A genome to the transcripts of each prolamin group was considerably less than the B and D genomes (Huo et al. 2018b).

The above data show that intact prolamin genes are characterized by differential expression between and within subgenomes. Although conserved cis-regulator motives have been detected within the promoter regions of different class of prolamins (Kawaura et al. 2012; Huo et al. 2017), it is likely that sequence variation at these regions, in term of the number/combination of cis-acting elements and SNP/ InDel, may impact the interaction with regulatory factors (Juhász et al. 2011; Ravel et al. 2014). For example, genetic variation in the promoter region of HMW-GS alleles has been associated with their different expression characteristics. In the cultivar Cheyenne (hexaploid bread), a tandem duplication of 53 bases was originally identified for the *Bx7* allele, and associated with its higher expression with respect to other subunits (Anderson and Greene 1989). Moreover, overexpression of the *Bx7* allele variant, *1Bx7*OE, was positively correlated with a 43 bp insertion found in the promoter region (Ragupathy et al. 2008).

These observations were further supported by a successive study that investigated the role of these genetic variants by using both transient expression assays in wheat endosperms and rice transgenic lines (Geng et al. 2014). In particular, the 43 bp insertion was proposed to serve as a strong enhancer to improve gene expression by affecting local chromatin structure and nucleosome positioning, whereas absence of the tandem 54 bp duplication, which contains the cereal-box that determinates seed-specific expression (Anderson and Greene 1989), led to loss of endosperm-specific expression (Geng et al. 2014). Guo et al. (2015) characterized the promoter region of the HMW-GS genes in 40 different wheat varieties, and found that the number of binding sites recognized by the wheat transcriptional factor TaGAMyb positively correlated with the abundance of HMW-GS proteins. Moreover, the different abundance of TaGAMyb binding sites among genes of the three subgenomes likely explains their final contribution at protein level in mature seeds. The authors also found that the y-type glutenin genes contained fewer TaGAMyb binding sites than the x-type glutenin genes, which again seem to explain the lower protein abundance of the y-type compared with the x-type (Guo et al. 2015). By comparing the promoters of two LMW-GS genes with contrasting expression patterns (i.e., a 21-fold difference between the highly expressed and the weakly expressed one), Qu et al. (2023) recently identified a sequence variation that could be associated with observed expression differences. Within the endosperm box, which confers endosperm-specific expression and is composed of the GCN4 motif and the adjacent Prolamin box, the weak *LMW-GS* gene has a T-C substitution located at the 5′ end of a GCN4 motif that alters the DNA binding affinity of the storage protein activator in the EMSA assay. In addition, variants were identified in additional regulatory elements, such as the TGACG and TCT motifs and the MYC and MYB recognition sites. Subgenome dominance in polyploid species has been evocated as a possible mechanism to explain the observed expression differences. However, the finding that even members of the same subgenome show different level of expression suggested that other factors may also be involved in this process, for example, epigenetic mechanisms (Huo et al. 2018a; Qu et al. 2023). In maize, epigenetic regulation of prolamin genes, and in particular of α-zeins, has been confirmed by several studies that have characterized these genes at both DNA methylation and chromatin level (Lund et al. 1995; Locatelli et al. 2009; Miclaus et al. 2011). In addition, parental expression in hybrids of a subset α-zeins has been shown to depend on the type of hybrid combination (Castelli et al. 2018). Available data suggest that epigenetic regulation is active also in wheat. For instance, suppression of the 5-methylcytosine DNA glycosylase DEMETER determined a reduction in LMW-GS and gliadin gene expression. DEMETER is an active element of the epigenetic process and acts by removing 5-methylcytosine from methylated sequences. Consequently, suppression of DEMETER in wheat endosperm avoids active demethylation of LMW- GS and gliadin genes, thus, reducing their expression capacity (Wen et al. 2012). These findings are supported by further studies that confirmed the occurrence of DNA methylation within the promoter regions of HMW-GS and γ-gliadin genes (Zhu et al. 2018; Zhou et al. 2022). Moreover, a correlation between gene expression and chromatin accessibility has been proved for the HMW-GS genes. Specifically, it was observed that TaGAMyb, which binds the HMW-GS gene promoter (see above), physically interacts with the wheat histone acetyltransferase TaGCN5 to establish a competent expression histone acetylation state at the HMW-GS genes. (Guo et al. 2015). This result seems to confirm previously observations in barley, where transgenic lines carrying the promoter region of two HMW-GS variants (*Bx7* and *By9*) fused to the *uidA* reported gene showed GUS expression after treatment with Trichostatin A, an agent that induces histone hyperacetylation (Geng et al. 2014). A final aspect to consider when evoking epigenetic mechanisms to explain gene expression is the general genomic landscape surrounding a gene sequence. Repetitive sequences, such as TEs or LTRs, are natural targets of epigenetic regulation, such as DNA methylation and repressive histone modifications, which occasionally can spread to neighbouring sequences, thereby altering their expression capabilities (Lauria and Rossi 2011). Considering the intimate coexistence between prolamin genes and sequences repeats, this condition may explain at least part the expression differences observed among members of this gene family: an aspect that is worthy of future studies.

4 Conclusion

Wheat plays a major role in humankind nutrition and major efforts are being made to improve wheat production worldwide. This crop supplies a relevant fraction of the total amount of calories produced by agriculture, but its success and widespread distribution are also due to the peculiar characteristics of its protein fraction. In this chapter, we have collected available information on the genomic organization of wheat prolamins and their expression patterns. Recently, the generation of high- quality sequences made possible a comparative analysis of subgenomes shedding new light on the evolution of prolamin loci. Before these technological advancements came into place, such analysis has been particularly challenging due to the complex genomic organization of these gene families and the specific characteristics of prolamin genes and loci. One main finding emerging from the comparative genomic analysis is that prolamin loci in different subgenomes have been subjected to varying levels of evolution, which have shaped the regions in term of copy number, collinearity and characteristic of repetitive elements present within the intergenic regions. It is anticipated that pangenomics studies, as the ones based on *de novo* transcriptome assembly recently performed in both durum and hexaploid wheat (Vendramin et al. 2019; White et al. 2024), will further expand the description of existing genetic variability in wheat and allow to refine our understanding of the dynamics responsible for the evolution of the prolamin gene family.

Information on the structure of prolamin loci and on the characteristics of individual genes is of great importance not only to understand the evolution of these loci and the factors controlling seed protein composition but also in view of improving flour technological and nutritional characteristics, and reducing the impact of CD at a global scale.

Acknowledgments We thank Stefania Masci for critical reading of the manuscript and useful suggestions. This work was carried out within the Agritech National Research Center and received funding from the European Union Next-GenerationEU (PIANO NAZIONALE DI RIPRESA E RESILIENZA (PNRR) MISSIONE 4 COMPONENTE 2, INVESTIMENTO 1.4 – D.D. 1032 17/06/2022, CN00000022).This manuscript reflects only the authors' views and opinions, neither the European Union nor the European Commission can be considered responsible for them.

References

Albani D, Hammond-Kosack MC, Smith C et al (1997) The wheat transcriptional activator SPA: a seed- specific bZIP protein that recognizes the GCN4-like motif in the bifactorial endosperm box of prolamin genes. Plant Cell 9:171–184. https://doi.org/10.1105/tpc.9.2.171

Altenbach SB, Vensel WH, DuPont FM (2010) Analysis of expressed sequence tags from a single wheat cultivar facilitates interpretation of tandem mass spectrometry data and discrimination of gamma gliadin proteins that may play different functional roles in flour. BMC Plant Biol 10:7. https://doi.org/10.1186/1471-2229-10-7

Altenbach SB, Tanaka CK, Hurkman WJ et al (2011) Differential effects of a post-anthesis fertilizer regimen on the wheat flour proteome determined by quantitative 2-DE. Proteome Sci 9:46. https://doi.org/10.1186/1477-5956-9-46

Altenbach SB, Chang H-C, Rowe MH et al (2020) Reducing the immunogenic potential of wheat flour: silencing of alpha gliadin genes in a U.S. wheat cultivar. Front Plant Sci 11. https://doi.org/10.3389/fpls.2020.00020

Anderson OD, Greene FC (1989) The characterization and comparative analysis of high-molecular- weight glutenin genes from genomes A and B of a hexaploid bread wheat. Theor Appl Genet 77:689–700. https://doi.org/10.1007/BF00261246

Anderson OD, Litts JC, Greene FC (1997) The α-gliadin gene family. I. Characterization of ten new wheat α-gliadin genomic clones, evidence for limited sequence conservation of flanking DNA, and Southern analysis of the gene family. Theor Appl Genet 95:50–58. https://doi.org/10.1007/s001220050531

Anderson O, Rausch C, Moullet O, Lagudah E (2003) The wheat D-genome HMW-glutenin locus: BAC sequencing, gene distribution, and retrotransposon clusters. Funct Integr Genomics 3:56–68. https://doi.org/10.1007/s10142-002-0069-z

Anderson OD, Dong L, Huo N, Gu YQ (2012) A new class of wheat gliadin genes and proteins. PLoS One 7(12):e52139. https://doi.org/10.1371/JOURNAL.PONE.0052139

Anderson OD, Huo N, Gu YQ (2013) The gene space in wheat: the complete γ-gliadin gene family from the wheat cultivar Chinese Spring. Funct Integr Genomics 13:261–273. https://doi.org/10.1007/s10142-013-0321-8

Bäumlein H, Nagy I, Villarroel R et al (1992) *Cis* -analysis of a seed protein gene promoter: the conservative RY repeat CATGCATG within the legumin box is essential for tissue-specific expression of a legumin gene. Plant J 2:233–239. https://doi.org/10.1046/j.1365-313X.1992.t01-45-00999.x

Blechl AE, Anderson OD (1996) Expression of a novel high-molecular-weight glutenin subunit gene in transgenic wheat. Nat Biotechnol 14:875–879. https://doi.org/10.1038/nbt0796-875

Camerlengo F, Sestili F, Silvestri M et al (2017) Production and molecular characterization of bread wheat lines with reduced amount of α-type gliadins. BMC Plant Biol 17:248. https://doi.org/10.1186/s12870-017-1211-3

Cassidy BG, Dvorak J, Anderson OD (1998) The wheat low-molecular-weight glutenin genes: characterization of six new genes and progress in understanding gene family structure. Theor Appl Genet 96:743–750. https://doi.org/10.1007/s001220050797

Castelli S, Mascheretti I, Cosentino C et al (2018) Uniparental and transgressive expression of α-zeins in maize endosperm of o2 hybrid lines. PLoS One 13:e0206993. https://doi.org/10.1371/journal.pone.0206993

Chen XY, Cao XY, Zhang YJ et al (2016) Genetic characterization of cysteine-rich type-b avenin-like protein coding genes in common wheat OPEN. Nat Publ Group 6. https://doi.org/10.1038/srep30692

Clarke BC, Phongkham T, Gianibelli M et al (2003) The characterisation and mapping of a family of LMW-gliadin genes: effects on dough properties and bread volume. Theor Appl Genet 106:629–635. https://doi.org/10.1007/s00122-002-1091-1

Cloutier S, Banks T, Nilmalgoda S (2005) Molecular understanding of wheat evolution at the Glu-B1 locus. In: proceedings of the international conference on plant genomics and biotechnology: chal lenges and opportunities, Raipur, India, p 40

D'Ovidio R, Masci S (2004) The low-molecular-weight glutenin subunits of wheat gluten. J Cereal Sci 39:321–339. https://doi.org/10.1016/j.jcs.2003.12.002

D'Ovidio R, Lafiandra D, Tanzarella OA et al (1991) Molecular characterization of bread wheat mutants lacking the entire cluster of chromosome 6A controlled gliadin components. J Cereal Sci 14:125–129. https://doi.org/10.1016/S0733-5210(09)80133-1

D'Ovidio R, Masci S, Porceddu E, Kasarda DD (1997) Duplication of the *Bx7* high-molecular-weight glutenin subunit gene in bread wheat (*Triticum aestivum* L.) cultivar 'Red River 68'. Plant Breed 116:525–531. https://doi.org/10.1111/j.1439-0523.1997.tb02184.x

De Bustos A, Rubio P, Jouve N (2001) Characterisation of two gene subunits on the 1R chromosome of rye as orthologs of each of the Glu-1 genes of hexaploid wheat. Theor Appl Genet 103:733–742. https://doi.org/10.1007/s001220100669

Dong L, Zhang X, Liu D et al (2010) New insights into the organization, recombination, expression and functional mechanism of low molecular weight Glutenin subunit genes in bread wheat. PLoS One 5:e13548. https://doi.org/10.1371/journal.pone.0013548

Dong L, Liu H, Zhang J et al (2015) Single-molecule real-time transcript sequencing facilitates common wheat genome annotation and grain transcriptome research. BMC Genomics 16:1039. https://doi.org/10.1186/s12864-015-2257-y

Dong L, Huo N, Wang Y et al (2016) Rapid evolutionary dynamics in a 2.8-Mb chromosomal region containing multiple prolamin and resistance gene families in *Aegilops tauschii*. Plant J 87:495–506. https://doi.org/10.1111/tpj.13214

Feldman M (2001) Origin of cultivated wheat. The world wheat book: a history of wheat breeding. pp 3–53

Gao S, Gu YQ, Wu J et al (2007) Rapid evolution and complex structural organization in genomic regions harboring multiple prolamin genes in the polyploid wheat genome. Plant Mol Biol 65:189–203. https://doi.org/10.1007/s11103-007-9208-1

Ge W, Gao Y, Xu S et al (2021) Genome-wide identification, characteristics and expression of the prolamin genes in Thinopyrum elongatum. BMC Genomics 22:864. https://doi.org/10.1186/s12864-021-08088-x

Geng Y, Pang B, Hao C et al (2014) Expression of wheat high molecular weight Glutenin subunit 1Bx is affected by large insertions and deletions located in the upstream flanking sequences. PLoS One 9:e105363. https://doi.org/10.1371/journal.pone.0105363

Gianibelli MC, Larroque OR, MacRitchie F, Wrigley CW (2001) Biochemical, genetic, and molecular characterization of wheat Glutenin and its component subunits. Cereal Chem 78:635–646. https://doi.org/10.1094/CCHEM.2001.78.6.635

Gil-Humanes J, Pistón F, Tollefsen S et al (2010) Effective shutdown in the expression of celiac disease- related wheat gliadin T-cell epitopes by RNA interference. Proc Natl Acad Sci 107:17023–17028. https://doi.org/10.1073/pnas.1007773107

Gu YQ, Coleman-Derr D, Kong X, Anderson OD (2004a) Rapid genome evolution revealed by comparative sequence analysis of orthologous regions from four Triticeae genomes. Plant Physiol 135:459–470. https://doi.org/10.1104/pp.103.038083

Gu YQ, Crossman C, Kong X et al (2004b) Genomic organization of the complex α-gliadin gene loci in wheat. Theor Appl Genet 109:648–657. https://doi.org/10.1007/s00122-004-1672-2

Gu YQ, Salse J, Coleman-Derr D et al (2006) Types and rates of sequence evolution at the high- molecular-weight Glutenin locus in Hexaploid wheat and its ancestral genomes. Genetics 174:1493–1504. https://doi.org/10.1534/genetics.106.060756

Gu YQ, Wanjugi H, Coleman-Derr D et al (2010) Conserved globulin gene across eight grass genomes identify fundamental units of the loci encoding seed storage proteins. Funct Integr Genomics 10:111–122. https://doi.org/10.1007/s10142-009-0135-x

Guo W, Yang H, Liu Y et al (2015) The wheat transcription factor Ta GAM yb recruits histone acetyltransferase and activates the expression of a high-molecular-weight glutenin subunit gene. Plant J 84:347–359. https://doi.org/10.1111/tpj.13003

Guzmán-López MH, Ruipérez V, Marín-Sanz M et al (2024) Identification of RNAi hypoallergic bread wheat lines for wheat-dependent exercise-induced anaphylaxis patients. Front Nutr 10. https://doi.org/10.3389/fnut.2023.1319888

Halstead-Nussloch G, Tanaka T, Copetti D et al (2021) Multiple wheat genomes reveal novel Gli-2 sublocus location and variation of celiac disease epitopes in duplicated α-gliadin genes. Front Plant Sci 12:715985. https://doi.org/10.3389/fpls.2021.715985

Harberd NP, Bartels D, Thompson RD (1985) Analysis of the gliadin multigene loci in bread wheat using nullisomic-tetrasomic lines. Mol Gen Genet 198:234–242. https://doi.org/10.1007/BF00383001

Hastie AR, Dong L, Smith A et al (2013) Rapid genome mapping in Nanochannel arrays for highly complete and accurate De novo sequence assembly of the complex Aegilops tauschii genome. PLoS One 8:e55864. https://doi.org/10.1371/journal.pone.0055864

Huang X-Q, Cloutier S (2008) Molecular characterization and genomic organization of low molecular weight glutenin subunit genes at the Glu-3 loci in hexaploid wheat (Triticum aestivum L.). Theor Appl Genet 116:953–966. https://doi.org/10.1007/s00122-008-0727-1

Huang S, Sirikhachornkit A, Su X et al (2002) Genes encoding plastid acetyl-CoA carboxylase and 3-phosphoglycerate kinase of the *Triticum / Aegilops* complex and the evolutionary history of polyploid wheat. Proc Natl Acad Sci 99:8133–8138. https://doi.org/10.1073/pnas.072223799

Hug N, Longman D, Cáceres JF (2016) Mechanism and regulation of the nonsense-mediated decay pathway. Nucleic Acids Res 44:1483–1495. https://doi.org/10.1093/nar/gkw010

Huo N, Dong L, Zhang S et al (2017) New insights into structural organization and gene duplication in a 1.75-Mb genomic region harboring the α-gliadin gene family in *Aegilops tauschii*, the source of wheat D genome. Plant J 92:571–583. https://doi.org/10.1111/tpj.13675

Huo N, Zhu T, Altenbach S et al (2018a) Dynamic evolution of α-gliadin Prolamin gene family in Homeologous genomes of Hexaploid wheat. Sci Rep 8:5181. https://doi.org/10.1038/s41598-018-23570-5

Huo N, Zhang S, Zhu T et al (2018b) Gene duplication and evolution dynamics in the Homeologous regions harboring multiple Prolamin and resistance gene families in Hexaploid wheat. Front Plant Sci 9:673. https://doi.org/10.3389/fpls.2018.00673

Huo N, Zhu T, Zhang S et al (2019) Rapid evolution of α-gliadin gene family revealed by analyzing Gli- 2 locus regions of wild emmer wheat. Funct Integr Genomics 19:993–1005. https://doi.org/10.1007/s10142-019-00686-z

Ikeda TM, Araki E, Fujita Y, Yano H (2006) Characterization of low-molecular-weight glutenin subunit genes and their protein products in common wheats. Theor Appl Genet 112:327–334. https://doi.org/10.1007/s00122-005-0131-z

Johal J, Gianibelli MC, Rahman S et al (2004) Characterization of low-molecular-weight glutenin genes in Aegilops tauschii. Theor Appl Genet 109:1028–1040. https://doi.org/10.1007/s00122-004-1711-z

Jouanin A, Borm T, Boyd LA et al (2019a) Development of the GlutEnSeq capture system for sequencing gluten gene families in hexaploid bread wheat with deletions or mutations induced by γ- irradiation or CRISPR/Cas9. J Cereal Sci 88:157–166. https://doi.org/10.1016/j.jcs.2019.04.008

Jouanin A, Schaart JG, Boyd LA et al (2019b) Outlook for coeliac disease patients: towards bread wheat with hypoimmunogenic gluten by gene editing of α- and γ-gliadin gene families. BMC Plant Biol 19:333. https://doi.org/10.1186/s12870-019-1889-5

Juhász A, Makai S, Sebestyén E et al (2011) Role of conserved non-coding regulatory elements in LMW Glutenin gene expression. PLoS One 6:e29501. https://doi.org/10.1371/journal.pone.0029501

Juhász A, Belova T, Florides CG et al (2018) Genome mapping of seed-borne allergens and immunoresponsive proteins in wheat. Sci Adv 4. https://doi.org/10.1126/sciadv.aar8602

Kawaura K, Mochida K, Ogihara Y (2005) Expression profile of two storage-protein gene families in Hexaploid wheat revealed by large-scale analysis of expressed sequence tags. Plant Physiol 139:1870–1880. https://doi.org/10.1104/pp.105.070722

Kawaura K, Wu J, Matsumoto T et al (2012) Genome change in wheat observed through the structure and expression of α/β-gliadin genes. Funct Integr Genomics 12:341–355. https://doi.org/10.1007/s10142-012-0269-0

Kong X-Y, Gu YQ, You FM et al (2004) Dynamics of the evolution of orthologous and paralogous portions of a complex locus region in two genomes of allopolyploid wheat. Plant Mol Biol 54:55–69. https://doi.org/10.1023/B:PLAN.0000028768.21587.dc

Lafiandra D, Kasarda DD, Morris R (1984) Chromosomal assignment of genes coding for the wheat gliadin protein components of the cultivars 'Cheyenne' and 'Chinese Spring' by two-dimensional (two-pH) electrophoresis. Theor Appl Genet 68:531–539. https://doi.org/10.1007/BF00285007

Lauria M, Rossi V (2011) Epigenetic control of gene regulation in plants. Bioch Biophys Acta 1809:369–378. https://doi.org/10.1016/j.bbagrm.2011.03.002

Lawrence GJ, Shepherd KW (1981) Chromosomal location of genes controlling seed proteins in species related to wheat. Theor Appl Genet 59:25–31. https://doi.org/10.1007/BF00275771

Levy AA, Feldman M (2022) Evolution and origin of bread wheat. Plant Cell 34:2549–2567. https://doi.org/10.1093/plcell/koac130

Lew E, Kuzmickyi DD, Kasarda DD (1992) Characterization of low molecular weight Glutenin subunits by reversed-phase high-performance liquid chromatography, sodium dodecyl sulfate- polyacrylamide gel electrophoresis, and N-terminal amino acid sequencing. Cereal Chem 69:508–515

Li Y, Fu J, Shen Q, Yang D (2020) High-molecular-weight Glutenin subunits: genetics, structures, and relation to end use qualities. Int J Mol Sci 22:184. https://doi.org/10.3390/ijms22010184

Lionetti E, Gatti S, Pulvirenti A, Catassi C (2015) Celiac disease from a global perspective. Best Pract Res Clin Gastroenterol 29:365–379. https://doi.org/10.1016/j.bpg.2015.05.004

Locatelli S, Piatti P, Motto M, Rossi V (2009) Chromatin and DNA modifications in the *Opaque2* – mediated regulation of gene transcription during maize endosperm development. Plant Cell 21:1410–1427. https://doi.org/10.1105/tpc.109.067256

Lund G, Ciceri P, Viotti A (1995) Maternal-specific demethylation and expression of specific alleles of zein genes in the endosperm of *Zea mays* L. Plant J 8:571–581. https://doi.org/10.1046/j.1365-313X.1995.8040571.x

Luo M-C, Gu YQ, You FM et al (2013) A 4-gigabase physical map unlocks the structure and evolution of the complex genome of *Aegilops tauschii,* the wheat D-genome progenitor. Proc Natl Acad Sci 110:7940–7945. https://doi.org/10.1073/pnas.1219082110

Luo M-C, Gu YQ, Puiu D et al (2017) Genome sequence of the progenitor of the wheat D genome Aegilops tauschii. Nature 551:498–502. https://doi.org/10.1038/nature24486

Ma F, Li M, Li T et al (2013) Overexpression of Avenin-like b proteins in bread wheat (Triticum aestivum L.) improves dough mixing properties by their incorporation into Glutenin polymers. PLoS One 8:e66758. https://doi.org/10.1371/journal.pone.0066758

Maccaferri M, Harris NS, Twardziok SO et al (2019) Durum wheat genome highlights past domestication signatures and future improvement targets. Nat Genet 51:885–895. https://doi.org/10.1038/s41588-019-0381-3

Marcussen T, Sandve SR, Heier L et al (2014) Ancient hybridizations among the ancestral genomes of bread wheat. Science 1979:345. https://doi.org/10.1126/science.1250092

Masci S, Rovelli L, Kasarda DD et al (2002) Characterisation and chromosomal localisation of C-type low-molecular-weight glutenin subunits in the bread wheat cultivar Chinese Spring. Theor Appl Genet 104:422–428. https://doi.org/10.1007/s001220100761

Massa AN, Wanjugi H, Deal KR et al (2011) Gene space dynamics during the evolution of *Aegilops tauschii, Brachypodium distachyon, Oryza sativa*, and *Sorghum bicolor* genomes. Mol Biol Evol 28:2537–2547. https://doi.org/10.1093/molbev/msr080

Mena M, Vicente-Carbajosa J, Schmidt RJ, Carbonero P (1998) An endosperm-specific DOF protein from barley, highly conserved in wheat, binds to and activates transcription from the prolamin-box of a native B-hordein promoter in barley endosperm. Plant J 16:53–62. https://doi.org/10.1046/j.1365-313x.1998.00275.x

Miclaus M, Xu J-H, Messing J (2011) Differential gene expression and Epiregulation of alpha Zein gene copies in maize haplotypes. PLoS Genet 7:e1002131. https://doi.org/10.1371/journal.pgen.1002131

Noma S, Kawaura K, Hayakawa K et al (2016) Comprehensive molecular characterization of the α/β- gliadin multigene family in hexaploid wheat. Mol Gen Genomics 291:65–77. https://doi.org/10.1007/s00438-015-1086-7

Ogihara Y, Mochida K, Kawaura K et al (2004) Construction of a full-length cDNA library from young spikelets of hexaploid wheat and its characterization by large-scale sequencing of expressed sequence tags. Genes Genet Syst 79:227–232. https://doi.org/10.1266/ggs.79.227

Ozuna CV, Iehisa JCM, Giménez MJ et al (2015) Diversification of the celiac disease α-gliadin complex in wheat: a 33-mer peptide with six overlapping epitopes, evolved following polyploidization. Plant J 82:794–805. https://doi.org/10.1111/tpj.12851

Paris R, Petruzzino G, Savino M et al (2021) Genome-wide identification, characterization and expression pattern analysis of the γ-gliadin gene family in the durum wheat (Triticum durum Desf.) cultivar Svevo. Genes (Basel) 12:1743. https://doi.org/10.3390/genes12111743

Payne PI (1987) Genetics of wheat storage proteins and the effect of allelic variation on bread-making quality. Annu Rev Plant Physiol 38:141–153. https://doi.org/10.1146/annurev.pp.38.060187.001041

Payne PI, Lawrence GJ (1983) Catalogue of alleles for the complex gene loci, Glu-A1, Glu-B1, and Glu-D1 which code for high-molecular-weight subunits of glutenin in hexaploid wheat. Cereal Res Commun 11(1):29–35. http://www.jstor.org/stable/23781365

Pistón F, Dorado G, Martín A, Barro F (2006) Cloning of nine γ-gliadin mRNAs (cDNAs) from wheat and the molecular characterization of comparative transcript levels of γ-gliadin subclasses. J Cereal Sci 43:120–128. https://doi.org/10.1016/j.jcs.2005.07.002

Pourmohammadi K, Abedi E, Hashemi SMB (2023) Gliadin and glutenin genomes and their effects on the technological aspect of wheat-based products. Curr Res Food Sci 7:100622. https://doi.org/10.1016/j.crfs.2023.100622

Qu G, Wang K, Mu J et al (2023) Identifying *cis* -acting elements associated with the high activity and endosperm specificity of the promoters of genes encoding low-molecular-weight Glutenin subunits in common wheat (*Triticum aestivum*). J Agric Food Chem. https://doi.org/10.1021/acs.jafc.3c04209

Ragupathy R, Naeem HA, Reimer E et al (2008) Evolutionary origin of the segmental duplication encompassing the wheat GLU-B1 locus encoding the overexpressed Bx7 (Bx7OE) high molecular weight glutenin subunit. Theor Appl Genet 116:283–296. https://doi.org/10.1007/s00122-007-0666-2

Rasheed A, Xia X, Yan Y et al (2014) Wheat seed storage proteins: advances in molecular genetics, diversity and breeding applications. J Cereal Sci 60:11–24. https://doi.org/10.1016/j.jcs.2014.01.020

Ravel C, Fiquet S, Boudet J et al (2014) Conserved cis-regulatory modules in promoters of genes encoding wheat high-molecular-weight glutenin subunits. Front Plant Sci 5. https://doi.org/10.3389/fpls.2014.00621

Reeves CD, Okita TW (1987) Analyses of α/β--type gliadin genes from diploid and hexaploid wheats. Gene 52:257–266. https://doi.org/10.1016/0378-1119(87)90052-7

Rooke L, Békés F, Fido R et al (1999) Overexpression of a gluten protein in transgenic wheat results in greatly increased dough strength. J Cereal Sci 30:115–120. https://doi.org/10.1006/jcrs.1999.0265

Sabelli PA, Shewry PR (1991) Characterization and organization of gene families at the Gli-1 loci of bread and durum wheats by restriction fragment analysis. Theor Appl Genet 83:209–216. https://doi.org/10.1007/BF00226253

Salentijn EM, Goryunova SV, Bas N et al (2009) Tetraploid and hexaploid wheat varieties reveal large differences in expression of alpha-gliadins from homoeologous Gli-2 loci. BMC Genomics 10:48. https://doi.org/10.1186/1471-2164-10-48

Salentijn EM, Esselink DG, Goryunova SV et al (2013) Quantitative and qualitative differences in celiac disease epitopes among durum wheat varieties identified through deep RNA-amplicon sequencing. BMC Genomics 14:905. https://doi.org/10.1186/1471-2164-14-905

Sánchez-León S, Gil-Humanes J, Ozuna CV et al (2018) Low-gluten, nontransgenic wheat engineered with CRISPR/Cas9. Plant Biotechnol J 16:902–910. https://doi.org/10.1111/pbi.12837

Sapone A, Bai JC, Ciacci C et al (2012) Spectrum of gluten-related disorders: consensus on new nomenclature and classification. BMC Med 10:13. https://doi.org/10.1186/1741-7015-10-13

Schaart JG, Salentijn EMJ, Goryunova SV et al (2021) Exploring the alpha-gliadin locus: the 33-mer peptide with six overlapping coeliac disease epitopes in *Triticum aestivum* is derived from a subgroup of *Aegilops tauschii*. Plant J 106:86–94. https://doi.org/10.1111/tpj.15147

Scherf KA, Koehler P, Wieser H (2016) Gluten and wheat sensitivities—an overview. J Cereal Sci 67:2–11. https://doi.org/10.1016/j.jcs.2015.07.008

Shewry PR, Halford NG (2002) Cereal seed storage proteins: structures, properties and role in grain utilization. J Exp Bot 53:947–958. https://doi.org/10.1093/jexbot/53.370.947

Shewry PR, Tatham AS (2016) Improving wheat to remove coeliac epitopes but retain functionality. J Cereal Sci 67:12–21. https://doi.org/10.1016/j.jcs.2015.06.005

Shewry PR, Tathamt AS (1990) The prolamin storage proteins of cereal seeds: structure and evolution. Biochem J 267:1–12

Shewry PR, Tatham AS, Kasarda DD (1992) Cereal proteins and coeliac disease. In: Marsh MN (ed) Coeliac disease. Blackwell, London, pp 305–348

Shewry PR, Halford NG, Tatham AS et al (2003) The high molecular weight subunits of wheat glutenin and their role in determining wheat processing properties. pp 219–302

Smulders MJM, Jouanin A, Schaart J et al (2015) Development of wheat varieties with reduced contents of celiac-immunogenic epitopes through conventional and GM strategies. Proceedings of the 28th meeting of the Working Group on Prolamin Analysis and Toxicity (P. Koehler, ed.), 25–27 September 2014, Nantes, France, pp 47–56. https://www.wgpat.com/proceeding_28th.html

Sollid LM, Qiao S-W, Anderson RP et al (2012) Nomenclature and listing of celiac disease relevant gluten T-cell epitopes restricted by HLA-DQ molecules. Immunogenetics 64:455–460. https://doi.org/10.1007/s00251-012-0599-z

van Herpen TW, Goryunova SV, van der Schoot J et al (2006) Alpha-gliadin genes from the A, B, and D genomes of wheat contain different sets of celiac disease epitopes. BMC Genomics 7:1. https://doi.org/10.1186/1471-2164-7-1

Vendramin V, Ormanbekova D, Scalabrin S et al (2019) Genomic tools for durum wheat breeding: de novo assembly of Svevo transcriptome and SNP discovery in elite germplasm. BMC Genomics 20:278. https://doi.org/10.1186/s12864-019-5645-x

Veraverbeke WS, Delcour JA (2002) Wheat protein composition and properties of wheat Glutenin in relation to Breadmaking functionality. Crit Rev Food Sci Nutr 42:179–208. https://doi.org/10.1080/10408690290825510

Walkowiak S, Gao L, Monat C et al (2020) Multiple wheat genomes reveal global variation in modern breeding. Nature 588:277–283. https://doi.org/10.1038/s41586-020-2961-x

Wang D-W, Li D, Wang J et al (2017) Genome-wide analysis of complex wheat gliadins, the dominant carriers of celiac disease epitopes. Sci Rep 7:44609. https://doi.org/10.1038/srep44609

Wang D, Li F, Cao S, Zhang K (2020) Genomic and functional genomics analyses of gluten proteins and prospect for simultaneous improvement of end-use and health-related traits in wheat. Theor Appl Genet 133:1521–1539. https://doi.org/10.1007/s00122-020-03557-5

Wen S, Wen N, Pang J et al (2012) Structural genes of wheat and barley 5-methylcytosine DNA glycosylases and their potential applications for human health. Proc Natl Acad Sci 109:20543–20548. https://doi.org/10.1073/pnas.1217927109

White B, Lux TM, Rusholme-Pilcher R et al (2024) De novo annotation of the wheat pan-genome reveals complexity and diversity within the hexaploid wheat pan-transcriptome. bioRxiv

Wicker T, Yahiaoui N, Guyot R et al (2003) Rapid genome divergence at orthologous low molecular weight Glutenin loci of the a and am genomes of wheat [W]. Plant Cell 15:1186–1197. https://doi.org/10.1105/tpc.011023

Wieser H (2007) Chemistry of gluten proteins. Food Microbiol 24:115–119. https://doi.org/10.1016/j.fm.2006.07.004

Wieser H, Bushuk W, MacRitchie F (2006) The polymeric glutenins. In: Wrigley C, Bekes F, Bushuk W (eds) Gliadin and Glutenin: the unique balance of wheat quality. St. Paul American Association of Cereal Chemistry, pp 213–240

Wrigley CW (1996) Giant proteins with flour power. Nature 381:738–739. https://doi.org/10.1038/381738a0

Xu J-H, Messing J (2009) Amplification of prolamin storage protein genes in different subfamilies of the Poaceae. Theor Appl Genet 119:1397–1412. https://doi.org/10.1007/s00122-009-1143-x

Zhang Y, Hu X, Islam S et al (2018) New insights into the evolution of wheat avenin-like proteins in wild emmer wheat (*Triticum dicoccoides*). Proc Natl Acad Sci 115:13312–13317. https://doi.org/10.1073/pnas.1812855115

Zhou Z, Liu C, Qin M et al (2022) Promoter DNA hypermethylation of *TaGli-γ-2.1* positively regulates gluten strength in bread wheat. J Adv Res 36:163–173. https://doi.org/10.1016/j.jare.2021.06.021

Zhu J, Fang L, Yu J et al (2018) 5-Azacytidine treatment and TaPBF-D over-expression increases glutenin accumulation within the wheat grain by hypomethylating the Glu-1 promoters. Theor Appl Genet 131:735–746. https://doi.org/10.1007/s00122-017-3032-z

Zimin AV, Puiu D, Hall R et al (2017a) The first near-complete assembly of the hexaploid bread wheat genome, *Triticum aestivum*. Gigascience 6. https://doi.org/10.1093/gigascience/gix097

Zimin AV, Puiu D, Luo M-C et al (2017b) Hybrid assembly of the large and highly repetitive genome of *Aegilops tauschii*, a progenitor of bread wheat, with the MaSuRCA mega-reads algorithm. Genome Res 27:787–792. https://doi.org/10.1101/gr.213405.116

The Potential of Genome Editing for Breeding Durum Wheat

Francesco Camerlengo

Abstract

The applications of genome editing (GE) technologies in wheat is constantly increasing, but less frequently applied to durum wheat compared to bread wheat. Durum wheat, which accounts for about 5–8% of the world wheat production, attracts less interest from breeders and researchers, likely due to the narrower diffusion of its cultivation. At the same time, it has good nutritional value and is adaptable to poor soils, therefore durum wheat could represent a source of genetic variability for introducing new traits and genes into bread wheat. Despite the challenges of applying GE technologies in wheat, which mostly arise from the transformation and regeneration efficiency and the high genomic redundancy due to polyploidy, many applications have been reported for yield enhancement, improvement of grain quality, and increased resilience to biotic/abiotic stress. Alternative methods, including CRISPR systems and delivery strategies, are being developed to overcome limitations such as slow protocols, genotype recalcitrance, and transgene integration in the host genome. The exploration of various CRISPR variants for different types of genetic modifications and fine regulation of gene expression allow a more precise and efficient site-directed mutagenesis of complex genomes, helping researchers and breeders modify homoeologous genes to produce specific phenotypes. GE, including CRISPR/Cas systems, offers the opportunity to deepen knowledge of the molecular mechanisms underlying wheat productivity, quality, and resilience to abiotic/biotic stress. Due to its accuracy and effectiveness, GE is deeply transforming research and breeding for agriculture with even more application for commercial crop/wheat improvement.

F. Camerlengo (✉)
Department of Agricultural and Food Sciences, Alma Mater Studiorum, University of Bologna, Bologna, Italy
e-mail: f.camerlengo@unibo.it

1 Introduction

Even though the cultivation of bread wheat (*Triticum aestivum*) is much more prevalent around the world, durum wheat (*Triticum turgidum*) is also a valuable species for human consumption and crop breeding. Durum wheat (tetraploid, $2n = 4x = 28$) and bread wheat (hexaploid, $2n = 6x = 42$) are two closely related species that can adapt differently to various environmental conditions. The hexaploid genome confers an evolutionary advantage particularly in terms of yield per hectare and a higher bread-making attitude, which may account for the greater spread of bread wheat, but durum wheat

R. Tuberosa et al. (eds.), *The Durum Wheat Genome*, Compendium of Plant Genomes,
https://doi.org/10.1007/978-3-031-84689-2_15

is more resilient than bread wheat to abiotic stress, therefore it is more suitable for poor marginal lands (Mastrangelo and Cattivelli 2021). From a breeding point of view, the durum wheat genome may also represent a source of genetic variability that can be used to introduce new traits and genes into bread wheat. Polyploidization, domestication, and breeding led to a genetic bottleneck of hexaploid wheat arising from few hybridization events and a restrictive human selection (Haudry et al. 2007). Tetraploid wheat such as durum may compensate a constrained gene pool of hexaploid wheat. In addition, durum wheat is receiving growing attention because it also represents a source of nutritional and healthy compounds like pigments scarcely present in bread wheat (Saini et al. 2023), besides its technological attitude for processing into a large variety of end products. Both species are polyploid and characterized by large genomes; this implies that most genes exist as two or more copies located on homoeologous chromosomes of each sub-genome. Sometimes, homoeologous genes are functionally redundant and therefore phenotypes controlled by recessive genes are rarely observed. However, homoeologous genes can have additive effects or can be preferentially expressed in different tissues. Controlling the action/influence of each gene to measure a phenotypic effect is very challenging and time-consuming to be achieved with traditional breeding methods. Genome editing (GE) technologies represent a toolbox that can help researcher and breeders to modify precisely homoeologous genes in order to produce a specific phenotype. GE technologies have been developed in the last two decades and include CRISPR/Cas (Clustered Regularly Interspaced Short Palindromic Repeats/CRISPR-associated protein) systems that rapidly became the most prevalent GE technology based on site-directed nucleases (SDNs) for specifically altering the genome of an organism. Furthermore, advancements in CRISPR systems and variants are adding new perspectives and achievements to model/shape plant genomes. To date, GE technologies have been successfully applied to most crops and plant species reducing the time needed to produce new genetic variability useful for functional genomics and breeding programs. Although GE technologies application in wheat is more challenging compared to other crops, many applications have been reported to date for yield enhancement, improvement of grain quality, and increased resilience to biotic/abiotic stress (Zhou et al. 2023; Ahmar et al. 2023) in bread wheat more than durum.

Different factors constrain successful GE application in wheat that mostly arise from the transformation and regeneration efficiency and the high genomic redundancy due to polyploidy. High wheat regeneration efficiencies depend on multiple factors that should be optimized on the basis of genotypes and transformation method (Hayta et al. 2021). The improvement in the efficiency of plant regeneration from tissue culture has been achieved through the expression of different plant morphological regulator genes in monocots species (Debernardi et al. 2020; Che et al. 2022; Johnson et al. 2023). In addition, new delivery strategies and transformation methods can overcome the limitations resulting from genotype dependence and tissue culture steps (Prestwich et al. 2024).

Due to its easiness, accuracy, and effectiveness, GE is deeply transforming research and breeding for agriculture with even more application for commercial crop improvement. In light of the findings and of the need to develop new cultivars to face the effects of climate change and the growing demand for food production, the development and the adoption/implementation of new breeding techniques such as GE along with the adoption of sustainable agronomic practices are required.

2 CRISPR-Cas Genome Editing

In contrast to the transgenic approach, GE technologies produce defined mutants directly targeting genomic regions in a DNA sequence-specific manner. The ability to introduce desired modifications in genomes is one of the major goals if not the most important to have been achieved in molecular biotechnology. Beginning with the discovery of meganucleases (endonucleases that recognize long stretches of DNA), researchers have been able to achieve targeted double-strand breaks (DSBs) in genomes for the first time. This discovery has led to a swift and continuous advancement in GE technologies based on site-directed nucleases, particularly over the past two decades. The progression of GE technologies was improved by the introduction of new tools that leverage the capacity of some eukaryotic proteins (zinc finger and transcription activator-like effector) to bind the DNA and rely on the ability of endonucleases to cut the double-strand helix. Two types of programmable nuclease proteins have been obtained by fusing the DNA cleavage domain of FokI endonuclease to DNA binding proteins: the zinc finger nucleases (ZFNs), and the transcription activator-like effector nuclease (TALEN). These two programmable site-specific nucleases are more flexible and easier to use compared to the previous meganuclease because they are able to recognize short DNA stretches and increase editing specificity (Adli 2018). Lastly, CRISPR-Cas systems were discovered as part of the adaptive immune system of bacteria and archaea. They allow the cleavage of invading DNA in a sequence-dependent manner, giving protection against bacterial phage and viruses. Generally, CRISPR-Cas systems are ribonucleoprotein (RNP) complexes that consist of a single-stranded RNA, which serves as a guide to identify the target location in the genome and therefore identified as guide RNA (gRNA), and of an RNA-guided endonuclease (Cas) able of cleaving DNA strands once reached the target site, thereby enabling the formation of DNA double-strand breaks (DSBs). CRISPR-Cas systems typically contain gene sequences for Cas proteins, a CRISPR array composed of "spacer" sequences inserted between short palindromic repeats and trans-activating CRISPR RNA (tracrRNA). The CRISPR array is processed into mature CRISPR RNA (crRNA) and often forms complexes with the tracrRNA that each have a single spacer conferring the specificity for the target sites on the foreign invading DNA and the binding affinity with Cas proteins. These enzymes are activated when combined with the tracrRNA/crRNA complex, then are guided to the target site and allowed to cleave the DNA. To distinguish between self and non-self target sequences, Cas proteins only recognize the complementary spacer sequence of the invading DNA that is adjacent to a protospacer adjacent motif (PAM) sequence (generally short DNA stretches of three to five nucleotides), which is not present in the CRISPR array. Notably, Jinek et al. (2012) engineered a single RNA chimera to mimic the ability of the tracrRNA/crRNA complex to direct the CRISPR-associated protein Cas9 to recognize and cleave specific DNA traits producing DSBs. Once a DSB is induced, different genome modifications can be achieved depending on the DNA repair pathways adopted by the cell, such as the non-homologous end-joining (NHEJ) mechanism or homologous recombination (HR) (Shrivastav et al. 2008). The NHEJ occurs more often than HR during DSBs repair. HR is restricted to late S/G2 phases in the cell cycle, whereas NHEJ functions throughout the entire cell cycle. Insertions or deletions (indels) are frequently caused by NHEJ and can result in frame-shift mutations and gene-knockouts. Alternatively, when a template with regions of homology to the sequence surrounding the DSB is available, the DSB can be repaired by HR; therefore, this mechanism can be exploited to introduce new DNA fragments in a specific location into genomes. Since then, engineered CRISPR systems allow site-directed mutagenesis of complex organisms relying on the ability to produce breaks into the DNA and to induce repair mechanisms to enable precise genetic modifications.

2.1 Natural and Engineered CRISPR Systems

Most of gene editing applications are performed by CRISPR-Cas9 but many other systems, naturally occurring in different organism, are continuously being discovered and engineered to be used as GE variants (Koonin et al. 2017). Based on Cas proteins and crRNA biogenesis and interference processes, CRISPR systems have been divided in two classes, and several types.

Primarily, class 1 CRISPR systems employ multi-Cas protein complexes for modifying DNA sequences whereas class 2 introduces targeted modification through one endonuclease protein (Koonin et al. 2017). To date, the most widely used as a genome editing tool is class 2 type II Cas9 from *Streptococcus pyogenes* (SpCas9), because of its high efficiency and its simple PAM sequence (5'-NGG-3′). Another class 2 highly efficient CRISPR variant, with many applications in crops like wheat (Li et al. 2023), is the Cas12a (also known as Cpf1). This variant has RNase III activity for crRNA biogenesis and its endonuclease generates DSBs with cohesive overhangs, compared with SpCas9, which produces blunt ends. The cohesive ends facilitate the insertion of a DNA fragment with complementary ends into the cleaved site using HR. The RNase III activity can be exploited for multiplex genome engineering via tandemly arrayed pre-crRNA-expressing constructs that produce multiple mature crRNAs processed by Cpf1 (Koonin et al. 2017). The Cas13 family, also belonging to class 2 system, consists of RNA-guided RNases able to cleave single-strand RNAs. The Cas13b has two RNase domains and causes RNA cleavage as well as secondary RNA degradation of minor RNA. Therefore, after recognition and cleavage of a target transcript, it leads to non-specific degradation of any nearby transcripts regardless of the spacer (Koonin et al. 2017). The recent progress in understanding CRISPR systems and Cas proteins along with the necessity to enhance editing precision, targeting fidelity, and packaging for delivering CRISPR components into different tissues and cell type, allowed to extend CRISPR tool capabilities via the re-engineering of Cas proteins. Introduction of point mutation in RuvC and HNH nuclease domains of Cas9, inactivating catalytic activity, can be exploited to modulate Cas activity. A point mutation in one of the two catalytic domains results in a nickase (nCas9) enzyme (D10A and H840A SpCas9) that combined with paired gRNAs can improve cleavage specificity. Simultaneous point mutation in both domains results in catalytically inactive dead Cas9 (dCas9) with complete loss of DNA cleavage activity. Dead Cas9, with point mutations in both of the nuclease domains, can be recruited by gRNAs but cannot cleave the target site. Various functional protein domains can be loaded on dCas9 and exploited for different applications; fusing dCas9 with transcription activation or repression domains enables gene regulation, and its ability to bind DNA without cutting can be used for gene tagging (Boti et al. 2023).

Two variants of SpCas9 were engineered and used in wheat to recognize the 5'-NG- 3' PAM instead of 5'-NGG- 3′ expanding the range of editable targets of the mostly used endonuclease (Wang et al. 2021). Walton et al. (1979) developed an engineered variant named SpRy targeting the genomic DNA independently from PAM restriction.

It has been shown that the development and application of new CRISPR systems recognizing distinct PAMs, employing diverse endonucleases or having different gRNAs structure and processing features, expands the range of editable loci in wheat genome potentially allowing any/every possible genomic applications (Wang et al. 2021).

2.2 Base Editing and Prime Editing

All these CRISPR variants are continually being explored in order to achieve different types of modification or to finely regulate gene expression. Most applications are aimed to knockout genes, but CRISPR may be used also for the insertion of new DNA sequences, up- or down-regulation of gene expression, DNA/RNA tagging, and base editing (BE). The latest advancements in CRISPR system have expanded the frontiers of GE in plants enabling targeted nucleotide modifications through base editing (BE) and reverse transcription of RNA sequences into plant genome using prime editing (PE).

BE enables to convert a single base into another without DSBs and DNA donor templates in order to alter one or several amino acids of the codified protein (Komor et al. 2016; Gaudelli et al. 2017). These systems are characterized by base editors composed of a DNA deaminase (cytosine or adenosine deaminase) fused to a nickase endonuclease (more frequently Cas9 of Cas12a) that produce only single-strand breaks. One or more gRNAs guide the Cas-deaminase fusion to the target site, where a single-stranded DNA R-loop is formed allowing access for the DNA deaminase (Anzalone et al. 2019, 2020). The base editors mostly used are cytosine base editors (CBEs), which catalyze the conversion of C/G base pairs to T/A base pairs; and adenine base editors (ABEs), which catalyze A/T-to-G/C conversions (Komor et al. 2016; Gaudelli et al. 2017). Although ABEs and CBEs are difficult to apply in polyploid crops due to the presence of homoeologous gene copies, improvements in base editing efficiency of DNA deaminase enzymes fostered their use for targeted point mutations at multiple endogenous loci in wheat (Zong et al. 2017; Li et al. 2018; Han et al. 2022).

Genome editing technologies that directly write new genetic information into a specified DNA site along the genomes is called prime editing and was introduced in 2019 by Anzalone et al. (Lin et al. 2020) who developed the technology in mammalian cells. This system consists of a nCas9 endonuclease fused to an engineered reverse transcriptase (RT), programmed with a prime editing guide RNA (pegRNA) that contains the spacer for the target site and the RNA template for the reverse transcriptase. The complex binds the target DNA, cuts only one strand, and promotes the reverse transcription of the RNA template into the target genomic locus. Prime editing in plants is affected by pegRNA primer-binding site length, RT template length, and nicking sgRNA position. Optimization of these factors increased the average efficiency of prime editing in wheat (Lin et al. 2020; Jin et al. 2021). Enhanced frequency of editing events at genomic sites in rice and wheat has been obtained using and modifying diverse RT enzymes and introducing nucleotide substitutions into the RT template. Jin et al. (2021) created a web-based tool (PlantPegDesigner) to design dual-pegRNAs and engineer prime editors with improved editing performance for prime editing in monocot plants. However, introducing DNA sequences into a target genomic locus is still limited by the size of the DNA fragment to be introduced. Recently, to overcome constraints about the length of introduced DNA, Sun et al. (2023) developed a new genome editing approach, named PrimeRoot, to insert DNA sequences precisely into the plant genome without DSB intermediates, combining prime editing with the highly active Cre-Lox66/Lox71 or FLP-FRT1 site-specific recombinase systems. PrimeRoot requires the use of prime editor to introduce short recombinase sites into the target genomic locus that are recognized and used by a recombinase to introduce a new DNA sequence from a donor template.

These engineered CRISPR systems have significantly broadened the application of CRISPR-Cas technologies, but they still require optimization for wheat.

3 Delivery Strategies of CRISPR Components

Diverse strategies have been developed to deliver CRISPR system components into plants through physical, chemical, biological, and in vivo methods. Traditional approaches including *Agrobacterium*-mediated transformation and biolistic/particle bombardment have been and are still being successfully used for the transformation of wheat explants. Otherwise, some alternative approaches are continuously developed mostly to overcome barriers for faster protocols, and avoid genotype recalcitrance and transgene integration in the host genome (Prestwich et al. 2024). These limitations often affect the transformation of monocotyledonous crops, especially wheat. The selection of a delivery method, such as *Agrobacterium*-mediated, biolistic, RNA virus-based, or *in planta*-based, depends on several factors. These include the species and the type of explant used for transformation, the specific CRISPR system that needs to be delivered, and whether a stable or transient transformation is required. Much progress has been made in the improvement of the transformation efficiency through the implementation of plasmid vectors with growth regulating factors and through the optimization of in vitro regeneration protocols. Recently, Debernardi et al. (2020) used a sequence encoding a fusion protein combining the growth-regulating factor 4 (GRF4) and its cofactor GRF-interacting factor 1 (GIF1) to increase the efficiency and the speed of regeneration in wheat, triticale, and rye. They also assessed the effect of GRF4–GIF1 chimera on different durum wheat cultivars obtaining a significant increase of regeneration frequency even for more recalcitrant genotypes. To demonstrate the capacity of this system to extend genome editing technologies to crops with low regeneration efficiency, a vector containing the GRF4–GIF1, Cas9 gene, and a gRNA targeting the wheat gene Q was successfully used for the *Agrobacterium*-mediated transformation of durum wheat immature embryos. Immature embryos are most often used as explants for genetic transformation of wheat through both *Agrobacterium*-mediated and biolistic transformation but the preparation of samples to be transformed requires laborious procedures and in vitro regeneration. The particle bombardment method was also used for "*in planta*" transformation of shoot apical meristem (SAM) (Hamada et al. 2018; Liu et al. 2021). The method developed by Hamada et al. (2018), named *in planta* particle bombardment (iPB), requires no tissue culture steps because gold particles coated with plasmid expressing CRISPR components are bombarded into the SAM exposed from dry imbibed seeds. The iPB method is also suitable for ribonucleoprotein (RNP) delivery intended to transiently express CRISPR components without transgene integration (Kumagai et al. 2022).

Recently, among biological methods, plant RNA virus-based vector systems are gaining more and more attention for the delivery of CRISPR components into plant overcoming genotype dependence. Direct inoculation of wheat leaves with barley stripe mosaic virus-based sgRNA delivery vector (BSMV-sg) was used to deliver gRNAs into Cas9-expressing wheat plants (Li et al. 2021b; Wang et al. 2022) in order to generate heritable gene editing. This system relies on the ability of viruses to spread across the plant cells and to deliver a DNA-based viral replicon containing gRNA for target sites.

To improve the capacity of editing multiple alleles at once in polyploid species, double-haploid technologies and haploid inducer are being used for editing wheat (Karmacharya et al. 2023; Goyal et al. 2024). These methods reduce the number of alleles to be edited by half and enable to avoid genotype dependency. Embryogenic microspores can easily be isolated in a relatively short time and used as explants for transformation. After transformation by transfection, the microspores are screened for gene editing events and then homozygous plants are generated after diploidization (Bhowmik et al. 2018). The microspore-based haploid strategy greatly enhances the efficiency of GE to produce homozygous mutants at the first generation. The use of intergeneric crosses between different crop species offers an alternative way to deliver CRISPR component into every potential genotype (Kelliher et al. 2019). Transgenic maize expressing *Cas9*, and gRNAs were used for intergenic crosses with durum and bread wheat cultivars (Budhagatapalli et al. 2020); mutations were successfully introduced in two key wheat genes, BRASSINOSTEROID-INSENSITIVE 1 (BRI1) and SEMI-DWARF 1 (SD1) which are involved in the regulation of plant height. After crosses, embryos were rescued in vitro and edited haploid plants selected. Once the zygote has been formed, the expression of *Cas9* and gRNAs derived from maize sperm cells allow mutations of target sites to be induced during mitosis. In the course of initial embryonic cell divisions, maize chromosomes are rapidly lost, leading to produce haploid wheat embryos. Lastly, haploid edited plants were subjected to double-haploid induction through colchicine treatment (Laurie and Bennett 1988). This method that couples exploiting haploid induction, targeted mutagenesis, and double-haploid production may overcome genotype dependence of wheat transformation and regeneration. Other delivery methods need to be developed and implemented in wheat. Nanomaterials have emerged as a promising candidate for delivery of nucleic acids and proteins into intact plant cells (Prestwich et al. 2024).

The chance to avoid the integration of vector DNA or portions of it harboring Cas and gRNAs sequences into plant genome makes CRISPR editing even more attractive especially for producing genome edited crops to be commercialized. Using classical transformation methods such as biolistic or *Agrobacterium*-mediated transformation, the integration of CRISPR components into the plant genome is highly probable. This elevates the risk of inducing undesired genetic alterations, with transgene integration and off-target mutation being the most significant ones. In addition, the legislation of some countries, among which those belonging to the European Union, identifies edited crops with exogenous DNA integration as unmarketable. The use of CRISPR/Cas9 ribonucleoproteins (RNPs) was one of the first approaches explored to this end. It relies on the use of pre-assembled RNPs (Cas9 and gRNA) delivered to plant cell and rapidly degraded. Because no foreign DNA is used in CRISPR/Cas9 RNP-mediated genome editing, the mutants obtained are completely transgene free (Liang et al. 2017). Editing wheat without integration of foreign DNA could be achieved also transiently expressing CRISPR/Cas9 components. This transient expression-based, genome-editing system, in the absence of herbicide or antibiotic selection, is able to produce transgene-free and homozygous mutants from the first generation after transformation in both bread and durum wheat (Zhang et al. 2016; Camerlengo et al. 2020). Otherwise, once an editing event is obtained, the CRISPR components should be segregated out through generations by back-crossing edited plants with the corresponding wild-type or elite cultivars, thereby obtaining transgene-free progeny with desired modification.

4 Applications of Genome Editing in Durum Wheat

GE technologies could assist breeders in achieving breeding objectives faster and more efficiently. In the perspective of feeding the developing world population and maintaining optimal yields even under adverse environmental conditions, enhancing crop yield remains one of the major traits to be improved. Specifically for wheat, the optimal yields should be accompanied by high technological quality and nutritional value of derived food. This would lead to produce end-products with improved health benefits while also reducing toxic, immunogenic, and antinutritional compounds.

Although the use of GE in wheat is steadily on the rise, these technologies are less frequently applied to durum wheat compared to bread wheat (Table 1), probably due its narrow cultivation that accounts for only 5–8% of the global wheat production, attracting less interest from breeders and researchers. This notwithstanding, its cultivation could be expanded to marginal lands using GE to modify genetic traits for broader adaptability and the possible genetic flow from tetraploid to hexaploid wheat would boost the improvement of bread wheat.

Functional genomic studies make large use of GE technologies, such as CRISPR-Cas9, to validate gene functions and to regulate metabolic pathways. However, these technologies provide breeders and researchers more options to address the development of new and improved crops to meet current environmental challenges (Nigro et al. 2024).

4.1 Genome Editing Contribution to Address New Food Challenges (Yield and Quality)

Enhancing grain yield continues to be a primary objective in crop breeding, especially durum wheat. This goal is closely tied to both the agricultural performance of the crops and the environmental conditions where they are grown. Nowadays, improving the yield of the major crops and adapting them to limited environmental conditions is essential to face the growing global demand for food. GE technologies are helping in the understanding of the molecular mechanisms at the basis of wheat productivity, with a focus on grain morphology, spike/floret fertility, and on a more efficient use of resources. Improving grain yield is challenging because it is affected by several quantitative traits and geno-

Table 1 Genome editing applications in durum wheat

Variety	Target gene(s)	Phenotypic traits	CRISPR/ system	Delivery method	Explant for transformation	Type of editing	References
D6 D7	*BRI1 SD1*	Plant height	CRISPR-Cas9	Pollination by cas9/guide RNA-transgenic maize and haploid induction	Floret	IN/DEL	[Budhagatapalli et al. 2020]
Shimai11 Yumai4	*GASR7*	Grain length and weight	CRISPR-Cas9	Biolistic	Embryo	IN/DEL	[Zhang et al. 2016]
Svevo	*CM3 - CM16*	Reduction of immunogenic potential	CRISPR-Cas9	Biolistic	Embryo	IN/DEL	[Camerlengo et al. 2020]
Kronos	*WAPO-A1*	Spikelet number	CRISPR-Cas9	Agrobacterium-mediated	Embryo	DEL	[Kuzay et al. 2022]
Don Pedro	*α-Gliadins*	Reduction of immunogenic potential	CRISPR-Cas9	Biolistic	Embryo	DEL	[Sánchez-León et al. 2018]
JIN8	*LFL2*	Interploidy hybridization	CRISPR-Cas9	Agrobacterium-mediated	Embryo	IN/DEL	[Yang et al. 2023]
Kronos	*ZIP4*	Homoeologous recombination	CRISPR-Cas9	Agrobacterium-mediated	Embryo	DEL	[Draeger et al. 2023]

type by environment interactions (Maccaferri et al. 2019). Spikelet number per spike (SNS) is considered one of the yield components with high heritability that can be improved. A significant and stable quantitative trait locus (QTL) for SNS was identified on chromosome 7AL (Kuzay et al. 2019). Among the genes within the QTL interval, the wheat ortholog of *APO1* (*WAPO1*) was identified as the causal gene underpinning this QTL. This was confirmed through functional validation obtained producing loss-of-function mutants and overexpressing transgenic lines. Kuzay et al. (2022) generated through CRISPR-Cas9 single and double mutants of *WAPO1* homeologs in durum wheat without the stable integration of CRISPR-Cas9 vector components detecting highly significant differences for SNS.

Grain size (both length and width) and weight influence wheat yield and are controlled by several genes, among which is *TaGASR7* (Dong et al. 2014). Zhang et al. (2016) introduced mutation in the *TtGASR7* homoeologous genes of two different durum and bread wheat varieties transiently expressing CRISPR/Cas9 components. The edited lines exhibited significantly elevated thousand kernel weight, because *GASR7* acts as a negative regulator of grain weight.

While GE in wheat has been used to target several significant agronomic traits, the primary goal of most wheat breeding programs is to develop new cultivars that not only have high yields but also high nutritional value and desirable technological properties of grain and flours (Langridge et al. 2022). The importance of nutrition and health factors has increased lately because many diseases have been linked to eating foods made from wheat. By altering the biosynthesis of protein, starch, and lipids, gene editing tools can improve the chemical makeup of wheat kernels. This also enables the production of wheat with better nutritional value, where minor components have a significant impact. Foods made from wheat are often associated with pathologies such as Celiac Disease (CD), allergies, and Non-Celiac Wheat Sensitivity (NCWS) (Scherf 2019). Prolamins, glutenins, and gliadins are proteins in the wheat grain endosperm that form the gluten matrix. They are the main causes of the onset of CD. Sánchez-León et al. (2018) applied the CRISPR/Cas9 technique to lower the amount of α-gliadins in durum and bread wheat kernels. They targeted shared regions among the gene family members, as α-gliadins are a large group of proteins with high similarity in their sequences. Otherwise, they observed that the other classes of gliadins (ω- and γ-) were also affected, but the mutant lines had less immunogenic peptides generally more prevalent in the α-gliadins. Certain protein families in the endosperm, such as α-amylase/trypsin inhibitors (ATIs), play a role in plant defense mechanisms and can trigger wheat allergies and NCWS in predisposed patients. In 2020, Camerlengo et al. (2020) employed CRISPR-Cas9 for the multiplex editing of genes encoding the ATI family members WTAI-CM3 and WTAI-CM16 in the elite durum wheat cultivar Svevo, producing wheat lines with a decreased quantity of ATIs and free of transgenes. The modified lines were entirely devoid of the target ATI subunits, which led to a substantial reduction in their potential to cause allergies. Also, the processing of wheat flour can create harmful and unhealthy substances. During the high-temperature processing of foods made from wheat flour, free asparagine transforms into acrylamide, a substance known to cause cancer. In 2021, Raffan et al. (2021) deactivated the asparagine synthase gene, *TaASN2*, in bread wheat varieties, which resulted in a decreased level of free asparagine in the grain. Zhang et al. (2018) reported an increase in the SDS (Sodium dodecyl sulfate) sedimentation volume of bread wheat lines in which they edited *TaGW2* homoeologous genes. Grain protein content and flour protein content were considerably higher in all the mutant lines, particularly in AAbbDD and AAbbdd genotypes. Glutenins and gliadins were also increased. Otherwise, they observed that the grains of double and triple mutants were morphologically wrinkled compared to the control plants and single mutants.

Many other modifications of chemical composition of the kernel determining technological and rheological quality have been obtained in bread wheat so far (Abe et al. 2019; Li et al.

2021a; Sun et al. 2022), but there are still few examples in durum wheat.

4.2 Genome Editing for Resilience to Climate Change (Stress and Genetic Flow)

Conventional farming practices depend on the use of chemical substances to prevent crop losses caused by pest infections. However, the widespread use of these chemicals can have serious implications for both human health and the environment. So far, creating wheat varieties resistant to pest diseases has been a significant objective in wheat breeding initiatives. To date, GE technologies have been applied to modify several genes in wheat to enhance its resistance to fungal infections. In 2014, Shan et al. (2014) were the first to report a successful experiment for wheat resistance to fungal infection using the CRISPR-Cas system, where they modified the *MLO* gene that produces the MILDEW-RESISTANCE LOCUS (MLO) protein. *MLO* causes susceptibility to powdery mildew in plants, therefore knockout of the *TaMLO* gene leads to improved powdery mildew tolerance. Several studies have targeted resistance to *Fusarium gramineaurm*, a fungus that causes one of the most harmful diseases in wheat. The lipoxygenase genes, *TaLpx* or *TaLox*, have been identified to be involved in wheat susceptibility to *Fusarium graminearum*. These enzymes are crucial in plant defense responses mediated by jasmonic acid, as they catalyze the hydrolysis of polyunsaturated fatty acids and trigger oxylipins biosynthesis. Zhang et al. (2016) produced durum and bread wheat plants with a mutated *TaLOX* at a frequency of 9.5%, with 44.7% of these being homozygous mutants. These plants showed a slight tolerance to *Fusarium graminearum* infection of spikes. Further gene targets for enhancing wheat resistance to *Fusarium gramineaurm* were edited in bread wheat but they have not yet been tested in durum wheat (Su et al. 2019; Brauer et al. 2020).

Factors like drought, extreme temperatures, and salinity could dramatically impact on crop performance (Kole et al. 2015).

GE technologies have been barely applied to target genes regulating cellular and molecular mechanisms of wheat responses to enhance abiotic stress resilience. Few findings indicate that the use of GE holds immense promise for altering physiological and morphological traits related to source-sink ratio resulting from water and resource assimilation. Knockout of *TaSal1* homoeologous genes perform better on media containing polyethylene glycol than wild-type plants (Abdallah et al. 2022). Drought resistance is also correlated with root system architecture that can be modified to improve uptake of nutrients and water required for the growth of the plant in different soil conditions. An *in planta Agrobacterium*-mediated CRISPR/Cas9 system was adopted to generate targeted mutations in *TaRPK1* homoeologous genes increasing root morphological traits such as length, depth, volume, and surface area (Abdul Rahim et al. 2024).

Such studies have demonstrated the potential of the application of GE to produce valuable traits for the genetic improvement not only of bread wheat but also of durum wheat. Continued research and breeding initiatives are essential to gain a deeper understanding of the genetic and physiological mechanisms of metabolic and signaling pathways involved in stress tolerance mechanisms (Trono and Pecchioni 2022). Identifying the gene regulatory networks that play a role in stress response and targeting them with gene editing tools will pave the way for the creation of new varieties that are both stress-resistant and high-yielding.

Durum wheat could be used to increase genetic variability of hexaploid wheat through the introgression of genes and/or genomic fragments that gain the improvement of grain quality, agronomic performance, and biotic/abiotic stress resilience (Mazzucotelli et al. 2024). During the history of co-evolution of tetraploid and hexaploid wheats, genomic introgressions through hybridization occurred mainly from tetraploids into hexaploids in which hexaploid wheat served as the maternal parent. Breeders employed maternal-excess hybridization/crosses (in which the female parent is the hexaploid wheat) more than paternal-excess hybridization/crosses (in which the female parent is the tetraploid wheat) to avoid poor seed set and low seed germination rates mostly due to deregulation of genomic imprinting that leads to dysfunctional endosperm. Otherwise, the inheritance of the tetraploid genome is lower in the progenesis of maternal- than paternal-excess crosses limiting the utilization efficiency of maternal-excess crosses for breeding programs. Yang et al. (2023) provided insights into the molecular mechanisms underlying the interploidy block of paternal-excess crosses during the history of wheat evolution. They found candidate genes for improving the utilization efficiency of paternal-excess crosses in wheat breeding programs; the endosperm-specific transcription factor gene *TaLFL2* (usually paternally expressed) contributes to paternal-excess interploidy hybridization barriers in wheat. The loss of function of this gene obtained through CRISPR-Cas9 decreases the expression of the *TaNAC019* nutrient accumulation regulator negatively altering the endosperm development as usually happens in paternal-excess crosses. These represent genomic targets for improving the efficiency of paternal-excess crosses in wheat breeding programs.

The homoeologous sub-genomes of bread and durum wheat are closely related sharing A and B sub-genomes with a similar gene content and order. Few *loci* have been reported to stabilize polyploid genomes during meiosis promoting cytologically diploid behavior of both tetraploid and hexaploidy wheat. Crossing over and recombination only occur between homologous chromosomes within each sub-genome, ensuring genome stability and fertility. In both tetraploid and hexaploid wheat, the major meiotic gene *ZIP4-B2* (*Ph1*) on chromosome 5B (Martín et al. 2017) promotes crossing over between homologous chromosomes and suppresses crossing over between homoeologous chromosomes. In addition to *ZIP4-B2*, there are further *ZIP4* alleles on group 3 chromosomes (*ZIP4-A1* and *ZIP4-B1*). Draeger et al. (2023) produced single, double, and triple *Ttzip4* through TILLING and CRISPR-Cas9 to determine the effect of each *ZIP4* gene on cross-over formation in the tetraploid wheat cultivar "Kronos." Their findings suggest that controlling *ZIP4* genes in tetraploid wheat could be useful in breeding programs for the introgression of genes from wild relatives and related species into elite wheat cultivars.

References

Abdallah NA, Elsharawy H, Abulela HA et al (2022) Multiplex CRISPR/Cas9-mediated genome editing to address drought tolerance in wheat. GM Crops Food 16. https://doi.org/10.1080/21645698.2022.2120313

Abdul Rahim A, Uzair M, Rehman N et al (2024) CRISPR/Cas9 mediated TaRPK1 root architecture gene mutagenesis confers enhanced wheat yield. J King Saud Univ Sci 36. https://doi.org/10.1016/j.jksus.2023.103063

Abe F, Haque E, Hisano H et al (2019) Genome-edited triple-recessive mutation alters seed dormancy in wheat. Cell Rep 28:1362–1369.e4. https://doi.org/10.1016/j.celrep.2019.06.090

Adli M (2018) The CRISPR tool kit for genome editing and beyond. Nat Commun 9:1911

Ahmar S, Hensel G, Gruszka D (2023) CRISPR/Cas9-mediated genome editing techniques and new breeding strategies in cereals—current status, improvements, and perspectives. Biotechnol Adv 69:108248

Anzalone AV, Randolph PB, Davis JR et al (2019) Search-and-replace genome editing without double-strand breaks or donor DNA. Nature 576:149–157. https://doi.org/10.1038/s41586-019-1711-4

Anzalone AV, Koblan LW, Liu DR (2020) Genome editing with CRISPR–Cas nucleases, base editors, transposases and prime editors. Nat Biotechnol 38:824–844

Bhowmik P, Ellison E, Polley B et al (2018) Targeted mutagenesis in wheat microspores using CRISPR/Cas9. Sci Rep 8. https://doi.org/10.1038/s41598-018-24690-8

Boti MA, Athanasopoulou K, Adamopoulos PG et al (2023) Recent advances in genome-engineering strategies. Genes (Basel) 14

Brauer EK, Balcerzak M, Rocheleau H et al (2020) Genome editing of a Deoxynivalenol-induced transcription factor confers resistance to Fusarium graminearum in wheat. Mol Plant-Microbe Interact 33:553–560. https://doi.org/10.1094/MPMI-11-19-0332-R

Budhagatapalli N, Halbach T, Hiekel S et al (2020) Site-directed mutagenesis in bread and durum wheat via pollination by cas9/guide RNA-transgenic maize used as haploidy inducer. Plant Biotechnol J 18:2376–2378. https://doi.org/10.1111/pbi.13415

Camerlengo F, Frittelli A, Sparks C et al (2020) CRISPR-Cas9 multiplex editing of the α-amylase/trypsin inhibitor genes to reduce allergen proteins in durum wheat. Front Sustain Food Syst 4. https://doi.org/10.3389/fsufs.2020.00104

Che P, Wu E, Simon MK et al (2022) *Wuschel2* enables highly efficient CRISPR/Cas-targeted genome editing during rapid de novo shoot regeneration in sorghum. Commun Biol 5. https://doi.org/10.1038/s42003-022-03308-w

Debernardi JM, Tricoli DM, Ercoli MF et al (2020) A GRF–GIF chimeric protein improves the regeneration efficiency of transgenic plants. Nat Biotechnol 38:1274–1279. https://doi.org/10.1038/s41587-020-0703-0

Dong L, Wang F, Liu T et al (2014) Natural variation of *TaGASR7-A1* affects grain length in common wheat under multiple cultivation conditions. Mol Breed 34:937–947. https://doi.org/10.1007/s11032-014-0087-2

Draeger TN, Rey MD, Hayta S et al (2023) *ZIP4* is required for normal progression of synapsis and for over 95% of crossovers in wheat meiosis. Front Plant Sci 14. https://doi.org/10.3389/fpls.2023.1189998

Gaudelli NM, Komor AC, Rees HA et al (2017) Programmable base editing of T to G C in genomic DNA without DNA cleavage. Nature 551:464–471. https://doi.org/10.1038/nature24644

Goyal L, Kaur M, Mandal M et al (2024) Potential gene editing targets for developing haploid inducer stocks in rice and wheat with high haploid induction frequency. 3 Biotech 14. https://doi.org/10.1007/s13205-023-03857-9

Hamada H, Liu Y, Nagira Y et al (2018) Biolistic-delivery-based transient CRISPR/Cas9 expression enables in planta genome editing in wheat. Sci Rep 8. https://doi.org/10.1038/s41598-018-32714-6

Han H, Wu Z, Zheng L et al (2022) Generation of a high-efficiency adenine base editor with TadA8e for developing wheat dinitroaniline-resistant germplasm. Crop J 10:368–374. https://doi.org/10.1016/j.cj.2021.08.006

Haudry A, Cenci A, Ravel C et al (2007) Grinding up wheat: a massive loss of nucleotide diversity since domestication. Mol Biol Evol 24:1506–1517. https://doi.org/10.1093/molbev/msm077

Hayta S, Smedley MA, Clarke M et al (2021) An efficient agrobacterium-mediated transformation protocol for Hexaploid and tetraploid wheat. Curr Protoc 1. https://doi.org/10.1002/cpz1.58

Jin S, Lin Q, Luo Y et al (2021) Genome-wide specificity of prime editors in plants. Nat Biotechnol 39:1292–1299. https://doi.org/10.1038/s41587-021-00891-x

Jinek M, Chylinski K, Fonfara I et al (2012) A Programmable Dual-RNA-Guided DNA Endonuclease in Adaptive Bacterial Immunity

Johnson K, Cao Chu U, Anthony G et al (2023) Rapid and highly efficient morphogenic gene-mediated hexaploid wheat transformation. Front Plant Sci 14. https://doi.org/10.3389/fpls.2023.1151762

Karmacharya A, Li D, Leng Y et al (2023) Targeting disease susceptibility genes in wheat through wide hybridization with maize expressing Cas9 and guide RNA. Mol Plant-Microbe Interact 36:554–557. https://doi.org/10.1094/MPMI-01-23-0004-SC

Kelliher T, Starr D, Su X et al (2019) One-step genome editing of elite crop germplasm during haploid induction. Nat Biotechnol 37:287–292. https://doi.org/10.1038/s41587-019-0038-x

Kole C, Muthamilarasan M, Henry R et al (2015) Application of genomics-assisted breeding for generation of climate resilient crops: progress and prospects. Front Plant Sci 6:563

Komor AC, Kim YB, Packer MS et al (2016) Programmable editing of a target base in genomic DNA without double-stranded DNA cleavage. Nature 533:420–424. https://doi.org/10.1038/nature17946

Koonin EV, Makarova KS, Zhang F (2017) Diversity, classification and evolution of CRISPR-Cas systems. Curr Opin Microbiol 37:67–78

Kumagai Y, Liu Y, Hamada H et al (2022) Introduction of a second "green revolution" mutation into wheat via in planta CRISPR/Cas9 delivery. Plant Physiol 188:1838–1842

Kuzay S, Xu Y, Zhang J et al (2019) Identification of a candidate gene for a QTL for spikelet number per spike on wheat chromosome arm 7AL by high-resolution genetic mapping. Theor Appl Genet 132:2689–2705. https://doi.org/10.1007/s00122-019-03382-5

Kuzay S, Lin H, Li C et al (2022) ++ is the causal gene of the 7AL QTL for spikelet number per spike in wheat. PLoS Genet 18. https://doi.org/10.1371/journal.pgen.1009747

Langridge P, Alaux M, Almeida NF et al (2022) Meeting the challenges facing wheat production: the strategic research agenda of the global wheat initiative. Agronomy 12(11):2767

Laurie DA, Bennett MD (1988) The production of haploid wheat plants from wheat • maize crosses

Li C, Zong Y, Wang Y et al (2018) Expanded base editing in rice and wheat using a Cas9-adenosine deaminase fusion. Genome Biol 19. https://doi.org/10.1186/s13059-018-1443-z

Li J, Jiao G, Sun Y et al (2021a) Modification of starch composition, structure and properties through editing of *TaSBEIIa* in both winter and spring wheat varieties by CRISPR/Cas9. Plant Biotechnol J 19:937–951. https://doi.org/10.1111/pbi.13519

Li T, Hu J, Sun Y et al (2021b) Highly efficient heritable genome editing in wheat using an RNA virus and bypassing tissue culture. Mol Plant 14:1787–1798. https://doi.org/10.1016/j.molp.2021.07.010

Li C, Chu W, Gill RA et al (2023) Computational tools and resources for CRISPR/Cas genome editing. Genomics Proteomics Bioinformatics 21:108–126

Liang Z, Chen K, Li T et al (2017) Efficient DNA-free genome editing of bread wheat using CRISPR/Cas9 ribonucleoprotein complexes. Nat Commun 8. https://doi.org/10.1038/ncomms14261

Lin Q, Zong Y, Xue C et al (2020) Prime genome editing in rice and wheat. Nat Biotechnol 38:582–585. https://doi.org/10.1038/s41587-020-0455-x

Liu Y, Luo W, Linghu Q et al (2021) In planta genome editing in commercial wheat varieties. Front Plant Sci 12. https://doi.org/10.3389/fpls.2021.648841

Maccaferri M, Harris NS, Twardziok SO et al (2019) Durum wheat genome highlights past domestica-

tion signatures and future improvement targets. Nat Genet 51:885–895. https://doi.org/10.1038/s41588-019-0381-3
Martín AC, Rey MD, Shaw P, Moore G (2017) Dual effect of the wheat *Ph1* locus on chromosome synapsis and crossover. Chromosoma 126:669–680. https://doi.org/10.1007/s00412-017-0630-0
Mastrangelo AM, Cattivelli L (2021) What makes bread and durum wheat different? Trends Plant Sci 26:677–684
Mazzucotelli E, Mastrangelo AM, Desiderio F et al (2024) Gene flow between tetraploid and Hexaploid wheat for breeding innovation. Wheat Genome:135–163
Nigro D, Smedley MA, Camerlengo F, Hayta S (2024) Using gene editing strategies for wheat improvement. In: A roadmap for plant genome editing. Springer Nature Switzerland, pp 183–201
Prestwich BD, Cardi T, Bakhsh A et al (2024) Novel delivery methods for CRISPR-based plant genome editing. In: A roadmap for plant genome editing. Springer Nature Switzerland, pp 41–67
Raffan S, Sparks C, Huttly A et al (2021) Wheat with greatly reduced accumulation of free asparagine in the grain, produced by CRISPR/Cas9 editing of asparagine synthetase gene *TaASN2*. Plant Biotechnol J 19:1602–1613. https://doi.org/10.1111/pbi.13573
Saini P, Kaur H, Tyagi V et al (2023) Nutritional value and end-use quality of durum wheat. Cereal Res Commun 51:283–294
Sánchez-León S, Gil-Humanes J, Ozuna CV et al (2018) Low-gluten, nontransgenic wheat engineered with CRISPR/Cas9. Plant Biotechnol J 16:902–910. https://doi.org/10.1111/pbi.12837
Scherf KA (2019) Immunoreactive cereal proteins in wheat allergy, non-celiac gluten/wheat sensitivity (NCGS) and celiac disease. Curr Opin Food Sci 25:35–41
Shan Q, Wang Y, Li J, Gao C (2014) Genome editing in rice and wheat using the CRISPR/Cas system. Nat Protoc 9:2395–2410. https://doi.org/10.1038/nprot.2014.157
Shrivastav M, De Haro LP, Nickoloff JA (2008) Regulation of DNA double-strand break repair pathway choice. Cell Res 18:134–147. https://doi.org/10.1038/cr.2007.111
Su Z, Bernardo A, Tian B et al (2019) A deletion mutation in TaHRC confers Fhb1 resistance to Fusarium head blight in wheat. Nat Genet 51:1099–1105. https://doi.org/10.1038/s41588-019-0425-8
Sun Z, Zhang M, An Y et al (2022) CRISPR/Cas9-mediated disruption of Xylanase inhibitor protein (XIP) gene improved the dough quality of common wheat. Front Plant Sci 13. https://doi.org/10.3389/fpls.2022.811668
Sun C, Lei Y, Li B et al (2023) Precise integration of large DNA sequences in plant genomes using PrimeRoot editors. Nat Biotechnol 42. https://doi.org/10.1038/s41587-023-01769-w
Trono D, Pecchioni N (2022) Candidate genes associated with abiotic stress response in plants as tools to engineer tolerance to drought, salinity and extreme temperatures in wheat: an overview. Plants 11
Walton RT, Christie KA, Whittaker MN (1979) Kleinstiver BP (2020) unconstrained genome targeting with near-PAMless engineered CRISPR-Cas9 variants. Science 368:290–296. https://doi.org/10.1126/science.aba8853
Wang W, Tian B, Pan Q et al (2021) Expanding the range of editable targets in the wheat genome using the variants of the Cas12a and Cas9 nucleases. Plant Biotechnol J 19:2428–2441. https://doi.org/10.1111/pbi.13669
Wang W, Yu Z, He F et al (2022) Multiplexed promoter and gene editing in wheat using a virus-based guide RNA delivery system. Plant Biotechnol J 20:2332–2341. https://doi.org/10.1111/pbi.13910
Yang G, Feng M, Yu K et al (2023) Paternally imprinted LATE-FLOWERING2 transcription factor contributes to paternal-excess interploidy hybridization barriers in wheat. J Integr Plant Biol 65. https://doi.org/10.1111/jipb.13574
Zhang Y, Liang Z, Zong Y et al (2016) Efficient and transgene-free genome editing in wheat through transient expression of CRISPR/Cas9 DNA or RNA. Nat Commun 7. https://doi.org/10.1038/ncomms12617
Zhang Y, Li D, Zhang D et al (2018) Analysis of the functions of TaGW2 homoeologs in wheat grain weight and protein content traits. Plant J 94:857–866. https://doi.org/10.1111/tpj.13903
Zhou X, Zhao Y, Ni P et al (2023) CRISPR-mediated acceleration of wheat improvement: advances and perspectives. J Genet Genomics 50:815–834
Zong Y, Wang Y, Li C et al (2017) Precise base editing in rice, wheat and maize with a Cas9-cytidine deaminase fusion. Nat Biotechnol 35:438–440. https://doi.org/10.1038/nbt.3811

Organelle Genomics in Durum Wheat: An Overview of Research Progress

Keshav Sharma and Shahryar Kianian

Abstract

Mitochondria, the organelles involved in cellular respiration and energy generation (adenosine triphosphate [ATP]), regulate essential cellular activities. They are primarily associated with major molecular functions, including cell division, growth, damage repair, and stress tolerance. The molecular mechanisms executing these functions deploy macromolecules such as hormones, metabolites, amino acids, proteins, reactive oxygen species (ROS), and superoxides and contribute to signaling pathways. Mitochondria play an important role in metabolic and stress-responsive signaling pathways, and the signaling molecules associated with those pathways include many, but salicylic acid, jasmonic acid, brassinosteroids, ethylene, auxin, gibberellin, and cytokinin are the most studied. As the powerhouses of the cell, mitochondria maintain homeostasis in the cellular environment through the production of ATP, generation of ROS, and scavenging the excess ROS produced in the cytoplasm. All these critical cell functions are regulated by the limited number of mitochondrial genes and the nucleus through retrograde and anterograde signaling. Cytonuclear communication regulates growth, development, and stress. Mitochondrial role in male sterility and its application to plant breeding have garnered researchers' attention to produce pure, cost-effective, and productive hybrids and combine desirable agronomic traits. Study of mitochondrial genes conducted in diploid, tetraploid, and hexaploid wheat provided important insights on its evolution, development, and defense. Mitochondria have a much smaller genome and have limited number of genes relative to the nucleus. They have repetitive DNA sequences with large introns and share some homology with the nuclear genome. This provided insight into the evolutionary theory that most mitochondrial genes were transferred to the nucleus over time. As major organelles have their own genome, mitochondria contribute to cell homeostasis and crop improvement. Although many details are yet to be resolved, we review here mitochondrial genes, structure, and function and some pioneering work on bread wheat and its wild relatives.

Keywords

Organelle · Mitochondria · Wheat · Alloplasmic · ROS · CMS · Genome

K. Sharma
Department of Plant Pathology, University of Minnesota, Saint Paul, MN, USA

S. Kianian (✉)
Department of Plant Pathology, University of Minnesota, Saint Paul, MN, USA

USDA ARS Cereal Disease Laboratory, Saint Paul, MN, USA
e-mail: shahryar.kianian@usda.gov

R. Tuberosa et al. (eds.), *The Durum Wheat Genome*, Compendium of Plant Genomes,
https://doi.org/10.1007/978-3-031-84689-2_16

1 Introduction

Genetic analysis of qualitative and quantitative traits in plants has justifiably focused on the nuclear genome with the organelle genome ignored. However, organellar genomes and their critical role have finally drawn the attention of breeders and scientists in recent years (Ogihara et al. 2005; Ghavami et al. 2014; Noyszewski et al. 2014; Soltani et al. 2016; Former et al. 2022; Fan et al. 2023). Mitochondrial genes control multiple cell functions and are known to have a significant effect on the generation of adenosine triphosphate (ATP), metabolism, stress, and disease resistance (Miquel et al. 1980; Taanman 1999; Henze and Martin 2003; McBride et al. 2006; Ryan and Hoogenraad 2007; Xiong et al. 2009; Gleason et al. 2011; Galluzzi et al. 2012; Zhang et al. 2015; Huang et al. 2016; Shao et al. 2017; Dopp et al. 2021; Gao et al. 2022). Mitochondrial genes also regulate nuclear genes through retrograde signaling and control genes associated with photosynthesis, metabolism, stress response, and organelle structure (Woodson and Chory 2008; Millar et al. 2011; Soltani et al. 2016; Suzuki 2023; Fábián et al. 2024). Cytoplasmic male sterility (CMS), the most widely explored trait controlled by mitochondrial genes, provides an opportunity to enhance cross-breeding for desirable agronomic traits (Saxena and Hingane 2015; Milner et al. 2020; Han et al. 2021; Melonek et al. 2021; Xu et al. 2023). Mitochondrial genes play a vital role in respiration, intercellular communication, cell signaling, hormone action, development, metabolism, defense responses, and programmed cell death (PCD) (Miquel et al. 1980; McBride et al. 2006; Ryan and Hoogenraad 2007; Zhang et al. 2015; Gibellini et al. 2015; Huang et al. 2016; Giacomello et al. 2020; Dache and Thierry 2023). Studies have shown that various diseases, including metabolic and degenerative ones, are also related to altered mitochondrial genes (Goto et al. 1990; Holt et al. 1988; Wallace 2005; Zhang et al. 2023). This suggests that mitochondria act as a biological hub providing a microenvironment for biosynthetic and catabolic pathways (Galluzzi et al. 2012; Le and Millar 2023). The evolution and lineage of mitochondria could reveal details on their roles (Gray et al. 1999; Gray 1999; Raval et al. 2023) but require more research on a broader range of functions in the cellular environment.

The mitochondria are semi-autonomous bodies believed to have originated from alpha-proteobacterium and a cyanobacterium (Gray et al. 1999, Gray 1999; Knoop 2004; Chandel 2021). During this evolutionary process, some ribosomal genes (rRNA, tRNA) and ribosomal proteins disappeared from it's genome to form the modern mitochondria (Gray et al. 1999; Gray 1999; Kubo et al. 1993; Handa 2003; Clifton et al. 2004; Ogihara et al. 2005; Tian et al. 2006; Warren and Sloan 2020). It is reported that most of the mitochondrial genes were transferred to the nucleus during evolution (Lang et al. 1999; Warren and Sloan 2020). Most often, mitochondrial genes are maternally inherited; however, rare paternal transfer events have been recorded (Kihara 1982; Woodson and Chory 2008; Kawaura et al. 2011; Ghavami et al. 2014). The wide range of mitochondrial genome size in plants resulted from additions and/or deletions and subsequent duplication of non-mitochondrial sequences, including nuclear, chloroplast, and viral DNA in intergenic regions (Cui et al. 2009; Liu et al. 2011; Warren and Sloan 2020). Typically, plant mitochondria are larger in size when compared to animal mitochondria, but the number of genes is limited (Cui et al. 2009; Noyszewski et al. 2014; Møller 2001). The reported mitochondrial size is 15–30 kb in animals and 200 kb–11.3 mb in plants (Burger et al. 2003; Sloan et al. 2012). The larger mitochondrial genome size in plants is due to increased non-coding sequences and repetitive DNA (Unseld et al. 1997). Several factors including DNA transfer events and mutations influence mitochondrial evolution in plants (Burger et al. 2003; Lynch et al. 2006; Cui et al. 2009; Warren and Sloan 2020), which differs significantly from that of their animal counterparts (Cui et al. 2009).

2 Organelle Genome Variation and Crossing in Wheat

Wheat is one of the major staple grain crops grown worldwide, covering more than 218 million ha by area (Giraldo et al. 2019). It is a significant crop in terms of world nutrient supply and economy (Giraldo et al. 2019; Sowell and Swearingen 2021; FAO 2022). Current hexaploid wheat evolved through a series of polyploidization events between the three nuclear donors (*Triticum urartu*-A genome, *Aegilops speltoides*-B genome, and *Aegilops tauschii*-D genome) and one cytoplasm donor *A. speltoides* (Golovnina et al. 2007; Tsunewaki 2009; Haider 2012; Pont et al. 2019). Since hexaploid wheat is compatible with many of the species in the *Triticum–Aegilops* group, this has given boundless opportunities for scientists to cross wheat as a male with plasmon donors, and the hybrids produced are in turn crossed as female to the hexaploid parents (Tsunewaki et al. 1996). The nuclear–cytoplasmic interactions in wheat are altered by replacing cytoplasm through this scheme with related species (Tsunewaki et al. 1996). During the crossing, as expected, replacement of wheat cytoplasm with closely related species has less effect than crossing to phylogenetically distant species (Tsunewaki et al. 1996). Thus, phylogenetic relationships determine species compatibility, which eventually leads to coevolution of nucleus and cytoplasm and compatible cyto-nuclear (CN) interactions (Allen 2004; Atienza et al. 2007; Crosatti et al. 2013).

3 The Mitochondria as an Autonomous Body

Mitochondria are double-membrane cell organelles separated from the cytosol by an outer membrane. The genetic information and matrix are packed inside its inner membrane called cristae (Taanman 1999; Chandel 2021). There is a high variation in genome structure, size, and genes present in higher plant mitochondria (Cui et al. 2009; Møller 2001). Compared to gymnosperms, angiosperm mitochondrial genes have been through an accelerated rate of evolution (Cui et al. 2009). The size of mitochondria and the number of genes vary widely in eukaryotes; however, number of active genes does not always vary with genome size (Burger et al. 2003; Alverson et al. 2011a, and b). Usually, the mitochondrial genome of plants is large, but the number of proteins coded remains like that of animals (Boore 1997; Burger et al. 2003). It remains unanswered how this extra DNA contributes to the cell and if it results from replication or duplication events (Nair 1993; Alverson et al. 2011a, and b). Compared to nucleus, the number of functional genes in mitochondria is less and is believed that most of the mitochondrial genes were lost or transferred to the nuclear genome during evolution (Taanman 1999; Woodson and Chory 2008; Adams and Palmer 2003). Although mitochondrial genes are limited, they are an integral part of its structure and major cellular functions such as respiration, oxidative phosphorylation, RNA maturation, transcription, translation, and protein import (Burger et al. 2003; Chandel 2021; Møller 2001). Most of the mitochondrial genes are conserved but their order differs significantly among species (Kubo et al. 2000; Satoh et al. 2004; Noyszewski et al. 2014).

Sequencing has shown that a substantial amount of mitochondrial genome is repetitive in nature (Ogihara et al. 2005; Alverson et al. 2011a, and b; Noyszewski et al. 2014; Wynn and Christensen 2019), supporting a theory that recombination events could have led to multiple isoforms (Lonsdale 1984; Palmer and Shields 1984; Kubo and Newton 2008; Sloan 2013; Kozik et al. 2019). Nucleotide and amino acid substitution rates vary within individual genes and across genes (Cui et al. 2009; Hu et al. 2023). DNA transfer events may have also occurred between mitochondria and chloroplasts as they contain some identical genes (Lonsdale et al. 1983: Li et al. 2024). Their sequence homology indicates that DNA transfer occurred among organelles (Lonsdale et al. 1985; 1983; Newton 1988; Li et al. 2024). The transfer events involve a cytoplasmic environment that offers transit and a buffer to the transferring of DNA. As the cytoplasmic genome is complex, the nuclear–cytoplasmic interaction is altered through hybridization events while producing alloplasmic lines that confer new phenotypes (Tsunewaki 1993; Tsunewaki et al. 1996; Tsunewaki 2009; Soltani et al. 2014). These hybridization events result in altered growth, development, and defense responses (Tsunewaki et al. 2002; Soltani et al. 2016; reviewed in Nakamura 2023). Also, the recombination and duplication events contribute to the genetic makeup and diversity of the organellar genome (Newton 1988; Noyszewski et al. 2014; Ghavami et al. 2014).

The structure of plant mitochondria is complex and results from extensive genomic rearrangements (Xiong et al. 2008; Soltani et al. 2014; Hu et al. 2023), and natural biological events such as fission and fusion that regulate the morphology, distribution, and function of mitochondria (reviewed in Chan 2006; Rakhmatullina et al. 2022). Although researchers have depicted them as a circular molecule in most plants, it is still unclear if it is circular or linear (Leaver and Gray 1982; Newton 1988; Nair 1993; Bendich 1993; Backert et al. 1997; Sloan 2013; Sloan et al. 2018; Kozik et al. 2019; Wu et al. 2022). Mitochondria are constantly fusing with neighboring mitochondria, and this initiates the fusion of both outer and inner membranes and the mitochondrial matrix (reviewed in Chan 2006; Giacomello et al. 2020). These mixing events induce variation in mitochondria within a brief period, making them semi-autonomous. Thus, mitochondrial morphology is shaped by balancing fission and fusion (reviewed in Chan 2006; Giacomello et al. 2020). The complexity in plant mitochondria results from their variable sequences and branched structures that include linear and circular molecules (Bendich 1993; Oldenburg and Bendich 2001; Alverson et al. 2011a; and b). The rate of internal DNA rearrangement in plant mitochondria is higher leading to structural changes, however rate of DNA mutation or nucleotide substitution is slow (Palmer and Herbon 1988; Knoop 2004). The highly repetitive sequences in mitochondrial genome facilitate recombination and rearrangements leading to complex genome organization (Lonsdale 1984; Bendich 1993, Ogihara et al. 2005; reviewed in Xiong et al. 2008; Sloan 2013; Hu et al. 2023).

4 Wheat Mitochondria

The mitochondrial genome of wheat ranges in size from 420 to 452 kb and consists of 80 genes with exons, 10–44 protein-coding genes, 122–149 ORFs (Open Reading Frames), 3–9 rRNAs, and 17–27 tRNAs (Ogihara et al. 2005; Noyszewski et al. 2014; Han et al. 2021; Hu et al. 2023). The size of the plant mitochondrial genome is variable and has larger introns and repetitive sequences (Kubo et al. 2000; Knoop 2004; Cui et al. 2009; Noyszewski et al. 2014; Hu et al. 2023). Frequent repeats are due to duplication and recombination events (Alverson et al. 2011a, 2011b; Davila et al. 2011; Tanaka et al. 2012, 2014; Hu et al. 2023). Each mitochondrion harbors a high genome copy number in the form of double-stranded DNA that replicates and expresses in this environment (Lang et al. 1999; Lohse et al. 2013; Noyszewski et al. 2014). The genes present in cereal mitochondria are similar but vary significantly in their genome structure and arrangements (Lonsdale 1984; Notsu et al. 2002; Clifton et al. 2004; Ogihara et al. 2005; Noyszewski et al. 2014). Although the wheat mitochondrial genome is smaller than other cereal crops, the mitochondrial gene map shows little synteny with other cereals such as rice and maize (Notsu et al. 2002; Clifton et al. 2004; Ogihara et al. 2005). Evolutionary analysis of protein-coding genes among maize, rice sorghum, and wheat shows changes in the nucleotide and amino acid sequences (Noyszewski et al. 2014). The structural variation among mitochondrial genes is induced by short repeat pairs (Ogihara et al. 2005). This indicates gene shuffling during speciation events (Ogihara et al. 2005). Intramolecular recombination and rearrangements induce changes and are responsible for the complex hybridization in the mitochondrial genome (Dewey et al. 1985). The mitochondrial genomes of wheat cultivars such as Chinese Yumai (CY) and Chinese Spring (CS) have a much higher similarity as measured by single nucleotide polymorphisms (SNPs) and presence/absence of small non-coding regions among them (Cui et al. 2009; Hu et al. 2023). During evolution, ribosomal genes such as rRNA and tRNA were lost, and changes in gene and amino acid sequence were observed in ATP synthase (Kubo et al. 1993; Handa 2003; Clifton et al. 2004; Ogihara et al. 2005; Tian et al. 2006; Cui et al. 2009). Variation in nucleotides and amino acids differ among individual genes, mitochondrial genomes, and plant lineages.

The NADH ubiquinone dehydrogenase complex and cytochrome-c biogenesis proteins have a slower substitution rate, whereas the ATP synthase family, cytochrome-c oxidases, and especially nad9 have a higher substitution rate (Cui et al. 2009). In tracking mitochondrial gene evolution, authors have indicated that gene loss or reacquisition in this organelle occurs via horizontal and vertical gene transfers (Bergthorsson et al. 2003; Richardson and Palmer 2007). As they harbor both nuclear and chloroplastic genes, horizontal transfers might induce significant variation (Bergthorsson et al. 2003; reviewed in Zardoya 2020). There is a remarkable difference in genetic recombination of the cytoplasmic and nuclear genomes. In sexual reproduction, nuclear genomes are hybrids, whereas plasmons are haploid, homoplasmic, and maternal (Hattori et al. 2002). The mitochondrial sequences in *Triticum* and *Aegilops* have significant differences in their genomes (Ogihara et al. 2005; Ghavami et al. 2014; Hu et al. 2023). The SNPs in the entire genome appear to be 40 within *Triticum* and 679 between *Triticum* and *Aegilop*s species (Ogihara et al. 2005; Ghavami et al. 2014). The location of genes and order vary among diploid *Aegilops tauschii*, tetraploid and hexaploid wheat, which indicates multiple rearrangements and shuffling have occurred over the time (Ghavami et al. 2014; Hu et al. 2023). Changes in gene sequence and arrangements were evident within various durum wheat lines and are significantly higher in alloplasmic than euplasmic lines (Noyszewski et al. 2014). This supports the hypothesis that nuclear genome substitution could have induced accelerated evolution in alloplasmic lines compared to parental lines (Ghavami et al. 2014).

Alloplasmic lines are the perfect system to explore organellar genome recombination and rearrangements (Tsunewaki 2009; Soltani et al. 2016). Variation in alloplasmic lines compared with their maternal genotypes indicates rearrangements and evolution due to mitochondrial genome shuffling (Ghavami et al. 2014). These accelerated evolutionary events in alloplasmic mitochondria add value to the cytoplasmic genome regarding how these changes could be applied (Allen et al. 2007; Bentolila and Stefanov 2012; Ghavami et al. 2014). The observed variation includes gene types, rearrangements, SNPs, additions, deletions, and changes in amino acids and proteins (Cui et al. 2009; Ogihara et al. 2005; Kawaura et al. 2011; Noyszewski et al. 2014; Ghavami et al. 2014). These changes in the cytoplasmic genome have varying effects on cyto-nuclear communication (Oda et al. 1992; Unseld et al. 1997; Kubo et al. 2000; Notsu et al. 2002; Handa 2003; Clifton et al. 2004; Sugiyama et al. 2005; Cui et al. 2009; Ogihara et al. 2005; Kawaura et al. 2011; Noyszewski et al. 2014; Ghavami et al. 2014; Nakamura 2023). Mitochondrial genome sequencing in wheat provided clues as to the sequence information, genome structure, homology, and synteny at various ploidy levels and compared to other organisms such as rice and maize (Ogihara et al. 2005; Cui et al. 2009; Noyszewski et al. 2014; Hu et al. 2023). Genes in mitochondria have been categorized into various complex systems. They are classified into ORF genes, various complexes (Complex I, Complex II, Complex III, Complex IV, and Complex V), cytochrome-c biosynthesis, SecY-independent transport, and ribosomal units and RNAs (Table 1) (Nair 1993; Notsu et al. 2002; Handa 2003; Clifton et al. 2004; Ogihara et al. 2005; Noyszewski et al. 2014; Hu et al. 2023). Every complex has unique genes regulating crucial functions in the cellular environment. Mutation events such as SNPs, additions, and deletions are also common in mitochondrial genes within and between species (Cui et al. 2009; Kawaura et al. 2011; Noyszewski et al. 2014). Several factors such as DNA recombination, rearrangements, additions, and deletions explain or cause mutations in the mitochondrial genome (Cui et al. 2009; Ogihara et al. 2005; Kawaura et al.

Table 1 Mitochondrial gene classification according to their functions and name

Gene function	Gene name
Complex I	*nadl, nad2, nad3, nad4, nad4L, nad5, nad6, nad7, nad9*
Complex II	–
Complex III	*Cob*
Complex IV	*Coxl, cox2, cox3*
Complex V	*atp1, atp4, atp6, atp8, atp9*
Cytochrome-c biogenesis	*ccmB, ccmFN*
SecY-independent transport	*mttB*
Ribosomal RNAs	*5 s rRNA, 18srRNA, 26srRNA*
Ribosomal protein small subunit	*Rps1, rps2, rps3, rps4, rps12, rps13*
Ribosomal protein large subunit	*rpl16, rpl5*
Open Reading frames	*Orf173, Orf240, Orf349, Orf359*

2011; Noyszewski et al. 2014; Ghavami et al. 2014; Gualberto and Newton, 2017; Hu et al. 2023). The recombination events involving strand exchange cause most of the DNA mutations (Gualberto and Newton, 2017; Hu et al. 2023), and among them, homologous recombination is more common compared with non-homologous end-joining (Gualberto and Newton, 2017). These events play a significant role in genome size and structural variation.

4.1 Gene Classification

Mitochondrial genes are classified into ORF genes, various complexes (Complex I, Complex II, Complex III, Complex IV, and Complex V), cytochrome-c biosynthesis, SecY-independent transport, and ribosomal units and RNAs (Nair 1993; Notsu et al. 2002; Handa 2003; Clifton et al. 2004; Ogihara et al. 2005; Noyszewski et al. 2014; Hu et al. 2023).

4.1.1 Complex I (NADH Dehydrogenase)

Complex I include NAD genes, a crucial element in cellular metabolism. There are five such genes: NAD 1, NAD 2, NAD 3, NAD 4, and NAD 5 (Kubo and Newton 2008; Møller 2001; Hu et al. 2023). The respiratory complex NADH–ubiquinone oxidoreductase is the electron transport chain (ETC) component (Fearnley and Walker 1992). In plants, this complex consists of nine nad (NADH) genes, that is two more than in mammalian mitochondria (Gäbler et al. 1994; Grohmann et al. 1994; Unseld et al. 1997). These genes are also found in the nuclear and chloroplast genomes (Rasmusson et al. 1998).

4.1.2 Complex II/Ubiquinone Oxidoreductase

Angiosperm Complex II consists of SDH3 and SDH4 (Kubo and Newton 2008); however, details on this complex are lacking in wheat mitochondria.

4.1.3 Complex III (Ubiquinol Cytochrome-c Reductase)

Complex III comprises COB genes and are identical in all *Triticum species* and show some variation in *Aegilops species* (Rasmusson et al. 1998). These genes are also found in nuclear and chloroplast genomes (Rasmusson et al. 1998).

4.1.4 Complex IV (Cytochrome-c Oxidase)

Complex IV consists of three genes: *COX1*, *COX2*, and *COX3*.

4.1.5 Complex V/Adenosine Triphosphate (ATP) Synthase

Complex V consists of ATP genes (*ATP1*, *ATP4*, *ATP6*, *ATP8*, and *ATP9*) associated with respiration and energy production.

4.1.6 Cytochrome-c Biosynthesis Genes

Several genes such as *ccmB*, *ccmFN*, *ccFC*, and *ccmC* fall under this category.

4.1.7 Ribosomal Proteins

The ribosomal proteins include rps genes. The wheat mitochondria include *RPS1*, *RPS2*, *RPS3*, *RPS*, *RPS7*, *RPS12*, *RPS13*, *RPS15*, *RPS16*, and *RPS19*.

5 Cyto-Nuclear Interaction

The cytoplasmic environment is diverse with intracellular liquid, macromolecules, ions, organic solutes, proteins, and membrane-bound organelles (Luby-Phelps 2000, 2013). Cytoplasm, the protein-rich matrix, provides a unique niche for the cellular components, traveling vesicles, and expressed proteins and helps regulate crucial functions such as intracellular transport and signaling, enzyme catalysis, and protein folding (Luby-Phelps 2013). The diversity of the matrix and its contribution to cellular balance can be validated by studying its interaction with the nucleus and observing their effects on plant phenotypes (Tsunewaki 2009; Ghavami et al. 2014). The mitochondrial alternative oxidase (AOX) genes involved in plant physiological processes induce ethylene-mediated drought tolerance and autophagy. Also, the RNAi of AOX genes altered reactive oxygen species (ROS) production and autophagy that reduced drought tolerance in tomatoes (Zhu et al. 2018). The cyto-nuclear (CN) interaction remains crucial in controlling major molecular functions and maintaining cell harmony (Tsunewaki 2009; Chen and Liu 2014; Chen et al. 2017; Møller 2001). The cytoplasmic genes interact with the nucleus for gene expression, and the nuclear genes depend on the organellar genomes for some critical gene functions; this emphasizes the importance of CN crosstalk and its role in growth, survival, and yield (Woodson and Chory 2008; Millar et al. 2011; Soltani et al. 2016; Møller 2001).

Organellar genes regulate signaling pathways, stress-related genes, phenotypes, and cyto-nuclear communications (Kwak et al. 2003; Sandhu et al. 2007; Xu et al. 2012; Crosatti et al. 2013; Virdi et al. 2015; Soltani et al. 2016; Shao et al. 2017; Welchen et al. 2021; Møller 2001). The majority of the organellar proteins are produced by the nuclear genome and are then transferred to organelle; likewise, some proteins produced in organelle alter the expression of nuclear genes associated with metabolic and photosynthetic pathways (Millar et al. 2011; Soltani et al. 2016; reviewed in Møller 2001). The process of organellar proteins regulating nuclear proteins is known as retrograde signaling (Millar et al. 2011; Møller 2001). The signaling molecules involved in this process include RNA, proteins, transcription factors, reactive oxygen species, metabolites, and others (Reviewed in Leister 2012; Møller 2001).

CN connections are maintained through generations and might be halted or modified by interspecific hybridizations. The traits observed through interspecific hybridization are believed to induce adaptive superiority leading to the evolution of new species (Levin 2003; Budar and Roux 2011). These hybridization events could yield desirable traits, but there is an equal or greater chance of undesirable traits (Busch and Maan 1978; Tsunewaki 1993). With a new combination of the nucleus from one parent and cytoplasm from another, alloplasmic lines have altered cyto-nuclear interactions (Tsunewaki et al. 2002; Tsunewaki 2009; Noyszewski et al. 2014). The alloplasmic lines produced via successive backcrossing reported altered cyto-nuclear crosstalk leading to cytoplasmic male sterility (CMS) lines (Hanson 1991). CMS is caused by both the mitochondrial and nuclear genes and takes several forms (Nair 1993; Cui et al. 1996; Chen and Liu 2014; Milner et al. 2020). However, a single factor could be enough to change fertility and phenotypes (Martín et al. 2008; Milner et al. 2020). Breeders have widely exploited CMS lines to produce hybrids for better agronomic traits and successive breeding experiments.

Breeding for resistance or quality traits involves shuffling cyto-nuclear combinations creating a novel cellular environment, a major step toward genome evolution (Tsunewaki 2009; Yıldırım et al. 2019). Alloplasmic lines possess novel interactions between the nucleus and cytoplasm and are invisible under normal conditions (Maan 1991; Tsunewaki et al. 1996; Tsunewaki 2009; Soltani et al. 2016). The effect of cytoplasm on plant physical traits and yield has been studied through modified cyto-nuclear interactions (Tsunewaki et al. 2002; Atienza et al. 2007; Tsunewaki 2009; Noyszewski et al. 2014). In alloplasmic lines, plants with improved vigor and resistance to various stresses were observed

(Mullaney 1981; Wu et al. 1998b; Voluevich and Buloichik 1992; Hou et al. 2000; Liu et al. 2002; Tsunewaki et al. 2002; Atienza et al. 2007; Tsunewaki 2009; Klimov et al. 2005; Ghavami et al. 2014; Soltani et al. 2016). An alloplasmic line of *Triticum durum* induced resistance to several wheat diseases, which indicates that cytoplasmic substitution could be exploited to enhance quality traits in plants (Ghavami et al. 2014). Alloplasmic wheat lines have shown higher gluten content without having significant damage on other agronomic traits (Atienza et al. 2007). Other traits such as life span, ear size, flowering, and dry matter weight are influenced by cyto-nuclear combinations (Wu et al. 1998a; Tsunewaki et al. 2002; Ghavami et al. 2014). Therefore, nuclear–cytoplasmic interaction is pivotal for crop life and quality improvement (Ghavami et al. 2014; Welchen et al. 2021).

6 Organellar Genes and Cytoplasmic Male Sterility in Plants

Organellar genes are the critical components of male fertility in plants. They are inherited maternally and control the molecular mechanism of pollen development and reproductive function. Failure to produce anthers or viable pollen leads to cytoplasmic male sterility (CMS) (Cui et al. 2009; Geng et al. 2018; Brownfield 2021). The CMS in plants is associated with the mitochondrial genes and their interaction with the nucleus, RNA editing, and accumulation of cytotoxic proteins (Chen and Liu 2014; Saxena and Hingane 2015; reviewed in Chen et al. 2017; Møller 2001). Genes associated with CMS result from the mitochondrial genome rearrangements that alter normal gene function and pollen fertility (Wang et al. 2006; Liu et al. 2011; Chen and Liu 2014; Geng et al. 2018). This phenotype can, in certain situations, be ameliorated by fertility restorer genes located in the nucleus (Chen and Liu 2014; reviewed in Chen et al. 2017). Factors leading to CMS can be categorized based on their origin, such as natural or spontaneous, interspecific crosses, and intrageneric crosses (Edwardson 1970). Notably, CMS occurs when the interaction between cytoplasm and nucleus is disturbed and leads to male sterility (Hanson and Bentolila 2004; Martín et al. 2018). Halting pollen fertility can help induce cross-hybridization (Han et al. 2021) particularly in self-pollinated crops. The CMS resulting from a new cyto-nuclear combination is known as alloplasmic male sterility (Hanson and Bentolila 2004) and the major force driving this phenotype is the cytoplasmic (mitochondrial) genes (Hanson and Bentolila 2004; Han et al. 2021; Møller 2001).

Cytoplasmic male sterility can be regulated by more than one gene (Chen and Liu 2014; Møller 2001). Comparative analysis of the differentially abundant proteins in CMS and maintainer (fertile) show that proteins associated with oxidative phosphorylation, glycolysis, tricarboxylic acid cycle, carbohydrate, and energy metabolism pathways are upregulated in the maintainer lines when compared to the CMS lines (Geng et al. 2018). The CMS is also governed by the abnormal CMS genes in mitochondria that produce cytotoxic proteins inhibiting proper growth of male gametophytic cells (Wang et al. 2006; Møller 2001). Some mitochondrial genes cause CMS by suppressing the male fertility in plants. The TALEN-mediated knockout of the CMS-associated gene *orf79* in Boro-Taichung type CMS rice and *orf125* in Kosena-type CMS rapeseed restored fertility in both crops (Kazama et al. 2019). Fertility restorer genes also regulate fertility by suppressing the function of CMS-associated gene in wheat. The fertility restorer genes *Rf1* and *Rf3* in wheat bind and cleave the CMS gene *orf279* transcript inhibiting the expression of CMS trait and restoring the pollen fertility (Melonek et al. 2021). The mitochondrial ETC-associated gene *atp8* carries CMS traits in various crops and shares common promoter and 5′UTR region with orf279 (Chen and Liu 2014; Melonek et al. 2021). Apart from genes, differences in the mitochondrial genome size have also been noted in CMS lines when compared with their maintainer lines. The genome of the CMS line (Ks3) in wheat has been reported to be larger than the maintainer lines (Km3) with some additional ORFs and repeat sequences (Liu et al.

2011). Most of the genes in Km3 and Ks3 are conserved with some difference in genes and sequences (Liu et al. 2011). Genomics analysis of the Km3 and Ks3 shows rpl5 is only present in Km3 lines and there is significant sequence difference in *at6*, *nad6*, and *nad9* genes and *rpl5* genes (Liu et al. 2011).

CMS can be regulated by some nuclear genes such as *MutS Homolog1*(*MSH1*) (Sandhu et al. 2007; Martín et al. 2008). Through transcriptome and RT-PCR analyses, the role of *MSH1* in plant cellular pathways has been validated and altering the function of MSH1 through RNAi seems to impact important cellular functions and pathways. Suppression of this gene through RNAi induced varying phenotypes, including CMS (Sandhu et al. 2007; Xu et al. 2012; Virdi et al. 2015; Shao et al. 2017). The male sterility in tobacco plants resulted from mitochondrial DNA rearrangements (Sandhu et al. 2007; Xu et al. 2012). Other phenotypes include reduced growth, branching, internodal length, variable leaf morphology, reduced stomatal density, and delayed flowering (Xu et al. 2012). Reduced growth relates to the changes in auxin and gibberellic acid metabolism and regulators associated with flowering and vernalization (Willige et al. 2011; Xu et al. 2012). These phenotypic changes in plants also demonstrate the interaction of mitochondria and chloroplasts in the cell. The disruption of *MSH1* and green-white variegation in the leaf are also associated with the changes in the chloroplast and its interaction with mitochondria (Xu et al. 2011). This modifies the development of chloroplast and redox status and induces tolerance to photooxidative light, thereby assisting in survival and reproduction (Xu et al. 2011).

7 Role of Mitochondria in Oxidative Stress

Unfavorable environmental conditions induce stress in plants affecting their growth, development, and productivity, which includes the production of reactive oxygen species (ROS) leading to oxidative stress (Møller 2001; Chaki et al. 2020). Oxidative stress is the condition of imbalance between oxidants and the antioxidants in the cell. When ROS are elevated, the production of antioxidants is reduced, and the electron transport chain (ETC) is inhibited (Blokhina and Fagerstedt 2010; Rigoulet et al. 2011). Cell organelles (mitochondria and chloroplasts) are the major sites for the ETC and generation of the ROS (Woodson and Chory 2008; Blokhina and Fagerstedt 2010; Rigoulet et al. 2011; Galluzzi et al. 2012; Gibellini et al. 2015) and plants tolerate these stresses by defending against oxidative stress at the cellular level (Pastore et al. 2007). In mitochondria, Complex I and Complex III produce most of the ROS (Pastore et al. 2007; Rigoulet et al. 2011; Galluzzi et al. 2012). ROS are also produced in peroxisomes and used as cellular weapons to stop the invasion from pathogens (Møller 2001). Plant mitochondria use several strategies such as skipping ROS production, detoxifying ROS, repairing ROS-induced molecular damage, and maintaining the ETC properly oxidized (Møller 2001). ROS as signaling molecules in cells affect stomatal movement, growth, development, and abiotic and biotic stress responses (Baxter et al. 2014; Yang et al. 2001). Unwanted or higher concentrations of ROS have a detrimental effect on overall growth and development; therefore, proper balance is critical (Møller 2001; Petrov et al. 2015). The wider functions of ROS involve a significant number of genes. In *Arabidopsis*, about 152 genes are reported to be involved (Mittler et al. 2004). The function of ROS is not solely dependent on its generation but also on its scavenging process. The localization, intensity, and longevity of ROS signaling depend on the cooperative action of ROS-producing and scavenging pathways (Mittler et al. 2004; Bailey-Serres and Mittler 2006). Maintaining ROS concentration at a lower level is critical for signaling pathways and proper balance in the cellular environment (Mittler et al. 2004). ROS are also vital for ABA signaling and the movement of stomatal guard cells (Mittler et al. 2004; Kwak et al. 2003). ROS production in durum wheat is controlled by several energy-dissipating systems such as alternative oxidase, plant uncoupling protein, and mitochondrial potassium channel (Pastore et al. 1999, 2000, 2001, 2007). ROS production is associated with various enzymes, including NADPH oxidases (Mittler et al. 2004; Bailey-Serres and Mittler 2006). Impairing the function of NADPH oxidase catalytic subunit genes *AtrbohD* and *AtrbohF* alters ABA signaling and stomatal closing (Kwak et al. 2003). The function of these mitochondrial genes is also critical for seed germination, and root and shoot growth. The mutants of these NADPH subunits impact ABA signaling and inhibit seed germination and root growth in *Arabidopsis* (Kwak et al. 2003; Yang et al. 2014). Therefore, the NADPH oxidases assist ABA for ROS production and defense (Lamb and Dixon 1997;Kwak et al. 2003).

8 Role of Mitochondria in Defense

Mitochondria conduct a wide range of cellular functions, including metabolism, growth, and resistance to abiotic and biotic stress (Mittler et al. 2004; Galluzzi et al. 2012; Baxter et al. 2014). Defense action of the organellar genome relates to the function of hormones and proteins regulated by mitochondrial genes and their interaction with the nucleus. The cyto-nuclear interaction regulates the expression of genes associated with photosynthesis, electron transport chain, male fertility, leaf metabolism, and retrograde signaling, and introducing alien cytoplasm alters these functions in wheat (Crosatti et al. 2013). The role of mitochondria in defense and disease resistance is equally important and has been documented (reviewed in Bohra et al. 2016). The defense action is initiated with stress responses such as oxidative burst and production of ROS (Glazebrook 2005). These ROS activate signaling pathways to induce hypersensitive cell death (HR) that blocks food supply to pathogens (Glazebrook 2005; Zurbriggen et al. 2010). Mitochondria generate ROS during stress and induce programmed cell death (PCD) (Zurbriggen et al. 2010). Application of bacterial elicitors in *Arabidopsis* resulted in induced ROS generation in mitochondria and reduced ATP (Zurbriggen et al. 2010). ROS production involves a broad molecular network that includes the plasma membrane, chloroplasts, and mitochondria (Zurbriggen et al. 2010). These ROS activate mitogen-activated protein kinase (MAPK) pathways that induce resistance against abiotic and biotic stresses (Nakagami et al. 2005; Zurbriggen et al. 2010). Abiotic stress is associated with environmental factors such as low temperature, low soil moisture, touch, and wounding (Ichimura et al. 2000). Biotic stress is caused by various pathogens, including fungi, bacteria, nematodes, and viruses (Wang et al. 2021; McNeece et al. 2019; Ma et al. 2021).

For plant defense, the initial step is critical and includes timely detection of the pathogen attack and immediate response. Initially, pathogens attack plants through different means and try to hijack the immune system (Jones and Dangl 2006). Plants use transmembrane pattern recognition receptors (PRR) and polymorphic nucleotide-binding Leucine-rich repeat (NB-LRR) to detect pathogen-activated molecular patterns (PAMPs) and respond to the pathogen attack (Jones and Dangl 2006). Pathogens later use effectors to dismantle the PAMP-triggered immunity (PTI) resulting in effector-triggered susceptibility. NB-LRR proteins identify these effectors, and plants use a more advanced form of defense response known as effector-triggered immunity (ETI) to induce resistance (Jones and Dangl 2006). The PTI and ETI involve a network of genes, including salicylic acid (SA) (Alvarez 2000). SA has a broad cellular network and activates mitogen-activated protein kinase (MAPK) signaling to induce resistance (Zhang and Klessig 1997; Ichimura et al. 2000; Yang et al. 2000; Zhang and Liu 2001; Nakagami et al. 2005; Wang et al. 2021; Ma et al. 2021). SA also activates the expression of glutathione S-transferase (GST) proteins located in the cytosol, mitochondria, and several other cell organelles and is involved in broad cellular functions (Moons 2005; Goto et al. 1990; reviewed in Raza 2011; Gullner et al. 2018). The glutathione S-transferase (GST) in mitochondria induces early defense responses in plants (Moons

2005; Gleason et al. 2011; Raza 2011; Gullner et al.2018). Overexpression of GST induced resistance against biotic and abiotic stress in plants (Hu et al. 2023; Sharma et al. 2014). In animal cells, overexpressed GST reduced cytotoxicity and prevented loss of mitochondrial membrane potential (Goto et al. 2009). Further, GST mutants in *Arabidopsis* altered the expression of stress-related genes, including SA, and showed higher susceptibility to bacterial and fungal pathogens (Gleason et al. 2011). The expression of GST modulated by SA, the function of the SA pathway, and SA-induced ROS production are controlled by mitochondrial Complex II (Gleason et al. 2011; Welchen et al. 2021). The alteration of SA expression through GST modification and SA expression controlled by mitochondrial genes highlights the importance of mitochondria in intramolecular functions and defense response.

9 Mitochondria in Cell Growth and Development

Mitochondria are a major site for respiration and energy generation and, as such, play a critical role in growth (Millar et al. 2011). They harbor most of the components essential for respiration. The electron transport chain inside the inner mitochondrial membrane is supported by its four complexes I, II, III, and IV. These complexes interact with the help of cytochrome-c and ubiquinone (Millar et al. 2011). Most of the mitochondrial proteins are synthesized in the nucleus; however, for a stable respiratory network, proper interaction with mitochondrial proteins is important (Millar et al. 2011). Mitochondria produce ATP through oxidative phosphorylation and modulate several biosynthesis pathways related to enzymes, vitamins, amino acids, and proteins (reviewed in Berkowitz et al. 2016). Mitochondrial genes contribute significantly to cell division, growth, and development of plants (Tsunewaki 2009; Bashir et al. 2011). Mitochondrial iron-transporter (MIT) genes have been found to induce growth in yeast and rice plants. Their mutants resulted in reduced growth symptoms such as lower chlorophyll content, leaf width, shoot length, and reduction in root and shoot dry weight. These MIT mutants also showed delayed flowering, reduction in tillers, and yield (Bashir et al. 2011). Mitochondria regulate several growths and stress-related hormones. The growth hormones auxin, gibberellin, cytokinin, and ethylene, as well as the defense hormones abscisic acid, salicylic acid, jasmonates, and brassinosteroids, interact with mitochondria to execute cellular functions and counter various stresses (Berkowitz et al. 2016).

Mitochondria as a regulator of growth and development in plants contribute to higher yield and desirable agronomic traits. Their function in the cellular environment has been studied through overexpression, mutation, and cross-hybridization. The mitochondria-localized heat shock cognate protein 70–1 (mtHSC70–1) regulates vegetative and reproductive growth. The mutants of mtHSC70–1 induce altered vegetative and reproductive growth resulting in abnormal cell division, cell elongation, root and shoot growth, flower morphology, anther development, and embryogenesis (Li et al. 2021). The abnormal growth of the embryo in mtHSC70–1 mutants was due to the altered expression of auxin biosynthesis and polar auxin transport genes. This was validated by applying auxin inhibitor to the wild-type plants, which induced retrograde signaling, showing similar phenotypes, along with altered auxin biosynthesis and polar auxin transport genes (Kerchev et al. 2014; Li et al. 2021). This shows that mitochondrial HSP70 protein regulates auxin-mediated embryogenesis (Kerchev et al. 2014; Li et al. 2021). Overexpression of *Brz-insensitive-long hypocotyls 2* (*BIL2*) gene related to the mitochondrial-localized DnaJ/Heat shock protein 40 (DnaJ/HSP40 protein) family induces cell elongation and root elongation, and increases the number of branches, inflorescences, and lateral roots (Bekh-Ochir et al. 2013). BIL2-RNAi mutants resulted in reduced branching, shorter hypocotyl, and a smaller inflorescence (Bekh-Ochir et al. 2013). The mitochondrial uncoupling protein1 (UCP1) gene overexpression in *Nicotiana tabacum* increased seed number and overall yield (Barreto et al. 2017). Mitochondrial genes coordinate with the nuclear genome for their expression and function, and their interaction is crucial for plant growth and development. The role of mitochondria and their role in cyto-nuclear interaction is observed by altering their interaction through interspecific hybridization and production of alloplasmic lines (Tsunewaki 2009; Soltani et al. 2016). Alloplasmic lines with variable phenotypes, including induced growth and development, have been obtained (Tsunewaki et al. 2002; Atienza et al. 2007; Tsunewaki 2009; Soltani et al. 2016). *Aegilops mutica* cytoplasm when crossed with *Triticum aestivum* cv. Chris nucleus resulted in alloplasmic *A. mutica* Chris with a higher growth rate, taller plants, and supernumerary spikelets (Soltani et al. 2016). The introduction of new cytoplasm in alloplasmic lines alters cyto-nuclear communication affecting metabolites and transcripts at the cellular environment (Crosatti et al. 2013; Soltani et al. 2016). Nuclear genes regulated by the organellar genome execute biological functions such as metabolism and photosynthesis through retrograde signaling (Millar et al. 2011; Soltani et al. 2016).

10 Future Perspectives and Challenges

Mitochondrion is the organelle of respiration and the unit that coordinates growth, development, and biotic stress response. When changes are introduced to mitochondrial genes, their interaction with the nuclear genome is altered and thus variation in vegetative growth, fertility, and stress response is observed. Male sterility has been the most widely exploited trait from this interaction, but important agronomic traits also need more attention. With advancements in genomics, proteomics, and next-generation sequencing, molecular engineering could be the next step to advance toward trait introgressions and analysis. Mitochondrial genes contributing to major cellular functions have been identified, and their interactions with other cytoplasmic genomes and nuclei have been studied. Mitochondria share some common sequences with chloroplast and nuclear genomes, indicating some clues as to their evolution and transfer of sequences among organelles (Filip and Skuza 2021). Plant organellar genome sequencing has increased in recent years. Similarly, molecular biology techniques such as gene cloning and editing could be exploited to further functional analysis of this important organelle. Mitochondrial genes are limited, randomly located in the genome, and have large introns. They have a highly repetitive sequence that facilitates rearrangements and the formation of sub-genomic circles through recombination (Lonsdale 1984). Although the structure of mitochondrial genomes is mostly portrayed as circular, evidence of a linear genome has also been presented. A clear answer on mitochondrial genome structure seems elusive.

References

Adams KL, Palmer JD (2003) Evolution of mitochondrial gene content: gene loss and transfer to the nucleus. Mol Phylogenet Evol 29(3):380–395

Allen JO (2004) Effect of teosinte cytoplasmic genomes on maize phenotype. Genetics 169(2):863–880

Allen CA, Van Der Giezen M, Allen JF (2007) Origin, function, and transmission of mitochondria. In: Origin of mitochondria and hydrogenosomes. Springer, Berlin, Heidelberg, pp 39–56

Alvarez ME (2000) Salicylic acid in the machinery of hypersensitive cell death and disease resistance. In: Programmed cell death in higher plants. Springer, pp 185–198

Alverson AJ, Rice DW, Dickinson S, Barry K, Palmer JD (2011a) Origins and recombination of the bacterial-sized multichromosomal mitochondrial genome of cucumber. Plant Cell 23(7):2499–2513

Alverson AJ, Zhuo S, Rice DW, Sloan DB, Palmer JD (2011b) The mitochondrial genome of the legume *Vigna radiata* and the analysis of recombination across short mitochondrial repeats. PLoS One 6(1):e16404

Atienza SG, Martin AC, Ramirez MC, Martin A, Ballesteros J (2007) Effects of *Hordeum chilense* cytoplasm on agronomic traits in common wheat. Plant Breed 126(1):5–8

Backert S, Nielsen BL, Börner T (1997) The mystery of the rings: structure and replication of mitochondrial genomes from higher plants. Trends Plant Sci 2(12):477–483

Bailey-Serres J, Mittler R (2006) The roles of reactive oxygen species in plant cells. Plant Physiol 141

Bashir K, Ishimaru Y, Nishizawa NK (2011) Identification and characterization of the major mitochon-

drial Fe transporter in rice. Plant Signal Behav 6(10):1591–1593
Baxter A, Mittler R, Suzuki N (2014) ROS as key players in plant stress signaling. J Exp Bot 65(5):1229–1240
Bekh-Ochir D, Shimada S, Yamagami A, Kanda S, Ogawa K, Nakazawa M, Matsui M, Sakuta M, Osada H, Asami T, Nakano T (2013) A novel mitochondrial DnaJ/Hsp40 family protein BIL2 promotes plant growth and resistance against environmental stress in brassinosteroid signaling. Planta 237(6):1509–1525
Bendich AJ (1993) Reaching for the ring: the study of mitochondrial genome structure. Curr Genet 24(4):279–290
Bentolila S, Stefanov S (2012) A reevaluation of rice mitochondrial evolution based on the complete sequence of male-fertile and male-sterile mitochondrial genomes. Plant Physiol 158(2):996–1017
Bergthorsson U, Adams KL, Thomason B, Palmer JD (2003) Widespread horizontal transfer of mitochondrial genes in flowering plants. Nature 424(6945):197–201
Berkowitz O, De Clercq I, Van Breusegem F, Whelan J (2016) Interaction between hormonal and mitochondrial signalling during growth, development and in plant defence responses. Plant Cell Environ 39(5):1127–1139
Blokhina O, Fagerstedt KV (2010) Reactive oxygen species and nitric oxide in plant mitochondria: origin and redundant regulatory systems. Physiol Plant 138(4):447–462
Bohra A, Jha UC, Adhimoolam P, Bisht D, Singh NP (2016) Cytoplasmic male sterility (CMS) in hybrid breeding in field crops. Plant Cell Rep 35(5):967–993
Budar F, Roux F (2011) The role of organelle genomes in plant adaptation: time to get to work! Plant Signal Behav 6(5):635–639
Burger G, Gray MW, Lang BF (2003) Mitochondrial genomes: anything goes. Trends Genet 19(12):709–716
Busch RH, Maan SS (1978) Effects of alien Cytoplasms on agronomic and bread-making traits of two spring wheat cultivars 1. Crop Sci 18(5):864–866
Chaki M, Begara-Morales JC, Barroso JB (2020) Oxidative stress in plants. Antioxidants (Basel) 9:481
Chan DC (2006) Mitochondria: dynamic organelles in disease, aging, and development. Cell 125(7):1241–1252
Chandel NS (2021) Mitochondria. Cold Spring Harb Perspect Biol 13(3):a040543
Chen L, Liu YG (2014) Male sterility and fertility restoration in crops. Annu Rev Plant Biol 65:579–606
Chen Z, Zhao N, Li S, Grover CE, Nie H, Wendel JF, Hua J (2017) Plant mitochondrial genome evolution and cytoplasmic male sterility. Crit Rev Plant Sci 36(1):55–69
Clifton SW, Minx P, Fauron CMR, Gibson M, Allen JO, Sun H, Thompson M, Barbazuk WB, Kanuganti S, Tayloe C, Meyer L, Wilson RK, Newton KJ (2004) Sequence and comparative analysis of the maize NB mitochondrial genome. Plant Physiol 136(3):3486–3503
Crosatti C, Quansah L, Marré C, Giusti L, Roncaglia E, Atienza SG, Cattivelli L, Fait A (2013) Cytoplasmic genome substitution in wheat affects the nuclear-cytoplasmic crosstalk leading to transcript and metabolite alterations. BMC Genomics 14(1):1–21
Cui X, Wise RP, Schnable PS (1996) The *rf2* nuclear restorer gene of male-sterile T-cytoplasm maize. Science 272(5266):1334–1336
Cui P, Liu H, Lin Q, Ding F, Zhuo G, Hu S, Liu D, Yang W, Zhan K, Zhang A, Yu J (2009) A complete mitochondrial genome of wheat (*Triticum aestivum* cv. Chinese Yumai), and fast evolving mitochondrial genes in higher plants. J Genet 88(3):299–307
Dache ZAA, Thierry AR (2023) Mitochondria-derived cell-to-cell communication. Cell Rep 42(7):112728
Davila JI, Arrieta-Montiel MP, Wamboldt Y, Cao J, Hagmann J, Shedge V, Xu YZ, Weigel D, Mackenzie SA (2011) Double-strand break repair processes drive evolution of the mitochondrial genome in Arabidopsis. BMC Biol 9(1):1–14
Dewey RE, Levings CS, Timothy DH (1985) Nucleotide sequence of ATPase subunit 6 gene of maize mitochondria. Plant Physiol 79(3):914–919
Dopp IJ, Yang X, Mackenzie SA (2021) A new take on organelle-mediated stress sensing in plants. New Phytol 230(6):2148–2153
Fábián A, Péntek BK, Soós V, Sági L (2024) Heat stress during male meiosis impairs cytoskeletal organization, spindle assembly and tapetum degeneration in wheat. Front Plant Sci 14:1314021
Fan F, Zhu YX, Wu MY, Yin WX, Li GQ, Hahn M, Hamada MS, Luo CX (2023) Mitochondrial inner membrane ABC transporter Bcmdl1 is involved in conidial germination, virulence, and resistance to anilinopyrimidine fungicides in *Botrytis cinerea*. Microbiol Spectr 11(4):e00108–e00123
FAO (2022). https://www.fao.org/worldfoodsituation/csdb/en/
Fearnley IM, Walker JE (1992) Conservation of sequences of subunits of mitochondrial complex I and their relationships with other proteins. Biochim Biophys Acta 1140(2):105–134
Filip E, Skuza L (2021) Horizontal gene transfer involving chloroplasts. Int J Mol Sci 22(9):4484
Gäbler L, Herz U, Brennicke A, Grohmann L, Liddell A, Leaver CJ, Schröder W (1994) The 42.5 kDa subunit of the NADH: ubiquinone oxidoreductase (complex I) in higher plants is encoded by the mitochondrial nad7 gene. Mol Gen Genet MGG 244(1):33–40
Galluzzi L, Kepp O, Trojel-Hansen C, Kroemer G (2012) Mitochondrial control of cellular life, stress, and death. Circ Res 111(9):1198–1207
Gao M, Zhao H, Zheng L, Zhang L, Peng Y, Ma W, Tian R, Yuan Y, Ma F, Li M, Ma B (2022) Overexpression of apple *Ma12*, a mitochondrial pyrophosphatase pump gene, leads to malic acid accumulation and the upregulation of malate dehydrogenase in tomato and apple calli. Horticult Res 9:uhab053
Geng X, Ye J, Yang X, Li S, Zhang L, Song X (2018) Identification of proteins involved in carbohydrate

metabolism and energy metabolism pathways and their regulation of cytoplasmic male sterility in wheat. Int J Mol Sci 19(2):324

Ghavami F, Bassi FM, Burciaga R, Soltani A, Noyszewski A, De Jimenez MKM, Gu YQ, Meinhardt S, Elias EM, Kianian PM, Mergoum M, Maan SS, Kianian SF (2014) Developing improved durum wheat germplasm by altering the cytoplasmic genomes. Options Méditerranéennes. Série A, Séminaires Méditerranéens 110:177–187

Giacomello M, Pyakurel A, Glytsou C, Scorrano L (2020) The cell biology of mitochondrial membrane dynamics. Nat Rev Mol Cell Biol 21(4):204–224

Gibellini L, Bianchini E, De Biasi S, Nasi M, Cossarizza A, Pinti M (2015) Natural compounds modulating mitochondrial functions. Evid Based Complement Alternat Med 2015

Giraldo P, Benavente E, Manzano-Agugliaro F, Gimenez E (2019) Worldwide research trends on wheat and barley: a bibliometric comparative analysis. Agronomy 9(7):352

Glazebrook J (2005) Contrasting mechanisms of defense against biotrophic and necrotrophic pathogens. Annu Rev Phytopathol 43:205–227

Golovnina KA, Glushkov SA, Blinov AG, Mayorov VI, Adkison LR, Goncharov NP (2007) Molecular phylogeny of the genus *Triticum* L. Plant Syst Evol 264(3):195–216

Goto YI, Koga Y, Horai S, Nonaka I (1990) Chronic progressive external ophthalmoplegia: a correlative study of mitochondrial DNA deletions and their phenotypic expression in muscle biopsies. J Neurol Sci 100(1–2):63–69

Gray MW (1999) Evolution of organellar genomes. Curr Opin Genet Dev 9(6):678–687

Gray MW, Burger G, Lang BF (1999) Mitochondrial evolution. Science 283(5407):1476–1481

Grohmann L, Thieck O, Herz U, Schrüoder W, Brennicke A (1994) Translation of nad9 mRNAs in mitochondria from Solanum tuberosum is restricted to completely edited transcripts. Nucleic Acids Res 22(16):3304–3311

Haider N (2012) Evidence for the origin of the B genome of bread wheat based on chloroplast DNA. Turk J Agric For 36(1):13–25

Han Y, Gao Y, Zhou H, Zhai X, Ding Q, Ma L (2021) Mitochondrial genes are involved in the fertility transformation of the thermosensitive male-sterile line YS3038 in wheat. Mol Breed 41(10):1–21

Handa H (2003) The complete nucleotide sequence and RNA editing content of the mitochondrial genome of rapeseed (*Brassica napus* L.): comparative analysis of the mitochondrial genomes of rapeseed and *Arabidopsis thaliana*. Nucleic Acids Res 31(20):5907–5916

Hanson MR (1991) Plant mitochondrial mutations and male sterility. Annu Rev Genet 25(1):461–486

Hanson MR, Bentolila S (2004) Interactions of mitochondrial and nuclear genes that affect male gametophyte development. Plant Cell 16(suppl_1):S154–S169

Henze K, Martin W (2003) Essence of mitochondria. Nature 426(6963):127–128

Holt IJ, Harding AE, Morgan-Hughes JA (1988) Deletions of muscle mitochondrial DNA in patients with mitochondrial myopathies. Nature 331(6158):717–719

Hou N, Wu YW, Liu CG, Zhang CL, Zhang Y (2000) Studies of salt tolerance of alloplasmic wheat. Yi Chuan xue bao=. Acta Genet Sin 27(4):325–330

Hu HL, Zhang F, Wang P, Lu FH (2023) Evolutionary genetics of wheat mitochondrial genomes. Crop J 11(6):1774–1781

Huang S, Van Aken O, Schwarzländer M, Belt K, Millar AH (2016) The roles of mitochondrial reactive oxygen species in cellular signaling and stress response in plants. Plant Physiol 171(3):1551–1559

Ichimura K, Mizoguchi T, Yoshida R, Yuasa T, Shinozaki K (2000) Various abiotic stresses rapidly activate *Arabidopsis* MAP kinases ATMPK4 and ATMPK6. Plant J 24(5):655–665

Jones JD, Dangl JL (2006) The plant immune system. Nature 444(7117):323–329

Kerchev PI, De Clercq I, Denecker J, Mühlenbock P, Kumpf R, Nguyen L, Audenaert D, Dejonghe W, Van Breusegem F (2014) Mitochondrial perturbation negatively affects auxin signaling. Mol Plant 7(7):1138–1150

Kihara H (1982) Importance of cytoplasm in plant genetics. Cytologia 47(3–4):435–450

Klimov VV, Astakhova NV, Alieva GP, Sal'nikov EB, Trunova TI, Morozova ZA, Semenov OG (2005) Effect of alien cytoplasm of goatgrass on biological and physiological properties of alloplasmic wheats. Biol Bull 32(3):234–239

Knoop V (2004) The mitochondrial DNA of land plants: peculiarities in phylogenetic perspective. Curr Genet 46(3):123–139

Kozik A, Rowan BA, Lavelle D, Berke L, Schranz ME, Michelmore RW, Christensen AC (2019) The alternative reality of plant mitochondrial DNA: one ring does not rule them all. PLoS Genet 15(8):e1008373

Kubo T, Newton KJ (2008) Angiosperm mitochondrial genomes and mutations. Mitochondrion 8(1):5–14

Kubo T, Mikami T, Kinoshita T (1993) The sugar beet mitochondrial genome contains an ORF sharing sequence homology with the gene for the 30 kDa subunit of bovine mitochondrial complex I. Mol Gen Genet MGG 241(3):479–482

Kubo T, Yamamoto MP, Mikami T (2000) The nad4L-orf25 gene cluster is conserved and expressed in sugar beet mitochondria. Theor Appl Genet 100(2):214–220

Kwak JM, Mori IC, Pei ZM, Leonhardt N, Torres MA, Dangl JL, Bloom RE, Bodde S, Jones JD, Schroeder JI (2003) NADPH oxidase *AtrbohD* and *AtrbohF* genes function in ROS-dependent ABA signaling in *Arabidopsis*. EMBO J 22(11):2623–2633

Lamb C, Dixon RA (1997) The oxidative burst in plant disease resistance. Annu Rev Plant Biol 48(1):251–275

Le XH, Millar AH (2023) The diversity of substrates for plant respiration and how to optimize their use. Plant Physiol 191(4):2133–2149

Leaver CJ, Gray MW (1982) Mitochondrial genome organization and expression in higher plants. Annu Rev Plant Physiol 33(1):373–402

Levin DA (2003) The cytoplasmic factor in plant speciation. Syst Bot 28(1):5–11

Li G, Li Z, Yang Z, Leshem Y, Shen Y, Men S (2021) Mitochondrial heat-shock cognate protein 70 contributes to auxin-mediated embryo development. Plant Physiol 186(2):1101–1121

Li G, Zhang H, Lin Z, Li H, Xu G, Xu Y, Ji R, Luo W, Qiu Y, Qiu S, Tang H (2024) Comparative analysis of chloroplast and mitochondrial genomes of sweet potato provides evidence of gene transfer. Sci Rep 14(1):4547

Liu CG, Wu YW, Hou H, Zhang C, Zhang Y, McIntosh RA (2002) Value and utilization of alloplasmic common wheats with *Aegilops crassa* cytoplasm. Plant Breed 121(5):407–410

Liu H, Cui P, Zhan K, Lin Q, Zhuo G, Guo X, Ding F, Yang W, Liu D, Hus S, Yu J, Zhang A (2011) Comparative analysis of mitochondrial genomes between a wheat K-type cytoplasmic male sterility (CMS) line and its maintainer line. BMC Genomics 12(1):1–14

Lohse M, Drechsel O, Kahlau S, Bock R (2013) OrganellarGenomeDRAW—a suite of tools for generating physical maps of plastid and mitochondrial genomes and visualizing expression data sets. Nucleic Acids Res 41(W1):W575–W581

Lonsdale DM (1984) A review of the structure and organization of the mitochondrial genome of higher plants. Plant Mol Biol 3(4):201–206

Lonsdale DM (1985) Movement of genetic material between the chloroplast and mitochondrion in higher plants. In: Genetic flux in plants. Springer, Vienna, pp 51–60

Lonsdale DM, Hodge TP, Howe CJ, Stem DB (1983) Maize mitochondrial DNA contains a sequence homologous to the ribulose-1, 5-bisphosphate carboxylase large subunit gene of chloroplast DNA. Cell 34(3):1007–1014

Lynch M, Koskella B, Schaack S (2006) Mutation pressure and the evolution of organelle genomic architecture. Science 311(5768):1727–1730

Ma Z, Qin G, Zhang Y, Liu C, Wei M, Cen Z, Yan Y, Luo T, Li Z, Liang H, Huang D, Deng G (2021) Bacterial leaf streak 1 encoding a mitogen-activated protein kinase confers resistance to bacterial leaf streak in rice. Plant J 107(4):1084–1101

Maan SS (1991) Nucleo-cytoplasmic genetics of wheat. In: Nuclear and Organellar genomes of wheat species (Proc. Dr. Kihara Mem. Int. Symp. Cytoplasm. Engineer. Wheat). Kihara Memorial Foundation, pp 175–194

Martín AC, Atienza SG, Ramírez MC, Barro F, Martín A (2008) Male fertility restoration of wheat in *Hordeum chilense* cytoplasm is associated with 6HchS chromosome addition. Aust J Agric Res 59(3):206–213

Martín AC, Castillo A, Atienza SG, Rodríguez-Suárez C (2018) A cytoplasmic male sterility (CMS) system in durum wheat. Mol Breed 38(7):1–9

McBride HM, Neuspiel M, Wasiak S (2006) Mitochondria: more than just a powerhouse. Curr Biol 16(14):R551–R560

McNeece BT, Sharma K, Lawrence GW, Lawrence KS, Klink VP (2019) The mitogen activated protein kinase (MAPK) gene family functions as a cohort during the Glycine max defense response to Heterodera glycines. Plant Physiol Biochem 137:25–41

Melonek J, Duarte J, Martin J, Beuf L, Murigneux A, Varenne P, Comadran J, Sebastien S, Levadoux S, Bernath-Levin K, Torney F, Pichon JP, Perez P, Small I (2021) The genetic basis of cytoplasmic male sterility and fertility restoration in wheat. Nat Commun 12(1):1–14

Millar AH, Whelan J, Soole KL, Day DA (2011) Organization and regulation of mitochondrial respiration in plants. Annu Rev Plant Biol 62:79–104

Milner MJ, Craze M, Bowden S, Bates R, Wallington EJ, Keeling A (2020) Identification of genes involved in male sterility in wheat (*Triticum aestivum* L.) which could be used in a genic hybrid breeding system. Plant Direct 4(3):e00201

Miquel J, Economos AC, Fleming J, Johnson JE Jr (1980) Mitochondrial role in cell aging. Exp Gerontol 15(6):575–591

Mittler R, Vanderauwera S, Gollery M, Van Breusegem F (2004) Reactive oxygen gene network of plants. Trends Plant Sci 9(10):490–498

Møller IM (2001) Plant mitochondria and oxidative stress: electron transport, NADPH turnover, and metabolism of reactive oxygen species. Annu Rev Plant Biol 52(1):561–591

Moons A (2005) Regulatory and functional interactions of plant growth regulators and plant glutathione S-transferases (GSTs). Vitamins Hormones 72:155–202

Mullaney EJ (1981) Increased resistance to *Septoria nodorum* in alloplasmic lines of a wheat cultivar. Phytopathology 71(2):244–244

Nair CKK (1993) Mitochondrial genome organization and cytoplasmic male sterility in plants. J Biosci 18(3):407–422

Nakagami H, Pitzschke A, Hirt H (2005) Emerging MAP kinase pathways in plant stress signalling. Trends Plant Sci 10(7):339–346

Nakamura C (2023) Nucleus-cytoplasm compatibility: a genetic system underlying allopolyploid speciation as exemplified in alloplasmic lines of wheat. Cytologia 88(3):189–195

Newton KJ (1988) Plant mitochondrial genomes: organization, expression and variation. Annu Rev Plant Physiol Plant Mol Biol 39(1):503–532

Notsu Y, Masood S, Nishikawa T, Kubo N, Akiduki G, Nakazono M, Hirai A, Kadowaki K (2002) The complete sequence of the rice (*Oryza sativa* L.) mitochondrial genome: frequent DNA sequence acquisition and loss during the evolution of flowering plants. Mol Gen Genomics 268(4):434–445

Noyszewski AK, Ghavami F, Alnemer LM, Soltani A, Gu YQ, Huo N, Meinhardt S, Kianian PM, Kianian

SF (2014) Accelerated evolution of the mitochondrial genome in an alloplasmic line of durum wheat. BMC Genomics 15(1):1–16

Oda K, Yamato K, Ohta E, Nakamura Y, Takemura M, Nozato N, Akashi K, Kanegae T, Ogura Y, Kohchi T, Ohyama K (1992) Gene organization deduced from the complete sequence of liverwort Marchantia polymorpha mitochondrial DNA: a primitive form of plant mitochondrial genome. J Mol Biol 223(1):1–7

Ogihara Y, Yamazaki Y, Murai K, Kanno A, Terachi T, Shiina T, Miyashita N, Nasuda S, Nakamura C, Mori N, Tsunewaki K (2005) Structural dynamics of cereal mitochondrial genomes as revealed by complete nucleotide sequencing of the wheat mitochondrial genome. Nucleic Acids Res 33(19):6235–6250

Oldenburg DJ, Bendich AJ (2001) Mitochondrial DNA from the liverwort *Marchantia polymorpha*: circularly permuted linear molecules, head-to-tail concatemers, and a 5′ protein. J Mol Biol 310(3):549–562

Palmer JD, Herbon LA (1988) Plant mitochondrial DNA evolved rapidly in structure, but slowly in sequence. J Mol Evol 28(1):87–97

Palmer JD, Shields CR (1984) Tripartite structure of the *Brassica campestris* mitochondrial genome. Nature 307(5950):437–440

Pastore D, Stoppelli MC, Di Fonzo N, Passarella S (1999) The existence of the K^+ channel in plant mitochondria. J Biol Chem 274(38):26683–26690

Pastore D, Fratianni A, Di Pede S, Passarella S (2000) Effects of fatty acids, nucleotides and reactive oxygen species on durum wheat mitochondria. FEBS Lett 470(1):88–92

Pastore D, Trono D, Laus MN, Di Fonzo N, Passarella S (2001) Alternative oxidase in durum wheat mitochondria. Activation by pyruvate, hydroxypyruvate and glyoxylate and physiological role. Plant Cell Physiol 42(12):1373–1382

Pastore D, Trono D, Laus MN, Di Fonzo N, Flagella Z (2007) Possible plant mitochondria involvement in cell adaptation to drought stress: a case study: durum wheat mitochondria. J Exp Bot 58(2):195–210

Pont C, Leroy T, Seidel M, Tondelli A, Duchemin W, Armisen D, Lang D, Bustos-kots D, Goue N, Balfourier F, Molnar-Lang M, Lage J, Kilian B, Ozkan H, Waite D, Dyer S, Latellier T, Alaux M, Wheat and Barley Legacy for Breeding Improvement (WHEALBI) Consortium, Russell J, Keller B, van Eeuwijk F, Spannagl M, Mayer KFX, Waugh R, Stein N, Cattivelli L, Haberer G, Charmet G, Salse J (2019) Tracing the ancestry of modern bread wheats. Nat Genet 51(5):905–911

Rakhmatullina D, Mazina A, Ponomareva A, Dmitrieva S, Beckett RP, Minibayeva F (2022) Mdivi-1 induced mitochondrial fusion as a potential mechanism to enhance stress tolerance in wheat. Life 12(9):1386

Rasmusson AG, Heiser V, Zabaleta E, Brennicke A, Grohmann L (1998) Physiological, biochemical and molecular aspects of mitochondrial complex I in plants. Biochim Biophys Acta 1364(2):101–111

Raval PK, Martin WF, Gould SB (2023) Mitochondrial evolution: gene shuffling, endosymbiosis, and signaling. Sci Adv 9(32):eadj4493

Richardson AO, Palmer JD (2007) Horizontal gene transfer in plants. J Exp Bot 58(1):1–9

Rigoulet M, Yoboue ED, Devin A (2011) Mitochondrial ROS generation and its regulation: mechanisms involved in H_2O_2 signaling. Antioxid Redox Signal 14(3):459–468

Ryan MT, Hoogenraad NJ (2007) Mitochondrial-nuclear communications. Annu Rev Biochem 76:701–722

Sandhu APS, Abdelnoor RV, Mackenzie SA (2007) Transgenic induction of mitochondrial rearrangements for cytoplasmic male sterility in crop plants. Proc Natl Acad Sci 104(6):1766–1770

Satoh M, Kubo T, Nishizawa S, Estiati A, Itchoda N, Mikami T (2004) The cytoplasmic male-sterile type and normal type mitochondrial genomes of sugar beet share the same complement of genes of known function but differ in the content of expressed ORFs. Mol Gen Genomics 272(3):247–256

Saxena KB, Hingane AJ (2015) Male sterility systems in major field crops and their potential role in crop improvement. In: Plant biology and biotechnology. Springer, New Delhi, pp 639–656

Shao MR, Raju SKK, Laurie JD, Sanchez R, Mackenzie SA (2017) Stress-responsive pathways and small RNA changes distinguish variable developmental phenotypes caused by MSH1 loss. BMC Plant Biol 17(1):1–14

Sloan DB (2013) One ring to rule them all? Genome sequencing provides new insights into the 'master circle'model of plant mitochondrial DNA structure. New Phytol 200(4):978–985

Sloan DB, Alverson AJ, Chuckalovcak JP, Wu M, McCauley DE, Palmer JD, Taylor DR (2012) Rapid evolution of enormous, multichromosomal genomes in flowering plant mitochondria with exceptionally high mutation rates. PLoS Biol 10(1):e1001241

Sloan DB, Warren JM, Williams AM, Wu Z, Abdel-Ghany SE, Chicco AJ, Havird JC (2018) Cytonuclear integration and co-evolution. Nat Rev Genet 19(10):635–648

Soltani A, Ghavami F, Mergoum M, Hegstad J, Noyszewski A, Meinhardt S, Kianian SF (2014) Analysis of ATP6 sequence diversity in the *Triticum–Aegilops* species group reveals the crucial role of rearrangement in mitochondrial genome evolution. Genome 57(5):279–288

Soltani A, Kumar A, Mergoum M, Pirseyedi SM, Hegstad JB, Mazaheri M, Kianian SF (2016) Novel nuclear-cytoplasmic interaction in wheat (*Triticum aestivum*) induces vigorous plants. Funct Integr Genomics 16(2):171–182

Sowell AR, Swearingen B (2021) Wheat outlook, WHS-21h, U.S. Department of Agriculture, Economic Research Service, August 16, 2021

Sugiyama Y, Watase Y, Nagase M, Makita N, Yagura S, Hirai A, Sugiura M (2005) The complete nucleotide sequence and multipartite organization of the tobacco mitochondrial genome: comparative analysis of

mitochondrial genomes in higher plants. Mol Genet Genom 272:603–615
Suzuki N (2023) Fine tuning of ROS, redox and energy regulatory systems associated with the functions of chloroplasts and mitochondria in plants under heat stress. Int J Mol Sci 24(2):1356
Taanman JW (1999) The mitochondrial genome: structure, transcription, translation and replication. Biochim Biophys Acta 1410(2):103–123
Tanaka Y, Tsuda M, Yasumoto K, Yamagishi H, Terachi T (2012) A complete mitochondrial genome sequence of Ogura-type male-sterile cytoplasm and its comparative analysis with that of normal cytoplasm in radish (*Raphanus sativus* L.). BMC Genomics 13(1):352
Tanaka Y, Tsuda M, Yasumoto K, Terachi T, Yamagishi H (2014) The complete mitochondrial genome sequence of *Brassica oleracea* and analysis of coexisting mitotypes. Curr Genet 60(4):277–284
Tian X, Zheng J, Hu S, Yu J (2006) The rice mitochondrial genomes and their variations. Plant Physiol 140(2):401–410
Tsunewaki K (1993) Genome-plasmon interactions in wheat. Japanese J Genet 68(1):1–34
Tsunewaki K (2009) Plasmon analysis in the *Triticum-Aegilops* complex. Breed Sci 59(5):455–470
Tsunewaki K, Wang GZ, Matsuoka Y (1996) Plasmon analysis of *Triticum* (wheat) and *Aegilops*. 1. Production of alloplasmic common wheats and their fertilities. Genes Genet Syst 71(5):293–311
Tsunewaki K, Wang GZ, Matsuoka Y (2002) Plasmon analysis of *Triticum* (wheat) and *Aegilops*. 2. Characterization and classification of 47 plasmons based on their effects on common wheat phenotype. Genes Genet Syst 77(6):409–427
Unseld M, Marienfeld JR, Brandt P, Brennicke A (1997) The mitochondrial genome of Arabidopsis thaliana contains 57 genes in 366,924 nucleotides. Nat Genet 15(1):57–61
Virdi KS, Laurie JD, Xu YZ, Yu J, Shao MR, Sanchez R, Kundariya H, Wang D, Riethoven JJM, Wamboldt Y, Arrieta-Montiel MP, Mackenzie SA (2015) *Arabidopsis* MSH1 mutation alters the epigenome and produces heritable changes in plant growth. Nat Commun 6(1):1–9
Voluevich EA, Buloichik AA (1992) Nuclear-cytoplasmic interactions in the resistance of wheat to fungal pathogens. II. Effects of cultivated and wild cereal cytoplasms on the expression of the genome of the Leningradka variety during interaction with the floury mildew pathogen. Soviet genetics
Wallace DC (2005) A mitochondrial paradigm of metabolic and degenerative diseases, aging, and cancer: a dawn for evolutionary medicine. Annu Rev Genet 39:359–407
Wang Z, Zou Y, Li X, Zhang Q, Chen L, Wu H, Su D, Chen Y, Guo J, Luo D, Liu YG (2006) Cytoplasmic male sterility of rice with boro II cytoplasm is caused by a cytotoxic peptide and is restored by two related PPR motif genes via distinct modes of mRNA silencing. Plant Cell 18(3):676–687
Wang K, Shao Z, Guo F, Wang K, Zhang Z (2021) The mitogen-activated protein kinase kinase TaMKK5 mediates immunity via the TaMKK5–TaMPK3–TaERF3 module. Plant Physiol 187:2323
Warren JM, Sloan DB (2020) Interchangeable parts: the evolutionarily dynamic tRNA population in plant mitochondria. Mitochondrion 52:144–156
Welchen E, Canal MV, Gras DE, Gonzalez DH (2021) Cross-talk between mitochondrial function, growth, and stress signalling pathways in plants. J Exp Bot 72(11):4102–4118
Willige BC, Isono E, Richter R, Zourelidou M, Schwechheimer C (2011) Gibberellin regulates PIN-FORMED abundance and is required for auxin transport–dependent growth and development in Arabidopsis thaliana. Plant Cell 23(6):2184–2195
Woodson JD, Chory J (2008) Coordination of gene expression between organellar and nuclear genomes. Nat Rev Genet 9(5):383–395
Wu Y, Zhang C, Liu C, Shuxin R, Yan Z (1998a) Breeding technology of alloplasmic wheat. Sci China Ser C Life Sci 41(5):449–458
Wu Y, Shuxin R, Chunguang L, Ning H, Cuilan Z, Yan Z (1998b) Influence of alloplasm on resistance to wheat scab. Chin Sci Bull 43(12):1038–1041
Wu ZQ, Liao XZ, Zhang XN, Tembrock LR, Broz A (2022) Genomic architectural variation of plant mitochondria—a review of multichromosomal structuring. J Syst Evol 60(1):160–168
Wynn EL, Christensen AC (2019) Repeats of unusual size in plant mitochondrial genomes: identification, incidence and evolution. G3 9(2):549–559
Xiong AS, Peng RH, Zhuang J, Gao F, Zhu B, Fu XY, Xue Y, Jin XF, Tian YS, Zhao W, Yao QH (2008) Gene duplication and transfer events in plant mitochondria genome. Biochem Biophys Res Commun 376(1):1–4
Xiong Y, Singh IN, Hall ED (2009) Tempol protection of spinal cord mitochondria from peroxynitrite-induced oxidative damage. Free Radic Res 43(6):604–612
Xu YZ, Arrieta-Montiel MP, Virdi KS, de Paula WB, Widhalm JR, Basset GJ, Davila JI, Elthon TE, Elowsky CG, Sato SJ, Clemente TE, Mackenzie SA (2011) MutS HOMOLOG1 is a nucleoid protein that alters mitochondrial and plastid properties and plant response to high light. Plant Cell 23(9):3428–3441
Xu YZ, Santamaria RDLR, Virdi KS, Arrieta-Montiel MP, Razvi F, Li S, Ren G, Yu B, Alexander D, Guo L, Feng X, Dweikat IM, Clemente TE, Mackenzie SA (2012) The chloroplast triggers developmental reprogramming when MUTS HOMOLOG1 is suppressed in plants. Plant Physiol 159(2):710–720
Xu Y, Yu D, Chen J, Duan M (2023) A review of rice male sterility types and their sterility mechanisms. Heliyon 9:e18204
Yang KY, Liu Y, Zhang S (2001) Activation of a mitogen-activated protein kinase pathway is involved in disease resistance in tobacco. Proc Natl Acad Sci 98(2):741–746
Yıldırım A, Sönmezoğlu ÖA, Sayaslan A, Kandemir N, Gökmen S (2019) Molecular breeding of durum wheat

cultivars for pasta quality. Qual Assurance Saf Crops Foods 11(1):15–21

Zardoya R (2020) Recent advances in understanding mitochondrial genome diversity. F1000Res:9

Zhang S, Klessig DF (1997) Salicylic acid activates a 48-kD MAP kinase in tobacco. Plant Cell 9(5):809–824

Zhang S, Liu Y (2001) Activation of salicylic acid–induced protein kinase, a mitogen-activated protein kinase, induces multiple defense responses in tobacco. Plant Cell 13(8):1877–1889

Zhang Y, Zhang S, Zhang G, Liu X, Wang C, Xu J (2015) Comparison of mitochondrial genomes provides insights into intron dynamics and evolution in the caterpillar fungus *Cordyceps militaris*. Fungal Genet Biol 77:95–107

Zhang L, Wu J, Zhu Z, He Y, Fang R (2023) Mitochondrion: a bridge linking aging and degenerative diseases. Life Sci 121666

Zhu T, Zou L, Li Y, Yao X, Xu F, Deng X, Zhang D, Lin H (2018) Mitochondrial alternative oxidase-dependent autophagy involved in ethylene-mediated drought tolerance in *Solanum lycopersicum*. Plant Biotechnol J 16(12):2063–2076

Zurbriggen MD, Carrillo N, Hajirezaei MR (2010) ROS signaling in the hypersensitive response: when, where and what for? Plant Signal Behav 5(4):393–396

Gleason C, Huang S, Thatcher LF, Foley RC, Anderson CR, Carroll AJ, Millar AH, Singh KB (2011) Mitochondrial complex ii has a key role in mitochondrial-derived reactive oxygen species influence on plant stress gene regulation and defense. Proc. Natl. Acad. Sci. USA. 108:10768–10773. https://doi.org/10.1073/pnas.1016060108

Lang BF, Gray MW, Burger G (1999) Mitochondrial genome evolution and the origin of eukaryotes. Annu Rev Genet 33:351–97. PMID: 10690412. https://doi.org/10.1146/annurev.genet.33.1.351

Kawaura K, Saeki A, Masumura T, Morita S, Ogihara Y (2011) Heteroplasmy and expression of mitochondrial genes in alloplasmic and euplasmic wheat. Genes Genet Syst. 86(4):249–55. https://doi.org/10.1266/ggs.86.249

Hattori N, Kitagawa K, Takumi S, Nakamura C (2002) Mitochondrial DNA heteroplasmy in wheat, Aegilops and their nucleus-cytoplasm hybrids. Genetics 160(4):1619–30. https://doi.org/10.1093/genetics/160.4.1619

Gulberto JM, Newton KJ (2017) Plant Mitchondrical Genomes: Dynamics and Mechanism of Mutation. Annu Rev Plant Biol. 68:225–252. https://doi.org/10.1146/annurev-arplant-043015-112232 Epub 2017 Feb 9.PMID: 28226235

Luby-Phelps K (2000) Cytoarchitecture and physical properties of cytoplasm: volume, viscosity, diffusion, intracellular surface area. Int Rev Cytol 192:189–221.https://doi.org/10.1016/s0074-7696(08)60527-6

Luby-Phelps K (2013) The physical chemistry of cytoplasm and its influence on cell function: an update. Mol Biol Cell. 24(17):2593-6. https://doi.org/10.1091/mbc.E12-08-0617

Leister D (2012) Retrograde signaling in plants: from simple to complex scenarios. Front Plant Sci. 19(3)135. https://doi.org/10.3389/fpls.2012.00135

Brownfield L (2021) Plant breeding: Revealing the secrets of cytoplasmic male sterility in wheat. Curr Biol. 7;31(11):R724–R726. https://doi.org/10.1016/j.cub.2021.04.026

Edwardson JR (1970) Cytoplasmic male sterility. BOI. Rev. 36:341–420.

Kazama T, Okuno M, Watari Y, Yanase S, Koizuka C, Tsuruta Y, Sugaya H, Toyoda A, Itoh T, Tsutsumi N, Toriyama K, Koizuka N, Arimura SI (2019) Curing cytoplasmic male sterility via TALEN-mediated mitochondrial genome editing. Nat Plants. 5(7):722–730. https://doi.org/10.1038/s41477-019-0459-z

Petrov V, Hille J, Mueller-Roeber B, Gechev TS (2015) ROS-mediated abiotic stress-induced programmed cell death in plants. Front Plant Sci. 18(6):69. https://doi.org/10.3389/fpls.2015.00069

Yang L, Zhang J, He J, Qin Y, Hua D, Duan Y, Chen Z, Gong Z (2014) ABA-mediated ROS in mitochondria regulate root meristem activity by controlling PLETHORA expression in Arabidopsis. PLoS Genet. 18;10(12):e1004791. https://doi.org/10.1371/journal.pgen.1004791

Raza H (2011) Dual localization of glutathione S-transferase in the cytosol and mitochondria: implications in oxidative stress, toxicity and disease. FEBS J. 8(22):4243–51. https://doi.org/10.1111/j.1742-4658.2011.08358.x

Gullner G, Komives T, Kiraly L, Schroder P (2018) Glutathione S-transferase enzymes in plant-pathogen interactions. Front Plant Sci. 9:1836. https://doi.org/10.3389/fpls.2018.01836

Sharma R, Sahoo A, Devendran R, Jain M (2014) Over-expression of a rice tau class glutathione s-transferase gene improves tolerance to salinity and oxidative stresses in Arabidopsis. PLoS One. 9(3):e92900. https://doi.org/10.1371/journal.pone.0092900

Barreto P, Yassitepe JECT, Wilson ZA, Arruda P (2017) Mitochondrial Uncoupling Protein 1 Overexpression Increases Yield in Nicotiana tabacum under Drought Stress by Improving Source and Sink Metabolism. Front Plant Sci. 8:1836. https://doi.org/10.3389/fpls.2017.01836

Filip E, Skuza L (2021) Horizontal gene transfer involving chloroplasts. Int J Mol Sci. 22(9):4484. https://doi.org/10.3390/ijms22094484

Boore JL (1997) Transmission of mitochondrial DNA-playing favorites? Bioessays. (9):751-3 https://doi.org/10.1002/bies.950190904

MIX
Papier aus verantwortungsvollen Quellen
Paper from responsible sources
FSC® C105338

If you have any concerns about our products,
you can contact us on
ProductSafety@springernature.com

In case Publisher is established outside the EU,
the EU authorized representative is:
Springer Nature Customer Service Center GmbH
Europaplatz 3, 69115 Heidelberg, Germany

Printed by Libri Plureos GmbH
in Hamburg, Germany